Optical Physics: Advanced Techniques and Applications

Optical Physics: Advanced Techniques and Applications

Editor: Vladimir Latinovic

NY RESEARCH PRESS
New York

Published by NY Research Press
118-35 Queens Blvd., Suite 400,
Forest Hills, NY 11375, USA
www.nyresearchpress.com

Optical Physics: Advanced Techniques and Applications
Edited by Vladimir Latinovic

International Standard Book Number: 978-1-63238-545-1 (Hardback)

Cataloging-in-Publication Data

Optical physics : advanced techniques and applications / edited by Vladimir Latinovic.
p. cm.
Includes bibliographical references and index.
ISBN 978-1-63238-545-1
1. Physical optics. 2. Optics. 3. Optical materials. I. Latinovic, Vladimir.
QC395.2 .068 2017
535.2--dc23

Printed in the United States of America.

Contents

Preface

This book unfolds the innovative aspects of the advanced techniques used in optical physics. It concentrates on the applications of optical physics, which aids in advancing communications, medicine, manufacturing, and various other sectors. The study of electromagnetic radiation is the primary focus of this field. This book is a compilation of chapters that discuss the most vital concepts and emerging trends in this area of study. The various advancements in optical physics are glanced at and their applications as well as ramifications are looked at in detail. Scientists, students and researchers actively engaged in the field of optical physics will find this book full of crucial and unexplored concepts.

This book has been the outcome of endless efforts put in by authors and researchers on various issues and topics within the field. The book is a comprehensive collection of significant researches that are addressed in a variety of chapters. It will surely enhance the knowledge of the field among readers across the globe.

It gives us an immense pleasure to thank our researchers and authors for their efforts to submit their piece of writing before the deadlines. Finally in the end, I would like to thank my family and colleagues who have been a great source of inspiration and support.

Editor

1

Electromechanical control of nitrogen-vacancy defect emission using graphene NEMS

Antoine Reserbat-Plantey[1,*], Kevin G. Schädler[1,*], Louis Gaudreau[1,*], Gabriele Navickaite[1], Johannes Güttinger[1], Darrick Chang[1], Costanza Toninelli[2], Adrian Bachtold[1] & Frank H.L. Koppens[1,3]

Despite recent progress in nano-optomechanics, active control of optical fields at the nanoscale has not been achieved with an on-chip nano-electromechanical system (NEMS) thus far. Here we present a new type of hybrid system, consisting of an on-chip graphene NEMS suspended a few tens of nanometres above nitrogen-vacancy centres (NVCs), which are stable single-photon emitters embedded in nanodiamonds. Electromechanical control of the photons emitted by the NVC is provided by electrostatic tuning of the graphene NEMS position, which is transduced to a modulation of NVC emission intensity. The optomechanical coupling between the graphene displacement and the NVC emission is based on near-field dipole–dipole interaction. This class of optomechanical coupling increases strongly for smaller distances, making it suitable for nanoscale devices. These achievements hold promise for selective control of emitter arrays on-chip, optical spectroscopy of individual nano-objects, integrated optomechanical information processing and open new avenues towards quantum optomechanics.

[1] ICFO-Institut de Ciencies Fotoniques, The Barcelona Institute of Science and Technology, Castelldefels, Barcelona 08860, Spain. [2] CNR-INO, Istituto Nazionale di Ottica, LENS Via Carrara 1, Sesto Fiorentino (FI) 50019, Italy. [3] ICREA – Institució Catalana de Recerça i Estudis Avancats, Barcelona, Spain.
* These authors contributed equally to this work. Correspondence and requests for materials should be addressed to A.B. (email: adrian.bachtold@icfo.eu) or to F.H.L.K. (email: frank.koppens@icfo.eu).

Active, *in situ* control of light at the nanoscale remains a challenge in modern physics and in nanophotonics in particular[1–3]. A promising approach is to take advantage of the technological maturity of nanoelectromechanical systems (NEMS) and to combine it with on-chip optics[4–6]. However, in scaling down the dimensions of such integrated devices, the coupling of a NEMS to optical fields becomes challenging. Despite recent advances in nano-optomechanical coupling[7–10], *in situ* control of light at the nanoscale with an on-chip NEMS has not been accomplished thus far. In this context, recent work has shown that graphene is an ideal platform for both nanophotonics[11–15] and nanomechanics[16–18].

Here we demonstrate a single device, combining these two platforms. In this device, the transduction between nanomotion and an optical field is due to a strong modification of an emitter's relaxation rate and light emission when graphene is placed in its near field[14,19–24], at nanometre-scale distances. The coupling strength increases strongly for shorter distances and is enhanced because of graphene's two-dimensional (2D) character and linear dispersion. As such, this near-field hybrid optomechanical coupling mechanism between graphene and a point dipole is intrinsically nanoscale in comparison with the evanescent coupling involving micron-scale cavities and waveguides in previous lines of work[4,25–27]. In addition, owing to its electromechanical properties, graphene NEMS can be actuated and deflected electrostatically over few tens of nanometres with modest voltages applied to a gate electrode[17,28–31]. The graphene motion can thus be used to modulate the light emission, while the emitted field can be used as a universal probe[19,20,23] of the graphene position.

The coupling between an emitter and graphene can manifest itself in various ways, such as Stark shift[32], dipolar-coupling-induced modification of the emission intensity[20–22,33] or energy (Casimir–Polder[34]) or as energy transfer to graphene plasmons[11,14]. In our experiment, we use the dipole–graphene coupling to control the nitrogen-vacancy centre (NVC) emission by the graphene displacement. This effect is due to non-radiative energy transfer (n-RET) and is mediated by dipolar interactions between the emitting point dipole and induced (lossy) dipoles in graphene, as shown schematically in Fig. 1a. As a consequence, it gives rise to a diverging decay rate[19] $\Gamma_G \propto d^{-4}$ of the emitter in the presence of graphene at a separation $d = 5$–50 nm. Therefore, the emission is reduced with decreasing graphene-emitter separation. Here graphene offers the advantage to be a 2D gapless broadband energy sink. First, the distance dependence of the energy transfer rate is governed by material-free parameters and wavelength, which can be exploited as a universal nanoruler[19,20,23]. Second, the enhanced dipolar coupling strength and stronger distance dependence (d^{-4} compared with d^{-3} for bulk materials[20]) makes the near-field dipolar interaction a more effective and divergent coupling mechanism between a graphene NEMS and a fluorescent emitter.

Results

Hybrid graphene–NVC optomechanical device. To harness near-field dipolar interactions for nano-optomechanical coupling, we propose and demonstrate a novel type of integrated hybrid device as shown in Fig. 1b. Our device consists of a four-layer graphene membrane designed to be suspended some 10–50 nm above a nanodiamond containing one or a few fluorescent NVCs[35,36], as shown in Fig. 1c,d. By applying a combination of a d.c. and a.c. voltage to the conducting graphene membrane relative to the doped silicon backgate, we control the graphene–NVC separation and can simultaneously drive the resonator at radio frequencies. Thus, our device enables electromechanical control of the NVC emission, and complementarily this emission is a transducer of the resonator's nanomotion.

Our hybrid devices are fabricated in arrays (10–100 devices per chip); however, we address them individually. The NVC fluorescence is monitored by a custom-made scanning confocal microscope (Fig. 1f), which simultaneously records the reflectance (Fig. 1e) from the device. These reflection measurements allow us to detect the nanomotion of the graphene resonators by interferometry (*cf.* Supplementary Notes 1, 2 and Supplementary Fig. 1), yielding a map of the mechanical resonance frequency f_m as shown in Fig. 1g. Here f_m depends on drum diameter and can vary between similar drums because of inhomogeneous strain and the presence of ripples as well as photothermal effects[37]. Together, Fig. 1f,g reveal the colocalization of the emitters and graphene resonators by optical measurements.

Electromechanical control of NVC emission. To quantitatively study and control the near-field interaction between the NVCs and the graphene resonator, we tune the graphene-emitter separation electrostatically. The membrane is attracted towards the NVCs by applying a potential difference V_g^{dc} between the backgate and the graphene drum. Optical interferometry measurements (*cf.* Supplementary Note 2 and Supplementary Fig. 2) show that the static deflection of the graphene scales as $\xi_{static} \propto (V_g^{dc})^2$, in agreement with electrostatic actuation. These measurements allow actuation calibration on the order of $1.2 \pm 0.1\,\mathrm{nm\,V^{-2}}$ for the sample shown here (*cf.* Supplementary Note 3). For a given value of V_g^{dc}, the graphene-emitter separation is $d_{G\text{–}NVC}(V_g^{dc}) = d_0 - \xi_{static}(V_g^{dc})$, where d_0 is the initial graphene-emitter separation extracted from the measured device topology (Fig. 1c). Such an electrostatic actuation provides *in situ* and stable control of the graphene deflection from its initial position to the point of contact with the nanodiamond (*cf.* Supplementary Fig. 3).

We use this *in situ* control to extract the separation between the graphene and a localized emitter, a quantity that is difficult to extract using far-field or local probe techniques. Furthermore, we experimentally verify that the graphene-emitter coupling is due to n-RET by measuring the NVC emission as a function of the membrane position. As shown in Fig. 2, we observe a nonlinear reduction of the NVC emission as the membrane is electrostatically deflected towards the nanodiamond. As introduced above, the emitter decay rate $\Gamma_G(d)$ has a nonlinear separation dependence, which induces a separation-dependent total emission strength given as[20] (black line in Fig. 2):

$$\phi_G = \frac{\phi_0}{1 + \frac{9\nu\alpha}{256\pi^3(\epsilon_{eq}+1)^2\left(\frac{\lambda}{d_{G\text{–}NVC}}\right)^4}} + \phi_{bg} \qquad (1)$$

Here, ϕ_0 is the emission of a single emitter in the absence of graphene, $\nu \in [1,2]$ takes into account the emitting dipole orientation, α is the fine structure constant, ϵ_{eq} is the equivalent relative permittivity of the separating medium (*cf.* Supplementary Note 2) and λ is the emission wavelength, which we take to be the peak wavelength of the NVC emission spectrum (*cf.* Supplementary Note 4 and Supplementary Fig. 4). We fit the data with the emission model shown above where we consider energy transfer from an individual NVC to graphene[19]. We consider this interacting emitter to be the NVC closest to the graphene, within a nanodiamond containing an ensemble of one to four NVCs[36]. Further, we assume NVCs embedded deeper within the nanodiamond to contribute to a fluorescence background ϕ_{bg} (visible in Fig. 2 for small separations), with negligible dependence on the graphene membrane position

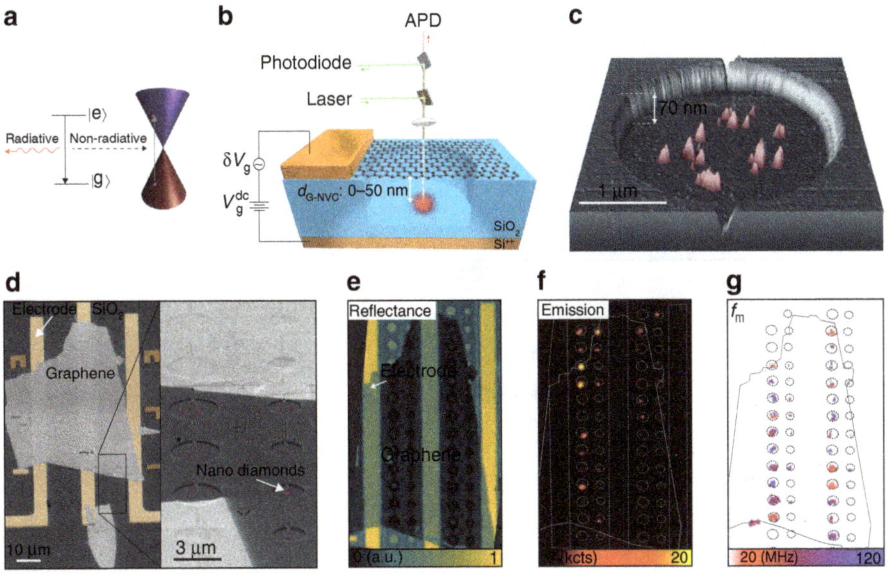

Figure 1 | Graphene-NVC hybrid optomechanical device. (**a**) Energy diagrams of an optical emitter and graphene at the K point of the Brillouin zone (Dirac cone). For small separations $d_{G\text{-}NVC} < 50$ nm, the relaxation of the excited emitter is predominantly due to near-field dipole-dipole interaction by excitation of electron-hole pairs in graphene. (**b**) Sketch of the hybrid optomechanical device. The graphene resonator is driven and displaced electrostatically by d.c. and a.c. voltages V_g^{dc} and δV_g, while its nanomotion is measured optically via the emitter using single-photon counters (APD) and by interferometry. (**c**) False colour AFM topology of nanodiamonds (red) deposited in the centre of a hole etched into SiO_2. (**d**) False colour scanning electronic micrograph (SEM) of arrays of hybrid graphene devices. The labelled hole corresponds to **c**. Graphene is closely suspended over the nanodiamonds (0–50 nm) and clamped at the edges of the holes by Vander Waals interactions. (**e**) False colour confocal reflection map of the same device shown in **c** at $T = 3$ K. At each laser position, both emission and a mechanical spectrum are recorded, thus providing a spatial map of the NVC emission (**f**), and the extracted mechanical resonance frequency f_m (**g**).

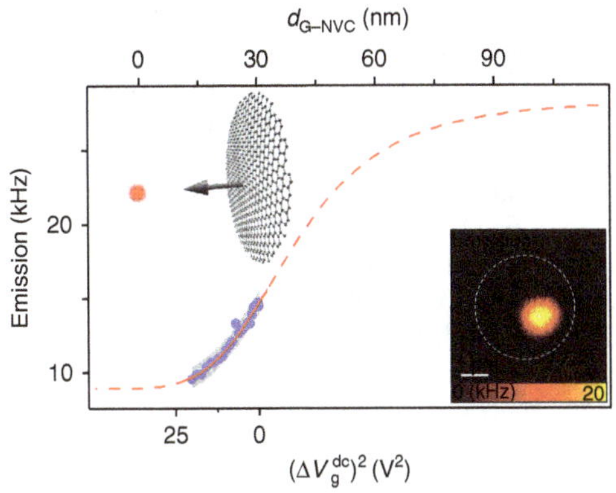

Figure 2 | *In situ* electromechanical control of dipolar coupling. Dependence of measured NVC emission intensity (blue) on the graphene-emitter separation $d_{G\text{-}NVC}$, controlled by electrostatic deflection of the membrane. With increasing $\left(\Delta V_g^{dc}\right)^2$, the membrane approaches the NVC, thereby reducing $d_{G\text{-}NVC}$ and quenching the NVC emission. Data can be fitted with an n-RET model (red), which also allows deflection calibration. For large separations, the extrapolated emission rate agrees with measured emission of nanodiamonds in the absence of graphene. The grey band is an uncertainty interval of width 1,500 Hz, estimated in detail in the Supplementary Note 6. Inset: optical emission map of the hybrid system showing localized NVC emission (dashed line: graphene resonator outline).

(*cf.* Supplementary Note 5 and Supplementary Fig. 5). This background fluorescence inhibits the use of radiative lifetime decay measurements as an alternative method to probe the n-RET separation dependence. However, good agreement of our data with the n-RET model shows that emission measurements provide an indirect optical probe of the graphene position. Given the assumptions described, the main contributions to the measurement uncertainty—shown in Fig. 2 as a grey band—are the broad emission spectrum as well as the uncertainty in the initial membrane position (*cf.* Supplementary Note 6). We remark that the observed emission reduction cannot be a result of an interferometric modulation of the excitation intensity because of graphene deflection. Indeed, for our device geometry (oxide thickness and hole depth) the interferometric effect would lead to an increase in the emission on decreasing $d_{G\text{-}NVC}$, in contrast to the measurements shown in Fig. 2 (*cf.* Supplementary Note 4 and Supplementary Fig. 3 for further discussion).

Nanomotion transduction to NVC emission. Our hybrid device enables high frequency and local control of individual emitters at subwavelength scales. To demonstrate this concept, we show optomechanical transduction of radiofrequency graphene resonator nanomotion to NVC emission, by performing time-resolved emission measurements. During a mechanical oscillation cycle, the graphene-emitter separation $d_{G\text{-}NVC}$ periodically varies with an amplitude δz, and the graphene position is imprinted on the NVC fluorescence. To observe this, we drive the resonator capacitively at frequency f_{drive} and simultaneously perform time-correlated single-photon counting of the emitted photons over a few mechanical periods. By repeating such synchronized acquisition, we obtain a histogram of photon arrival times modulated at f_{drive}, as shown in Fig. 3a. As such, this transduction mechanism involves three successive steps: (i) the initial electromechanical actuation of the membrane, followed by (ii) a quasi-instantaneous optomechanical transduction due to n-RET ($\frac{c}{d_{G\text{-}NVC}} \gg f_m$, c being the speed of light) and finally (iii) a conversion into a time-resolved electronic signal at the single-photon-counting module.

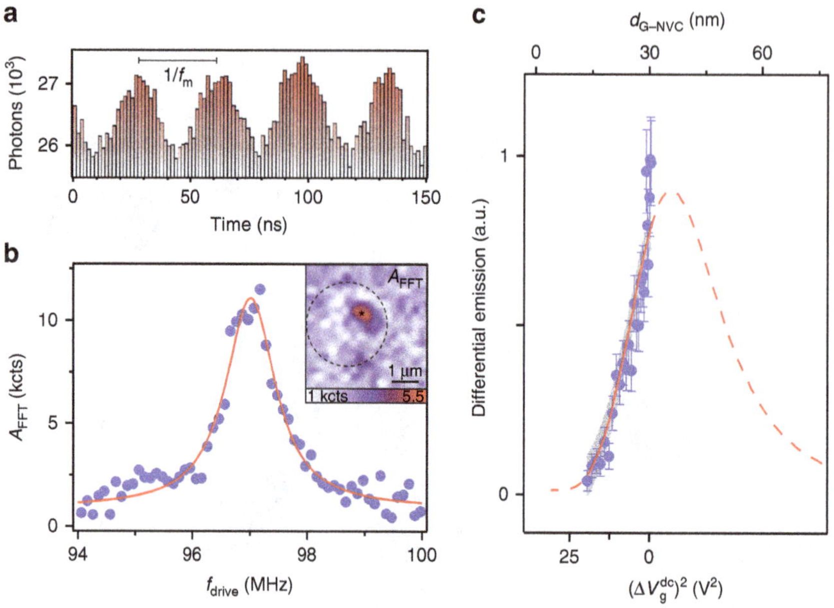

Figure 3 | Graphene nanomotion transduction to NVC emission. (**a**) Time trace of NVC emission (bars) modulated by a driven graphene membrane oscillating at f_m in its near field. Distance-dependent dipolar emitter-graphene coupling imprints the nanomotion of the graphene membrane on the emission. (**b**) Mechanical resonance obtained from differential oscillating NVC emission. Each point corresponds to the amplitude of the Fourier component at $f = f_{drive}$ of the emission time trace (as in **a**). A Lorentzian fit of the data (red line) yields the same value for f_m as obtained independently by optical interferometry. Inset: spatial map of Fourier component at $f = f_{drive}$, localized around the emitter position (*). The dashed circle indicates the graphene drum. (**c**) Dependence of measured (blue) and extrapolated (red) differential NVC emission intensity on the graphene-emitter separation d_{G-NVC}, normalized by the membrane's resonant oscillation amplitude A_m (which increases with V_g, thus reducing the measurement uncertainty for small separations d_{G-NVC}). The grey band is an uncertainty interval (*cf.* Supplementary Note 6).

Additional interferometric measurements enable calibration of the driven oscillation amplitude to be ~ 1 nm at resonance (*cf.* Supplementary Note 3 and Supplementary Fig. 4). The optomechanical transduction step is linear for $\delta z << d_{G-NVC}$, resulting in the observed sinusoidal emission modulation despite the nonlinear dependence of emission on d_{G-NVC}.

To reveal the mechanical spectrum of the graphene resonator in the NVC emission, we extract the modulation depth $A_{FFT}(f_{drive})$ defined as the Fourier component of the emission time traces at different frequencies f_{drive}. At mechanical resonance, both the amplitude of motion $\delta z(f_m)$ and $A_{FFT}(f_m)$ are greatest. Indeed, by sweeping f_{drive} through f_m (independently measured by interferometry), we can reconstruct the mechanical spectrum of the graphene resonator (Fig. 3b) through the near-field transduction mechanism.

This transduction strength can be tuned *in situ*. To this end, we record $A_{FFT}(f_m)$ while varying the stationary separation d_{G-NVC}, as shown in Fig. 3c. Here the differential emission $\Delta A_{FFT} = A_{FFT}(f_m)/A_r(f_m)$ is the measured emission modulation amplitude $A_{FFT}(f_m)$, normalized to the resonant oscillation amplitude $A_r(f_m)$ as obtained from interferometry. This normalization is necessary to compensate the increase in A_r with increasing backgate voltage as $\delta z(f_{drive}) \propto \chi_m V_g^{dc} V_g^{ac}(f_{drive})$, where χ_m is the mechanical susceptibility. The measurements shown in Fig. 3c reveal that ΔA_{FFT} diminishes with increasing V_g^{dc}. Indeed, while the near-field interaction diverges with decreasing d_{G-NVC}, the observed emission, and thus the transduced signal, is quenched. Our data can be fitted by the derivative of NVC emission with respect to d_{G-NVC}, as expected from equation (1). We find that the largest transduction would be obtained for a separation of 35 ± 3 nm. These results, summarized in Fig. 3c, show that we achieve active control of the optomechanical coupling strength by tuning the separation between a local emitter and a vibrating graphene NEMS.

Finally, we explore the localized nature of the near-field coupling mechanism in the normal plane through time-resolved emission from our hybrid device over the whole area of the driven resonator. A spatial map of $A_{FFT}(f_m)$ (inset of Fig. 3b) shows a stronger signal at the NVC site, as expected for such n-RET coupling between graphene and a point-like emitter. This observation highlights the intrinsic localization of the interaction, confined within a subwavelength volume. As such, a small ensemble of emitters can be used as a local transducer of the graphene NEMS motion, which enables eigenmode shape reconstruction. On the other hand, driving the NEMS at higher-order mechanical modes, with subwavelength spatial modulations, would allow addressing and coupling to individual emitters distributed over separations unresolvable with far-field optics.

Discussion

Our device also holds promise for dissipative optomechanics experiments. To quantify the dissipative coupling demonstrated here, we use an established formalism[38,39] to extract a dimensionless dissipative coupling strength $\widetilde{B} = \partial_z \Gamma_G z_{zpm} \Gamma_0^{-1} \sim 10^{-5}$ and the dissipative optomechanical coupling rate $\partial_z \Gamma_G = 6\,\mathrm{MHz\,nm}^{-1}$, where Γ_G is the decay rate of the emitter in presence of graphene, $z_{zpm} \sim 65$ fm is the zero-point motion of the resonato and, $\Gamma_0 \sim 30$ MHz is the intrinsic decay rate of an NVC in a nanodiamond[40] (*cf.* Supplementary Note 7). This compares favourably with existing dissipative optomechanical devices such as a microdisk coupled to an optical waveguide[41], a graphene resonator coupled to a microsphere optical resonator[25] or a photonic crystal nanocavity[42]. The dissipative coupling quantified here takes into account spectral broadening of the emitter we selected. We remark that Fourier limited linewidth emitters such as molecules[43] or NVC's in bulk diamonds[44], in nanocrystals or implanted close to a surface[45] are readily available.

In conclusion, we have realized a novel device comprising a graphene NEMS dissipatively coupled to NVCs by near-field dipolar coupling. Our work offers interesting perspectives for lock-in detection of weak fluorescence signals, NEMS position sensing, electromechanical control of emitters on-chip and fast electromechanical light modulation at the single-photon level. On the basis of the system presented, we envision a similar device that harnesses vacuum forces[34], inducing a divergent and dispersive optomechanical coupling between a quantum emitter and a 2D system. For instance, using a semiconducting 2D resonator (for example, MoS_2) instead of graphene allows to supress the n-RET dissipative coupling, so that the coupling is purely dispersive. A hybrid system combining such resonators ($f_m \sim 10$–100 MHz) placed at 20–40 nm from a lifetime-limited linewidth ($\Gamma_0/2\pi \sim 50$ MHz) emitter, such as a dibenzoterrylene (DBT) molecule[43], would feature a large dispersive coupling strength[34] of $g_0/2\pi \sim 10$–100 MHz. Such systems may approach the ultrastrong coupling regime ($g_0/2\pi > f_m$, $\Gamma_0/2\pi$). Importantly, as discussed in ref. 34, this type of system may enable significant squeezing of mechanical motion at room temperature and position detection on timescales that are short compared with f_m^{-1} because of its extremely large displacement sensitivity. This would then enable coherent manipulation of mechanical and optical degrees of freedom at the level of single quantum[7,46,47].

Methods

Fabrication. Initially, we pattern electrodes on the surface of a 285-nm-thick SiO_2 layer thermally grown on a p-doped Si chip using electron beam lithography (EBL) and thermal evaporation of Au. Holes of 80–100 nm depth and with diameters ranging from 2 to 5 μm are patterned using EBL and subsequent reactive ion etching. Air escape trenches of 150 nm width are etched between adjacent holes to avoid buckling of the subsequently transferred graphene membrane when the device is placed in vacuum. Using a mask exposed by EBL, we deposit nanodiamonds selectively within the etched holes by spin-drop casting[48]. We use a commercially available suspension of ~40 nm diameter nanodiamonds, which contain one to four embedded NVCs. Graphite is exfoliated mechanically using commercial polydimethylsiloxane (PDMS) sheets. Large, few-layer graphene flakes on the PDMS are identified under an optical microscope by contrast measurement. Identified flakes (*cf.* Supplementary Note 1 and Supplementary Fig. 1) are then transferred to target hole arrays using a three-axis micromanipulator[49], forming graphene membranes closely suspended above nanodiamonds. This dry stamp approach has the advantage of high spatial transfer precision and avoids contamination by process liquids.

Electrical device actuation and optical readout. Measurements are performed under vacuum in a cryostat. Data shown in Fig. 1e–g are taken at 3 K and data shown in Figs 2 and 3 are taken at 300 K at 10^{-6} mbar. Our hybrid devices are actuated electrically using a waveform generator and a low-noise voltage source to provide a.c. and d.c. voltage signals. We use a custom-built confocal microscope to locally illuminate the device with 532-nm laser light and simultaneously readout both the reflected and emitted light. The reflection signal is detected using a fast photodiode for the high-frequency reflection component and a powermeter for the static reflection component. We read out the high-frequency photodiode signal with a lock-in amplifier or spectrum analyser to study the resonator mechanics. The emitted light is detected by an avalanche photodetector (APD) combined with 532-nm notch and dichroic filters. Time-resolved emission traces are recorded by feeding the APD signal to a time-correlated single-photon counter, triggered at the radiofrequency driving voltage provided by the waveform generator. The electrostatic actuation is kept low (typically, 1–10 μV) to remain in the linear mechanical regime[18]. Data shown in Fig. 1f are obtained by interferometric detection of the electrostatically driven motion of graphene drums (*cf.* Supplementary Note 2). Graphene motion is detected as a modulation of the reflected light. The drums are driven at a frequency *f*, and the optical reflection modulation amplitude *A*(*f*) at this frequency is read out via a lock-in type of measurement. For each position of the laser, we fit the recorded spectra and extract relevant parameters (f_m, *Q*, amplitude). For most resonators, we do not observe higher-mode resonances within our measurement frequency range, for which reason we attribute our extracted peak frequency to be the mechanical resonance of the fundamental mechanical mode.

Extraction of dissipative optomechanical coupling strength. The detection sensitivity of our readout scheme is $\partial_z\phi = \partial\phi/\partial\Gamma_G.\partial_z\Gamma_G$, where ϕ is the number of emitted photons per second. From Fig. 2, one can extract $\partial_z\phi = 300\,\text{Hz nm}^{-1}$ for a separation of 30 nm and then evaluate $|\partial\phi/\partial\Gamma_G| = 5\times10^{-5}$ using a background-corrected expression for ϕ. Together with the experimental value for $\partial_z\phi$, this yields $\partial_z\Gamma_G = 6\times10^6\,\text{Hz nm}^{-1}$. Given a zero-point motion amplitude $z_{ZPM} = 65$ fm, the dimensionless dissipative coupling strength $B = \partial_z\Gamma_G.z_{ZPM}.\Gamma_0^{-1} = 1.3\times10^{-5}$ is extracted following the definition of Elste *et al.*[38] and Weiss *et al.*[39]. See Supplementary Note 7 for more details.

References

1. Almeida, V. R., Barrios, C. A., Panepucci, R. R. & Lipson, M. All-optical control of light on a silicon chip. *Nature* **431,** 1081–1084 (2004).
2. Rotenberg, N. & Kuipers, L. Mapping nanoscale light fields. *Nat. Photon.* **8,** 919–926 (2014).
3. Hsieh, P. *et al.* Photon transport enhanced by transverse Anderson localization in disordered superlattices. *Nat. Phys.* **11,** 17–19 (2015).
4. Chan, J. *et al.* Laser cooling of a nanomechanical oscillator into its quantum ground state. *Nature* **498,** 18 (2011).
5. Li, M. *et al.* Harnessing optical forces in integrated photonic circuits. *Nature* **456,** 480–484 (2008).
6. Pernice, W. H. P. Circuit optomechanics: concepts and materials. *IEEE Trans. Ultrason. Ferroelectr. Freq. Control* **61,** 1889–1898 (2014).
7. Yeo, I. *et al.* Strain-mediated coupling in a quantum dot-mechanical oscillator hybrid system. *Nat. Nanotechnol.* **9,** 106–110 (2014).
8. Teissier, J., Barfuss, A., Appel, P., Neu, E. & Maletinsky, P. Strain coupling of a nitrogen-vacancy center spin to a diamond mechanical oscillator. *Phys. Rev. Lett.* **113,** 020503 (2014).
9. Montinaro, M., Wüst, G. & Munsch, M. Quantum dot opto-mechanics in a fully self-assembled nanowire. *Nano Lett.* **14,** 4454–4460 (2014).
10. Ovartchaiyapong, P., Lee, K. W., Myers, B. A. & Jayich, A. C. B. Dynamic strain-mediated coupling of a single diamond spin to a mechanical resonator. *Nat. Commun.* **5,** 4429 (2014).
11. Koppens, F. H. L., Chang, D. E. & García De Abajo, F. J. Graphene plasmonics: a platform for strong light-matter interactions. *Nano Lett.* **11,** 3370–3377 (2011).
12. Vakil, A. & Engheta, N. Transformation optics using graphene. *Science* **332,** 1291–1294 (2011).
13. Liu, M. *et al.* A graphene-based broadband optical modulator. *Nature* **474,** 64–67 (2011).
14. Nikitin, A. Y., Guinea, F., Garcia-Vidal, F. J. & Martin-Moreno, L. Fields radiated by a nanoemitter in a graphene sheet. *Phys. Rev. B* **84,** 195446 (2011).
15. Grigorenko, A. N., Polini, M. & Novoselov, K. S. Graphene plasmonics. *Nat. Photon.* **6,** 749–758 (2012).
16. Bunch, J. S. *et al.* Electromechanical resonators from graphene sheets. *Science* **315,** 490–493 (2007).
17. Chen, C. *et al.* Performance of monolayer graphene nanomechanical resonators with electrical readout. *Nat. Nanotechnol.* **4,** 861–867 (2009).
18. Eichler, A. *et al.* Nonlinear damping in mechanical resonators made from carbon nanotubes and graphene. *Nat. Nanotechnol.* **6,** 339–342 (2011).
19. Gómez-Santos, G. & Stauber, T. Fluorescence quenching in graphene: a fundamental ruler and evidence for transverse plasmons. *Phys. Rev. B* **84,** 165438 (2011).
20. Gaudreau, L. *et al.* Universal distance-scaling of nonradiative energy transfer to graphene. *Nano Lett.* **13,** 2030–2035 (2013).
21. Tisler, J. *et al.* Single defect center scanning near-field optical microscopy on graphene. *Nano Lett.* **13,** 3152–3156 (2013).
22. Federspiel, F. *et al.* Distance dependence of the energy transfer rate from a single semiconductor nanostructure to graphene. *Nano Lett.* **15,** 1252–1258 (2015).
23. Mazzamuto, G. *et al.* Single-molecule study for a graphene-based nano-position sensor. *New J. Phys.* **16,** 113007 (2014).
24. Brenneis, A. *et al.* Ultrafast electronic read-out of diamond NV centers coupled to graphene. *Nat. Nanotechnol.* **10,** 135–139 (2015).
25. Cole, R. M. *et al.* Evanescent-field optical readout of graphene mechanical motion at room temperature. *Phys. Rev. Appl.* **3,** 1–7 (2015).
26. Rath, P., Khasminskaya, S., Nebel, C., Wild, C. & Pernice, W. H. P. Diamond-integrated optomechanical circuits. *Nat. Commun.* **4,** 1690 (2013).
27. Anetsberger, G. *et al.* Near-field cavity optomechanics with nanomechanical oscillators. *Nat. Phys.* **5,** 909–914 (2009).
28. Weber, P., Güttinger, J., Tsioutsios, I., Chang, D. E. & Bachtold, A. Coupling graphene mechanical resonators to superconducting microwave cavities. *Nano Lett.* **14,** 2854–2860 (2014).
29. Singh, V. *et al.* Probing thermal expansion of graphene and modal dispersion at low-temperature using graphene nanoelectromechanical systems resonators. *Nanotechnology* **21,** 165204 (2010).
30. Bao, W. *et al.* *In situ* observation of electrostatic and thermal manipulation of suspended graphene membranes. *Nano Lett.* **12,** 5470–5474 (2012).
31. Wong, C.-L., Annamalai, M., Wang, Z.-Q. & Palaniapan, M. Characterization of nanomechanical graphene drum structures. *J. Micromech. Microeng.* **20,** 115029 (2010).

32. Puller, V., Lounis, B. & Pistolesi, F. Single molecule detection of nanomechanical motion. *Phys. Rev. Lett.* **110,** 125501 (2013).
33. Tielrooij, K. J. *et al.* Electrical control of optical emitter relaxation pathways enabled by graphene. *Nat. Phys.* **11,** 281–287 (2015).
34. Muschik, C. A. *et al.* Harnessing vacuum forces for quantum sensing of graphene motion. *Phys. Rev. Lett.* **112,** 223601 (2014).
35. Doherty, M. W. *et al.* The nitrogen-vacancy colour centre in diamond. *Phys. Rep.* **528,** 1–45 (2013).
36. Beams, R. *et al.* Nanoscale fluorescence lifetime imaging of an optical antenna with a single diamond NV center. *Nano Lett.* **13,** 3807–3811 (2013).
37. Barton, R. *et al.* Photothermal self-oscillation and laser cooling of graphene optomechanical systems. *Nano Lett.* **12,** 4681–4686 (2012).
38. Elste, F., Girvin, S. M. & Clerk, A. A. Quantum noise interference and backaction cooling in cavity nanomechanics. *Phys. Rev. Lett.* **102,** 1–5 (2009).
39. Weiss, T., Bruder, C. & Nunnenkamp, A. Strong-coupling effects in dissipatively coupled optomechanical systems. *New J. Phys.* **15,** 045017 (2013).
40. Mohtashami, A. & Koenderink, A. F. Suitability of nanodiamond nitrogen-vacancy centers for spontaneous emission control experiments. *New J. Phys.* **15,** 043017 (2013).
41. Li, M., Pernice, W. H. P. & Tang, H. X. Reactive cavity optical force on microdisk-coupled nanomechanical beam waveguides. *Phys. Rev. Lett.* **103,** 223901 (2009).
42. Wu, M. *et al.* Dissipative and dispersive optomechanics in a nanocavity torque sensor. *Phys. Rev. X* **4,** 21052 (2014).
43. Nicolet, A. A. L. *et al.* Single dibenzoterrylene molecules in an anthracene crystal: main insertion sites. *ChemPhysChem* **8,** 1929–1936 (2007).
44. Tamarat, P. *et al.* Stark shift control of single optical centers in diamond. *Phys. Rev. Lett.* **97,** 083002 (2006).
45. Chu, Y. *et al.* Coherent optical transitions in implanted nitrogen vacancy centers. *Nano Lett.* **14,** 1982–1986 (2014).
46. Rabl, P. *et al.* A quantum spin transducer based on nanoelectromechanical resonator arrays. *Nat. Phys.* **6,** 602–608 (2010).
47. Arcizet, O. *et al.* A single nitrogen-vacancy defect coupled to a nanomechanical oscillator. *Nat. Phys.* **7,** 879–883 (2011).
48. Bermúdez-Ureña, E. *et al.* Coupling of individual quantum emitters to channel plasmons. *Nat. Commun.* **6,** 7883 (2015).
49. Castellanos-Gomez, A. *et al.* Deterministic transfer of two-dimensional materials by all-dry viscoelastic stamping. *2D Mater.* **1,** 011002 (2014).

Acknowledgements

We thank O. Arcizet, E. Bermúdez-Ureña, J. Bertelot, V. Bouchiat, F. Dubin, J. Moser, C. Muschik, M. Lewenstein, M. Lundeberg, J. Osmond, K. Tielrooij, I. Tsioutsios and P. Weber for discussions and help with the experiments. K.G.S. is supported by the Erasmus Mundus Doctorate Program Europhotonics (Grant No. 159224-1-2009-1-FR-ERA MUNDUS-EMJD). L.G. acknowledges financial support from Marie-Curie International Fellowship COFUND and ICFOnest programme. C.T. acknowledges support from the MIUR programme Atom-based Nanotechnology and from the Ente Cassa di Risparmio di Firenze with the project GRANCASSA. A.B. acknowledges supports by the ERC starting grant 279278 (CarbonNEMS), the Spanish MEC (MAT2012-31338) associated to FEDER and the EE Graphene Flagship (contract no. 604391). F.H.L.K. acknowledges support by Fundacio Cellex Barcelona, the ERC Career integration grant (294056, GRANOP), the ERC starting grant (307806, CarbonLight), the Government of Catalonia through the SGR grant (2014-SGR-1535), the Mineco grants Ramón y Cajal (RYC-2012-12281) and Plan Nacional (FIS2013-47161-P), and support by the EC under the Graphene Flagship (contract no. CNECT-ICT-604391).

Author contributions

K.G.S. made the samples with the support of A.R.-P., L.G. and G.N. A.R.-P. built the set-up in collaboration with K.G.S. and support from J.G. and L.G. K.G.S. led the time-resolved oscillation measurements with the support of A.R.-P. and L.G. A.R.-P. led the interferometric motion detection measurements with the support of K.G.S. and L.G. L.G. built the analysis tools with the help of K.G.S. Analysis of the measurements was performed by A.R.-P., K.G.S. and L.G. A.R.-P., K.G.S, A.B. and F.H.L.K. wrote the manuscript with critical comments from all authors. A.R.-P., K.G.S., A.B. and F.H.L.K. conceived the experiment. F.H.L.K. supervised the work.

Additional information

Competing financial interests: The authors declare no competing financial interests.

2

Collective atomic scattering and motional effects in a dense coherent medium

S.L. Bromley[1], B. Zhu[1], M. Bishof[1,†], X. Zhang[1,†], T. Bothwell[1], J. Schachenmayer[1], T.L. Nicholson[1,†], R. Kaiser[2], S.F. Yelin[3,4], M.D. Lukin[4], A.M. Rey[1] & J. Ye[1]

We investigate collective emission from coherently driven ultracold ^{88}Sr atoms. We perform two sets of experiments using a strong and weak transition that are insensitive and sensitive, respectively, to atomic motion at 1 μK. We observe highly directional forward emission with a peak intensity that is enhanced, for the strong transition, by $>10^3$ compared with that in the transverse direction. This is accompanied by substantial broadening of spectral lines. For the weak transition, the forward enhancement is substantially reduced due to motion. Meanwhile, a density-dependent frequency shift of the weak transition ($\sim$10% of the natural linewidth) is observed. In contrast, this shift is suppressed to $<$1% of the natural linewidth for the strong transition. Along the transverse direction, we observe strong polarization dependences of the fluorescence intensity and line broadening for both transitions. The measurements are reproduced with a theoretical model treating the atoms as coherent, interacting radiating dipoles.

[1] JILA, NIST and Department of Physics, University of Colorado, 440 UCB, Boulder, Colorado 80309, USA. [2] Université de Nice Sophia Antipolis, CNRS, Institut Non-Linéaire de Nice, UMR 7335, F-06560 Valbonne, France. [3] Department of Physics, University of Connecticut, Storrs, Connecticut 06269, USA. [4] Department of Physics, Harvard University, Cambridge, Massachussetts 02138, USA. † Present addresses: Physics Division, Argonne National Laboratory, Argonne, Illinois 60439, USA (M.B.); International Center for Quantum Materials, School of Physics, Peking University, Beijing 100871, China (X.Z.); Center for Ultracold Atoms, Massachusetts Institute of Technology, Cambridge, Massachusetts 02139, USA (T.L.N.). Correspondence and requests for materials should be addressed to A.M.R. (email: arey@jilau1.colorado.edu) or to J.Y. (email: Ye@jila.colorado.edu).

Understanding interactions between light and matter in a dense atomic medium is a long-standing problem in physical science[1,2] since the seminal work of Dicke[3]. In addition to their fundamental importance in optical physics, such interactions play a central role in enabling a range of new quantum technologies including optical lattice atomic clocks[4] and quantum networks[5].

The key ingredient in a dense sample is dipole–dipole interactions that arise from the exchange of virtual photons with dispersive and radiative contributions, and their relative magnitude varies between the near-field and far-field regimes. The dispersive (real) part is responsible for collective level shifts and the radiative (imaginary) part gives rise to line broadening and collective superradiant emission[6–8]. Intense theoretical efforts have been undertaken over many years, to treat the complex interplay between the dispersive and radiative dynamics[9–18]. However, experimental demonstrations that provide a complete picture to clarify these interactions have been elusive.

Collective level shifts and line broadening arising from the real and imaginary parts of dipole–dipole interactions have recently been observed in both atomic[19–23] and condensed matter[24] systems. The modification of radiative decay dynamics at low excitation levels has also been observed using short probe pulses[25–28], and interaction effects were manifested in coherent backscattering[29,30]. Although simple models of incoherent radiation transport have often been used to describe light propagation through opaque media[31,32] and radiation trapping in laser cooling of dense atomic samples[33], coherent effects arising from atom–atom interactions, which are necessary to capture correlated many-body quantum behaviour induced by dipolar exchange, are beginning to play a central role. For example, the dipole–dipole interaction is responsible for the observed dipolar blockade and collective excitations in Rydberg atoms[34–41]; it may also place a limit to the accuracy of an optical lattice clock and will require non-trivial lattice geometries to overcome the resulting frequency shift[42]. Previous theoretical efforts have already shown that physical conditions such as finite sample size, sample geometry and the simultaneous presence of dispersive and radiative parts can play crucial roles in atomic emission[10–13,43–45].

In this work we use millions of Sr atoms in optically thick ensembles, taking advantage of the unique level structure of Sr to address motional effects, to study these radiative and dispersive parts simultaneously. We demonstrate that a single, self-consistent, microscopic theory model can provide a unifying picture for the majority of our observations. These understandings can help underpin emerging applications based on many-body quantum science, such as lattice-based optical atomic clocks[4,46,47], quantum nonlinear optics[39], quantum simulations[48] and atomic ensemble-based quantum memories[49].

Results

Experimental setup. Bosonic alkaline-earth atoms with zero nuclear spin have simple atomic structure compared with the more complex hyperfine structure present in typical alkali metal atoms that complicates the modelling and interpretation of the experimental observations. For example, ^{88}Sr atoms have both a strong $^1S_0 - {}^1P_1$ blue transition ($\lambda = 461$ nm) and a spin-forbidden weak $^1S_0 - {}^3P_1$ red transition ($\lambda = 689$ nm), with a strict four-level geometry (Fig. 1a). When the atoms are cooled to a temperature of $\sim 1\,\mu$K, Doppler broadening at 461 nm is ~ 55 kHz, which is almost three orders of magnitude smaller than the blue transition natural linewidth, $\Gamma = 32$ MHz. To an excellent approximation, atomic motion is negligible for atomic coherence prepared by the 461-nm light. To the contrary, the red transition with a natural linewidth $\Gamma = 7.5$ kHz is strongly affected by atomic motion. By comparing the behaviours of the same atomic ensemble observed at these two different wavelengths (Fig. 1b), we can thus collect clear signatures of motional effects on coherent scattering and dipolar coupling[50,51].

We use the experimental scheme shown in Fig. 1a, to perform a comprehensive set of measurements of fluorescence intensity emitted by a dense sample of ^{88}Sr atoms. The sample is released from the trap and then illuminated with a weak probe laser. We vary the atomic density, cloud geometry, observation direction and polarization state of the laser field, and we report the system characteristics using three key parameters as follows: the peak scattered intensity, the linewidth broadening and the line centre shift. For example, along the forward and transverse directions we observe different values of intensity and linewidth broadening, as well as their dependence on light polarization (see Fig. 1c).

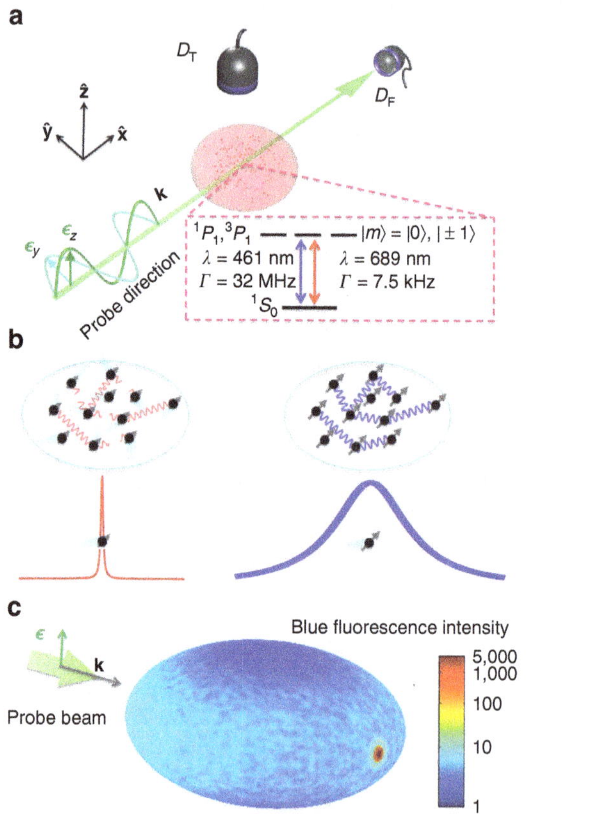

Figure 1 | The experimental scheme and concept. (**a**) We weakly excite the strontium atoms with a linearly polarized probe beam and measure the fluorescence with two detectors: one in the forward direction, $\hat{\mathbf{x}}$, and the other almost in the perpendicular direction, $\hat{\mathbf{z}}$. We probe two different $J = 0$ to $J' = 1$ transitions. The first transition is a $^1S_0 - {}^1P_1$ blue transition with a natural linewidth of $\Gamma = 32$ MHz and the second is a $^1S_0 - {}^3P_1$ red transition with $\Gamma = 7.5$ kHz. (**b**) In the coherent dipole model, photons are shared between atoms. When the Doppler broadened linewidth becomes comparable to the natural linewidth, dephasing must be considered. At our $\sim 1\,\mu$K temperatures the Doppler broadening is ≈ 40 kHz, meaning motional effects are important only for the red transition. (**c**) The three-dimensional intensity distribution predicted for a blue probe beam. The coupled dipole model predicts a strong 10^3 enhancement of the forward intensity compared with other directions and a finite fluorescence along a direction parallel to the incident polarization. The speckled pattern is due to randomly positioned atoms and can be removed by averaging over multiple atom configurations.

We also observe motional effects on the red transition in contrast to the same measurements on the blue transition.

In the experiment, up to 20 million ^{88}Sr atoms are cooled to $\sim 1\,\mu$K in a two-stage magneto-optical trap, the first based on the blue transition and the second on the red transition. The atomic cloud is then released from the magneto-optical trap and allowed to expand for a variable time of flight (TOF), which allows us to control its optical depth and density. They are subsequently illuminated for 50(100) μs with a large-size probe beam resonant with the blue (red) transition (Fig. 2a). The resulting scattered light is measured with two detectors far away from the cloud (see Fig. 1a). One detector is along the forward direction $\hat{\mathbf{x}}$ (detector D_{F}) and the other along the transverse direction $\hat{\mathbf{z}}$ (detector D_{T}, offset by $\sim 10°$). For a short TOF, the atomic cloud is anisotropic and has an approximately Gaussian distribution with an aspect ratio of $R_x{:}R_y{:}R_z = 2{:}2{:}1$, where $R_{\{x,y,z\}}$ are the root-mean-squared radii. We define OD as the on resonance optical depth of the cloud, $\mathrm{OD} = \frac{3N}{2(kR_\perp)^2}$, where $R_\perp$ depends on the direction of observation with $R_{\perp,\mathrm{T}} = R_x = R_y$ and $R_{\perp,\mathrm{F}} = (R_z R_y)^{1/2}$ for the transverse and forward directions respectively, N is the atom number and k is the laser wavevector for the atomic transition (see Supplementary Note 1).

Forward observations. The coherent effect manifests itself most clearly in the forward direction (Fig. 2). To separate the forward fluorescence from the probe beam, we focus the probe with a lens (L_1) after it has passed through the atomic cloud and then block it with a beam stopping blade, which can be translated perpendicular to the probe beam (Fig. 2a inset). The same lens (L_1) also collimates the atomic fluorescence so that it can be imaged onto D_{F}. The position of the beam stopper can be used to vary the angular range of collected fluorescence, characterized by the angle (θ) between $\hat{\mathbf{x}}$ and the edge of the beam stopper (see Methods). Using the maximum atom number available in the experiment, the measured intensity $I_{x,0}(\theta)$ is normalized to that collected at $\theta_{\max} = 7.5$ mRad. Both the blue (square) and red (triangle) transition results are displayed in Fig. 2a. For the blue transition, we observe a 1,000-fold enhancement of the normalized intensity for $\theta < 0.5$ mRad. The enhancement is also present for the red transition, but it is reduced by nearly two orders of magnitude at small θ due to the motional effect. On the other hand, the wider angular area of enhancement is attributed to the longer wavelength of the red transition. The forward intensity strongly depends on the atom number. In Fig. 2b, we present measurements of the forward intensity I_x versus the transverse intensity I_z at a fixed $\theta = 2$ mRad for different atom numbers. The intensities are normalized to those obtained at the peak atom number as used in Fig. 2a. To the first-order approximation, the transverse fluorescence intensity scales linearly with the atom number. Hence, the forward intensity of both the blue and red transitions scales approximately with the atom number squared.

In the forward direction, we have also investigated the linewidth broadening of the blue transition as a function of the atomic OD. By scanning the probe frequency across resonance, we extract the fluorescence linewidth, which is found to be determined primarily by the OD of the cloud (open squares in Fig. 2c). For the range of $0 < \mathrm{OD} < 20$, the lineshape is Lorentzian (see insets); however, the observed lineshape starts to flatten at the centre for $\mathrm{OD} > 20$. We have also varied the atom number by a factor of four, and to an excellent approximation the linewidth data are observed to collapse to the same curve when plotted as a function of OD (open triangles).

Transverse observations. For independent emitters, the forward fluorescence should have no dependence on the probe beam polarization; however, the transverse fluorescence (along $\hat{\mathbf{z}}$) should be highly sensitive to the probe polarization and it is even classically forbidden if the probe is $\hat{\mathbf{z}}$ polarized. However, multiple scattering processes with dipolar interactions can completely modify this picture by redistributing the atomic population in the three excited magnetic states and thus scrambling the polarization of the emitted fluorescence. Consequently, even for a $\hat{\mathbf{z}}$-polarized probe there should be a finite emission along $\hat{\mathbf{z}}$ (see Fig. 1c), with an intensity that increases with increasing OD. Our experimental investigation of the fluorescence properties along the transverse direction is summarized in Fig. 3. Under the same OD along $\hat{\mathbf{z}}$, the $\hat{\mathbf{y}}$-polarized probe beam (square) gives rise to a much more broadened lineshape for the blue transition than the $\hat{\mathbf{z}}$-polarized probe beam does (triangle), as shown in Fig. 3a. Meanwhile, the peak intensity ratio of $I_{\mathrm{ypol}}/I_{\mathrm{zpol}}$ decreases significantly with an increasing OD, indicating the rapidly rising fluorescence with

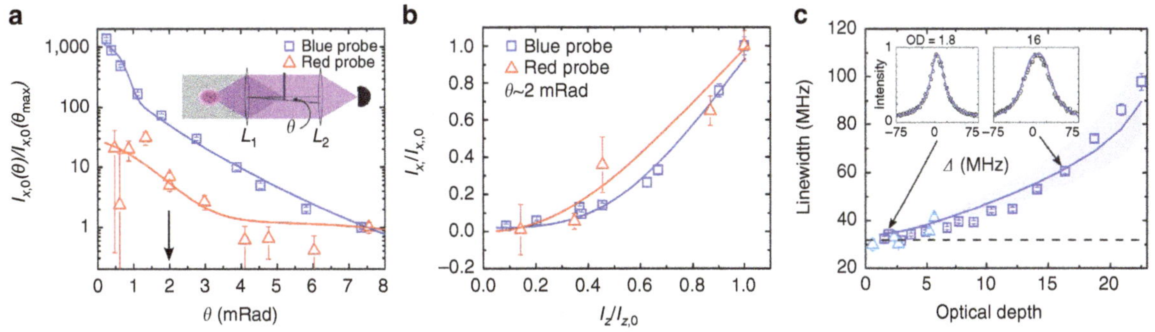

Figure 2 | Forward scattering. (**a**) Comparison of forward scattering intensity versus angle using a red and blue probe beam. We use the setup shown in the inset, to block the probe beam. After interacting with the atoms the probe beam is focused using a lens, which also collimates the fluorescence from the atoms. We block the probe beam using a beam stopper, which we translate perpendicular to the probe beam, to change the angular range of fluorescence collected by the detector, characterized by the angle (θ) between $\hat{\mathbf{x}}$ and the edge of the beam stopper (see Methods). The measured intensity, $I_{x,0}(\theta)$, for each probe beam is normalized to the intensity at $\theta_{\max} = 7.5$ mRad. The dephasing caused by motion reduces the forward intensity peak for the red transition. (**b**) Comparison of intensity in the forward direction, I_x, versus intensity in the transverse direction, I_z. Both are varied by changing N. All measurements are made at $\theta = 2$ mRad (arrow in **a**) and normalized to the intensity, $I_{x,0}$, for the atom number used in **a**. (**c**) Linewidth broadening in the forward direction measured by scanning the blue probe beam detuning, Δ, across resonance. Example lineshapes for different ODs are shown in the inset. Two different atom numbers are used, $N = 1.7(2) \times 10^7$ (blue squares) and $N/4$ (cyan triangles). The dashed line represents Γ for reference. All solid curves are based on the full theory of coupled dipoles and the band in **c** is for a $\pm 20\%$ atom number uncertainty. All error bars are for statistical uncertainties.

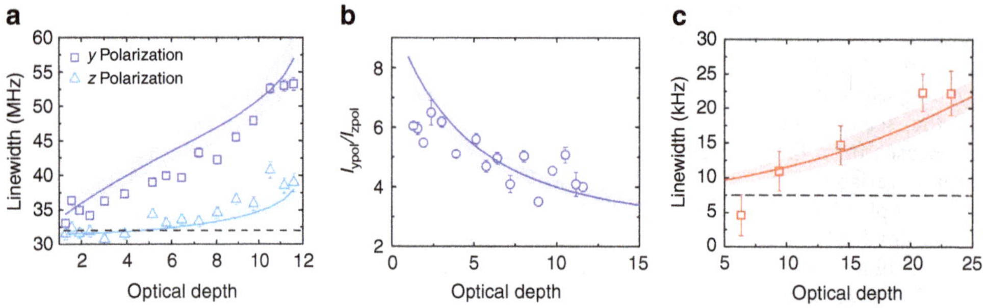

Figure 3 | Transverse scattering. (**a**) Linewidth broadening for the blue transition in the transverse direction for $\hat{\mathbf{y}}$ polarization (open squares) and $\hat{\mathbf{z}}$ polarization (open triangles). (**b**) Intensity ratio, I_{ypol}/I_{zpol}, of $\hat{\mathbf{y}}$ polarization to $\hat{\mathbf{z}}$ polarization measured in the transverse direction when a blue probe beam is used. For low optical depths single particle scattering is dominant and for single particle scattering almost zero intensity is predicted for $\hat{\mathbf{z}}$-polarized fluorescence, as this polarization points directly into the detector. (**c**) Linewidth broadening for the red transition in the transverse direction for $\hat{\mathbf{y}}$-polarized light, showing a similar trend to the blue transition. This transition is more sensitive to magnetic fields; thus, a large magnetic field is applied to probe only the $m=0$ to $m'=0$ transition. All solid curves are based on the full theory of coupled dipoles and the band in **a**, **b** and **c** is for a ±20% atom number uncertainty. All error bars are for statistical uncertainties.

a $\hat{\mathbf{z}}$-polarized probe when OD increases (Fig. 3b). For the red transition, the existence of Doppler broadening requires the lineshape data to be fitted to a Voigt profile. With the Doppler linewidth Δ_D fixed from the thermal velocity measured in free expansion, the Voigt profile determines the line centre and the Lorentzian linewidth with the Gaussian linewidth determined by the temperature. Figure 3c displays the Lorentzian linewidth obtained with a $\hat{\mathbf{y}}$-polarized red probe showing a strong increase of the linewidth with OD.

Spectral broadening and shift. To a good approximation, the dependence of the linewidth on OD along the forward and transverse directions (for the classically allowed $\hat{\mathbf{y}}$ polarization in the single scattering limit) is similar. However, owing to the anisotropic aspect ratio of the cloud, for the same TOF, the OD is lower along $\hat{\mathbf{z}}$ than along $\hat{\mathbf{x}}$. This is responsible for the smaller broadenings measured along $\hat{\mathbf{z}}$ than along $\hat{\mathbf{x}}$. The classically forbidden polarization direction, on the other hand, exhibits a different scaling with OD, which is understandable given that the emission in this case comes only from multiple scattering events with dipolar interactions. The transverse linewidth broadening for the red transition is similar to that of the blue, and it does not depend sensitively on motional effects. This behaviour is in stark contrast to another important observation: the shift of the transition centre frequency. Figure 4 contrasts the linecentre frequency shift observed for $^1S_0-{}^1P_1$ (square) and $^1S_0-{}^3P_1$ (triangle, with original data reported in ref. 52 and see Supplementary Fig. 1). The blue transition frequency shift is consistent with zero at the level of 0.004Γ using an atomic density of $10^{12}\,\text{cm}^{-3}$. However, the measured density shift for the red transition (normalized to the transition linewidth) is more than one order of magnitude larger. This density-related frequency shift significantly exceeds the predicted value based on general S-matrix calculations of s-wave collisions[52] (2.18×10^{-10} Hz cm^3 if the unitary limit is used).

Theory model. Before we turn to a microscopic model to obtain a full and consistent understanding of all these related experimental observations, we note that semiclassical models[53] treating the atomic cloud as a continuous medium of an appropriate refractive index can give an intuitive explanation of the linewidth broadening in the forward direction. Classically, an incoming electric field is attenuated as it propagates through the medium according to the Beer–Lambert law and the forward fluorescence intensity is determined by the same mechanism.

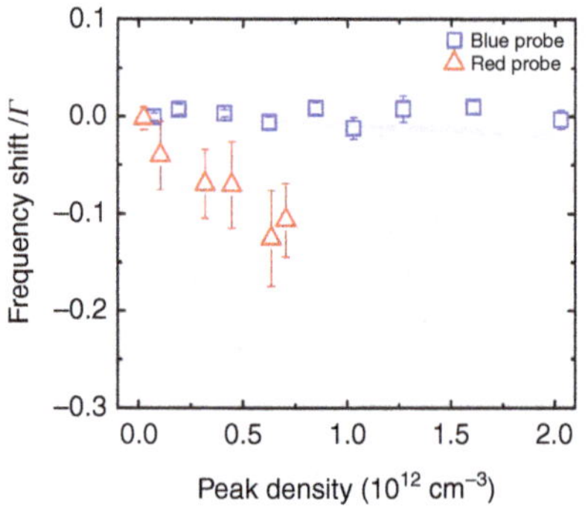

Figure 4 | Frequency shift. Comparison of frequency shift normalized to the corresponding natural linewidth for the blue and red transitions. The blue frequency shift is consistent with 0–0.004 of Γ at an atomic density of $10^{12}\,\text{cm}^{-3}$. The red shift, on the other hand, shows more than 0.1Γ at densities up to $0.7\times10^{12}\,\text{cm}^{-3}$.

This simple semiclassical model recovers the linear dependence of the forward width for small OD and predicts a nonlinear dependence of the linewidth for large OD and a flattening of the line centre. However, we find that this semiclassical approach cannot provide explanations for most aspects of the experimental observations.

The full microscopic model builds on a set of coherently coupled dipoles. Here, each four-level atom is treated as a discrete radiating dipole located at a frozen position, but coupled with retarded dipole radiation, and it is driven with a weak incident laser beam. The atomic ensemble follows the Gaussian distribution observed in the experiment with the appropriate aspect ratio. By solving the master equation in the steady state, we find that the coherence, $b_j^\alpha=\mathrm{Tr}[|g\rangle\langle e_j^\alpha|\hat{\rho}]$, of atom j, located at $\mathbf{r}_j$ is modified by other atoms as[18,54–60]:

$$b_j^\alpha=\frac{\Omega^\xi e^{i\mathbf{k}\cdot\mathbf{r}_j}}{2(\Delta^\alpha+i\frac{\Gamma}{2})}\delta_{\alpha,\xi}+\Gamma\sum_{\alpha',m\neq j}\frac{G_{\alpha,\alpha'}(\mathbf{r}_j-\mathbf{r}_m)}{(i\Delta^\alpha-\frac{\Gamma}{2})}b_m^{\alpha'}. \quad (1)$$

Here, $|g\rangle={}^1S_0$, $|e^\alpha\rangle$ corresponds to the three excited states of $^1P_1$ or $^3P_1$, with $\alpha\in\{x,\ y,\ z\}$ representing the Cartesian states. In addition, $\hat{\rho}$ is the reduced density matrix of the atoms and $\delta_{\gamma,\gamma'}$ is the Kronecker Delta. The driving laser's linear polarization state ξ is along $\hat{\mathbf{y}}$ or $\hat{\mathbf{z}}$, with wavevector $\mathbf{k}$ along $\hat{\mathbf{x}}$, Rabi frequency Ω^ξ and detuned by Δ^α from the $|g\rangle\rightarrow|e^\alpha\rangle$ transition. The function $G_{\alpha,\alpha'}(\mathbf{r})$ accounts for the retarded pairwise dipolar interactions and is given by[18,48,56] $G_{\alpha,\alpha'}(\mathbf{r}) = -i\frac{3}{4}\frac{e^{ikr}}{kr}[(\delta_{\alpha,\alpha'} - \hat{r}_\alpha\hat{r}_{\alpha'}) + (\delta_{\alpha,\alpha'} - 3\hat{r}_\alpha\hat{r}_{\alpha'})(\frac{i}{kr} - \frac{1}{(kr)^2})]$. The fluorescence intensity $I(\mathbf{r}_s) = \langle \hat{\mathbf{E}}^{(+)}(\mathbf{r}_s)\hat{\mathbf{E}}^{(-)}(\mathbf{r}_s)\rangle$, detected at position $\mathbf{r}_s$, can be determined[17,18] as a function of b_j^α,

$$I(\mathbf{r}_s) \approx \frac{9\hbar^2\Gamma^2}{16k^2\mu^2 r_s^2}\sum_{j,m}\left[\mathbf{b}_j \cdot \mathbf{b}_m^* - (\mathbf{b}_j \cdot \hat{\mathbf{r}}_s)(\mathbf{b}_m^* \cdot \hat{\mathbf{r}}_s)\right]e^{i\mathbf{k}_s\cdot(\mathbf{r}_j - \mathbf{r}_m)}, \quad (2)$$

with μ is the atomic transition dipole moment and $\mathbf{k}_s = k\hat{\mathbf{r}}_s$.

Discussion

To understand the forward enhancement we first consider non-interacting atoms under the zeroth order approximation. The atomic coherence is driven only by the probe field that imprints its phase and polarization onto the atoms: $b_j^{\alpha(0)} = \frac{i\delta_{\alpha,\xi}\Omega^\xi e^{i\mathbf{k}\cdot\mathbf{r}_j}/2}{i\Delta - \Gamma/2}$, where $\Delta = \Delta^\alpha$. The corresponding intensity, $I(\mathbf{r}_s) = \frac{9\hbar^2\Gamma^2}{16k^2\mu^2 r_s^2}\frac{|\Omega^\xi|^2}{4(\Delta^2 + \Gamma^2/4)}(N + N^2 e^{-|\mathbf{k}_s - \mathbf{k}_0|^2 R_{\perp,F}^2})$ has a Lorentzian profile. It also exhibits an N^2 scaling and an enhanced forward emission lobe, with an angular width given by the ratio between the transition wavelength and the transverse size of the sample $\Delta\theta \sim 1/(kR_{\perp,F})$. The forward lobe reflects the constructive interference of the coherently emitted radiation stimulated by the laser. Outside the coherent lobe the constructive interference is quickly reduced due to the random position of atoms[28,59,61]. The longer wavelength of the red transition corresponds to a wider angular width of the forward lobe for the red fluorescence.

Simple considerations can also give rise to a qualitative understanding of atomic motion-related effects on forward enhancement. Again for the red transition, the Doppler effect introduces random phases accumulated by $\delta\phi \sim kv/\Gamma$. Here, v is the thermal velocity. The dephasing reduces coherent photon emission and gives rise to a net suppression of the forward emission intensity. The suppression becomes stronger with Δ_D/Γ, with $\Delta_D = \sqrt{\frac{k_B T}{8m\lambda^2 \ln 2}}$ the Doppler width. Such a suppression is clearly observed for the red transition.

To address the linewidth broadening we now consider atoms coupled by dipolar interactions, which tend to emit collectively in an optically dense cloud. The collective emission manifests itself with a broader fluorescence linewidth. Moving to the first-order approximation, we note that the atomic coherence acquires contributions not only from the probe beam but also from the surrounding atoms, with $b_j^\alpha \sim b_j^{\alpha(0)} + b_j^{\alpha(1)}$. Here, $b_j^{\alpha(1)} = \frac{i\Omega^\xi\Gamma/2}{(i\Delta - \Gamma/2)^2}K_{\alpha,\xi}^j e^{ikx_j}$ and $K_{\alpha,\alpha'}^j = \sum_{m\neq j} G_{\alpha,\alpha'}(\mathbf{r}_j - \mathbf{r}_m)e^{ik(x_m - x_j)}$. For a relatively dilute cloud with average interparticle distance $\bar{r} \gg 1/k$, the far-field interactions dominate; thus, higher-order terms beyond $1/r$ can be neglected. Dipolar interactions modify the fluorescence lineshape, with consequences of both a frequency shift that depends on the cloud peak density n_0 and a line broadening that is proportional to OD: $\Delta \to \Delta + \bar{\Delta}$ and $\Gamma \to \Gamma + \bar{\Gamma}$, with $\bar{\Delta} = -\frac{3\sqrt{2}\pi n_0 k^{-3}}{16}\Gamma$ and $\bar{\Gamma} = \frac{\text{OD}}{4}\Gamma$. Thus, the first-order approximation provides an intuitive picture about the role of dipolar effects on the lineshape.

However, in a cloud with an increasingly large OD, dipolar interactions are stronger and multiple scattering processes become relevant. The first-order perturbative analysis then breaks down[62–64]. The full solution of equation (1) based on the coherent coupled dipole model becomes necessary to account for multiple scattering processes (see Methods). The first signatures arise from the forward fluorescence intensity, where its naive N^2 scaling is reduced with an increasing atom number as a consequence of multiple scattering processes. This effect is observed in both red and blue calculations, and it is expected to be more pronounced on the red transition due to its longer wavelength. However, atomic motion leads to a lower effective OD, which tends to suppress multiple scattering processes and thus helps to partially recover the collective enhancement. The solid lines in Fig. 2a,b represent such quantitative theory calculations for both transitions, which agree with the experiment.

Meanwhile, for the linewidth broadening observed in the forward direction, it becomes evident that the scaling of the linewidth versus OD turns nonlinear at large values of OD. The experimental data falls within the shaded area in Fig. 2c, which represents the full solution with a 20% uncertainty in the experimental atom number. Multiple scattering processes are also key to the explanation of the measured fluorescence along the transverse direction, especially for the classically forbidden polarization $\hat{\mathbf{z}}$. Indeed, for both intensity and linewidth broadening observed in the transverse direction, under either $\hat{\mathbf{y}}$ or $\hat{\mathbf{z}}$ probe polarization, the full model (shown as shaded areas in both Fig. 3a,b) reproduces well the experimental results on ${}^1S_0 - {}^1P_1$. Taking into account motional dephasing (see Supplementary Note 2), the transverse broadening for ${}^1S_0 - {}^3P_1$ is also well reproduced as shown in Fig. 3c.

So far, we have shown the observed effects on linewidth and fluorescence intensity are uniquely determined by OD. However, following the arguments discussed above, the frequency shift arising from the dipolar coupling is expected to scale with atomic density, $|\bar{\Delta}|/\Gamma \propto n_0 k^{-3}$, which includes both the collective Lamb shift and the Lorentz–Lorenz shift[50,65]. For our experimental density, this effect is $\lesssim 10^{-3}$, which is consistent with the observed frequency shift for the blue transition (Fig. 4). (It is noteworthy that the role of multiple scattering processes is to further suppress this frequency shift mechanism[50].) In contrast, for the red transition the measured density shift (normalized to Γ) is significantly larger than what is predicted from the current treatment of interacting dipoles; it is also much bigger than the unitarity limit of s-wave scattering. Qualitatively, we expect that as the atoms move and approach each other, the long-lived ground-excited state coherence in the red transition can be significantly modified by the collisional process and open higher partial wave channels. We can thus expect a larger collisional phase shift. This process can be further complicated by atomic recoil, light forces and Doppler dephasing[66].

We have shown that a coherent dipole model describes light scattering in a dense atomic medium with collective effects and multiple scatterings. The model captures the quantitative features of the experimental observations. Motional effects, as manifested in dephasing, can be captured in the model as well. Our results provide useful guides for further developments of optical atomic clocks and other applications involving dense atomic ensembles.

Methods

Coherent dipole model. Here we present the derivation of equation (1). The fundamental assumption is to treat the atoms as frozen during the interrogation. This is an excellent approximation if $\hbar\Gamma$ is much faster than other energy scales in the problem. The latter condition is always satisfied in the case of the blue transition. For the $J=0$ to $J=1$ configuration exhibited by ^{88}Sr, we can label the $J=0$ ground state as $|g\rangle$ and the excited $J=1$ states using a Cartesian basis $|e^z\rangle = |0\rangle$, $|e^x\rangle = (|-1\rangle - |+1\rangle)/\sqrt{2}$, $|e^y\rangle = i(|-1\rangle + |+1\rangle)/\sqrt{2}$. Here, the $|0, \pm 1\rangle$ states are the standard angular momentum ones. In the Cartesian basis, the vector transition operator for the j atom located at $\mathbf{r}_j$ can be written as $\mathbf{b}_j^\alpha = \hat{x}\hat{b}_j^x + \hat{y}\hat{b}_j^y + \hat{z}\hat{b}_j^z$. Here $\hat{b}_j^\alpha = |g\rangle_j\langle e^\alpha|$. On this basis, the master equation governing the evolution of the reduced density matrix of the N atom ensemble, $\hat{\rho}$, in the presence of an incident laser beam with linear polarization ξ can be written as[18]:

$$\frac{d\hat{\rho}}{dt} = -\frac{i}{2}\sum_{j,\alpha}\delta_{\alpha,\xi}\left[\Omega_j\hat{b}_j^{\alpha\dagger} + \Omega_j^*\hat{b}_j^{\alpha}, \hat{\rho}\right] - i\sum_{\substack{j,m\neq j\\ \alpha,\beta}}\left[g_{jm}^{\alpha\beta}\hat{b}_j^{\alpha\dagger}\hat{b}_j^{\beta}, \hat{\rho}\right] + i\sum_{j,\alpha}\Delta^{\alpha}[\hat{b}_j^{\alpha\dagger}\hat{b}_j^{\alpha}, \hat{\rho}] + \sum_{\substack{j,m\\ \alpha,\beta}}f_{jm}^{\alpha\beta}\left(2\hat{b}_j^{\alpha}\hat{\rho}\hat{b}_m^{\beta\dagger} - \left\{\hat{b}_j^{\alpha\dagger}\hat{b}_m^{\beta}, \hat{\rho}\right\}\right), \tag{3}$$

where $\Omega_j = \Omega^{\xi}e^{i\mathbf{k}\cdot\mathbf{r}_j}$ is the Rabi frequecy of the incident field, polarized along ξ ($\hat{\xi}\cdot\mathbf{k}=0$) and detuned by Δ^{α} from the atomic transition $|g\rangle \rightarrow |e^{\alpha}\rangle$. The parameters $g_{jm}^{\alpha\beta} = g_{\alpha,\beta}(\mathbf{r}_j - \mathbf{r}_m)$ and $f_{jm}^{\alpha\beta} = f_{\alpha,\beta}(\mathbf{r}_j - \mathbf{r}_m)$ are the components of the elastic and inelastic dipolar interactions between a pair of atoms at position $\mathbf{r}_j$ and $\mathbf{r}_m$, respectively, and are given by

$$g_{\alpha,\beta}(\mathbf{r}) = \frac{3\Gamma}{4}\left[\left(y_0(kr) - \frac{y_1(kr)}{kr}\right)\delta_{\alpha,\beta} + y_2(kr)\hat{r}_{\alpha}\hat{r}_{\beta}\right], \tag{4}$$

$$f_{\alpha,\beta}(\mathbf{r}) = \frac{3\Gamma}{4}\left[\left(j_0(kr) - \frac{j_1(kr)}{kr}\right)\delta_{\alpha,\beta} + j_2(kr)\hat{r}_{\alpha}\hat{r}_{\beta}\right], \tag{5}$$

where $r = |\mathbf{r}| = |\mathbf{r}_j - \mathbf{r}_m|$, $y_n(x)$, $j_n(x)$ are the spherical Bessel functions of the second and first kind, respectively. Here, also α, $\beta = x$, y or z represent Cartesian components. The symbol $\delta_{\gamma,\gamma'}$ is the Kronecker Delta. In the low-intensity limit, we can project the density matrix into a state space including the ground state $|G\rangle \equiv |g_1, g_2, ...g_N\rangle$ and states with only one excitation[57–59] such as $|j\alpha\rangle \equiv |g_1, ...e_j^{\alpha}, ...g_N\rangle$. In this reduced space, the relevant equations of motion simplify to

$$\frac{d\rho_{j\alpha,j\alpha}}{dt} = -\frac{i}{2}\left(\Omega_j\delta_{\alpha,\xi}\rho_{G,j\alpha} - \Omega_j^*\delta_{\alpha,\xi}\rho_{j\alpha,G}\right) - i\sum_{m\neq j,\beta}g_{jm}^{\alpha\beta}\left(\rho_{m\beta,j\alpha} - \rho_{j\alpha,m\beta}\right) - \sum_{m\neq j,\beta}f_{jm}^{\alpha\beta}\left(\rho_{m\beta,j\alpha} + \rho_{j\alpha,m\beta}\right) - \Gamma\rho_{j\alpha,j\alpha}, \tag{6}$$

$$\frac{d\rho_{j\alpha,G}}{dt} = -\frac{i}{2}\left(\Omega_j\delta_{\alpha,\xi}\rho_{G,G} - \sum_m\Omega_m\rho_{j\alpha,m\xi}\right) - i\sum_{m\neq j,\beta}\left(g_{jm}^{\alpha\beta} - if_{jm}^{\alpha\beta}\right)\rho_{m\beta,G} + \left(i\Delta^{\alpha} - \frac{\Gamma}{2}\right)\rho_{j\alpha,G}, \tag{7}$$

$$\frac{d\rho_{j\alpha,m\beta}}{dt} = -\frac{i}{2}\left(\Omega_j\delta_{\alpha,\xi}\rho_{G,m\beta} - \Omega_m^*\delta_{\xi,\beta}\rho_{j\alpha,G}\right) + i(\Delta^{\alpha} - \Delta^{\beta})\rho_{j\alpha,m\beta} - i\left(\sum_{l\neq j,\nu}\rho_{l\nu,m\beta}g_{jl}^{\alpha,\nu} - \sum_{l\neq m,\nu}\rho_{j\alpha,l\nu}g_{lm}^{\beta,\nu}\right) - \Gamma\rho_{j\alpha,m\beta} - \left(\sum_{l\neq j,\nu}\rho_{l\nu,m\beta}f_{jl}^{\alpha\nu} + \sum_{l\neq m,\nu}\rho_{j\alpha,l\nu}f_{lm}^{\beta\nu}\right), \tag{8}$$

$$\frac{d\rho_{G,G}}{dt} = -\frac{i}{2}\left(\sum_{j,\alpha}\Omega_j^*\rho_{j\alpha,G} - \Omega_j\rho_{G,j\alpha}\right) + \Gamma(1 - \rho_{G,G}) + \sum_{\substack{m,j\neq m\\ \alpha,\beta}}f_{jm}^{\alpha\beta}\left(\rho_{j\alpha,m\beta} + \rho_{m\beta,j\alpha}\right). \tag{9}$$

where $\rho_{G,G} = \mathrm{Tr}[\hat{\rho}|G\rangle\langle G|]$, $\rho_{j\alpha,m\beta} = \mathrm{Tr}\left[\hat{\rho}\left(\hat{b}_m^{\beta\dagger}\hat{b}_j^{\alpha}\right)\right]$ and $\rho_{j\alpha,G} = \mathrm{Tr}\left[\hat{b}_j^{\alpha}\hat{\rho}\right]$.

As we are interested in the situation of a weak probe limit, $\Omega^{\xi} \ll \Gamma$, we expand the density matrix in successive orders of Ω^{ξ}/Γ, $\hat{\rho} = \hat{\rho}^{(0)} + \hat{\rho}^{(1)} + \hat{\rho}^{(2)} + \ldots$, and keep the first-order terms. At this level of approximation, $\rho_{G,G} = 1$, $\rho_{j\alpha,m\beta} = 0$ and only the optical coherences $b_j^{\alpha} \equiv \rho_{j\alpha,G}$ evolve in time accordingly to the following set of linear equations:

$$\frac{db_j^{\alpha}}{dt} = \left(i\Delta^{\alpha} - \frac{\Gamma}{2}\right)b_j^{\alpha} - \frac{i}{2}\Omega_j\delta_{\alpha,\xi} - i\sum_{m\neq j,\beta}\left(g_{jm}^{\alpha\beta} - if_{jm}^{\alpha\beta}\right)b_m^{\beta}. \tag{10}$$

Here, $G_{\alpha,\beta}(\mathbf{r}) = (f_{\alpha,\beta}(\mathbf{r}) + ig_{\alpha,\beta}(\mathbf{r}))/\Gamma$. The steady-state solution can be obtained by setting $\frac{db_j^{\alpha}}{dt} = 0$ and then solving the subsequent $3N$ linear equations.

Measure the enhancement of forward fluorescence. To measure the scattered light in the forward direction, we use the setup shown in the inset of Fig. 2a, to tightly focus and block the probe beam, while still collecting most of the atomic fluorescence on the CCD (charge-coupled device) camera. We focus the probe beam, after it interacts with the atoms, to a small spot with 15 μm root-mean-squared radius and block it using a beam stopping blade. We then translate the beam stopper perpendicular to the probe beam by a distance Δx from our reference point of $x = 0$, which we define as the position of the beam stopper where we see the greatest fluorescence without saturating the CCD camera with the probe beam. As only the forward direction is particularly sensitive to positional changes, we convert this change in position to a change in angle simply using $\theta = \arctan\frac{\Delta x}{15\mathrm{cm}}$, where the first lens with a 15-cm focal length collimates the fluorescence. In numerical calculations, the CCD camera is simulated as a ring area centred around the forward direction and the average intensity collected over the ring is determined. The external radius is set to be large enough to reach the angular region outside the interference cone and the inner angular radius θ_{sim}, simulating the blocking of the signal by the beam stopper, is varied accordingly to the experiment. To account for the difference between σ_{sim} and the experiment cloud size, θ_{sim} is rescaled so that we satisfy the experimental observation that at θ_{max} the enhancement factor drops to 1.

References

1. Gross, M. & Haroche, S. Superradiance: an essay on the theory of collective spontaneous emission. *Phys. Rep.* **93**, 301–396 (1982).
2. Andreev, A. V., Emel'yanov, V. I. & Il'inskii, Y. A. Collective spontaneous emission (Dicke superradiance). *Sov. Phys. Usp.* **23**, 493–514 (1980).
3. Dicke, R. H. Coherence in spontaneous radiation processes. *Phys. Rev.* **93**, 99–110 (1954).
4. Bloom, B. J. *et al.* A new generation of atomic clocks: accuracy and stability at the 10^{-18} level. *Nature* **506**, 71–75 (2014).
5. Kimble, H. J. The quantum internet. *Nature* **453**, 1023–1030 (2008).
6. Skribanowitz, N., Herman, I. P., MacGillivray, J. C. & Feld, M. S. Observation of Dicke superradiance in optically pumped hf gas. *Phys. Rev. Lett.* **30**, 309–312 (1973).
7. Pavolini, D., Crubellier, A., Pillet, P., Cabaret, L. & Liberman, S. Experimental evidence for subradiance. *Phys. Rev. Lett.* **54**, 1917–1920 (1985).
8. Wang, T. *et al.* Superradiance in ultracold rydberg gases. *Phys. Rev. A* **75**, 033802 (2007).
9. Agarwal, G. S., Saxena, R., Narducci, L. M., Feng, D. H. & Gilmore, R. Analytical solution for the spectrum of resonance fluorescence of a cooperative system of two atoms and the existence of additional sidebands. *Phys. Rev. A* **21**, 257–259 (1980).
10. Friedberg, R. & Hartmann, S. Superradiant stability in specially shaped small samples. *Opt. Commun.* **10**, 298–301 (1974).
11. Friedberg, R. & Hartmann, S. R. Temporal evolution of superradiance in a small sphere. *Phys. Rev. A* **10**, 1728–1739 (1974).
12. Lewenstein, M. & Rzazewski, K. Collective radiation and the near-zone field. *J. Phys. A Math. Gen* **13**, 743–756 (1980).
13. Steudel, H. & Richter, T. Radiation properties of a continuously pumped two-atom system. *Ann. Phys. (Berlin)* **490**, 122–136 (1978).
14. Rzażewski, K. & Żakowicz, W. Initial value problem for two oscillators interacting with electromagnetic field. *J. Math. Phys.* **21**, 378–388 (1980).
15. Ruostekoski, J. & Javanainen, J. Quantum field theory of cooperative atom response: low light intensity. *Phys. Rev. A* **55**, 513–526 (1997).
16. Scully, M. O. Collective Lamb shift in single photon dicke superradiance. *Phys. Rev. Lett.* **102**, 143601 (2009).
17. Lehmberg, R. H. Radiation from an *N*-atom system. I. General formalism. *Phys. Rev. A* **2**, 883–888 (1970).
18. James, D. F. V. Frequency shifts in spontaneous emission from two interacting atoms. *Phys. Rev. A* **47**, 1336–1346 (1993).
19. Keaveney, J. *et al.* Cooperative lamb shift in an atomic vapor layer of nanometer thickness. *Phys. Rev. Lett.* **108**, 173601 (2012).
20. Meir, Z., Schwartz, O., Shahmoon, E., Oron, D. & Ozeri, R. Cooperative lamb shift in a mesoscopic atomic array. *Phys. Rev. Lett.* **113**, 193002 (2014).
21. Jennewein, S., Sortais, Y. R. P., Greffet, J.-J. & Browaeys, A. Propagation of Light Through Small Clouds of Cold Interacting Atoms. http://arxiv.org/abs/1511.08527 (2015).
22. Jennewein, S. *et al.* Observation of the Failure of Lorentz Local Field Theory in the Optical Response of Dense and Cold Atomic Systems. http://arxiv.org/abs/1510.08041 (2015).
23. Jenkins, S. D. *et al.* Optical Resonance Shifts in the Fluorescence Imaging of Thermal and Cold Rubidium Atomic Gases. http://arxiv.org/abs/1602.01037 (2016).
24. Röhlsberger, R., Schlage, K., Sahoo, B., Couet, S. & Rüffer, R. Collective lamb shift in single-photon superradiance. *Science* **328**, 1248–1251 (2010).
25. Balik, S., Win, A. L., Havey, M. D., Sokolov, I. M. & Kupriyanov, D. V. Near-resonance light scattering from a high-density ultracold atomic ^{87}Rb gas. *Phys. Rev. A* **87**, 053817 (2013).
26. Chalony, M., Pierrat, R., Delande, D. & Wilkowski, D. Coherent flash of light emitted by a cold atomic cloud. *Phys. Rev. A* **84**, 011401 (2011).
27. Guerin, W., Araujo, M. O. & Kaiser, R. Subradiance in a large cloud of cold atoms. *Phys. Rev. Lett.* **116**, 083601 (2016).
28. Kwong, C. C. *et al.* Cooperative emission of a coherent superflash of light. *Phys. Rev. Lett.* **113**, 223601 (2014).
29. Labeyrie, G. *et al.* Coherent backscattering of light by cold atoms. *Phys. Rev. Lett.* **83**, 5266–5269 (1999).
30. Kulatunga, P. *et al.* Measurement of correlated multiple light scattering in ultracold atomic ^{85}Rb. *Phys. Rev. A* **68**, 033816 (2003).

31. de Vries, P., van Coevorden, D. V. & Lagendijk, A. Point scatterers for classical waves. *Rev. Mod. Phys.* **70,** 447–466 (1998).
32. van Rossum, M. C. W. & Nieuwenhuizen, T. M. Multiple scattering of classical waves: microscopy, mesoscopy, and diffusion. *Rev. Mod. Phys.* **71,** 313–371 (1999).
33. Walker, T., Sesko, D. & Wieman, C. Collective behaviour of optically trapped neutral atoms. *Phys. Rev. Lett.* **64,** 408–411 (1990).
34. Heidemann, R. *et al.* Evidence for coherent collective Rydberg excitation in the strong blockade regime. *Phys. Rev. Lett.* **99,** 163601 (2007).
35. Lukin, M. D. *et al.* Dipole blockade and quantum information processing in mesoscopic atomic ensembles. *Phys. Rev. Lett.* **87,** 037901 (2001).
36. Saffman, M., Walker, T. G. & Mølmer, K. Quantum information with Rydberg atoms. *Rev. Mod. Phys.* **82,** 2313–2363 (2010).
37. Lahaye, T., Menotti, C., Santos, L., Lewenstein, M. & Pfau, T. The physics of dipolar bosonic quantum gases. *Rep. Prog. Phys.* **72,** 126401 (2009).
38. Löw, R. *et al.* An experimental and theoretical guide to strongly interacting Rydberg gases. *J. Phys. B At. Mol. Opt. Phys.* **45,** 113001 (2012).
39. Peyronel, T. *et al.* Quantum nonlinear optics with single photons enabled by strongly interacting atoms. *Nature* **488,** 57–60 (2012).
40. Ravets, S. *et al.* Coherent dipole-dipole coupling between two single Rydberg atoms at an electrically-tuned förster resonance. *Nat. Phys.* **10,** 914–917 (2014).
41. Günter, G. *et al.* Observing the dynamics of dipole-mediated energy transport by interaction-enhanced imaging. *Science* **342,** 954–956 (2013).
42. Chang, D. E., Ye, J. & Lukin, M. D. Controlling dipole-dipole frequency shifts in a lattice-based optical atomic clock. *Phys. Rev. A* **69,** 023810 (2004).
43. Kuś, M. & Wódkiewicz, K. Two-atom resonance fluorescence. *Phys. Rev. A* **23,** 853–857 (1981).
44. Żakowicz, W. Superradiant decay of a small spherical system of harmonic oscillators. *Phys. Rev. A* **17,** 343–352 (1978).
45. Sutherland, R. T. & Robicheaux, F. Coherent forward broadening in cold atom clouds. *Phys. Rev. A* **93,** 023407 (2016).
46. Nicholson, T. L. *et al.* Systematic evaluation of an atomic clock at 2×10^{-18} total uncertainty. *Nat. Commun.* **6,** 6896 (2015).
47. Ushijima, I., Takamoto, M., Das, M., Ohkubo, T. & Katori, H. Cryogenic optical lattice clocks. *Nat. Photon.* **9,** 185–189 (2015).
48. Olmos, B. *et al.* Long-range interacting many-body systems with alkaline-earth-metal atoms. *Phys. Rev. Lett.* **110,** 143602 (2013).
49. Specht, H. P. *et al.* A single-atom quantum memory. *Nature* **473,** 190–193 (2011).
50. Javanainen, J., Ruostekoski, J., Li, Y. & Yoo, S.-M. Shifts of a resonance line in a dense atomic sample. *Phys. Rev. Lett.* **112,** 113603 (2014).
51. Labeyrie, G., Delande, D., Kaiser, R. & Miniatura, C. Light transport in cold atoms and thermal decoherence. *Phys. Rev. Lett.* **97,** 013004 (2006).
52. Ido, T. *et al.* Precision spectroscopy and density-dependent frequency shifts in ultracold Sr. *Phys. Rev. Lett.* **94,** 153001 (2005).
53. Lagendijk, A. & Van Tiggelen, B. A. Resonant multiple scattering of light. *Phys. Rep* **270,** 143–215 (1996).
54. Morice, O., Castin, Y. & Dalibard, J. Refractive index of a dilute Bose gas. *Phys. Rev. A* **51,** 3896–3901 (1995).
55. Rouabah, M.-T. *et al.* Coherence effects in scattering order expansion of light by atomic clouds. *J. Opt. Soc. Am. A* **31,** 1031–1039 (2014).
56. Chomaz, L., Corman, L., Yefsah, T., Desbuquois, R. & Dalibard, J. Absorption imaging of a quasi-two-dimensional gas: a multiple scattering analysis. *New J. Phys.* **14,** 055001 (2012).
57. Bienaimé, T., Petruzzo, M., Bigerni, D., Piovella, N. & Kaiser, R. Atom and photon measurement in cooperative scattering by cold atoms. *J. Mod. Opt.* **58,** 1942–1950 (2011).
58. Bienaimé, T., Bachelard, R., Piovella, N. & Kaiser, R. Cooperativity in light scattering by cold atoms. *Fortschr. Phys.* **61,** 377–392 (2013).
59. Bienaimé, T. *et al.* Interplay between radiation pressure force and scattered light intensity in the cooperative scattering by cold atoms. *J. Mod. Opt.* **61,** 18–24 (2014).
60. Müller, C. A. & Miniatura, C. Multiple scattering of light by atoms with internal degeneracy. *J. Phys. A Math. Gen.* **35,** 10163–10188 (2002).
61. Allen, L. & Eberly, J. H. *Dover Books on Physics* (Dover Publications, 1987).
62. Pellegrino, J. *et al.* Observation of suppression of light scattering induced by dipole-dipole interactions in a cold-atom ensemble. *Phys. Rev. Lett.* **113,** 133602 (2014).
63. Fleischhauer, M. & Yelin, S. F. Radiative atom-atom interactions in optically dense media: quantum corrections to the lorentz-lorenz formula. *Phys. Rev. A* **59,** 2427–2441 (1999).
64. Lin, G.-D. & Yelin, S. F. in *Advances in Atomic, Molecular, and Optical Physics* Vol 61 (eds Berman, P., Arimondo, E. & Lin, C.) Ch. 6, 295–329 (Academic Press, 2012).
65. Friedberg, R., Hartmann, S. & Manassah, J. Frequency shifts in emission and absorption by resonant systems ot two-level atoms. *Phys. Rep.* **7,** 101–179 (1973).
66. Julienne, P. S. Cold binary atomic collisions in a light field. *J. Res. Natl Inst. Stand. Technol.* **101,** 487–503 (1996).

Acknowledgements
We are grateful to Paul Julienne, Chris Greene, John Cooper and Murray Holland for their important insights and stimulating discussions. This research is supported by NIST, NSF Physics Frontier Center at JILA, AFOSR, AFOSR-MURI, ARO, DARPA QuASAR, NSF Center for Ultracold Atoms at Harvard-MIT, ITAMP and ANR-14-CE26-0032.

Author contributions
S.L.B., M.B., X.Z., T.B., T.L.N. and J.Y. contributed to the executions of the experiments. B.Z., J.S., R.K., S.F.Y., M.D.L. and A.M.R. contributed to the development of the theory model. All authors discussed the results, contributed to the data analysis and worked together on the manuscript.

Additional information

Competing financial interests: The authors declare no competing financial interests.

3

Broadband single-molecule excitation spectroscopy

Lukasz Piatkowski[1], Esther Gellings[1] & Niek F. van Hulst[1,2]

Over the past 25 years, single-molecule spectroscopy has developed into a widely used tool in multiple disciplines of science. The diversity of routinely recorded emission spectra does underpin the strength of the single-molecule approach in resolving the heterogeneity and dynamics, otherwise hidden in the ensemble. In early cryogenic studies single molecules were identified by their distinct excitation spectra, yet measuring excitation spectra at room temperature remains challenging. Here we present a broadband Fourier approach that allows rapid recording of excitation spectra of individual molecules under ambient conditions and that is robust against blinking and bleaching. Applying the method we show that the excitation spectra of individual molecules exhibit an extreme distribution of solvatochromic shifts and distinct spectral shapes. Importantly, we demonstrate that the sensitivity and speed of the broadband technique is comparable to that of emission spectroscopy putting both techniques side-by-side in single-molecule spectroscopy.

[1]ICFO—Institut de Ciencies Fotoniques, The Barcelona Institute of Science and Technology, 08860 Castelldefels (Barcelona), Spain. [2]ICREA—Institució Catalana de Recerca i Estudis Avançats, 08010 Barcelona, Spain. Correspondence and requests for materials should be addressed to L.P. (email: Lukasz.Piatkowski@icfo.eu) or to N.F.v.H. (email: Niek.vanHulst@icfo.eu).

Optical spectroscopy is a primary analysis tool underlying almost any field of science; absorption, emission and excitation spectra are routinely recorded on bulk samples. The advent of single-molecule detection pushed spectroscopy to a next level: probing molecules one by one unravels inhomogeneities and dynamical processes that are otherwise hidden in the ensemble average. Intrinsic molecular diversity and distinct interactions of molecules with their nanoenvironment lead to wide distributions of their spectral properties. Pursuing a single molecule in time reveals discrete dynamics such as blinking[1] and spectral diffusion[2–6]. To date single molecules, quantum dots, nanoparticles and proteins are detected and tracked with wide-ranging applications in molecular biology, polymer chemistry, nanoscopy and so on.

Historically, the first single molecules were detected through absorption[7] and fluorescence excitation[8] at cryogenic temperatures—conditions leading to high photostability and large absorption cross-section have allowed the recording of both excitation and emission spectra of individual molecules[5,6,9–12]. Far superior signal-to-noise ratios, however, are obtainable with the background-free fluorescence-based detection as opposed to absorption-based spectroscopy; as a result detection of single molecules through their fluorescence has established as the predominant method of choice.

The vast majority of applications in biology and chemistry involve the study of molecules in their natural state that is under ambient conditions. At room temperature, however, the significantly broadened absorption lines combined with a substantial reduction of the photostability and absorption cross-section demands more sensitive fluorescence detection methods and spatial separation of individual molecules. The introduction of near-field imaging[13] and non-invasive confocal microscopy[14,15] have guaranteed the necessary sensitivity and spatial resolution, however, the limited number of photons emitted before photobleaching still puts a major constraint on all room temperature single-molecule spectroscopy techniques. Therefore, only time and detection efficient spectroscopic schemes are feasible, limiting room temperature experiments largely to the recording of emission spectra[15–18].

In recent efforts to go beyond single-molecule emission spectroscopy, individual molecules have been detected using photothermal contrast[19], scattered light[20,21], plasmonic structures[22] and even direct detection of a single molecule through absorption has been demonstrated at room temperature[23–25]. Although very interesting, these approaches are far from routine and their main drawback is that they typically lack spectral information.

Previously, Stopel *et al.*[26] determined the Stokes shift of individual molecules recording both excitation and emission spectrum at room temperature. Against the general believe, they found that the Stokes shift varies between individual molecules. In this study, the excitation spectra were recorded by serial scanning of the narrowband excitation wavelength derived from a white-light continuum. A disadvantage of this approach, however, is that blinking and bleaching of the molecule compete with the sequential scanning, which may result in an incomplete excitation spectrum. Recently Weigel *et al.*[27] were the first to close a coherent control loop on a single molecule, showing that essentially the excitation spectrum is probed.

The ability to routinely perform single-molecule excitation spectroscopy under ambient conditions side-by-side to the already well-established single-molecule emission spectroscopy would be highly valuable. The crucial difference between the two spectroscopies is that emission spectra probe the final spontaneous, nanosecond decay to the ground state and its vibrational progression, while excitation spectra explore the excited state, its vibrational manifold and any intermediate short-lived (ps) states towards the decay. The excitation spectra are particularly sensitive to coupling between molecules or the nanoenvironment of the single molecules in general. Having both spectra at our disposal, a complete picture of the spectral characteristics and excited state dynamics of the molecule emerges.

Beyond the fundamental properties of individual molecules, such as assymetry in absorption–emission spectra, origins of blinking–bleaching dynamics and intra/intermolecular inter-system-crossings, excitation spectra provide valuable information in applied fields such as environmental sciences, medicine and material sciences, just to name a few. For example excitation spectra are used to determine the nature, relative quantities and composition of chromophores in plants, photosynthetic units and coral reef matter[28,29]. Even in the studies on photodynamic therapy for cancer treatment, excitation spectra are used to non-invasively determine *in vivo* the penetration depth of drugs in skin[30]. However, in such applications the multichromophoric composition leads to complex photophysics including energy transfer, self-quenching, reabsorption and reemission, complicating the multicomponent analysis of excitation spectra. Clearly the ability to measure excitation spectra of molecules in extremely small volumes and at interfaces is desirable. A novel approach to single-molecule excitation spectroscopy at ambient conditions would provide sufficient sensitivity to probe even monolayers of such samples with diffraction-limited spatial resolution.

Here we demonstrate a fast and efficient broadband excitation spectroscopy method to record single-molecule excitation spectra at room temperature. Our results uncover heterogeneities in the excitation spectra with unexpectedly large spectral shifts over 100 nm between individual molecules, which have remained beyond observation so far. Exploiting an ultra-broadband laser, we probe the entire excitation spectrum at once while recording the resulting fluorescence response. Different excitation wavelengths are sampled when scanning a time delay between two interfering broadband (over 100 nm) laser pulses. The time-dependent fluorescence response then yields the excitation spectrum through a Fourier transformation. Conceptually, our method is a pulsed single-molecule approach, in contrast to earlier continuous-wave Fourier excitation spectroscopy on bulk samples[31,32]. An important advantage of interferometric excitation over scanned narrowband excitation is that in the presence of blinking and even bleaching (provided that the main part of the interferogram around $\Delta t = 0$ is largely intact) full spectral information is still contained in the measured (though incomplete) interferogram, with no significant effect on the quality of the measured excitation spectrum. The insensitivity of the interferometric approach to fluorescence intensity fluctuations and spectral jumps has also been shown for interferometric detection of the emission spectra of individual molecules at cryogenic temperatures[33]. Moreover, the coherence and fs-resolution of the interfering broadband pulse-pair can be exploited to control the excited state of individual molecules before fluorescence emission[34].

Results

Inhomogeneity of single molecules. All our measurements of the excitation and emission spectra were performed on single quaterrylene diimide (QDI) molecules derived from a rylene dye family[35,36]. Rylene dyes have been extensively used in single-molecule experiments because of their extraordinary brightness and photostability[23,33,37–39]. The absorption spectrum of a solution of QDI in toluene peaks around 750 nm, as shown in Fig. 1b (solid, black line), whereas the emission

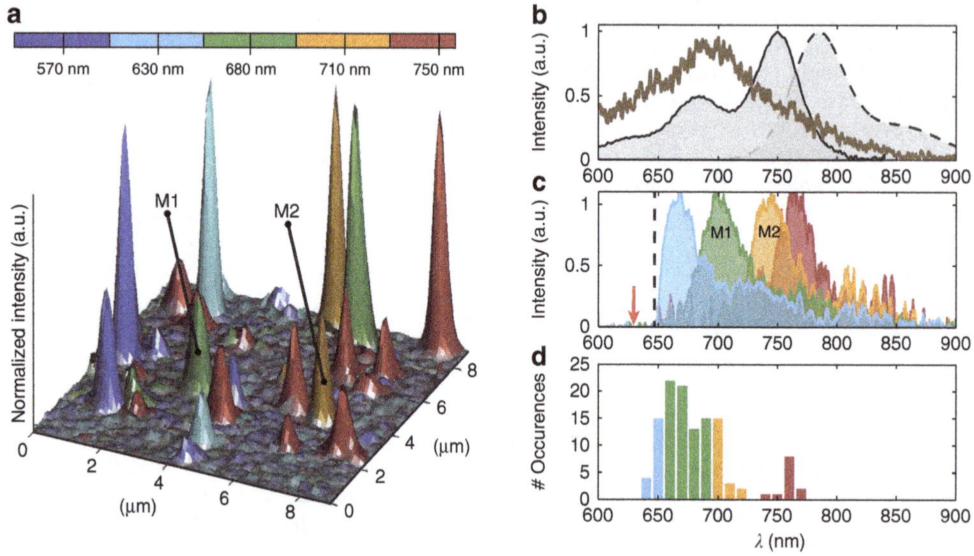

Figure 1 | Heterogeneity of single molecule emission. (**a**) A composite confocal image of dispersed QDI molecules in a PMMA matrix constructed from a series of five normalized images recorded at different excitation wavelengths. The assigned colour corresponds to appearance of the molecules in the image for a specific excitation wavelength. The height of the peaks reflects the relative fluorescence intensity between the molecules. (**b**) Absorption (black solid line) and emission (black dashed line) spectra of a solution of QDI in PMMA/toluene mixture. The brown solid line indicates the absorption spectrum of QDI molecules embedded in a solidified PMMA matrix. (**c**) Exemplary emission spectra associated with molecules absorbing at different wavelengths. Molecules M1 and M2 correspond to the two molecules from the composite confocal image in **a**. The red arrow indicates the excitation wavelength (633 nm), the black dashed line indicates the cutoff wavelength of the long-pass filter. (**d**) A histogram representing the positions of the maxima of the emission spectra for all measured molecules. The assigned colours correspond to the excitation wavelengths indicated in panel (**a**).

spectrum of the same QDI solution (upon excitation at 633 nm) is Stokes shifted to 780 nm (black dashed line).

We first turn our attention to the emission spectra of individual QDI molecules, which exhibit intriguing properties[40]. In Fig. 1a we show a composite confocal fluorescence image of QDI molecules embedded in a polymethyl methacrylate (PMMA) matrix. The image is constructed by superimposing five individually recorded images of the same area using different excitation (detection lower limit) wavelengths: 570 nm (610 nm); 630 nm (648 nm); 680 nm (730 nm); 710 nm (740 nm); and 750 nm (778 nm). Before the composition, images were normalized to unity. The height of the peaks reflects the relative fluorescence intensity between the molecules. Each molecule was labelled with a different colour according to the spectral range, in which it starts absorbing and emitting light. Molecules excited at the longest wavelength (750 nm) are labelled red; molecules that only start emitting when excited at 710 nm are labelled orange and so on. The efficient excitation of QDI molecules at short excitation wavelengths (around and below 650 nm) is surprising, as the absorption of QDI solution in this spectral range is very low (below 20% of the maximum). We have found, however, that the QDI absorption spectrum undergoes a significant blueshift (hypsochromic shift) and modification of the spectral shape, upon embedding the molecules in a PMMA matrix (see brown spectrum in Fig. 1b and Supplementary Note 1). The QDI/PMMA absorption spectrum is broadened and lacks the vibrational progression feature, which is clearly resolved in the solution absorption spectrum. The strong spectral shifts imply that to understand the spectral properties of QDI in PMMA, it is necessary to disentangle the excitation spectra of individual molecules.

To assess the spectral variation, we measured a total of 122 single-molecule emission spectra at excitation (detection) wavelength of 633 nm (above 655 nm). The emission spectra of four typical molecules representing the light blue, green, orange and red type molecules in the composite confocal image (Fig. 1a) are shown in Fig. 1c. The main emission peak position of the emission spectra was determined with a Gaussian fit, and the distribution of peak positions is plotted in Fig. 1d.

The presented data evidently shows that the emission spectra of individual QDI molecules in the PMMA matrix exhibit a remarkably large distribution of spectral shifts towards lower wavelengths with respect to the emission spectrum of the QDI ensemble solution spectrum. Moreover, it underlines that the maximum fluorescence peak can shift by more than 100 nm compared with the ensemble solution which has a maximum peak at 780 nm (see Fig. 1b, black, dashed line). The confocal image (Fig. 1a) illustrates that a significant amount of molecules only becomes visible when changing the wavelength by more than 150 from 750 to 570 nm. Conversely, the majority of molecules, emitting upon 750 nm excitation are not visible anymore when excited at 570 nm. The absorption of individual molecules shift spectrally more than the width of the QDI solution absorption spectrum (roughly 160 nm) and, thus, will remain unnoticed when the excitation wavelength is chosen solely based on the solution spectrum. This risk of selective detection is clearly observed in Fig. 1d. Both shape and position of the detected fluorescence emission distribution depend on the choice of excitation wavelength. These findings clearly indicate that a broad range of excitation wavelength is necessary to probe the entire distribution of emission spectra. The majority of single-molecule detection methods rely on the acquisition of fluorescence upon narrowband excitation with just a single wavelength, which limits and biases the quantitative analysis of spectral distributions in single-molecule experiments.

Although quite large, the observed spectral shifts and spectral variability are not uncommon in single-molecule emission spectra. In fact, analogous effects have been reported for a number of other molecules, including perylene diimide (molecular analogue of QDI) embedded in polyvinyl alcohol (PVA), for which ~40-nm wide distribution of emission spectra has been found because of the heterogeneity of the polymer matrix sites as well as twisted conformations of the core[41,42]. It has also been shown that changes in the photophysical form of the green fluorescent protein in polyvinyl alcohol lead to a spectral variations of nearly 100 nm due to π-stacking interactions, extended chromophoric π-systems and photoactivation[43]. Comparable spectral shifts have been measured in single carbocyanine dye molecules adsorbed on bare glass along with strong alterations in the shape of the vibronic bands[44].

Clearly, the dramatic variability of the observed emission spectra should be reflected in the excitation spectra, while differences give clues on the nature of the heterogeneity. In the following, we, thus, focus on the challenge to measure the excitation spectra of individual molecules to address their large spectral variability and directly compare excitation and emission spectra.

Broadband excitation spectroscopy. The experimental set-up is sketched in Fig. 2a (for details see Methods). The broadband laser excitation spectrum is derived from a Ti:sapphire oscillator and covers the wavelength range 655–770 nm, with 12 fs pulse duration. After propagating through a Mach–Zehnder interferometer including a delay line with a sub-fs interpulse delay precision, the interfering laser pulses with interpulse delay Δt excite a single molecule whose fluorescence response is detected. The total fluorescence intensity $F(\omega)$ is directly proportional to both the molecular excitation probability (absorption cross-section) $F_{\mathrm{QDI}}(\omega)$ and the laser spectral power $F_{\mathrm{laser}}(\omega)$ at excitation frequency ω. Each spectrum $F(\omega)$ has a Fourier-related time-dependent interferogram $H(\Delta t)$. A different combination of wavelengths is sampled at each delay line position and information on the excitation spectrum (that is, absorption cross-section) is contained in Δt dependent fluorescence intensity.

When exciting a molecule in its linear optical response regime, the measured fluorescence response $H(\Delta t)$ is the convolution of the laser interferometric autocorrelation function $H_{\mathrm{laser}}(\Delta t)$ and the interferogram of the excitation spectrum $H_{\mathrm{QDI}}(\Delta t)$ of the molecule: $H(\Delta t) \propto (H_{\mathrm{laser}} \otimes H_{\mathrm{QDI}})(\Delta t)$. Using the convolution theorem, the Fourier transformation of $H(\Delta t)$ turns into the product of the laser spectrum and QDI excitation spectrum: $F(\omega) \propto F_{\mathrm{laser}}(\omega) \cdot F_{\mathrm{QDI}}(\omega)$. Dividing $F(\omega)$ by the laser spectrum $F_{\mathrm{laser}}(\omega)$, we directly obtain the excitation spectrum of the molecule.

In Fig. 2b we show three examples of the time-dependent fluorescence response of single QDI molecules. Differences in the extent of the interference and beating in the interferogram recordings are a clear sign that each of the molecules interacts differently with the broadband excitation laser. Molecule M4 (red) shows a beating pattern on a time scale of ~25 fs, a clear signature of the presence of a superposition of (at least) two distinct frequency bands. The other two interferograms (molecule M1 and M2) clearly lack this feature. The corresponding Fourier transforms are shown in Fig. 2c. As expected, the red spectrum shows two bands whereas the other two (green, blue) show only one. Since the laser spectrum was identical in all three cases, the clear disparities between these spectra indicate differences in the excitation spectra of the molecules. The narrow spectra of M1 and M2 are either caused by a very narrow excitation spectrum of the molecule or the molecular excitation spectrum being shifted towards lower wavelengths, out of our laser spectral window.

Single-molecule excitation spectra. We have measured the fluorescence response of 25 molecules along with interferograms of the excitation laser. The excellent photostability of the QDI molecules enabled the recording of several consecutive interferograms for most molecules (a total of 95 individual interferograms).

In Fig. 3 we show distinct excitation spectra for five single QDI molecules (M1:M5). In Fig. 3a1–a5, we present the Fourier transformations of the fluorescence response (that is the product spectrum $F(\omega)$) measured on each individual molecule (solid line, green) and the laser spectrum measured aside the molecule (shaded area, blue). Dividing the green product spectra by the blue-laser spectra directly yields the excitation spectra of the molecules, which are presented in Fig. 3b1–b5 (solid line, red). For comparison, we also show the ensemble solution absorption spectrum (shaded area, grey), which has been spectrally offset from the ensemble position to match the measured single-molecule excitation spectra. Here we assume that only the absolute spectral position, and not the vibrational progression of individual QDI molecules undergoes significant change under encapsulation of the molecules in the PMMA matrix. The

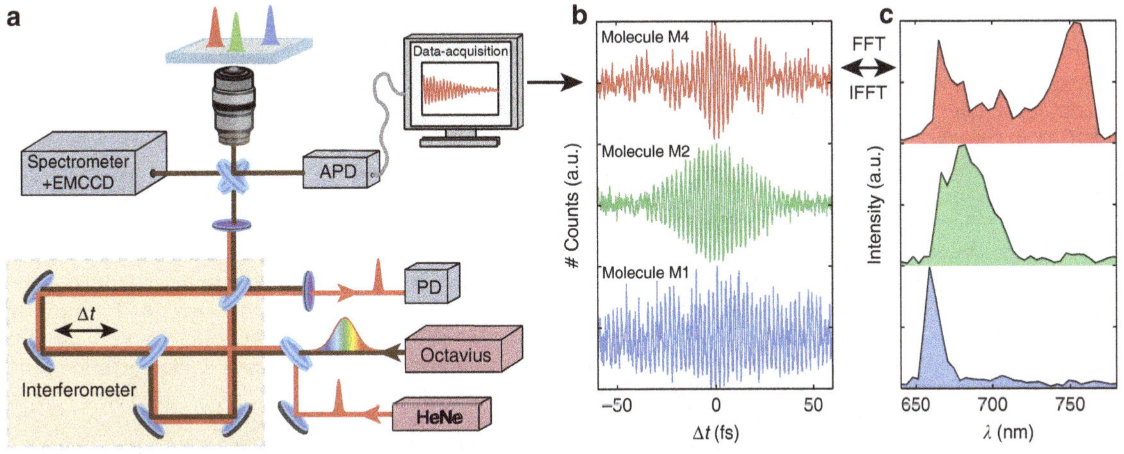

Figure 2 | The concept of broadband single-molecule excitation spectroscopy. (**a**) Schematic representations of the experimental set-up: a broadband fs laser (Octavius) is split in two branches, and a molecule of choice excited by the pulse-pair, while scanning the time delay Δt. (**b**) Three typical fluorescence interferograms of individual QDI molecules (M1, M2 and M4), each exhibiting a specific delay time-dependent fluorescence response and (**c**) their corresponding Fourier transformations. For full fluorescence interferograms see Supplementary Fig. 1. The typical contrast between the background count level and the fluorescence signal is 1/5–1/8 (see Supplementary Fig. 1). FFT, fast Fourier transform; IFFT, inverse fast Fourier transform

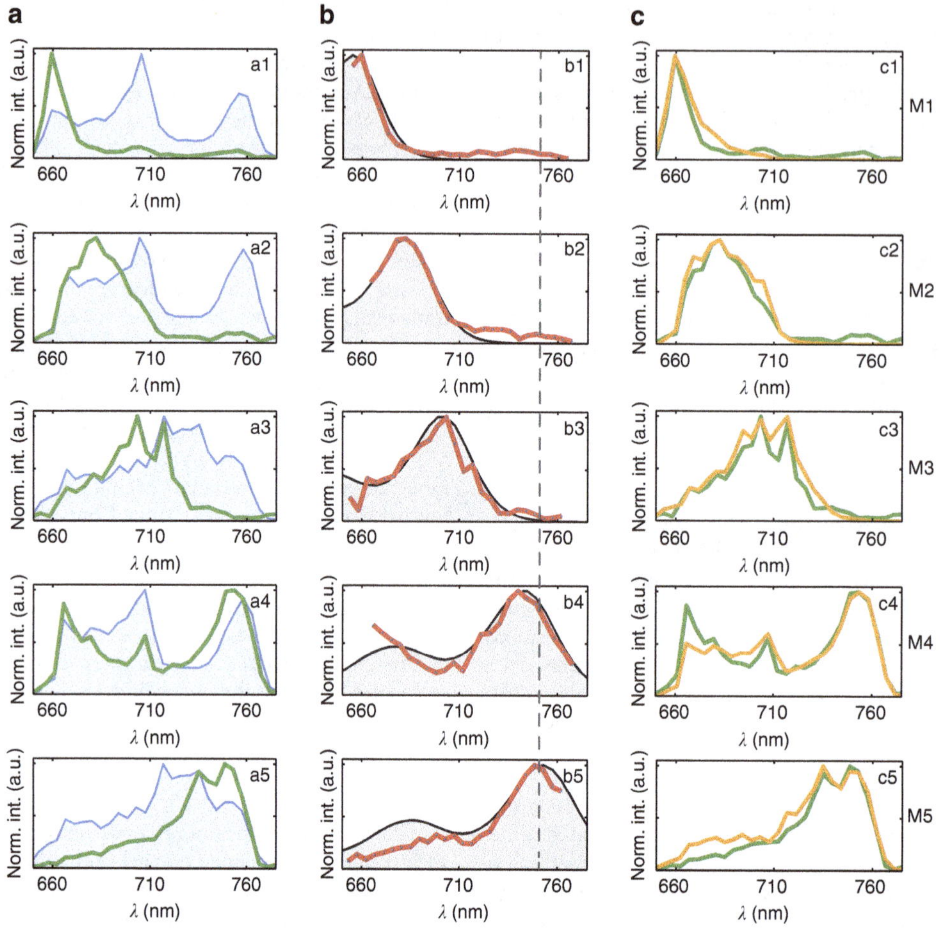

Figure 3 | Excitation spectra of single molecules at room temperature. (**a**) Experimental spectra of single molecules (green lines) for a series of five molecules (M1:M5), together with the laser spectrum (blue, shaded). (**b**) Single-molecule excitation spectra (red lines) obtained by dividing the experimental single-molecule product spectra by the corresponding laser spectra. Shaded grey spectra in **b** represent the ensemble absorption spectra shifted accordingly for direct comparison. The dashed line indicates the position of a maximum of the QDI ensemble absorption spectrum at 750 nm. (**c**) Single-molecule product spectra (green lines) compared with reconstructed theoretical spectra (orange lines) obtained by multiplying the appropriately shifted ensemble QDI absorption spectrum (grey, shaded) by the experimental laser spectra (blue, shaded).

excitation spectra of another 6 individual QDI molecules (M6:M11) are shown in Supplementary Fig. 2.

We find that the measured single-molecule excitation spectra are strongly shifted compared with the ensemble absorption spectrum. Interestingly, we can, however, reproduce the experimental product spectrum by multiplying the measured laser spectrum by the appropriately spectrally shifted QDI ensemble solution spectrum. The comparison between the measured (solid line, green) and reproduced (solid line, orange) single-molecule product spectrum is shown in Fig. 3c. The two curves agree quite well for all the measured molecules.

As can be seen in Fig. 3a, we have used two different shapes of laser spectra for the measurements (M1, M2 and M4—first shape; M3 and M5—second shape). We thus confirm that our approach and its sensitivity do not depend on the spectral shape of the laser. Each of the 25 molecules we measured exhibited a different excitation spectrum. While we observed hypsochromic shifts of up to 100 nm (see M1), we did not observe any molecules red-shifted by more than a few nm (see M5) with respect to the ensemble solution spectrum. The use of broadband excitation allows us to directly sample the distribution of single-molecule spectra over a 100 nm broad band, which with narrowband excitation would be difficult to probe. Because the measured excitation spectra of the molecules with small hypsochromic shifts (that is, large overlap with our laser spectrum) indeed resemble the ensemble solution absorption spectrum, we expect that the more (above 100 nm) blue-shifted molecules exhibit similar excitation spectrum containing the vibrational progression. For M1, M2 and M3, however, the vibrational bands are outside the spectral window of our broadband excitation laser.

Variations in the shape of the excitation spectra. Upon closer inspection, our results do display more spectral variability than just the spectral shifts shown in Fig. 3. The single-molecule excitation spectra exhibit a variety of inter-vibronic band distances, spectral band widths and relative peak intensities. In Fig. 4a, we demonstrate three typical excitation spectra for which the main electronic transition band (0′–0″, number indicating vibrational level, single/double apostrophe noting ground/excited electronic state) overlap the best. Differences in the relative peak intensities are clearly visible. Moreover, changes in the distance between the two vibronic transitions (0′–0″ and 0′–1″) are noticeable in the spectra. Similar spectral variations can also be found in the emission spectra, which for comparison we show in Fig. 4b. These observations illustrate that we have achieved sufficient sensitivity in the measured excitation spectra to probe the effect of the nanoenvironment on the vibrational modes of individual molecules.

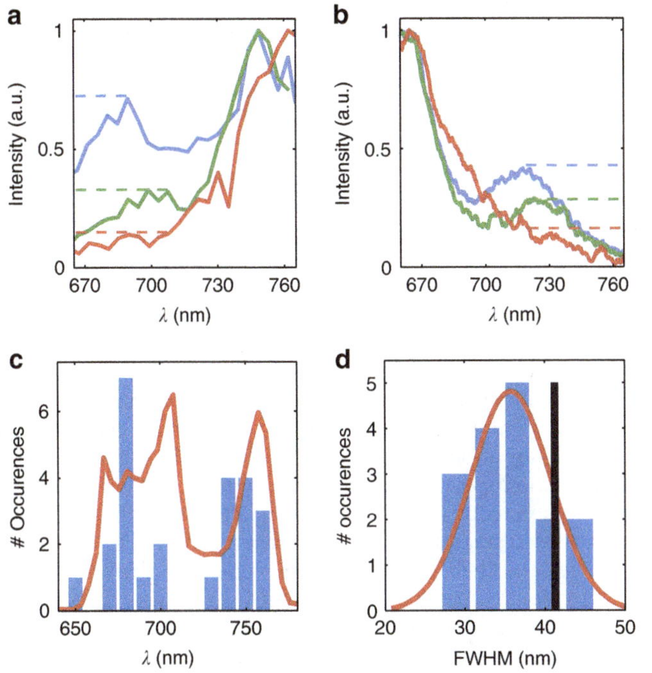

Figure 4 | Heterogeneity of the excitation and emission spectra. Comparison of three single-molecule excitation (**a**) and emission (**b**) spectra showed with red, green and blue lines. Both types of spectra show a similar spread of relative peak intensities and inter-vibronic band distances (see corresponding colours). (**c**) A histogram presenting spread of the 0′-0″ transistion positions of all measured excitation spectra along with the laser spectrum (red). (**d**) Distribution of the 0′-0″ transition peak widths along with a fitted Gaussian spread function. The vertical black line indicates the width of the transition in the bulk solution spectrum.

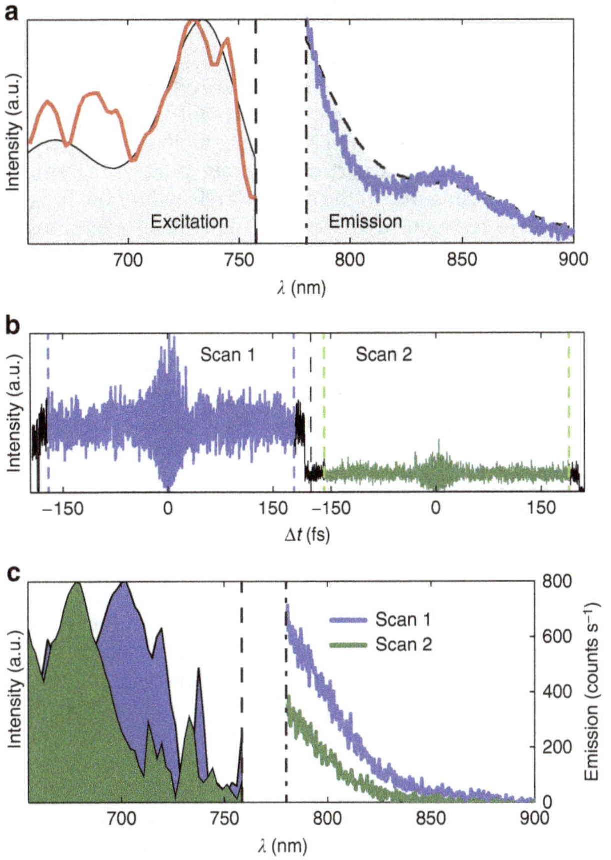

Figure 5 | Simultaneous detection of emission and excitation spectra of individual molecules. (**a**) Exemplary excitation (red) and emission (blue) spectra measured simultaneously on the same molecule. For comparison we plot the QDI solution absorption and emission spectra (shaded grey) separated by the solution Stokes shift. (**b**) Two (blue and green) consecutive fluorescence interferogram scans measured on the same molecule. Interferogram sections marked with a black line are the parts ommited in the Fourier transformation analysis. (**c**) The corresponding excitation spectra alongside the simultaneously recorded emission spectra for each (blue and green) of the two scans. Vertical dashed and dash-dotted lines in **a** and **c** indicate cutoff wavelengths of the laser and long-pass filter, respectively.

We determined a spread of the main transition positions for all the measured excitation spectra in Fig. 4c. We find that the 25 excitation spectra are rather homogeneously distributed across the laser spectrum (shown in red). As the fluorescence is detected above 785 nm, less emission is collected for molecules that are very blue-shifted compared with those absorbing at longer wavelengths, making the former underrepresented in our choice of excitation range. Furthermore, the shape of the laser spectrum may also have an indirect effect on the measured distribution—in case the laser spectrum would be more intense on the red side than on the blue side, the molecules absorbing on the blue side would be even further underrepresented as they would appear even dimmer in the confocal image. This once again pinpoints the importance of the right choice of excitation wavelength, detection range and even intensity distribution of the laser spectrum. We did not find any correlation between the position of the main electronic transition band and its spectral width. In forthcoming experiments, utilizing a broader laser excitation spectrum, our approach should allow us to study possible correlations between the spectral position of the excitation spectrum and the spectral properties of the vibrational progression (separation, intensity and widths).

For completeness, we have analysed the width of the main transition band (0′-0″) for 16 out of 25 measured excitation spectra (those for which the main transition band was completely overlapping with the spectral window of the laser) by fitting a Gaussian profile. The result is shown in a form of a histogram in Fig. 4d. We found that the average width of the main electronic transition band for the measured single molecules is 35.5 nm, which is nearly 20% narrower (roughly 6 nm) than that of the ensemble absorption band (41.5 nm, indicated with vertical black line). The reduced width of the main electronic transition band for the single molecules clearly indicates that we are resolving the inhomogeneous broadening of the ensemble. The observed widths for individual molecules, however, might still be broadened due to spectral diffusion occurring during the acquisition time of the interferogram.

The observed variations in the excitation spectra are not because of random fluctuations in the experiment, or low signal-to-noise ratio as the variations between successively acquired excitation spectra on the same molecule are typically much smaller than differences in excitation spectra between different molecules.

Simultaneous detection of excitation and emission spectra. Finally, we modified the experimental set-up to allow simultaneous detection of excitation and emission spectra of the same molecule at room temperature. To this end we placed a 50/50 beamsplitter in the fluorescence detection path and used half of the fluorescence signal for the detection of excitation spectra with the APD and the other half for the detection of emission spectra with the electron multiplying charge-coupled device (EMCCD) camera. In Fig. 5a we show the simultaneously recorded excitation (red) and emission (blue) spectrum of an individual QDI molecule. For comparison we also plot the QDI solution absorption and emission spectra (shaded grey). We spectrally offsetted both solution spectra to match the single-molecule spectra without changing the Stokes shift separation of the solution spectra. Both solution and single-molecule spectra match each other quite well. A part of the main peak of the emission spectrum is missing due to the cutoff wavelength of the long-pass filter used to filter out the laser light. For future experiments it might be advantageous to use a long-pass filter with a sharper cutoff slope and to move the cutoff point towards the middle of the Stokes separation.

QDI molecules are generally stable in their fluorescence emission; however, on a few occassions we observed jumps in the fluorescence intensity while recording fluorescence interferograms. On even fewer occassions such intensity fluctuations occurred towards the start or end of the recorded interferogram. A nice example of such discrete jump is shown in Fig. 5b. The investigated molecule clearly emits photons at distinct fluorescence levels. When analysing the two scans separately, the excitation spectra on the left side of Fig. 5c are obtained. They clearly have the main excitation peak at distinct spectral positions, separated by $\sim$20 nm. The corresponding, simultaneously acquired emission spectra are plotted alongside for both scans, showing similar shape but different intensity.

As this particular molecule absorbed and emitted more towards the blue side, we could not draw any direct conclusions on the spectral shift or position of the emission spectra, other than that we only detected the slope of the second emission band. However, we were able to compare the ratio in measured total fluorescence intensity to the ratio in excitation efficiency due to the spectral shift of the molecules' absorption spectrum. Taking the integrated product spectrum (which is the molecule's excitation spectrum times the laser spectrum), we found a ratio of 1/1.4 in excitation efficiency between the two scans. The ratio of fluorescence intensity between the two scans, however, is much larger and amounts to 1/4. It is thus clear that the emission spectrum must shift along with the excitation spectrum towards the blue, this way moving further out of the detection window (determined by the long-pass filter) and accounting for the lower detected fluorescence in the interferogram scans.

Discussion

We have demonstrated that our broadband Fourier excitation approach effectively captures excitation spectra of individual molecules at room temperature. We, thus, put forward yet another way of detecting and investigating single molecules, right along with absorption, emission and scattering. The single-molecule excitation spectra allow us to resolve the intrinsic and environmentally induced inhomogeneities in the excited state potential of single molecules. These are inaccessible through emission spectroscopy (or Raman scattering techniques), which only probe the ground state potential. Furthermore, we have shown that the interplay between the excitation wavelength and detection spectral window may obscure the real extent of spectral inhomogeneity among single molecules, as indicated recently in the single-molecule experiment at cryogenic temperature[45]. One might fail to detect a significant fraction of the spectrally distributed molecules. This further corroborates the need of broadband excitation spectroscopy for the study of the single-molecule static and dynamic properties and their quantitative analysis of both emission and excitation spectroscopy.

The presented broadband excitation spectroscopy technique has the potential to contribute to the understanding of many dynamical processes such as intersystem crossing which typically affects the excitation spectrum but not the emission and absorption spectra. Consequently, we can probe the mechanisms and spectral dependence of intra- and intermolecular intersystem crossing, which so far has mainly been investigated at cryogenic conditions through statistical analysis of the blinking dynamics[46–50] and through fluorescence-detected magnetic resonance spectroscopy[51–53]. Similarly, our approach offers new insights into single-molecule blinking dynamics. For single-molecule emission spectra acquired with a narrow excitation bandwidth, it is very challenging to differentiate between different blinking mechanisms like spectral jumps outside the excitation window and transitions to dark states. As we demonstrated, the broadband excitation spectroscopy approach allows us to follow the excitation spectrum in time and thus to verify correlations between blinking and spectral changes on the excitation side. We presented a proof-of-principal experiment, which demonstrates the feasibility of simultaneous acquisition of both the emission and excitation spectra and ambient conditions. We believe that this advancement in the single-molecule spectroscopy will allow for detailed studies on the interaction between individual molecules and their environment as well as for characterization of molecules in more complex systems. Finally, the technique will prove useful to follow slow-occurring chemical reactions in time through changes in the molecule's excitation spectrum both in solution and on the single-molecule level.

We note that our experimental approach is based on the concept of Fourier transform spectroscopy, however, with two important differences with respect to already established techniques. Firstly, commercially available FT spectrometers do not offer single-molecule sensitivity and do not allow for active control over excitation processes. As we use a coherent broadband light source we can manipulate the phase, time delay and chirp of the pulses. Transform limited pulses (in case of our set-up 12 fs) offer a time resolution that can be used to coherently probe femtosecond dynamics and (de)coherence of single molecules[54,55]. Using a laser with a sufficiently broad spectrum (exceeding the width of the absorption spectrum) it is possible to extract the dephasing time directly from the measured interferograms. Secondly, using two interfering excitation beams it should be possible to measure the excitation spectrum of non-fluorescent (single) molecules by detecting stimulated emission photons, a detection scheme which has already shown to reach nearly single-molecule sensitivity[56]. This would give access to the information normally obtained through fluorescence, but from non-fluorescent molecules.

On a technical note, the presented technique requires a broadband laser source and a simple (Mach–Zehnder) interferometer. However, it does not require any other modifications to a fluorescence detection scheme and thus is compatible with any confocal single-molecule detection set-up. It is worth noting that acquiring an excitation spectrum or an emission spectrum of a single molecule typically requires a similar number of photocounts and thus can be acquired in a comparable time. To obtain the presented excitation spectra of single molecules (with 4 nm resolution), a single interferogram was acquired for $\sim$120 s (see Methods). By reducing the sampling frequency and/or the data range (which decreases spectral resolution and spectral range) it is possible to record an interferogram within roughly 10 s without significantly affecting the shape of the measured excitation spectrum. The effect of reducing the resolution of the measured interferogram or its temporal range on the shape and quality of the excitation spectrum for molecule M4 (see Fig. 3), is shown in Supplementary Fig. 3. The acquisition time to obtain a reasonable fluorescence emission spectrum with the same excitation power is typically 2–10 s, depending on the brightness of the molecule. The few tens of seconds needed to measure an excitation spectrum of single molecules is comparable with the survival time of many biologically relevant fluorescing molecules. Therefore excitation spectra of molecules with low quantum effciency and photostability, like light harvesting complexes, is within reach.

Methods

Broadband excitation spectroscopy. The experimental set-up is schematically depicted in Fig. 2a. We used a broadband titanium-sapphire laser (Octavius-85M, Thorlabs) operating at 85 MHz and tuned to a central wavelength of 710 nm with a bandwidth of about 120 nm. The laser pulses were split into two parts in a Mach–Zehnder-type interferometer consisting of two identical 50/50 beamsplitters (Semrock) and a mechanical delay line (NRT 100/M, Thorlabs), which we used to control the path difference between the two arms of the interferometer. The interfering pulse pairs were propagated collinearly into an inverted microscope (Observer D1, Zeiss). A reference HeNe laser beam was propagated through the same interferometer, separated from the Ti:sapphire light using two band-pass filters (617 ± 36, 632 ± 11 nm, Semrock) and detected with a photodiode (PDA36A, Thorlabs). The derived reference interferogram was used to precisely determine the optical path difference between the interfering broadband pulse-pair. Before entering the microscope, the HeNe light was filtered out using a long-pass filter (635 nm LP, Semrock). In the microscope, the broadband Octavius pulses were reflected from a 50/50 beamsplitter and focused to a diffraction-limited spot on the sample with a high numerical aperture objective (1.3NA, × 100, Zeiss Fluar). The sample was placed on a piezo-controlled stage (Mad City Labs) allowing precise positioning of the molecules in the focal spot. The fluorescence from the sample was collected in reflection through the same objective and beamsplitter and sent either to a spectrometer equipped with an EMCCD camera for spectral detection (Newton, Andor) or to an avalanche photodiode (Perkin-Elmer) that allowed confocal optical imaging of the sample. The fluorescence was separated from the laser light using two long-pass filters (780 nm LP and 785 nm LP, Semrock). The laser interferograms were detected with a diffuser placed in front of an APD.

The experiment was performed as follows: first a confocal image of the sample was recorded. A molecule was placed in the focal spot, its position was optimized based on the intensity of the fluorescence signal and the interferogram scans were recorded until the molecule photobleached. Finally, the long-pass filter was replaced with the diffuser, the molecule was translated out of the focal spot and a series of reference laser interferograms was recorded.

Detection efficiency. The laser power was set to roughly 1 μW for a single beam at the sample position, which corresponds to around 450 W cm^{-2}. Considering transmission of optical elements in the detection path we were able to detect approximately 35% of the emitted fluorescence photons. For a typical QDI molecule this yielded 2–3 kcounts s^{-1}. The delay line was scanned over 60 μm with an interpulse delay ranging from − 200 to + 200 fs. The delay line velocity was set to 0.5 μm s^{-1}, which yielded a total single interferogram acquisition time of 120 s. The acquisition time was set to 10 ms, which including software data handling time, resulted in 12–13 measurement points per oscillation period of the broadband laser (corresponding to sampling rate of $\sim 5 \times 10^{15}$ Hz or a measurement point every 0.15 fs). The resulting typical resolution in the frequency domain was 4.5 nm.

Emission spectra. For the emission spectra, single molecules were excited by the HeNe laser (at 633 nm) using excitation intensity of 1–2 μW (450–900 W cm^{-2}). The fluorescence was separated from the excitation light using a notch filter (633/25 nm) for the HeNe laser or a long-pass filter (690 nm cutoff) for the Ti:sapphire laser. Typically 20 emission spectra were recorded in a series with integration time of 11 s/spectrum. The gain of the charge-coupled device camera was set to 200. The emission spectrum of the QDI solution (10^{-6} M in 1% w/v toluene/PMMA) shown in Fig. 1b (shaded grey) was measured using HeNe excitation in combination with a 635 nm LP filter.

Absorption spectra. The absorption spectrum of QDI (10^{-6} M) in toluene/PMMA (1% w/v) solution (Fig. 1b, shaded grey) was measured using a commercial spectrophotometer (NanoDrop 2000, Thermo Scientific). The absorption spectra as a function of evaporation (shown in Fig. 1b, also see Supplementary Fig. 4) were measured using the microscope's built-in halogen lamp. Light transmitted through the sample was collected through the objective and detected with the EMCCD camera. Typically 40 μl of solution (10^{-5} M) was placed on top of a microscope cover slip and a kinetic series of 8,000 transmitted light spectra was acquired, with integration time of 1 s per spectrum.

Multicolour excitation. Confocal fluorescence images at different excitation wavelengths were recorded using a 5 nm broad excitation bands (in combination with a long-pass filter for detection): 570 nm (610 nm); 630 nm (648 nm); 680 nm (730 nm); 710 nm (740 nm); and 750 nm (778 nm); derived from an ultra-broadband laser (SuperK, NKT).

Sample preparation. The QDI molecules were obtained from the Müllen group (Max Planck Institute for Polymer Research, Mainz, Germany). The samples were prepared by spin-coating a solution of QDI molecules at a roughly nM concentration in PMMA/toluene mixture (1% w/v). Approximately 50 μl of the solution was spin-coated on a #1 microscope cover slip for 60 s, at a spinning rate of 2,000 r.p.m. Before sample deposition, microscope cover slips were cleaned by leaving them in a piranha solution (1:2 ratio hydrogen peroxide to sulfuric acid) for about 30 min, then rinsing with deionized water and blow drying with nitrogen. We found this procedure to yield no or very little contamination on the cover slips.

References

1. Dickson, R. M., Cubitt, A. B., Tsien, R. Y. & Moerner, W. E. On/off blinking and switching behaviour of single molecules of green fluorescent protein. *Nature* **388,** 355–358 (1997).
2. Basché, T. & Moerner, W. E. Optical modification of a single impurity molecule in a solid. *Nature* **355,** 335–337 (1992).
3. Zumbusch, A., Fleury, L., Brown, R., Bernard, J. & Orrit, M. Probing individual two-level systems in a polymer by correlation of single molecule fluorescence. *Phys. Rev. Lett.* **70,** 3584–3587 (1993).
4. Ambrose, W. P. & Moerner, W. E. Fluorescence spectroscopy and spectral diffusion of single impurity molecules in a crystal. *Nature* **349,** 225–227 (1991).
5. Myers, A. B., Tchénio, P., Zgierski, M. Z. & Moerner, W. E. Vibronic spectroscopy of individual molecules in solids. *J. Phys. Chem.* **98,** 10377–10390 (1994).
6. Kiraz, A., Ehrl, M., Bräuchle, C. & Zumbusch, A. Low temperature single molecule spectroscopy using vibronic excitation and dispersed fluorescence detection. *J. Chem. Phys.* **118,** 10821–10824 (2003).
7. Moerner, W. E. & Kador, L. Optical detection and spectroscopy of single molecules in a solid. *Phys. Rev. Lett.* **62,** 2535–2538 (1989).
8. Orrit, M. & Bernard, J. Single pentacene molecules detected by fluorescence excitation in a p-terphenyl crystal. *Phys. Rev. Lett.* **65,** 2716–2719 (1990).
9. Feist, F. A., Tommaseo, G. & Basché, T. Observation of very narrow linewidths in the fluorescence excitation spectra of single conjugated polymer chains at 1.2 K. *Phys. Rev. Lett.* **98,** 208301–208305 (2007).
10. Feist, F. A. & Basché, T. Fluorescence excitation and emission spectroscopy on single MEH-PPV chains at low temperature. *J. Phys. Chem. B* **112,** 9700–9708 (2008).
11. Tchénio, P., Myers, A. B. & Moerner, W. E. Vibrational analysis of the dispersed fluorescence from single molecules of terrylene in polyethylene. *Chem. Phys. Lett.* **213,** 325–332 (1993).
12. Tchénio, P., Myers, A. B. & Moerner, W. E. Dispersed fluorescence spectra of single molecules of pentacene in p-terphenyl. *J. Phys. Chem.* **97,** 2491–2493 (1993).
13. Betzig, E. & Chichester, R. J. Single molecules observed by near-field scanning optical microscopy. *Science* **262,** 1422–1425 (1993).
14. Nie, S., Chiu, D. T. & Zare, R. N. Probing individual molecules with confocal fluorescence microscopy. *Science* **266,** 1018–1021 (1994).
15. Trautman, J. K. & Macklin, J. J. Time-resolved spectroscopy of single molecules using near-field and far-field optics. *Chem. Phys.* **205,** 221–229 (1996).
16. Xie, X. S. & Trautman, J. K. Optical studies of single molecules at room temperature. *Annu. Rev. Phys. Chem.* **49,** 441–480 (1998).
17. Trautman, J. K., Macklin, J. J., Brus, L. E. & Betzig, E. Near-field spectroscopy of single molecules at room temperature. *Nature* **369,** 40–42 (1994).
18. Xie, X. S. Single molecule spectroscopy and dynamics. *Acc. Chem. Res.* **29,** 598–606 (1996).
19. Gaiduk, A., Yorulmaz, M., Ruijgrok, P. V. & Orrit, M. Room-temperature detection of a single molecule's absorption by photothermal contrast. *Science* **330,** 353–356 (2010).
20. Piliarik, M. & Sandoghdar, V. Direct optical sensing of single unlabelled proteins and super-resolution imaging of their binding sites. *Nat. Commun.* **5,** 4495 (2014).
21. Ortega-Arroyo, J. & Kukura, P. Interferometric scattering microscopy (iSCAT): new frontiers in ultrafast and ultrasensitive optical microscopy. *Phys. Chem. Chem. Phys.* **14,** 15625–15636 (2012).
22. Zijlstra, P., Paulo, P. M. R. & Orrit, M. Optical detection of single non-absorbing molecules using the surface plasmon resonance of a gold nanorod. *Nat. Nanotechnol.* **7,** 379–382 (2012).
23. Celebrano, M., Kukura, P., Renn, A. & Sandoghdar, V. Single-molecule imaging by optical absorption. *Nat. Photon.* **5,** 95–98 (2011).
24. Kukura, P., Celebrano, M., Renn, A. & Sandoghdar, V. Single-molecule sensitivity in optical absorption at room temperature. *J. Phys. Chem. Lett.* **1,** 3323–3327 (2010).
25. Chong, S., Min, W. & Xie, X. S. Ground-state depletion microscopy: detection sensitivity of single-molecule optical absorption at room temperature. *J. Phys. Chem. Lett.* **1,** 3316–3322 (2010).
26. Stopel, M. H. W., Blum, C. & Subramaniam, V. Excitation spectra and stokes shift measurements of single organic dyes at room temperature. *J. Phys. Chem. Lett.* **5,** 3259–3264 (2014).

27. Weigel, A., Sebesta, A. & Kukura, P. Shaped and feedback-controlled excitation of single molecules in the weak-field limit. *J. Phys. Chem. Lett.* **6**, 4032–4037 (2015).
28. Pfündel, E. & Baake, E. A quantitive description of fluorescence excitation spectra in intact bean leaves greened under intermittent light. *Photosynth. Res.* **26**, 19–28 (1990).
29. Matthews, B. J. H., Jones, A. C., Theodoru, N. K. & Tudhope, A. W. Excitation-emission-matrix fluorescence spectroscopy applied to humic acid bands in coral reefs. *Mar. Chem.* **55**, 317–332 (1996).
30. Juzenas, P., Juzeniene, A., Kaalhus, O., Iani, V. & Moan, J. Noninvasive fluorescence excitation spectroscopy during application of 5-aminolevulinic acid in vivo. *Photochem. Photobiol. Sci.* **1**, 745–748 (2002).
31. Hirschberg, J. G. *et al.* Interferometric measurement of fluorescence excitation spectra. *Appl. Opt.* **37**, 1953–1957 (1998).
32. Paul, R., Steiner, A. & Gemperlein, R. Spectral sensitivity of *Calliphora erythrocephala* and other insect species studied with Fourier interferometric stimulation (FIS). *J. Comp. Physiol. A* **158**, 669–680 (1986).
33. Korlacki, R., Steiner, M., Qian, H., Hartschuh, A. & Meixner, A. J. Optical fourier transform spectroscopy of a single-walled carbon nanotube and single molecules. *Chem. Phys. Chem.* **8**, 1049–1055 (2007).
34. Dantus, M. & Lozovoy, V. V. Experimental coherent laser control of physicochemical processes. *Chem. Rev.* **104**, 1813–1859 (2004).
35. Chen, L., Li, C. & Müllen, K. Beyond perylene diimides: synthesis, assembly and function of higher rylene chromophores. *J. Mater. Chem. C* **2**, 1938–1956 (2014).
36. Geerts, Y. *et al.* Quaterrylenebis(dicarboximide)s: near infrared absorbing and emitting dyes. *J. Mater. Chem.* **8**, 2357–2369 (1998).
37. Hildner, R., Brinks, D. & van Hulst, N. F. Femtosecond coherence and quantum control of single molecules at room temperature. *Nature Phys.* **7**, 172–177 (2011).
38. Hwang, J., Fejer, M. M. & Moerner, W. E. Scanning interferometric microscopy for the detection of ultrasmall phase shifts in condensed matter. *Phys. Rev. A* **73**, 021802 (2006).
39. Brinks, D. *et al.* Visualizing and controlling vibrational wave packets of single molecules. *Nature* **465**, 905–908 (2010).
40. Piatkowski, L., Gellings, E. & van Hulst, N. F. Multicolour single molecule emission and excitation spectroscopy reveals extensive spectral shifts. *Faraday Discuss.* **184**, 207–220 (2015).
41. Margineanu, A. *et al.* Photophysics of a water – soluble rylene dye: comparison with other fluorescent molecules for biological applications. *J. Phys. Chem. B* **108**, 12242–12251 (2004).
42. Hofkens, J. *et al.* Conformational rearrangements in and twisting of a single molecule. *Chem. Phys. Lett.* **333**, 255–263 (2001).
43. Blum, C., Meixner, A. J. & Subramaniam, V. Room temperature spectrally resolved single-molecule spectroscopy reveals new spectral forms and photophysical versatility of *Aequorea* green fluorescent protein variants. *Biophysical J.* **87**, 4172–4179 (2004).
44. Weston, K. D., Carson, P. J., Metiu, H. & Buratto, S. K. Room-temperature fluorescence characteristics of single dye molecules adsorbed on a glass surface. *J. Chem. Phys.* **109**, 7474–7485 (1998).
45. Kunz, R. *et al.* Single-molecule spectroscopy unmasks the lowest exciton state of the B850 assembly in LH2 from Rps. acidophila. *Biophys. J.* **106**, 2008–2016 (2014).
46. Kol'chenko, M. A., Kozankiewicz, B., Nicolet, A. & Orrit, M. Intersystem crossing mechanisms and single molecule fluorescence: terrylene in anthracene crystals. *Opt. Spectrosc.* **98**, 681–686 (2005).
47. Kozankiewicz, B. & Orrit, M. Single-molecule photophysics, from cryogenic to ambient conditions. *Chem. Soc. Rev.* **43**, 1029–1043 (2014).
48. Mais, S. *et al.* Terrylenediimide: A novel fluorophore for single-molecule spectroscopy and microscopy from 1.4 K to room temperature. *J. Phys. Chem. A* **101**, 8435–8440 (1997).
49. Kummer, S., Basché, T. & Bräuchle, C. Terrylene in p-terphenyl: a novel single crystalline system for single molecule spectroscopy at low temperatures. *Chem. Phys. Lett.* **229**, 309–316 (1994).
50. Lang, E. *et al.* Comparison of the photophysical parameters for three perylene bisimide derivatives by single-molecule spectroscopy. *Chem. Phys. Chem.* **8**, 1487–1496 (2007).
51. Brouwer, A. C. J., Groenen, E. J. J. & Schmidt, J. Detecting magnetic resonance through quantum jumps of single molecules. *Phys. Rev. Lett.* **80**, 3944–3947 (1998).
52. Brouwer, A. C. J., Köhler, J., van Oijen, A. M., Groenen, E. J. J. & Schmidt, J. Single-molecules fluorescence autocorrelation experiments on pentacene: the dependence of intersystem crossing on isotopic composition. *J. Chem. Phys.* **110**, 9151–9159 (1999).
53. Brown, R., Wrachtrup, J., Orrit, M., Bernard, J. & von Borczyskowski, C. Kinetics of optically detected magnetic resonance of single molecules. *J. Chem. Phys.* **100**, 7182–7191 (1994).
54. Scherer, N. F. *et al.* Fluorescence-detected wave packet interferometry: time resolved molecular spectroscopy with sequences of femtosecond phase-locked pulses. *J. Chem. Phys.* **95**, 1487–1512 (1991).
55. Milota, F., Sperling, J., Szöcs, V., Tortschanoff, A. & Kauffmann, H. F. Correlation of femtosecond wave packets and fluorescence interference in a conjugated polymer: towards the measurement of site homogeneous dephasing. *J. Chem. Phys.* **120**, 9870–9884 (2004).
56. Min, W. *et al.* Imaging chromophores with undetectable fluorescence by stimulated emission microscopy. *Nature* **461**, 1105–1109 (2009).

Acknowledgements

We thank Yves L. A. Rezus and Daan Brinks for critical reading of the manuscript. L.P. acknowledges financial support from the Marie-Curie International Fellowship COFUND and the ICFOnest program. E.G. acknowledges financial support from the Erasmus + program. This research was supported by the European Commission (ERC Advanced Grant 247330-NanoAntennas), Spanish MINECO (PlanNacional project FIS2012-35527, network FIS2014-55563-REDC and Severo Ochoa grant SEV2015-0522), the Catalan AGAUR (2014 SGR01540) and Fundació CELLEX (Barcelona).

Author contributions

L.P. and N.F.v.H. designed the experiment. E.G. and L.P. performed the experiments and data analysis. E.G., L.P. and N.F.v.H. wrote the manuscript.

Additional information

Competing financial interests: The authors declare no competing financial interests.

4

Photoacoustics of single laser-trapped nanodroplets for the direct observation of nanofocusing in aerosol photokinetics

Johannes W. Cremer[1], Klemens M. Thaler[2], Christoph Haisch[2] & Ruth Signorell[1]

Photochemistry taking place in atmospheric aerosol droplets has a significant impact on the Earth's climate. Nanofocusing of electromagnetic radiation inside aerosols plays a crucial role in their absorption behaviour, since the radiation flux inside the droplet strongly affects the activation rate of photochemically active species. However, size-dependent nanofocusing effects in the photokinetics of small aerosols have escaped direct observation due to the inability to measure absorption signatures from single droplets. Here we show that photoacoustic measurements on optically trapped single nanodroplets provide a direct, broadly applicable method to measure absorption with attolitre sensitivity. We demonstrate for a model aerosol that the photolysis is accelerated by an order of magnitude in the sub-micron to micron size range, compared with larger droplets. The versatility of our technique promises broad applicability to absorption studies of aerosol particles, such as atmospheric aerosols where quantitative photokinetic data are critical for climate predictions.

[1] Department of Chemistry and Applied Biosciences, Laboratory of Physical Chemistry, ETH Zurich, Vladimir-Prelog-Weg 2, CH-8093 Zurich, Switzerland.
[2] Laboratory for Applied Laser Spectroscopy, Chair of Analytical Chemistry, Technical University of Munich, Marchioninistrasse 17, D-81377 Munich, Germany. Correspondence and requests for materials should be addressed to R.S. (email: rsignorell@ethz.ch).

Understanding fundamental processes that govern the reaction dynamics of gas phase, aerosol and cloud processes is crucial for the advancement of global atmospheric chemistry modelling[1–14]. Much of the chemistry occurring in the Earth's atmosphere is driven by sunlight. Photochemical reactions, in which aerosol particles or droplets act as the active reaction medium, can be highly complex because they are influenced by optical phenomena, transport properties and surface effects[2]. Optical phenomena play a fundamental role in light-initiated particle processes since the radiation flux within the particles determines the activation rate of the photochemically active species.

Focusing of electromagnetic radiation inside small particles leads to an enhancement of the overall light intensity, compared with the intensity of the incident radiation and to structuring and localization of the internal optical fields[15–23]. These phenomena depend strongly on the particle size, the particle composition and the wavelength of electromagnetic radiation. The fundamental influence of the enhanced electromagnetic energy density on the rate of photochemical reactions in micro- and nanodroplets has been recognized and calculations have provided limited evidence for enhanced photochemical rates[24–27]. Experimental results remain inconclusive concerning the influence of light enhancement on the kinetics, mainly because direct observation of the actual photoactive step was not possible[23,28–34]. The observation of size-dependent effects in ensembles of aerosol or emulsion droplets is often hindered because the droplet size distribution cannot be varied and determined with the necessary accuracy. However, even single-droplet techniques have so far not provided size-dependent photolysis rates because the direct measurement of the population decay of the photoactive substance was not possible.

Elastic light scattering is sensitive enough to allow measurements on single sub-micron-sized droplets, but the information content is not specific enough to extract size-dependent rates. Raman spectroscopy, by contrast, could provide specific information but it comes with the disadvantage of low sensitivity (long averaging times), which would make its application to study processes in single submicrometre droplets where nanofocusing becomes important very challenging. Single-droplet fluorescence studies require a fluorescing compound, which strongly restricts its applicability. Furthermore, the fluorescence depletion is not always a reliable measure of the population decay of the photoactive species because of varying quenching efficiencies. The recently presented cavity ring-down studies on single droplets provide information on the extinction but not directly on the droplet absorption[35,36]. Even in combination with light scattering measurements, the determination of rates in nanodroplets is likely prohibited by the uncertainty of the derived absorption.

This study reports the direct observation of light nanofocusing on the photokinetics in nanometre- to micron-sized droplets in the ultraviolet/visible (UV/VIS) range. To this end, we introduce single-droplet photoacoustic (PA) absorption spectroscopy, allowing the direct detection of the population decay of the photoactive substance. PA spectroscopy has been successfully used for the investigation of ensembles of aerosol particles[37–40], but its applicability to single aerosol particle studies has been controversial and has not previously been realised experimentally. Here we demonstrate the feasibility of single-droplet PA spectroscopy in combination with laser trapping, and provide direct experimental evidence for the size-dependence of the photolysis rate in model aerosol droplets due to nanofocusing effects. The results are compared with simulations using classical cavity electrodynamics.

Results

Principle of single-droplet PAs. The two experimental set-ups, using a microphone and a quartz tuning fork, respectively, for resonant single-droplet PA measurements, are sketched in Fig. 1a,b (see Methods). For the droplet absorption experiments, we use a $\lambda = 445$ nm excitation laser (Nichia laser diode NDB7112E) of variable power (0.3–40 mW) modulated at the resonance frequency of the PA-resonator and the tuning fork, respectively. The resonance frequency and the Q-factor of the PA-resonator and the tuning fork are 3.97 kHz and $\sim$8.9, and 32.7 kHz and $\sim$8,000, respectively (Supplementary Fig. 1). The power of the excitation laser is recorded by a power meter after passing the PA cell and the tuning fork, respectively (Supplementary Fig. 2). The amplified PA signals are averaged over either 30 ms or 200 ms. For single-particle trapping, we use a counter-propagating optical tweezer built from a continuous laser beam of $\lambda = 660$ nm of $\sim$200 mW (Laser Quantum, Opus 660) (Supplementary Fig. 2). Such multiple beam optical traps allow trapping of sub-micron droplets, and combine the advantage of a comparatively simple set-up with high trapping stability and tight particle confinement ($<$100 nm)[41–43]. Droplets are trapped by gradient forces pointing towards the trap centre for all translational degrees of freedom. Single droplets are captured in the trap centre from a plume of aerosol generated by a nebulizer (see Methods). The droplet size is determined from laser light elastically scattered by the droplet[41,44] (see Methods, Supplementary Fig. 3). Figure 1e shows an example for light scattering measurements for droplet sizing. In the microphone set-up (Fig. 1a), the trap centre is located in the middle of the PA-resonator above the microphone[45]. The trapping and excitation lasers enter and exit the cell through wide-band, anti-reflective windows coated for the respective wavelengths. The CMOS camera for particle imaging and light scattering measurements is placed perpendicular to the excitation and trapping laser. The aerosol inlet and outlet are on the side of the PA cell outside the resonator.

In the tuning fork set-up (Fig. 1b), the droplet is trapped between the tines of the fork with collinearly aligned excitation laser and trapping laser beams. The CMOS camera for particle imaging and light scattering measurements is placed opposite the tines of the fork. Figure 1c,d shows images of a single droplet trapped in between the tines and of a droplet ensemble flowing through the tines, respectively. The principal attractiveness of the tuning fork derives from its high detection sensitivity (very high Q-factors) and low sensitivity to environmental acoustic noise[46]. In our set-up, we mainly profit from the ease of combining it with laser trapping and light scattering measurements, as well as from the fact that it is chemically inert.

PA response of a single droplet. Figure 2 provides typical noise levels, background signals and a proof for single-droplet detection. Figure 2a illustrates the noise level and the background signal for the empty trap with the trapping laser on. With the excitation laser off ($-5\,s < t < 0\,s$), the background signal (average) and noise level (1 s.d.) are $\sim 2.2 \pm 1.2\,\mu V$. Once the excitation laser is turned on at time $t = 0$ s, a background signal of $BS = 5.3\,\mu V$ with a noise level of $NL = 1.7\,\mu V$ is recorded. The background signal is caused by excitation laser light scattered from the cell walls and hitting the microphone. Blocking the trapping laser (that is, disabling the trap) leaves the noise, as well as the background unchanged.

Figure 2b shows the same as Fig. 2a but with a single VIS441/tetraethylene glycol (TEG) solution droplet in the trap. The PA signal reaches a maximum (S_{max}) just after the excitation laser is turned on and then decreases exponentially as the VIS441 absorber undergoes photolysis. In Fig. 2b, the trap is disabled at

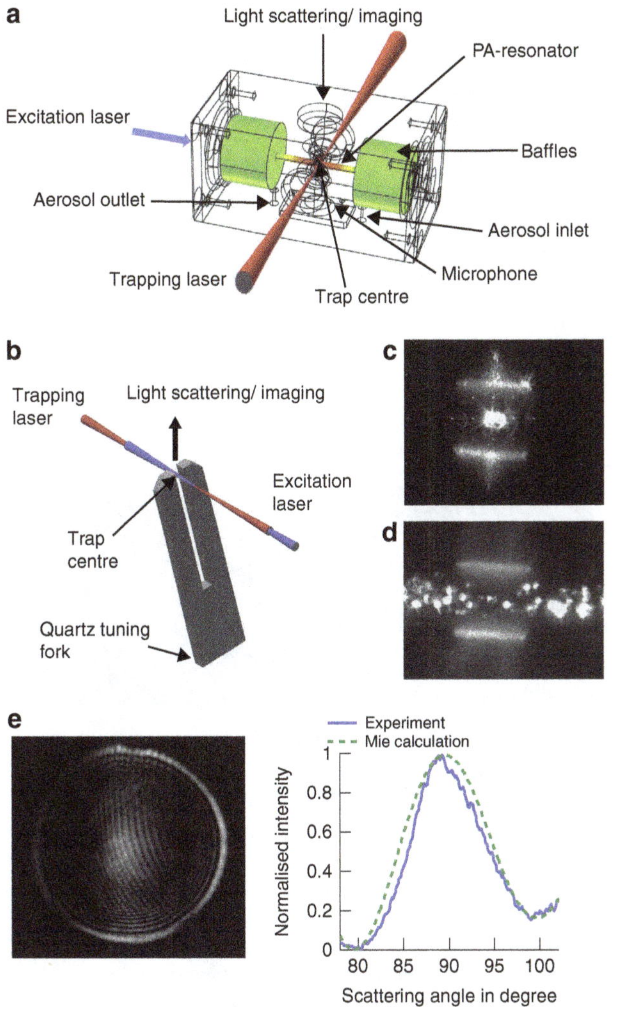

Figure 1 | Sketch of the experimental single-droplet PA set-ups. (**a**) Microphone set-up with PA-resonator, excitation laser, trapping laser and light scattering measurements. The colours in the PA cell indicate that the acoustic mode has its maximum amplitude (red) in the vicinity of the microphone and a value close to zero (green) in the region of the acoustical baffles. (**b**) Tuning fork set-up. (**c**) Snapshot of a single droplet trapped between the tines of the tuning fork (view from top). Note that the droplet ($\sim 1\,\mu m$) is much smaller than the detection volume between the tines ($\sim 0.3 \times 0.34 \times 2$ mm). (**d**) Snapshot of an ensemble of droplets flowing in between the tines from left to right. (**e**) Light scattering image as recorded by the CMOS camera (left) with experimental and calculated phase function (right) for a droplet with a radius of $a = 530$ nm.

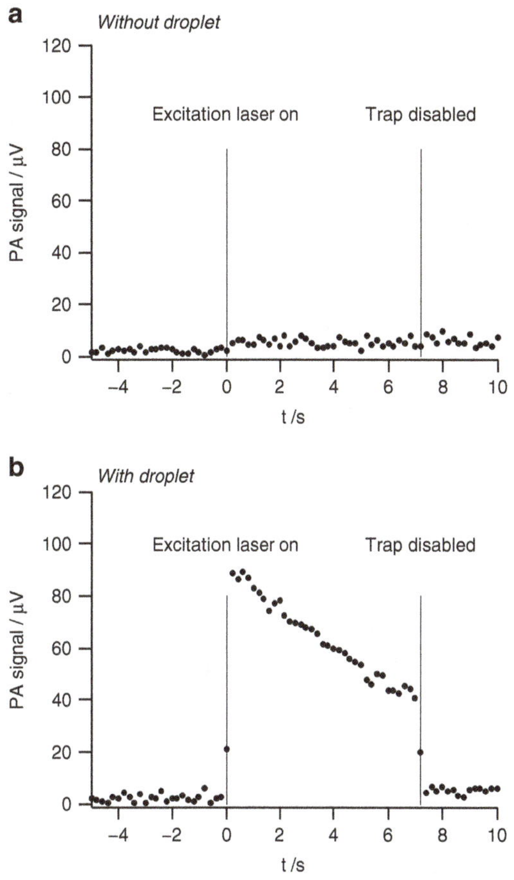

Figure 2 | Typical noise levels and background signals for the microphone set-up. (**a**) For the empty trap, identical noise levels and background signals are recorded for a pure TEG solvent droplet in the trap (data not shown). (**b**) With a VIS441/TEG solution droplet in the trap. At $t = 0$ s, the excitation laser (445 nm) is switched on and at $t = 7$ s the trapping laser is switched off.

$t = 7$ s which leads to the immediate loss of the droplet and to a decrease of the PA signal to the background signal BS. This proves that the signal between $0\,s < t < 7\,s$ indeed comes from a single droplet. The signal to noise ratio $\frac{S}{NL} = \frac{S_{max} - BS}{NL}$ depends on the power P of the excitation laser, the concentration of the solution and the droplet size. Figure 3 shows exemplary experimental data for solution droplets of different size excited with different laser powers P. With the tuning fork set-up, we find an improvement in the $\frac{S}{NL}$ ratio of a factor of ~ 3 compared with the microphone set-up. Note that the PA signal is caused by absorptive heating of the droplet and subsequent heat transfer to the surroundings. Evaporation of the solvent does not occur.

Detection limit. The minimum absorbance detectable with the single-particle PA set-ups at a given power of the excitation laser can be estimated from the PA signal and the single-droplet absorbance assuming that the noise level NL is the detection limit (see Methods, Calculation of droplet absorbance). As an example, we use the PA signal at $t = 0$ (S_{max}) of the 530 nm droplet shown in Fig. 3a recorded at a laser power of 2.8 mW with an averaging time of 200 ms. The refractive index of this droplet, $n + ik = 1.463 + i \cdot 0.0062$ (Methods and Supplementary Figs 4 and 5), yields an absorption cross-section of $C_{abs} = 0.22\,\mu m^2$. $(S_{max} - BS) = 33.3\,\mu V$ corresponds to an equivalent absorbance of $A = 1.8 \times 10^{-5}$. From the measured NL of the 530 nm droplet of $NL = 1.7\,\mu V$, we derive a minimum absorbance $A_{min} = 9 \times 10^{-7}$ detectable with the microphone set-up. The improvement in $\frac{S}{NL}$ by a factor of ~ 3 for the tuning fork set-up reduces the detection limit to $A_{min} \sim 3 \times 10^{-7}$, a minimum detectable absorption coefficient of $\alpha_{min} = 0.0074 \times 10^{-6}\,m^{-1}$ or a minimum detectable absorption cross-section $C_{abs,min} = 0.0037\,\mu m^2$ (laser power of 2.8 mW and averaging time of 200 ms) (see Methods equations (2)–(4)). The equivalent particle radius of 146 nm corresponds to a probe volume of 13 al. This far exceeds the performance of typical spectrometers ($A_{min} \sim 10^{-3}$–10^{-4}), and is at least comparable to the most sensitive laser spectroscopic absorption measurements for macroscopic probe volumes. Note that in our set-up, this sensitivity is achieved for small (attolitre) probe volumes and very short ($<< 1$ s) measurement times. Both can be further improved by increasing the laser power.

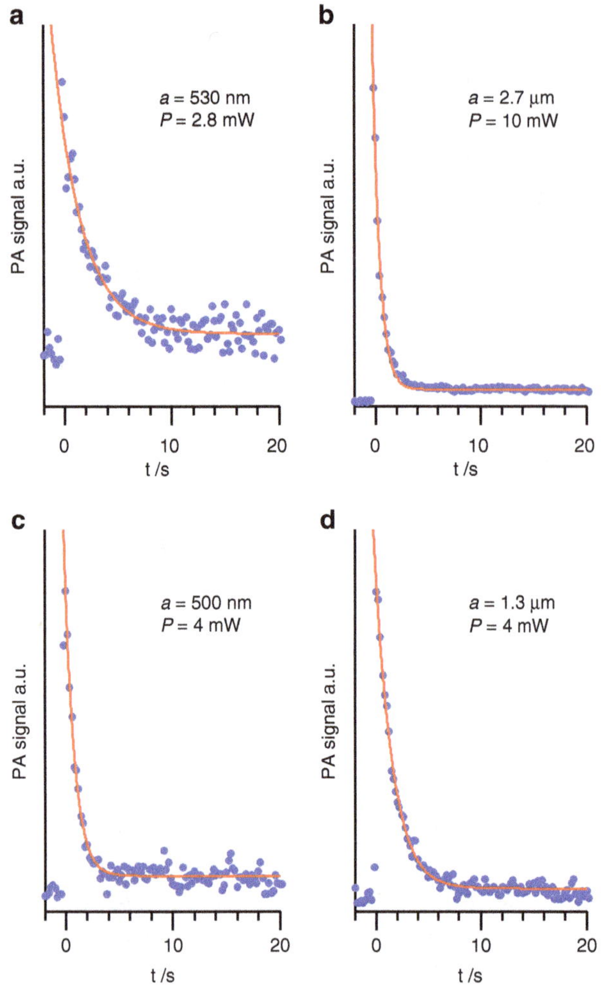

Figure 3 | Exemplary PA signals of VIS441/TEG solution droplets as a function of time. The decay of the signal is caused by photolysis of the solute. The experimental data (blue dots) are recorded for different droplet radii *a* and different power *P* of the excitation laser. (**a**,**b**) Recorded with the microphone set-up. (**c**,**d**) Recorded with the tuning fork setup. The red lines are fits to the experimental data providing experimental first half-lives $t_{1/2}$ (see Fig. 4).

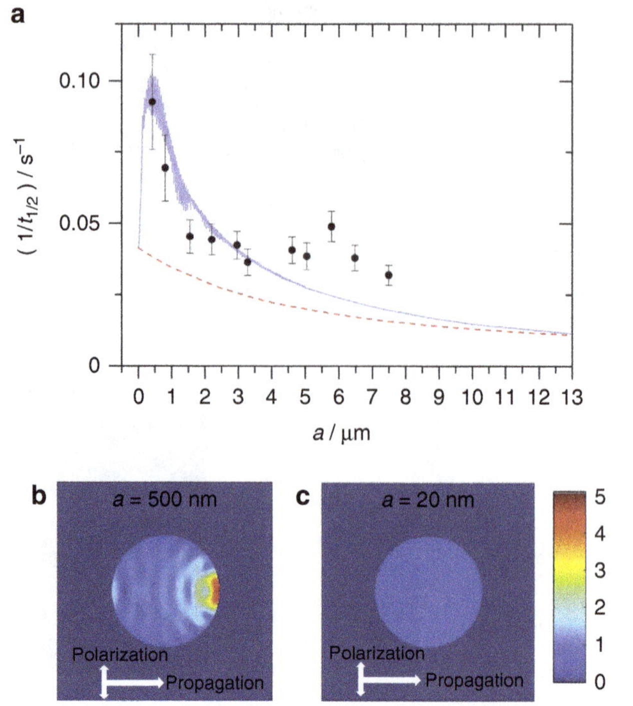

Figure 4 | Size-dependent photokinetics. (**a**) Inverse first half-lives $\frac{1}{t_{1/2}}$ as a function of the droplet radius for a laser power of 1 mW. Black circles: statistically evaluated experimental data. Error bars show 1 s.d. Full blue line: model prediction including nanofocusing and scattering effects. The calculations are for a quantum yield of 7×10^{-6}. Dashed red line: model prediction for a hypothetical bulk limit, that is, excluding contributions from nanofocusing and scattering. Distribution of the light intensity inside droplets at $t = 0$ s for (**b**) a 0.5 µm droplet and (**c**) a 20 nm droplet. The colour scheme is relative to an incident light intensity of 1.

Size-dependent photokinetics. The photokinetics in small droplets do not follow simple pseudo first order kinetics because the light intensity distribution inside the droplets is time dependent; that is, because of the concentration dependence of the nanofocusing. Therefore, we use the same initial concentration for all experiments so that the first half-life can be used as a measure for the size-dependence of the photokinetics. With our PA set-up, we directly measure the decay in absorption resulting from the population decay of the photoactive substance. Diffusion is so fast in the droplets ($\sim 10\,\mu m^2\,s^{-1}$) that concentration gradients cannot build up and homogeneous concentrations for the solute can be assumed at all times. A model for the droplet photokinetics under these conditions is provided in Methods (Calculation of droplet photokinetics).

The model prediction (full blue line in Fig. 4a) shows a strong droplet size-dependence with a maximum in the inverse first half-life $\frac{1}{t_{1/2}}$ at a droplet radius of ~0.5 µm. Pronounced increases in $\frac{1}{t_{1/2}}$ and fluctuations due to resonances are observed over the droplet size range from ~50 nm to ~1.2 µm. In this size range, the increase of the rate is caused by the enhancement of the internal electromagnetic field intensity through focusing of the light inside the droplet. Figure 4b shows the distribution of the light intensity inside a ~0.5 µm droplet at $t = 0$ s. The enhancement of the overall intensity and the local variation of the intensity are pronounced. The inverse half-life at a droplet radius of ~0.5 µm is increased by a factor of ~2.5 compared with the infinitely small droplet limit. The kinetics in these small droplets is no longer influenced by nanofocusing inside the droplet or light scattering by the droplet as visualized by the internal field intensity in Fig. 4c. The inverse half-life of larger droplets (>6 µm) exhibits only a weak size-dependence but decreases continuously (towards zero for infinitely large particles). The rate for these large droplets is determined by the balance between the decay of the absorber and the rise of the decay rate with time. As the photolysis proceeds, the penetration depth of the light and hence the internal field intensity increases. As in the case of very small droplets, nanofocusing is not important for very large droplets. Large droplets essentially represent the behaviour of thin bulk films with the same effective thickness as the droplet. The $\frac{1}{t_{1/2}}$ increases by a factor of ~10 for a 0.5 µm compared with a 13 µm droplet, which implies a substantial increase in the rate of sub-micron-size droplets relative to bulk. The dashed red line in Fig. 4a simulates the behaviour of a hypothetical droplet excluding the influence of nanofocusing and light scattering but still accounting for the droplet-size-dependent absorption (see Methods, Calculation of droplet photokinetics). This curve represents bulk behaviour. The comparison with the full blue line clearly shows the pronounced influence of light focusing on the rate in the sub-micron to micron size range.

The statistically evaluated experimental first half-lives (black circles in Fig. 4a) are determined from time-dependent PA measurements (see Methods, Statistical analysis). The experimental results clearly follow the size-trend predicted by the model (full blue line). The pronounced maximum of $\frac{1}{t_{1/2}}$ at a droplet size of $\sim 0.5\,\mu m$ is clearly visible even though the data scatter notably below $\sim 1\,\mu m$ (Supplementary Fig. 6) mainly because of the uncertainty in the droplet size determination (Methods, Droplet Sizing). Our experimental data show somewhat higher values of the inverse half-life for larger droplets than the model prediction. Deviations from the model assumptions including modified PA response in large droplets[47] could potentially account for this. We have recently introduced a broad-band scattering method for accurate sizing of submicrometre particles, which will allow us in future to significantly reduce the size uncertainty in the submicrometre range (unpublished data). However, already at the current level of accuracy, the data in Fig. 4a provide the first direct observation of the strong influence of nanofocusing of light on the photokinetics in droplets.

Discussion

The experimental results in Fig. 4 confirm a strong size-dependence of the rate of photochemical reactions in droplets. This optical phenomenon shows the most pronounced effect in the submicrometre to micrometre droplet size range for electromagnetic radiation in the UV to VIS range, that is, for the relevant frequency range in atmospheric processes. Classical cavity electrodynamics provides a semi-quantitative description of the kinetics for our ideal model system. The photokinetics of our model system is representative of typical atmospheric aerosols; that is, of typical optical properties of these particles. For example, similar quantitative results are predicted for aqueous droplets (Supplementary Fig. 7). The acceleration of the kinetics we find in the visible range is predicted to be even more pronounced in the UV range of the solar spectrum (Supplementary Fig. 7). Many aerosol particles are non-spherical. However, for particles with different shapes but the same volumes one finds quantitatively similar nanofocusing effects as for droplets. Nanofocusing also affects surface reactions since the strong intensity enhancement in forward direction shown for the internal field in Fig. 4b extends to the external field near the surface (not shown). The ability to measure and thus quantify the kinetics of the light-induced step in photochemical reactions in aerosol particles is of fundamental importance for atmospheric chemistry, where chemical processes are largely driven by sunlight. The diverse and complex processes (for example, transport and surface phenomena) in atmospheric aerosol particles require direct measurement methods as the one introduced here because simple models are of limited applicability.

The introduction of single-droplet PA spectroscopy in the present study finally makes the direct observation of the photoactive step possible. Single-droplet PA was previously deemed not feasible because of sensitivity and background issues. Here we demonstrate the viability of this new method and its very high sensitivity ($C_{abs,min} = 0.0037\,\mu m^2$) enabling studies even of single nanodroplets (10 al). PA spectroscopy provides a general absorption method that can be used in any frequency range. The combination with laser trapping lets us follow the evolution of individual droplets under controlled conditions over extended periods of time (up to several days). This versatility enables fundamental studies on many different droplet systems relevant to atmospheric and technical processes. The investigation of droplet photokinetics is just one example where this new broadly applicable single-droplet method can make an important contribution.

Methods

PA measurements with microphone. The PA cell is made of brass and consists of a longitudinal PA-resonator (length 40 mm, diameter 4 mm), which is connected to two buffer volumes with acoustical baffles for sound insulation (Fig. 1a)[45]. A sensitive microphone (EK 23029, Knowles) is used with a custom-made preamplifier. The output signal is recorded by a lock-in amplifier (Stanford, SR 830).

PA measurements with tuning fork. The distance between the two tines of the tuning fork (Q 32.768 kHz TC 38, AURIS) is 300 µm. Each tine has a width of 600 µm, a thickness of 340 µm and a length of 3.8 mm. The quartz tuning fork acts as the resonant acoustic transducer, which generates an electric signal on resonant excitation by an acoustic wave due to the piezoelectric effect[46]. The signal recording is identical to the microphone set-up except for the more precise reference frequency adapted to the higher Q-factor.

Aerosol generation and materials. To study photokinetics in single droplets, solutions of the photoactive dye VIS441 (Cyanine dye with formula $NaC_{17}H_{25}N_3O_5S_3$ and molar mass 470, QCR solutions) in TEG solvent (ACROS organics, 99.5%) are nebulized with a medical nebulizer (Pari, PARI Boy SX). A concentration of $4.55\,g\,l^{-1}$ VIS441 in TEG is used. For measurements on pure solvent droplets, pure TEG is nebulized. The Supplementary Fig. 4 shows an UV/VIS spectrum of a bulk solution of VIS441 in TEG and of pure TEG solvent, respectively.

Droplet sizing. The particle size is determined from excitation laser light scattered elastically by the droplet. The scattered light intensity is collected for scattering angles between 76.5° and 103.5° and focused onto a CMOS camera (Thorlabs, DCC1645C, 1280 × 1024 pixels) using a camera objective (Super Carenar, focal length = 50 mm, *f*-number = 1.7). The particle size is retrieved by fitting calculated phase functions to experimental ones using Mie theory[41,44]. The sizing of sub-micron-sized droplets is difficult because only few fringes are left in the scattering pattern (for example, Fig. 1e). Larger particles exhibit brighter scattering images and many more fringes (Supplementary Fig. 3), which makes sizing easier. We estimate uncertainties in the droplet radius of about half the wavelength.

Calculation of droplet absorbance. The PA signal S is assumed to be proportional to the power P_{abs} absorbed by the droplet, which is located at the centre of a Gaussian excitation laser beam (beam waist radius of 87 µm and cross-section $q_L = 11{,}889\,\mu m^2$) with incident power P

$$S \propto P_{abs} = I_0 \cdot C_{abs} = P \cdot \frac{C_{abs}}{q_L} \quad (1)$$

I_0 is the intensity incident on the droplet and C_{abs} is the droplet's absorption cross-section. The equivalent absorbance A due to absorption is given by

$$A = \ln\left(\frac{P}{P - P_{abs}}\right) = \ln\left(\frac{q_L}{q_L - C_{abs}}\right) \approx \frac{C_{abs}}{q_L} \quad (2)$$

For a single droplet in the PA cell, the equivalent absorption coefficient is given by,

$$\alpha = \frac{C_{abs}}{V_{res}} \quad (3)$$

where $V_{res} = 0.5\,cm^3$ is the volume of the PA-resonator. The absorption cross-section of the droplet is calculated from the Mie theory[44] with a refractive index of the surroundings equal to 1:

$$C_{abs} = \frac{2\pi}{\varepsilon \cdot \mu \cdot \omega^2}\left[\sum_j (2j+1)\left(\mathrm{Re}\{a_j(x,m) + b_j(x,m)\} - |a_j(x,m)|^2 - |b_j(x,m)|^2\right)\right] \quad (4)$$

Here a_n and b_n are the scattering coefficients, $x = \frac{2\pi a}{\lambda}$ is the size parameter, a is the droplet radius, λ its wavelength of light in vacuum, ω is the angular frequency of the light, ε and μ are the permittivity and the permeability, respectively, of the droplet, and $m = n + ik$ is the droplet's complex index of refraction at the wavelengths of the excitation laser ($\lambda = 445$ nm). The latter is determined from UV/VIS absorption and refractometric measurements of VIS441/TEG bulk solutions and a pure TEG solution using Kramers–Kronig inversion. The refractive index of the VIS441/TEG solution for a dye concentration $4.55\,g\,l^{-1}$ and the pure TEG solvent are $n + ik = 1.463 + i \cdot 0.0062$ and $n_s + ik_s = 1.460 + i \cdot 0.0000$, respectively. The refractive index of the VIS441/TEG solution (dye concentration $4.55\,g\,l^{-1}$) in the UV/VIS range is provided in the Supplementary Fig. 5. For other dye concentrations (see photokinetics), it is assumed to depend linearly on the dye concentration.

Calculation of droplet photokinetics. The droplet photokinetics is described by the following rate equation

$$\frac{dN}{dt} = -p \cdot \frac{I(r)}{h\nu} \cdot \sigma(r) \cdot N(r) \quad (5)$$

with the number density of reactant molecules N, Planck's constant h, excitation laser frequency ν, molecular absorption cross-section σ and the quantum yield

p. Here $\boldsymbol{r}$ denotes the location within the droplet and I is the local field intensity. Both I and σ depend on the complex index of refraction, which in turn depends on the number density N, so that the rate law is no longer pseudo first order. The power absorbed by the droplet is given by the rate of absorption integrated over the droplet's volume V

$$P_{abs}(t) = I_0 \cdot C_{abs}(N) = -hv \cdot \int p^{-1} \frac{dN}{dt} dV \quad (6)$$

Assuming fast diffusion, that is, $N \neq N(\boldsymbol{r})$, we obtain:

$$\frac{dN}{dt} = -f \cdot V^{-1} C_{abs}(N) \quad (7)$$

where $f = p \cdot I_0/hv$ is the product of incident photon flux and reaction probability. Equation (7) is integrated using a 4th order Runge–Kutta method with the time-dependent PA signal given by equation (1). The corresponding inverse first half-lives of the PA signal as a function of droplet radius are shown as a full blue line in Fig. 4a.

To illustrate the effect of nanofocusing, we compare the above model (full blue line in Fig. 4a) with a model that neglects the influence of nanofocusing (dashed red line in Fig. 4a). This model is obtained from equations (5) and (6)

$$C_{abs} = N(t) \int \frac{I(\boldsymbol{r})}{I_0} \sigma(\boldsymbol{r}) dV \quad (8)$$

by inserting the small particle limit[44] for σ

$$\sigma V^{-1} = \frac{\pi}{\lambda} \mathrm{Im}\left\{\frac{-18}{m^2+2}\right\} \quad (9)$$

and a Beer–Lambert expression for the intensity distribution within the particle

$$I(\boldsymbol{r}) = I_0 \cdot \exp\left\{-\frac{4\pi k}{\lambda} \ell(\boldsymbol{r})\right\} \quad (10)$$

where $\ell(\boldsymbol{r}) = r\cos\theta + \sqrt{a^2 - (r\sin\theta)^2}$ is the absorption path length at distance r from the centre of the particle and at polar angle θ relative to the incident beam direction.

Statistical analysis of experimental photolysis data. To account for the uncertainties both in the particle radii and in the decay half-lives of the experimental data set (Supplementary Fig. 6), we perform a two-step maximum likelihood analysis. First, the distribution of particle radii $D(a) = \sum g_i(a)$ is analysed assuming a normally distributed error for the size determination,

$$g_i(a) = \frac{\exp\{-(a-a_i)^2/2\sigma_a^2\}}{\sigma_a\sqrt{2\pi}} \quad (11)$$

with a constant s.d. of $\sigma_a = 220$ nm. The local extrema in D at a_k divide the size range into sections with a lower and an upper half for each cluster of data, which are combined into a single section for isolated data points. For each section, we finally obtain the most probable values for particle radius and the inverse half-life as weighted averages over the particles with weights given by,

$$w_{i,k} = \sigma_{t,i}^{-2} \int_{a_{k-1}}^{a_k} g_i(a)da \Big/ \sum_i \sigma_{t,i}^{-2} \int_{a_{k-1}}^{a_k} g_i(a)da \quad (12)$$

This implies normally distributed errors for the experimental inverse half-lives with s.d. $\sigma_{t,i}$ ranging from about 10% for the most accurate measurements to about 50% for measurements with $\frac{S}{NL} << 10$ (typically small particles). The error bars in Fig. 4a were obtained by s.e. propagation.

References

1. Nie, W. *et al.* Polluted dust promotes new particle formation and growth. *Sci. Rep.* **4,** 1–6 (2014).
2. George, C., Ammann, M., D'Anna, B., Donaldson, D. J. & Nizkorodov, S. A. Heterogeneous photochemistry in the atmosphere. *Chem. Rev.* **115,** 4218–4258 (2015).
3. Liu, F., Beames, J. M., Petit, A. S., McCoy, A. B. & Lester, M. I. Infrared-driven unimolecular reaction of CH_3CHOO Criegee intermediates to OH radical products. *Science* **345,** 1596–1598 (2014).
4. Reed Harris, A. E. *et al.* Photochemical kinetics of pyruvic acid in aqueous solution. *J. Phys. Chem. A* **118,** 8505–8516 (2014).
5. Finlayson-Pitts, B. J. Chlorine chronicles. *Nat. Chem.* **5,** 724 (2013).
6. Monge, M. E. *et al.* Alternative pathway for atmospheric particles growth. *Proc. Natl Acad. Sci. USA* **109,** 6840–6844 (2012).
7. Jimenez, J. L. *et al.* Evolution of organic aerosols in the atmosphere. *Science* **326,** 1525–1529 (2009).
8. Lack, D. A. *et al.* Relative humidity dependence of light absorption by mineral dust after long-range atmospheric transport from the Sahara. *Geophys. Res. Lett.* **36,** L24805 (2009).
9. Laskin, A. *et al.* Reactions at interfaces as a source of sulfate formation in sea-salt particles. *Science* **301,** 340–344 (2003).
10. Vaida, V., Kjaergaard, H. G., Hintze, P. E. & Donaldson, D. J. Photolysis of sulfuric acid vapor by visible solar radiation. *Science* **299,** 1566–1568 (2003).
11. Jacobson, M. Z. Strong radiative heating due to the mixing state of black carbon in atmospheric aerosols. *Nature* **409,** 695–697 (2001).
12. Lelieveld, J. & Crutzen, P. J. Influences of cloud photochemical processes on tropospheric ozone. *Nature* **343,** 227–233 (1990).
13. Crutzen, P. J. & Arnold, F. Nitric acid cloud formation in the cold Antarctic stratosphere: a major cause for the springtime 'ozone hole'. *Nature* **324,** 651–655 (1986).
14. Tyndall, J. On the blue colour of the sky, and the polarization of light. *Proc. R. Soc.* **37,** 384–394 (1869).
15. Benincasa, D. S., Barber, P. W., Zhang, J.-Z., Hsieh, W.-F. & Chang, R. K. Spatial distribution of the internal and near-field intensities of large cylindrical and spherical scatterers. *Appl. Opt.* **26,** 1348–1356 (1987).
16. Symes, R., Sayer, R. M. & Reid, J. P. Cavity enhanced droplet spectroscopy: principles, perspectives and prospects. *Phys. Chem. Chem. Phys.* **6,** 474–487 (2004).
17. Cappa, C. D., Wilson, K. R., Messer, B. M., Saykally, R. J. & Cohen, R. C. Optical cavity resonances in water micro-droplets: implications for shortwave cloud forcing. *Geophys. Res. Lett.* **31,** L10205 (2004).
18. Brem, B. T., Gonzalez, F. C. M., Meyers, S. R., Bond, T. C. & Rood, M. J. Laboratory-measured optical properties of inorganic and organic aerosols at relative humidities up to 95%. *Aerosol. Sci. Technol.* **46,** 178–190 (2012).
19. Preston, T. C. & Signorell, R. Vibron and phonon hybridization in dielectric nanostructures. *Proc. Natl Acad. Sci. USA* **108,** 5532–5536 (2011).
20. Preston, T. C. & Signorell, R. From plasmon spectra of metallic to vibron spectra of dielectric nanoparticles. *Acc. Chem. Res.* **45,** 1501–1510 (2012).
21. Hickstein, D. D. *et al.* Mapping nanoscale absorption of femtosecond laser pulses using plasma explosion imaging. *Am. Chem. Soc. Nano* **8,** 8810–8818 (2014).
22. Goldmann, M., Miguel-Sánchez, J., West, A. H. C., Yoder, B. L. & Signorell, R. Electron mean free path from angle-dependent photoelectron spectroscopy of aerosol particles. *J. Chem. Phys.* **142,** 224304 (2015).
23. Ruth Signorell & Jonathan, P. Reid (eds) *Fundamentals and Applications in Aerosol Spectroscopy* (CRC Press, 2011).
24. Mayer, B. & Madronich, S. Actinic flux and photolysis in water droplets: Mie calculations and geometrical optics limit. *Atmos. Chem. Phys.* **4,** 2241–2250 (2004).
25. Ray, A. K. & Bhanti, D. D. Effect of optical resonances on photochemical reactions in microdroplets. *Appl. Opt.* **36,** 2663–2674 (1997).
26. Ruggaber, A. *et al.* Modelling of radiation quantities and photolysis frequencies in the aqueous phase in the troposphere. *Atmos. Environ.* **31,** 3137–3150 (1997).
27. Bott, A. & Zdunkowski, W. Electromagnetic energy within dielectric spheres. *J. Opt. Soc. Am. A* **4,** 1361–1365 (1987).
28. Penconi, M. *et al.* The use of chemical actinometry for the evaluation of the light absorption efficiency in scattering photopolymerizable miniemulsions. *Photochem. Photobiol. Sci.* **14,** 308–319 (2015).
29. Nissenson, P., Knox, C. J. H., Finlayson-Pitts, B. J., Phillips, L. F. & Dabdub, D. Enhanced photolysis in aerosols: evidence for important surface effects. *Phys. Chem. Chem. Phys.* **8,** 4700–4710 (2006).
30. Mills, C. T., Rowland, G. A., Westergren, J. & Phillips, L. F. Quantum yields of CO_2 and SO_2 formation from 193 nm photo-oxidation of CO in a sulfuric acid aerosol. *J. Photochem. Photobiol. A* **93,** 83–87 (1996).
31. Kitagawa, F. & Kitamura, N. A laser trapping-spectroscopy study on the photocyanation of perylene across a single micrometre-sized oil droplets/water interface: droplet-size effects on photoreaction quantum yield. *Phys. Chem. Chem. Phys.* **4,** 4495–4503 (2002).
32. Barnes, M. D., Whitten, W. B. & Ramsey, J. M. Enhanced fluorescence yields through cavity quantum-electrodynamic effects in microdroplets. *J. Opt. Soc. Am. B* **11,** 1297–1304 (1994).
33. Taflin, D. C. & Davis, E. J. A study of aerosol chemical reactions by optical resonance spectroscopy. *J. Aerosol. Sci.* **21,** 73–86 (1990).
34. Ward, T. L., Zhang, S. H., Allen, T. & Davis, E. J. Photochemical polymerization of acrylamide aerosol particles. *J. Colloid Interface Sci.* **118,** 343–355 (1987).
35. Sanford, T. J., Murphy, D. M., Thomson, D. S. & Fox, R. W. Albedo measurements and optical sizing of single aerosol particles. *Aerosol. Sci. Technol.* **42,** 958–969 (2008).
36. Cotterell, M. I., Mason, B. J., Preston, T. C., Orr-Ewing, A. J. & Reid, J. P. Optical extinction efficiency measurements on fine and accumulation mode aerosol using single particle cavity ring-down spectroscopy. *Phys. Chem. Chem. Phys.* **17,** 15843–15856 (2015).
37. Haisch, C. Photoacoustic spectroscopy for analytical measurements. *Meas. Sci. Technol.* **23,** 012001 (2012).

38. Haisch, C., Menzenbach, P., Bladt, H. & Niessner, R. A wide spectral range photoacoustic aerosol absorption spectrometer. *Anal. Chem.* **84,** 8941–8945 (2012).
39. Lack, D. A. *et al.* Aircraft instrument for comprehensive characterization of aerosol optical properties, part 2: black and brown carbon absorption and absorption enhancement measured with photo acoustic spectroscopy. *Aerosol. Sci. Technol.* **46,** 555–568 (2012).
40. Gyawali, M. *et al.* Photoacoustic optical properties at UV, VIS, and near IR wavelengths for laboratory generated and winter time ambient urban aerosols. *Atmos. Chem. Phys.* **12,** 2587–2601 (2012).
41. David, G. *et al.* Stability of aerosol droplets in Bessel beam optical traps under constant and pulsed external forces. *J. Chem. Phys.* **142,** 154506 (2015).
42. Thanopulos, I., Luckhaus, D., Preston, T. C. & Signorell, R. Dynamics of submicron aerosol droplets in a robust optical trap formed by multiple Bessel beams. *J. Appl. Phys.* **115,** 154304 (2014).
43. Li, T., Kheifets, S., Medellin, D. & Raizen, M. G. Measurement of the instantaneous velocity of a brownian particle. *Science* **328,** 1673–1675 (2010).
44. Bohren, C. F. & Huffman, D. R. *Absorption and Scattering of Light by Small Particles* (John Wiley & Sons, 1998).
45. Beck, H. A., Niessner, R. & Haisch, C. Development and characterization of a mobile photoacoustic sensor for on-line soot emission monitoring in diesel exhaust gas. *Anal. Bioanal. Chem.* **375,** 1136–1143 (2003).
46. Kosterev, A. A., Tittel, F. K., Serebryakov, D. V., Malinovsky, A. L. & Morozov, I. V. Applications of quartz tuning forks in spectroscopic gas sensing. *Rev. Sci. Instrum.* **76,** 043105 (2005).
47. Raspet, R., Slaton, W. V., Arnott, W. P. & Moosmüller, H. Evaporation-condensation effects on resonant photoacoustics of volatile aerosols. *J. Atmos. Oceanic Technol.* **20,** 685–695 (2003).

Acknowledgements

This work was supported by the Swiss National Science Foundation (SNSF grant nr. 200020_159205), ETH Zurich and the TUM International Graduate School of Science and Engineering (IGSSE). We are very grateful to Dr David Luckhaus for support in the data analysis and to Prof. Markus Sigrist for helpful discussions concerning the PA set-up. We would like to thank Guido Grassi and Daniel Zindel from the analytical service at ETH and the electronic and mechanical workshops at ETH and TUM for their technical support.

Author contributions

J.W.C. implemented the experimental set-up and performed the measurements. K.M.T. and C.H. designed the PA cell and contributed to the experimental set-up and initial test measurements. J.W.C. and R.S. analysed the data. R.S. conceived the project, performed the calculations and wrote the manuscript.

Additional information

Competing financial interests: The authors declare no competing financial interests.

Spatial control of chemical processes on nanostructures through nano-localized water heating

Calum Jack[1], Affar S. Karimullah[1,2], Ryan Tullius[1], Larousse Khosravi Khorashad[3], Marion Rodier[1], Brian Fitzpatrick[1], Laurence D. Barron[1], Nikolaj Gadegaard[2], Adrian J. Lapthorn[1], Vincent M. Rotello[4], Graeme Cooke[1], Alexander O. Govorov[3] & Malcolm Kadodwala[1]

Optimal performance of nanophotonic devices, including sensors and solar cells, requires maximizing the interaction between light and matter. This efficiency is optimized when active moieties are localized in areas where electromagnetic (EM) fields are confined. Confinement of matter in these 'hotspots' has previously been accomplished through inefficient 'top-down' methods. Here we report a rapid 'bottom-up' approach to functionalize selective regions of plasmonic nanostructures that uses nano-localized heating of the surrounding water induced by pulsed laser irradiation. This localized heating is exploited in a chemical protection/deprotection strategy to allow selective regions of a nanostructure to be chemically modified. As an exemplar, we use the strategy to enhance the biosensing capabilities of a chiral plasmonic substrate. This novel spatially selective functionalization strategy provides new opportunities for efficient high-throughput control of chemistry on the nanoscale over macroscopic areas for device fabrication.

[1]School of Chemistry, University of Glasgow, Joseph Black Building, Glasgow G12 8QQ, UK. [2]School of Engineering, University of Glasgow, Rankine Building, Glasgow G12 8LT, UK. [3]Department of Physics and Astronomy, Ohio University, Athens, Ohio 45701, USA. [4]Department of Chemistry, University of Massachusetts, 710 North Pleasant Street, Amherst, Massachusetts 01003, USA. Correspondence and requests for materials should be addressed to A.S.K. (email: Affar.Karimullah@glasgow.ac.uk) or to M.K.(email: Malcolm.Kadodwala@glasgow.ac.uk).

Incorporating molecular functionality into engineered nanomaterials is a requirement for the development of new sensing[1–3], photovoltaic[4] and optical technologies[5–8]. Maximizing coupling between nanostructures and molecular species requires spatial control of surface chemistry on the nanoscale. Currently, there are a variety of techniques that can be used to spatially control chemical functionality on nanostructured substrates such as dip-pen lithography[9], inkjet printing[10] and direct laser patterning[11,12]. However, all these approaches are in essence 'top-down' methods for micrometre-scale substrate areas and each features limitations such as time-consuming processing and difficulties in achieving sub-50 nm resolutions. Furthermore, they require complex alignment procedures to be overlaid onto fabricated nanostructures.

In this study we show a high-throughput 'bottom-up' approach that uses the nano-localized heating of a liquid (water) surrounding a nanostructure, to spatially direct a protection/deprotection strategy[13]. This novel strategy enables molecular materials to be placed in selective regions of a nanostructure. The protection step involves the chemical passivation of the gold structure with a self-assembled monolayer (SAM) of a long-chain polyethylene glycol (PEG) thiol. The spatially selective deprotection step then is the thermally driven structural transformation of the PEG SAM.

Structurally complex plasmonic nanostructures can produce a range of electromagnetic (EM) fields that not only have different spatial extents but also differing intrinsic properties. Consequently, for sensing purposes, it would be advantageous to selectively place material in locations occupied by EM fields with the desired properties. To this end, we have employed the nano-localized heating-driven protection/deprotection strategy to functionalize chiral plasmonic nanostructures, to enable selective positioning of proteins in a region with EM fields of enhanced chiral asymmetry. The placement of proteins in regions with fields of optimal properties provides enhanced biosensing performance, enabling attomole ($\sim 6.0 \times 10^5$ molecules) rather than femtomole ($\sim 6.0 \times 10^8$ molecules) detection levels.

Results

Modelling of nano-localized thermoplasmonic water heating. To date, it is believed that spatially localized chemistry on individual nanostructures cannot be achieved through thermoplasmonic phenomena, although thermal gradients can be generated in ensembles of nanoparticles separated by a poorly conducting medium[14–16]. This assumption is based on the fact that although electric fields, created by plasmon excitation, may be highly localized, efficient thermal diffusion in metal nanostructures results in rapid and uniform temperature increase throughout a particular structure in less than a nanosecond. This rapid heat dissipation leads to a uniform temperature at the nanostructure surface and hence spatial uniformity of surface chemical/physical processes. In this study we demonstrate a new phenomenon, which leads to spatially localized thermally driven chemistry, allowing nanoscale precision placement of biomaterial on a nanostructure surface (Fig. 1). Localized chemistry in this study is the result of significant temperature gradients in the surrounding water and not thermal gradients across the nanostructure surface itself.

We demonstrate this new 'thermoplasmonic' effect using a chiral plasmonic nanomaterial formed using injection moulding, creating nanopatterned indentations in a polycarbonate slide (polycarbonate template) and coating the surface with a continuous Au film. The nanostructure in particular is a 'shuriken' structure (Fig. 2a), which is either left or right handed and is referred to as a templated plasmonic substrate (TPS)[17]. These structures are 500 nm end to end and have a pitch of 700 nm from centre to centre. In contrast to traditional electron beam or photolithography, the TPS offer low-cost high-throughput fabrication, effectively a disposable consumable ideally suited for technological exploitation. The TPS are incorporated into a liquid cell with a total volume of *ca.* 100 µl. Interfaces that will subsequently be referred to as the front face, back face, lower and upper surfaces are defined in Fig. 2 (further details in Supplementary Fig. 1). In our approach, the back face of the TPS was irradiated using a 1,064-nm Nd:YAG laser with an 8-ns pulse.

On irradiation, the plasmonic structure absorbs the energy based on its resonance condition and the spatial regions of the nanostructure with high current density will then act as heat sources[18,19]. To understand this pivotal step, we have modelled the spatially resolved time dynamics for thermal behaviour of the nanostructure and the surrounding water for an 8-ns laser pulse at a fluence of $15\,\mathrm{mJ\,cm^{-2}}$. Unlike plasmonic heating with femtosecond pulses[20], the gold film, owing to its high thermal conductivity, does not generate large thermal gradients (Fig. 3a) with nanosecond laser pulse durations[16,19]. Although the plasmonic fields maybe highly localized, the heat generated diffuses in the structure over such time scales (Supplementary Figs 3–6). With thermal diffusivity almost three orders of magnitude smaller than gold, the surrounding water does generate thermal gradients, developing three regions of varying thermal behaviour by the time the 8-ns pulse ends. The water at the top surface has a low average temperature but the two regions of water within the indentation itself have high temperatures over significant distances. The water in the central

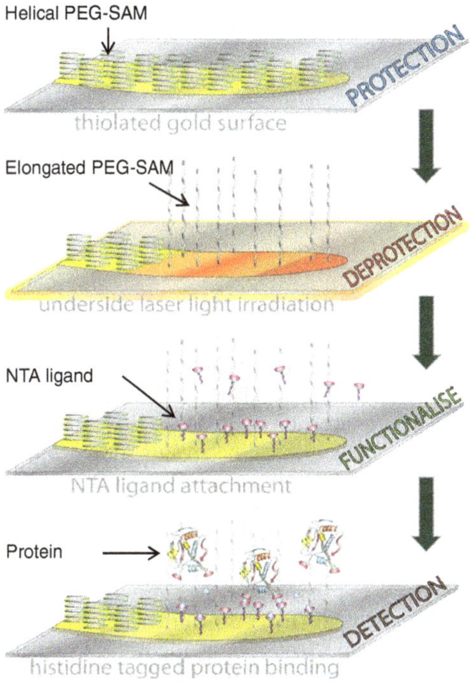

Figure 1 | Pictorial representation of the steps in the protection/deprotection strategy. The nanostructure is first protected using a thermally responsive PEG-SAM (helical form). The second step is the deprotection where the nanostructure is irradiated by a nanosecond pulse laser and the subsequent localized water temperatures will cause a transition in the PEG-SAM to its elongated form. The elongated PEG-SAM will not inhibit the NTA ligands from attaching to the surface, thereby functionalizing the selective regions. Proteins can then be positioned in the particular regions for detection.

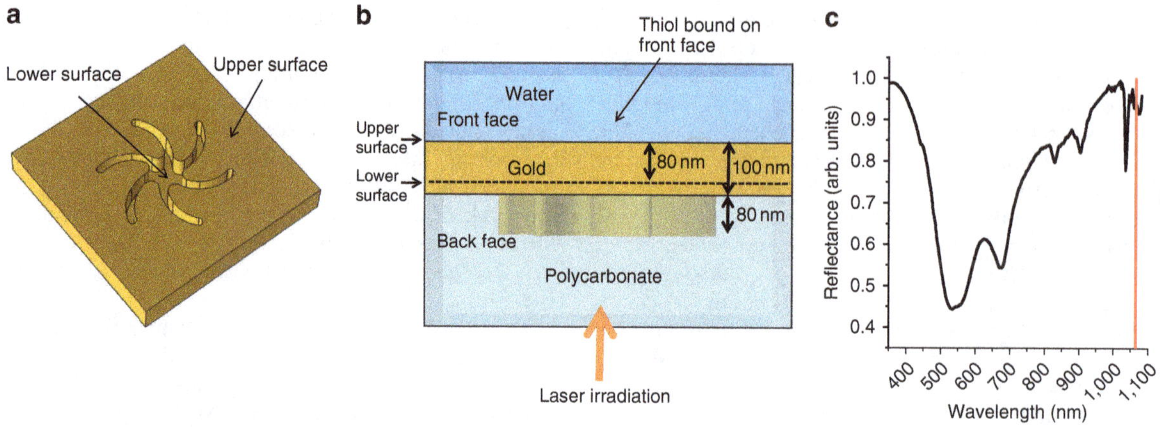

Figure 2 | TPS experimental geometry and reflectivity. (**a**) Perspective view of the front face of the gold layer on a TPS. (**b**) Side view of the TPS, which is irradiated using a nanosecond laser through the polycarbonate. (**c**) Reflectivity of the TPS measured through the polycarbonate. The red line indicates the wavelength of the nanosecond laser.

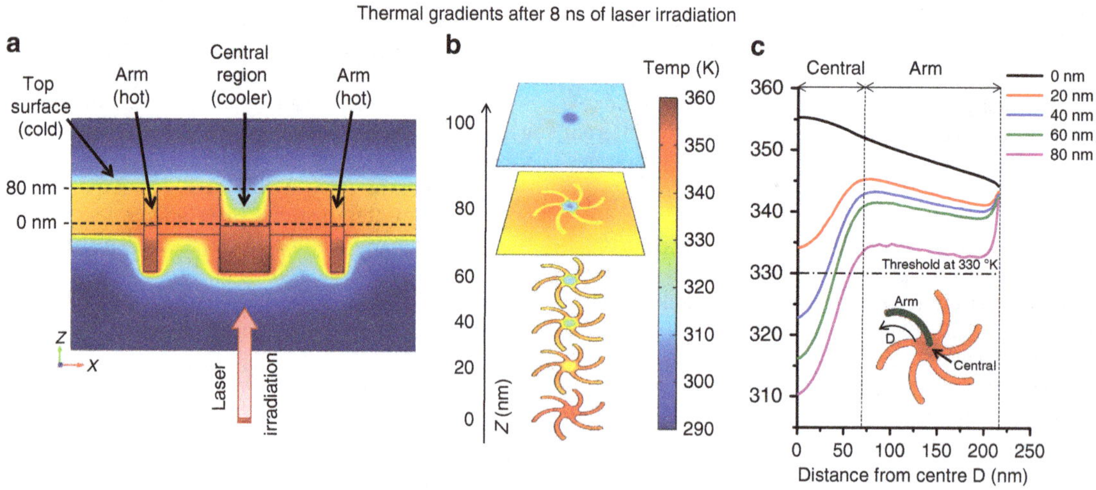

Figure 3 | Surface plot of the thermal gradients at 8 ns (pulse end). (**a**) Side view showing gold and water regions and (**b**) stacked surface plots of water regions for increasing height *Z*. (**c**) Graph showing the temperature changes along a parametric curve passing along an arm to the centre (as shown by green line in inset) of the nanostructure for varying height *Z*. The horizontal line at 330 K represents the threshold for the deprotection step discussed.

region of the indentation is cooler than that in the arms with increasing distance from the surface and the water temperature in the arms is relatively high and uniform with increasing height (Fig. 3a–c and Supplementary Fig. 3).

The time-dependant thermal behaviour shows that during the laser pulse, the water shows a steep increase in the average temperature for the two regions (Fig. 4) with the arm sections achieving higher temperatures than the central region, reaching a maximum at the end of the pulse (8 ns). This leads to thermal gradients in the water across the nanostructure that exist up to 12 ns after the pulse ends (that is, the arms are significantly hotter than the central region). After 12 ns, the water thermal gradients have completely disappeared and the water achieves thermal homogeneity with the metal as well. Equivalent thermal gradients do not exist in the metal (Fig. 4c). Thus, our simulations show thermal gradients in the water surrounding the nanostructure during the 8-ns laser pulse and a subsequent 4 ns afterwards. This phenomena derives from the disparity in the thermal diffusivities of water and Au. It is fundamentally different to the previously reported focusing of light into mesoscale volumes at water surfaces by Au nanoparticles, which creates a nanobubble of steam surrounding the resonantly heated nanoparticle surface[21].

Nano-localized thermoplasmonic chemical functionalization. The first step for our patterning process is the deposition of a protective SAM. To achieve spatially selective chemistry, we have used a thermally responsive SAM composed of a high-molecular-weight (6 kDa) PEG methyl ether thiol (PEG-thiol) polymer. The SAMs were formed by depositing PEG-thiol from solution onto the TPS; these substrates will subsequently be referred to as PEG-TPSs. The macromolecular structure of PEG displays a temperature dependency, which has been exploited previously to produce thermally responsive materials[22,23]. The macromolecular structure of PEG is governed by the conformation adopted by the $-(CH_2-CH_2-O)-$ subunits. At low temperatures or in polar environments a gauche conformation is favoured, whereas at higher temperatures or in non-polar environments the *trans* form is observed[24,25]. Hence, PEG thiols adopt a compact ordered helical structure within SAM at low temperature and in aqueous solutions, which inhibit the adsorption of biomaterials. Driven by the conformational transition of the $-(CH_2-CH_2-O)-$ subunits, PEG-SAM can undergo a transition from the helical state to an elongated form, which does not inhibit the adsorption of biomaterials[26]. To achieve this transition, Shima *et al.*[22] have shown that the PEG-SAM must be exposed to

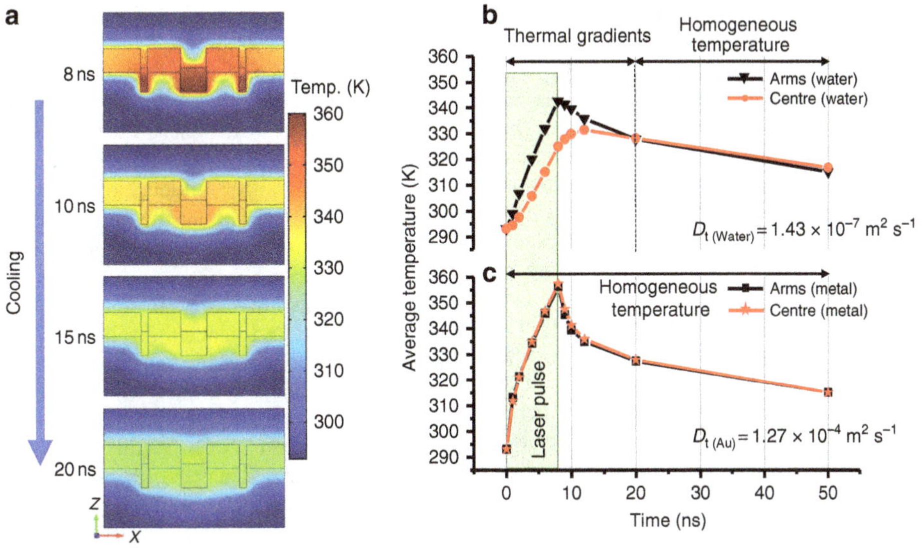

Figure 4 | Thermal behaviour of water on irradiation. (**a**) Simulated spatially resolved temperature dynamics after pulse (8 ns) laser irradiation. (**b**) Average temperature of water (10 nm < Z < 70 nm) and (**c**) metal (Z < 0 nm) in the arm and centre regions. The values D_t are the thermal diffusivity of water and Au.

temperatures over ~330 K, along the entire length of its elongated structure, ~35 nm (ref. 27).

The sensitivity of the wavelength of a plasmonic resonance to the refractive index of the near field enables change in the structure of the PEG-SAM to be monitored spectroscopically. We measured the shifts in the optical rotatory dispersion (ORD) of our left-handed plasmonic shurikens from the front face[17]. The helical and elongated *trans* form SAM have thickness of ~13 and ~35 nm, respectively[27]. Consequently, it would be expected that a transition from the helical to elongated *trans* form results in a significant increase in the thickness of the PEG-SAM layer and hence will cause a red shift in the plasmonic resonance. Indeed, when the PEG-TPSs are exposed either to water at 358 K (Fig. 5a) or a less polar solvent such as 1-butanol (see Supplementary Fig. 7), an irreversible red shift of +5.0 ± 0.2 nm in the wavelength of the plasmonic resonance in ORD spectra is in fact observed. We attribute the irreversible nature of the transition to the kinetic effects of lateral interactions (such as intertwining) of the elongated *trans* PEG within the SAM that occur when temperatures are between 323 K (see Supplementary Fig. 7) and 358 K (Fig. 5a). When a PEG-TPS is heated to 358 K in air (for 48 h), no discernible shift within error (−0.2 ± 0.2 nm) is observed (Supplementary Fig. 7). Hence, heating a PEG-TPS in air only, causes no change, indicating that the process is water mediated (Supplementary Fig. 7D)[27].

The PEG-TPS were irradiated using a pulsed laser from the back and the energy dependence of the transition was evaluated. A plain TPS with no PEG-SAM shows no change (Fig. 5b) when irradiated using 8 ns pulse laser (15 mJ cm^{-2} fluence), indicating that the laser irradiation causes no deformation to the nanostructure. When a PEG-TPS is irradiated using 400 μs pulse laser, no observable changes occur (Fig. 5c) even at the highest fluence (20 mJ cm^{-2}) used. When the PEG-TPS is irradiated using 8 ns pulse laser (15 mJ cm^{-2} fluence), a small shift of +1.0 ± 0.2 nm is observed, indicating that a small fraction of the PEG-SAM has been changed (Fig. 5d). The level of red shift saturates within 60 s of laser exposure with further radiation causing no measureable change (Supplementary Fig. 7). When a substrate that has had its PEG transformed to the elongated state using 1-butanol is irradiated with nanosecond pulses, no spectral changes are observed (Supplementary Fig. 7). These data clearly

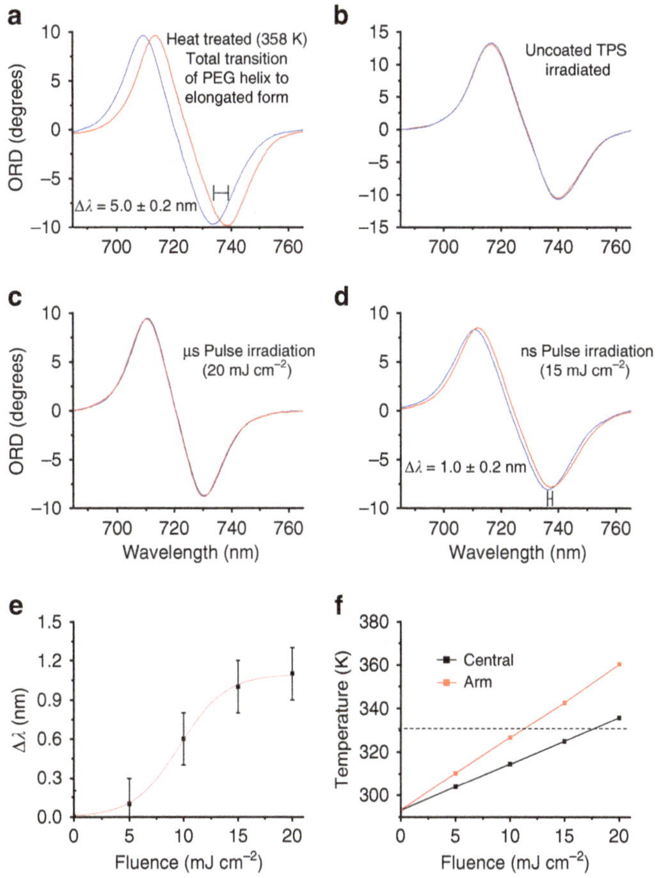

Figure 5 | TPS nanostructure ORD measurements. (**a**) PEG-TPS heat treated at 358 K. (**b**) A plain TPS irradiated with 8 ns pulse laser; PEG-TPS irradiated using (**c**) 400 μs pulse laser and (**d**) 8 ns pulse laser. (**e**) Fluence dependence of the resonance shift and (**f**) simulation results for average temperature within the nano indentation for varying fluence.

demonstrate that nanosecond laser heating causes only a partial yet irreversible change to the PEG-SAM. Furthermore, the effects of laser irradiation show dependency on the fluence, displaying stepwise behaviour (Fig. 5e). Nanosecond and microsecond pulses can be considered to be quasi continuous, as they have significantly longer durations than the electron–phonon relaxation time[28]. Consequently, they only generate linear hot electron-driven phenomenon rather than the nonlinear phenomenon associated with femtosecond pulses, which arise through high electron temperatures. Although microsecond pulses would generate smaller field intensities, they would occur for longer time scales. Thus, the time integrated total flux of hot electrons will be identical for both nano and microsecond pulses[29]. Hence, the experimental observations are consistent with a thermal process rather than either a direct optical or hot-electron-mediated excitation, as both would have no dependence on pulse duration and a linear dependency on fluence[30].

The small red shift (1.0 ± 0.2 nm) induced by nanosecond pulse laser irradiation compared with the shift (5.0 ± 0.2 nm) due to heating at 358 K indicates that only a small fraction ($\sim 20\%$) of the total area in the near fields of the nanostructures of the PEG-SAM undergoes a transition to the elongated state. This limited transformation is consistent with our hypothesis that thermal gradients within the surrounding water can direct spatially selective chemistry. The irreversible transformation is a kinetically limited process that consequently requires the water temperature to be above the threshold for a sufficient time period, to enable the elongated PEG-thiol to intertwine with each other. Earlier studies of the dynamic relaxation for polymers in solution with similar molecular weight have shown this time period to be $\sim$5–10 ns (ref. 31). For shorter time periods, the transformation would be reversible. The results from our simulation show that at the top surface, the temperatures in the water are not high enough to achieve the conformational transition. Hence, the PEG-SAM here will remain in its helical conformation over the entire surface. At the central cooler region of the structure, the average water temperature barely crosses the threshold for < 3 ns. The PEG-SAM here will not experience the required temperatures for a long-enough period and hence will also remain in a helical conformation. However, the arm region, is at a significantly higher average temperature (~ 350 K) and above the threshold for ~ 12 ns. Only in the arms is the average water temperature higher than the threshold for long enough to achieve the irreversible transformation.

The dependence of the resonance shift on the laser fluence is also in agreement with our simulation results on the dependence of temperatures on laser fluence (Fig. 5f). The average temperature in the arm only increases above the threshold for the PEG transition for a fluence of > 12 mJ cm^{-2} and just about crosses that threshold in the centre at the highest fluence of 20 mJ cm^{-2}, indicating that substantial transition of PEG-SAM will occur only in the arms.

To confirm the spatial localization of the PEG-SAM transformation and hence the validity of using thermal gradients in water to illicit spatially selective chemistry, we have collected atomic force microscopy (AFM) images. The significant height difference between the helical and the elongated *trans* form (~ 22 nm) of the PEG can readily be detected by AFM. In Fig. 6 are AFM images collected from PEG-TPS before and after irradiation with nanosecond pulses. The AFM images show that after laser irradiation there is an average reduction in the depth of the arms by 20–30 nm in the region at the end of the arms. This decrease in depth is an expected change in SAM thickness on going from helical to elongated *trans*-PEG. The spatial localization of the change in SAM structure is also supported by scanning electron microscopy, which show change in contrast at the end of the arms post irradiation (see Supplementary Fig. 8).

Enhanced sensitivity of TPS spectroscopy to biomaterials. The elongated *trans*-PEG-SAM in the arms should allow adsorption of biomolecules, whereas the helical PEG-SAM will inhibit adsorption, essentially creating 'nanopockets' in the arms. This ability of the PEG-SAM to selectively vary spatial chemistry provides a powerful tool for enhancing the functionality of nanomaterials. As an exemplar of the potential applications, we used the nanopockets created on the TPSs, to achieve detection of proteins at the attomole sensitivity. Specifically, the chiral evanescent fields that occupy the nanopockets were used to perform a form of polarimetry, plasmonic polarimetry[32], which measures the asymmetry of the effective refractive indices of chiral media on the handedness of the applied chiral evanescent field[2]. The asymmetries in refractive indices of chiral media can be parameterized by $\Delta\Delta\lambda = \Delta\lambda_R - \Delta\lambda_L$, where $\Delta\lambda_{R/L}$ is the shift in

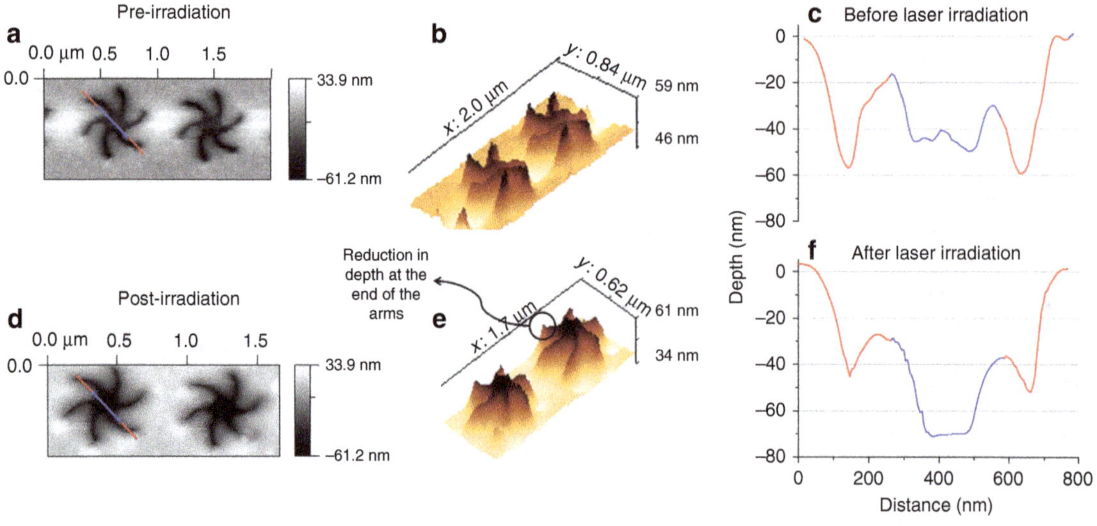

Figure 6 | AFM of pre and post irradiation of the PEG-TPS. (**a**,**d**) The height plots, (**b**,**e**) the inverted 3D plots and (**c**,**f**) the height data of along the lines shown in **a** and **d**. The red and blue colours correspond to the distances marked in the SEM(**a**,**d**).

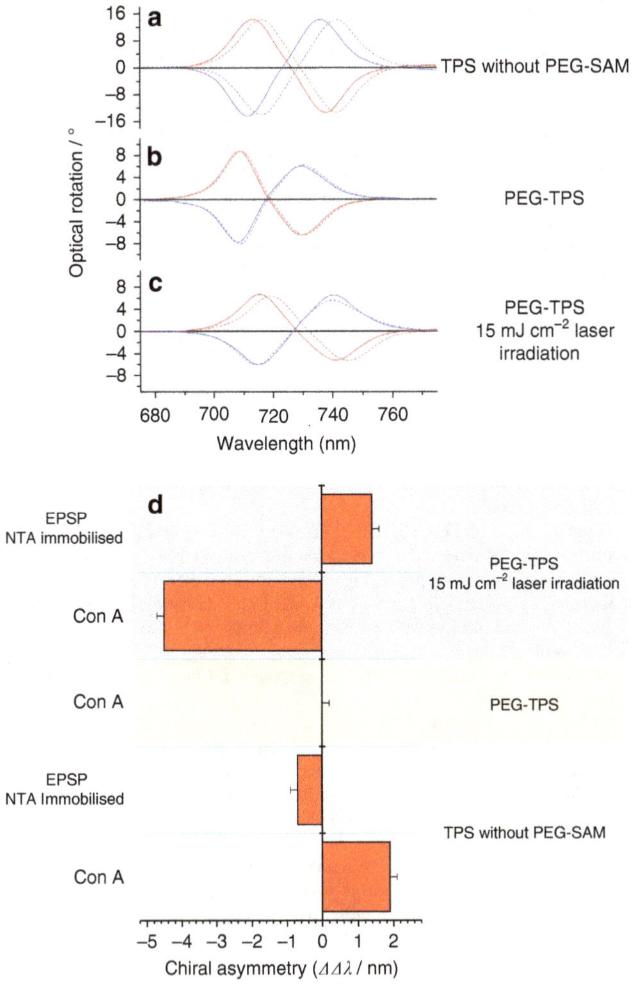

Figure 7 | ORD shifts from buffer (solid) to ConA (dotted). (**a**) TPS, (**b**) PEG-TPS but not irradiated and (**c**) irradiated PEG-TPS. Blue lines represent right-handed structures and red represents left-handed structures. (**d**) $\Delta\Delta\lambda$ results for ConA and EPSPS protein experiments.

the wavelengths of the ORD resonances induced by the adsorption of a chiral medium. The $\Delta\Delta\lambda$ parameter is analogous to the optical rotation measured with conventional optical polarimetry. We performed plasmonic polarimetry using bare PEG-TPSs and laser-irradiated PEG-TPSs. The ORD spectra collected for the left- and right-handed forms of the three different types of TPSs in buffer (reference value) and solution of a model protein, Concanavalin A (ConA), and the $\Delta\Delta\lambda$ values extracted from these spectra for the peak labelled are shown in Fig. 7d. ConA was chosen as an exemplar system, because previous work has shown that it displays a large plasmonic polarimetry response[2].

As expected, given that the helical PEG structure inhibits the adsorption of biomaterials, the ConA cannot be detected using non-irradiated PEG-TPSs. In contrast, $\Delta\Delta\lambda$ values of 1.9 and − 4.5 nm are observed for the bare and laser-irradiated substrates, respectively. The change in sign of the $\Delta\Delta\lambda$ values on going from bare to laser-irradiated PEG-TPSs can be correlated with the sign of the overall integrated optical chirality of areas available for protein adsorption on both substrates (Supplementary Fig. 2). In the case of the bare substrate, protein will be exposed to evanescent fields with an overall negative integrated optical chirality. In contrast, for the laser-irradiated PEG-TPS, ConA will only be adsorbed in the arms of the structure and thus will experience fields with an overall positive optical chirality. The $\Delta\Delta\lambda$ parameter decreases with decreasing laser fluence and hence smaller areas of thermally transformed PEG-SAM, as might be expected (see Supplementary Fig. 9).

In a second series of experiments, we functionalized the deprotected arms using a procedure first proposed by Sigal *et al.*[33], which enables recombinant histidine-tagged (His-tagged) proteins to be selectively immobilized to SAMs featuring a nitrilotriacetic acid (NTA) group co-ordinated to Ni(II). In particular, we have co-self-assembled a thiol-functionalized NTA derivative (NTA-thiol) and triethylene glycol mono-11-mercaptoundecyl ether (EG-thiol) spacer unit to create binding sites for the His-tagged protein 5-enolpyruvylshikimate-3-phosphate (EPSPS) synthase (see Supplementary Fig. 10) within the nanopockets. For comparison, we functionalized a bare TPS with the NTA-thiol/EG-thiol and immobilized the EPSPS across the whole surface. The $\Delta\Delta\lambda$ values obtained for EPSPS immobilized across the whole of the TPS and just in the nanopockets were − 0.7 and 1.4 nm, respectively. The behaviour of EPSPS is qualitatively the same as for ConA, with laser-processed substrates having $\Delta\Delta\lambda$ values that are larger and of the opposite sign to those obtained when protein covers the entire surface.

In our experiments, we measure a single array of our nanostructures, which is roughly a square with 300 μm sides. Assuming that the irreversible transformation of PEG takes place in the arms of all nanostructures, we estimate that the total deprotected area in a single array would be $6.65 \times 10^3\ \mu m^2$ (calculations in Supplementary Note 7). We only detect surface-bound molecules and hence can consider the total number of particles bound on our deprotected areas as those being detected. In the case of EPSPS, the molecules are His-tagged and chemically bound to the surface with no proteins present in the buffer during measurements. We estimate that for a plain TPS, the EPSPS contributing to the measured $\Delta\Delta\lambda$ values would be 1.46 femtomoles, whereas only 108 attomoles ($\sim 6.5 \times 10^7$ protein molecules) of EPSPS protein are being detected using our protection/deprotection strategy (calculations in Supplementary Note 7). In the case of ConA, we estimate that only 80 amol ($\sim 4.8 \times 10^7$ protein molecules) contribute to the measured $\Delta\Delta\lambda$ values (see Supplementary Note 7). Hence, a functionalized PEG-TPS allows us to detect attomole ($\sim 6.0 \times 10^5$ molecules) quantities and with enhanced performance from plasmonic polarimetry.

Discussion

In conclusion, we demonstrate a novel strategy to achieve spatially selective surface functionalization on a nanostructure. It is based on a protection/deprotection strategy, which is rapid and can chemically modify macroscopic areas of substrates. The spatially selective deprotection step is driven by the localized water heating-induced unravelling of an initially helical protective thiol in selective areas of a nanostructure. The unravelled regions are both less densely packed, thus can be subsequently functionalized with another thiol and do not inhibit biomolecule adsorption. The strategy has allowed us to enhance the sensitivity of a biosensor enabling analyte molecules to be selectively located in regions with EM fields of high net chiral asymmetry. Using thermal gradients in water (or any solvent) to control surface chemistry of nanoscale regions over macroscopic areas is a versatile route for the high throughput production of advanced functional materials.

Methods

Fabrication of TPSs. The templated substrates were prepared by a combination of high-resolution electron beam lithography and injection moulding. In brief, clean silicon substrates were coated with 80 nm of PMMA (Elvacite 2041, Lucite International) and exposed in a Vistec VB6 UHR EWF lithography tool operating at 100 kV. After exposure, the substrates were developed and submitted for electroplating, where a 300-μm-thick nickel shim was formed[34]. The shim was trimmed and mounted in a custom-made tool capable of manufacturing ASA standard polymer slides. An Engel Victory Tech 28 tons injection moulding machine was used in fully automatic production mode in the manufacture of the polymer slides using polycarbonate (Makrolon DP2015) as feedstock. Polycarbonate is used as a substrate material, because it is known to have the best ability to replicate the nanofeatures and is commonly used in the industry for optical storage media[35]. This process enables us to make more than 200 substrates per hour. The injection-moulded substrates have the chiral nanostructures imparted in the plastic surface and are subsequently covered by a continuous 100-nm Au film to complete the TPS process.

PEG-SAM functionalization and ConA solutions. PEG-thiol (Sigma-Aldrich) was used to make a 833-μM solution in 95% ethanol and the TPS was allowed to self-assemble onto the nanostructure for 18 h. After being removed from solution, the TPSs were washed with water and dried under nitrogen. ConA (Sigma-Aldrich) was prepared in solutions at a concentration of 1 mg ml $^{-1}$ (37.7 μM) using a 10-mM Tris/HCl buffer at pH 7.4. Surface plasmon resonance measurements to determine surface protein coverage were done using a Biacore 2000 Instrument (GE Lifesciences).

Preparation of the NTA-thiol and His-tagged EPSPS samples. NTA-thiol was purchased from Prochimia and EG-thiol was purchased from Sigma-Aldrich. The preparation for the NTA-thiol/EG-thiol monolayer was performed similar to that described by Sigal *et al.*[33].

An existing His-tagged *Escherica coli aroA* gene cloned into a pET 22b vector was transformed and overexpressed in BL21 star cells. One litre of liquid Luria broth media was used to grow the cells and protein expression was induced with isopropyl β-D-1-thiogalactopyranoside as an optical density of 0.6 at 600 nm. The cells were harvested 4 h after induction by centrifugation and resuspended in 50 mM Tris/HCl buffer pH 7.5. The cells were lysed using sonication and centrifuged at 25 K to remove all insoluble matter. The protein was purified from the cell lysate by NTA nickel affinity chromatography with the EPSPS protein eluted using 300 mM imidazole, 0.5 M NaCl in 20 mM Tris/HCl buffer pH 7.5, giving a protein solution with a concentration of 434.8 μM. This resulted in a yield of over 50 mg of protein. The protein was concentrated by centrifugal concentration and buffer exchanged by dialysis to make a final solution of 20 mg ml $^{-1}$ EPSPS in 50 mM Tris/HCl buffer pH 7.5. The purity of the protein was assessed by SDS–PAGE and the enzyme activity was assessed using a phosphate release assay[36].

HEPES-buffered saline (HBS) used in the EPSPS experiments were 10 mM HEPES and 150 mM NaCl in water adjusted to pH 7.4.

The EPSPS solution was made using EPSP synthase in 50 mM Tris/HCl buffer of pH 7.5 with a concentration of 4 mg ml $^{-1}$.

After SAMs were fabricated from NTA-thiol/EG-thiol, measurements were taken using HBS for buffer values. The EPSP solution was left for 2 h and then rinsed with HBS before measurements were taken.

EM field simulations. Numerical simulations of EM fields and thermal heat transfer were performed using a commercial finite-element package (COMSOL v4.4, Wave optics module with Multiphysics and a heat transfer module). Permittivity values for gold were taken from Palik's optical constants[37]. Drude broadening was applied using the method described by Kuzyk *et al.*[38]. Earlier work by Bouillard *et al.*[39] shows that the variation in dielectric properties over the temperatures associated to our simulation are insignificant ($<0.2\%$) and have hence been neglected. Periodic boundary conditions were used to emulate the array of nanostructures. Linear, polarized EM wave was applied at normal incidence through the polycarbonate substrate onto the structure. A subsequent heat transfer module was then used with the total heat dissipation from the EM model used as the heat source. A time-dependant function was applied to the total heat dissipation to create a heat source that would replicate a temporally square-shaped laser pulse as the source for EM heating. We used the appropriate thermal conductivities, heat capacity and density for the dielectrics and the metals. For the values used and further information on the simulations, see Supplementary Method and Supplementary Table 1.

ORD and laser irradiation. We have used a custom-made polarimeter that measures the reflected light from our samples. It uses a tungsten halogen light source (Thorlabs), Glan-Thompson polarizers (Thorlabs) and a $\times 10$ objective (Olympus). The samples are positioned with the help of a camera (Thorlabs, DCC1645C) and the spectrum is measured using a compact spectrometer (Ocean optics USB4000). Using Stokes methods, we can measure the intensity of light at four angles of the analyser and calculate the optical rotation dispersion of our chiral plasmonic arrays.

A nanosecond (8 ns)-pulsed Nd:YAG laser (Spectra Physics Quanta Ray) operating at a 10-Hz repetition rate was used for the irradiation. The sample was irradiated at normal incidence S-polarized light, using an unfocussed beam with an area of 1 cm ^{2}, for 1 min. Fluence was varied where stated. For all protein experiments, the fluence used was 15 mJ cm $^{-2}$.

References

1. Adato, R. & Altug, H. In-situ ultra-sensitive infrared absorption spectroscopy of biomolecule interactions in real time with plasmonic nanoantennas. *Nat. Commun.* **4**, 2154–2163 (2013).
2. Hendry, E. *et al.* Ultrasensitive detection and characterization of biomolecules using superchiral fields. *Nat. Nanotechnol.* **5**, 783–787 (2010).
3. Liu, N., Hentschel, M., Weiss, T., Alivisatos, A. P. & Giessen, H. Three-dimensional plasmon rulers. *Science* **332**, 1407–1410 (2011).
4. Atwater, H. A. & Polman, A. Plasmonics for improved photovoltaic devices. *Nat. Mater.* **9**, 205–213 (2010).
5. Pendry, J. B., Aubry, A., Smith, D. R. & Maier, S. A. Transformation optics and subwavelength control of light. *Science* **337**, 549–552 (2012).
6. Schurig, D. *et al.* Metamaterial electromagnetic cloak at microwave frequencies. *Science* **314**, 977–980 (2006).
7. Smith, D. R., Pendry, J. B. & Wiltshire, M. C. K. Metamaterials and negative refractive index. *Science* **305**, 788–792 (2004).
8. Gansel, J. K. *et al.* Gold helix photonic metamaterial as broadband circular polarizer. *Science* **325**, 1513–1515 (2009).
9. Salaita, K., Wang, Y. & Mirkin, C. A. Applications of dip-pen nanolithography. *Nat. Nanotechnol.* **2**, 145–155 (2007).
10. Park, J.-U. *et al.* High-resolution electrohydrodynamic jet printing. *Nat. Mater.* **6**, 782–789 (2007).
11. Shadnam, M. R., Kirkwood, S. E., Fedosejevs, R. & Amirfazli, A. Thermo-kinetics study of laser-induced desorption of self-assembled monolayers from gold: case of laser micropatterning. *J. Phys. Chem. B* **109**, 11996–12002 (2005).
12. Slater, J. H., Miller, J. S., Yu, S. S. & West, J. L. Fabrication of multifaceted micropatterned surfaces with laser scanning lithography. *Adv. Funct. Mater.* **21**, 2876–2888 (2011).
13. Jarowicki, K. & Kocienski, P. Protecting groups. *J. Chem. Soc. Perkin Trans.* **1**, 4005–4037 (1998).
14. Baldwin, C. L., Bigelow, N. W. & Masiello, D. J. Thermal signatures of plasmonic fano interferences: toward the achievement of nanolocalized temperature manipulation. *J. Phys. Chem. Lett.* **5**, 1347–1354 (2014).
15. Sanchot, A. *et al.* Plasmonic nanoparticle networks for light and heat concentration. *ACS Nano* **6**, 3434–3440 (2012).
16. Baffou, G., Quidant, R. & García de Abajo, F. J. Nanoscale control of optical heating in complex plasmonic systems. *ACS Nano* **4**, 709–716 (2010).
17. Karimullah, A. S. *et al.* Disposable plasmonics: plastic templated plasmonic metamaterials with tunable chirality. *Adv. Mater.* **27**, 5610–5616 (2015).
18. Govorov, A. O. & Richardson, H. H. Generating heat with metal nanoparticles. *Nano Today* **2**, 30–38 (2007).
19. Baffou, G. & Quidant, R. Thermo-plasmonics: using metallic nanostructures as nano-sources of heat. *Laser Photon. Rev.* **7**, 171–187 (2013).
20. Liu, L. *et al.* Highly localized heat generation by femtosecond laser induced plasmon excitation in Ag nanowires. *Appl. Phys. Lett.* **102**, 1–5 (2013).
21. Neumann, O. *et al.* Solar vapor generation enabled by nanoparticles. *ACS Nano* **7**, 42–49 (2013).
22. Shima, T. *et al.* Thermally driven polymorphic transition prompting a naked-eye-detectable bending and straightening motion of single crystals. *Angew. Chem. Int. Ed.* **53**, 7173–7178 (2014).
23. Love, J. C., Estroff, L. A., Kriebel, J. K., Nuzzo, R. G. & Whitesides, G. M. Self-assembled monolayers of thiolates on metals as a form of nanotechnology. *Chem. Rev.* **105**, 1103–1170 (2005).
24. Bjoerling, M., Karlstroem, G. & Linse, P. Conformational adaption of poly(ethylene oxide): a carbon-13 NMR study. *J. Phys. Chem.* **95**, 6706–6709 (1991).
25. Matsuura, H. & Fukuhara, K. Conformational analysis of poly(oxyethylene) chain in aqueous solution as a hydrophilic moiety of nonionic surfactants. *J. Mol. Struct.* **126**, 251–260 (1985).
26. Harder, P., Grunze, M. & Dahint, R. Molecular conformation in oligo (ethylene glycol)-terminated self-assembled monolayers on gold and silver surfaces determines their ability to resist protein adsorption. *J. Phys. Chem. B* **102**, 426–436 (1998).
27. Norman, A. I., Yiwei, F., Ho, D. L. & Greer, S. C. Folding and unfolding of polymer helices in solution. *Macromolecules* **40**, 2559–2567 (2007).
28. Gadzuk, J. W. The road to hot electron photochemistry at surfaces: a personal recollection. *J. Chem. Phys.* **137**, 091703 (2012).
29. Zhou, X.-L., Zhu, X.-Y. & White, J. M. Photochemistry at adsorbate/metal interfaces. *Surf. Sci. Rep.* **13**, 73–220 (1991).
30. Cao, L., Barsic, D. N., Guichard, A. R. & Brongersma, M. L. Plasmon-assisted local temperature control to pattern individual semiconductor nanowires and carbon nanotubes. *Nano Lett.* **7**, 3523–3527 (2007).
31. Bauer, D. R., Brauman, I. & Pecora, R. Depolarized Rayleigh spectroscopy studies of relaxation processes of polystyrenes in solution. *Macromolecules* **8**, 443–451 (1975).
32. Tullius, R. *et al.* 'Superchiral' spectroscopy: detection of protein higher order hierarchical structure with chiral plasmonic nanostructures. *J. Am. Chem. Soc.* **137**, 8380–8383 (2015).

33. Sigal, G. B., Bamdad, C., Barberis, A., Strominger, J. & Whitesides, G. M. A self-assembled monolayer for the binding and study of histidine-tagged proteins by surface plasmon resonance. *Anal. Chem.* **68,** 490–497 (1996).
34. Gadegaard, N., Mosler, S. & Larsen, N. B. Biomimetic polymer nanostructures by injection molding. *Macromol. Mater. Eng.* **288,** 76–83 (2003).
35. Monkkonen, K. *et al.* Replication of sub-micron features using amorphous thermoplastics. *Polym. Eng. Sci.* **42,** 1600–1608 (2002).
36. Oliveira, J. S., Mendes, M. A., Palma, M. S., Basso, L. A. & Santos, D. S. One-step purification of 5-enolpyruvylshikimate-3-phosphate synthase enzyme from *Mycobacterium tuberculosis. Protein Expr. Purif.* **28,** 287–292 (2003).
37. Palik, E. D. *Handbook of Optical Constants of Solids* (Elsevier Science, 1998).
38. Kuzyk, A. *et al.* Reconfigurable 3D plasmonic metamolecules. *Nat. Mater.* **13,** 1–5 (2014).
39. Bouillard, J. S. G., Dickson, W., O'Connor, D. P., Wurtz, G. A. & Zayats, A. V. Low-temperature plasmonics of metallic nanostructures. *Nano Lett.* **12,** 1561–1565 (2012).

Acknowledgements

We acknowledge financial support from the Engineering and Physical Sciences Research Council (EPSRC EP/K034936/1), National Science Foundation (NSF grant CHE-1307021) and JSPS Core to Core. A.K. thanks the Leverhulme Trust and technical support from the James Watt Nanofabrication Centre (JWNC). R.T. and C.J. thank the EPSRC for the award of scholarships. L.K.K. and A.O.G. were supported by the Volkswagen Foundation (Germany).

Author contributions

C.J., R.T., A.S.K. and M.R. performed the experimental work. A.S.K., L.K.K. and A.O.G. carried out COMSOL calculations for the EM fields and the local temperature maps. A.J.L. supervised protein purification and immobilization. B.F. facilitated the SAM formation. N.G. developed the fabrication process of the injection-moulded substrates. A.S.K., M.K., L.K.K. and A.O.G. discussed the mechanism of temperature increase and the origins of the observed chemical effects. M.K., A.S.K., L.D.B., A.J.L., V.R., G.C. and N.G. wrote the paper. M.K. conceived and designed the experiment.

Additional information

Competing financial interests: The authors declare no competing financial interests.

6

A series connection architecture for large-area organic photovoltaic modules with a 7.5% module efficiency

Soonil Hong[1,2], Hongkyu Kang[2,3], Geunjin Kim[1,2], Seongyu Lee[1,2], Seok Kim[1,2], Jong-Hoon Lee[1,2], Jinho Lee[2,4], Minjin Yi[3], Junghwan Kim[2,3], Hyungcheol Back[1,2], Jae-Ryoung Kim[3] & Kwanghee Lee[1,2,3,4]

The fabrication of organic photovoltaic modules via printing techniques has been the greatest challenge for their commercial manufacture. Current module architecture, which is based on a monolithic geometry consisting of serially interconnecting stripe-patterned subcells with finite widths, requires highly sophisticated patterning processes that significantly increase the complexity of printing production lines and cause serious reductions in module efficiency due to so-called aperture loss in series connection regions. Herein we demonstrate an innovative module structure that can simultaneously reduce both patterning processes and aperture loss. By using a charge recombination feature that occurs at contacts between electron- and hole-transport layers, we devise a series connection method that facilitates module fabrication without patterning the charge transport layers. With the successive deposition of component layers using slot-die and doctor-blade printing techniques, we achieve a high module efficiency reaching 7.5% with area of 4.15 cm^2.

[1] School of Materials Science and Engineering, Gwangju Institute of Science and Technology, Gwangju 61005, Republic of Korea. [2] Heeger Center for Advanced Materials, Gwangju Institute of Science and Technology, Gwangju 61005, Republic of Korea. [3] Research Institute for Solar and Sustainable Energies, Gwangju Institute of Science and Technology, Gwangju 61005, Republic of Korea. [4] Department of Nanobio Materials and Electronics, Gwangju Institute of Science and Technology, Gwangju 61005, Republic of Korea. Correspondence and requests for materials should be addressed to H.K. (email: gemk@gist.ac.kr) or to K.L. (email: klee@gist.ac.kr).

Bulk heterojunction solar cells, which are built on photoactive nanocomposites of electron-donating and electron-withdrawing organic semiconductors, are good candidates to be a ubiquitous renewable energy source that allows for integration with portable and wearable electronic applications[1,2]. Moreover, these organic solar cells (OSCs) are considered representative of the research field of printed electronics, because the solution processability of organic semiconductors enables cost-efficient, high-volume/throughput printing production with roll-to-roll manufacturing facility[3–8]. To realize this photovoltaic technology, research has focused mainly on enhancing device performance, extending device lifetime and developing up-scaling techniques for transitioning from small-area laboratory-scale devices to large-area industrial-scale modules[3–14]. Although the impressive progress made in the past two decades has led to considerable improvements in both the efficiency and operational stability of OSCs, the fabrication of large-area printed modules still suffers from significantly reduced power conversion efficiencies (PCEs), amounting to less than half the efficiencies of small-sized laboratory cells.

The current module architecture possesses an inherent weakness with regard to area loss, so-called aperture loss, which is well known to be a major contributor to the drastic performance degradation observed in large-area printed OSC modules[6–8]. The module geometry is based on a monolithic structure composed of several serially interconnecting subcells that are patterned into stripes with sufficiently narrow widths ($W \sim 10$ mm) to enable the sheet resistance of transparent electrodes to be neglected. However, such monolithic interconnections inevitably produce area loss to ensure contact areas for series connections between subcells. Furthermore, because of the low (millimetre scale) patterning resolutions of current printing techniques, using these techniques to create regularly spaced stripe-patterned subcells worsens unwanted area loss, resulting in very poor module PCEs with low geometric fill factors (FF, ratios between photoactive and total areas). Despite intense research efforts to reduce aperture loss by using laser ablation or metal-filament patterning techniques, only a few methods for realizing high-efficiency printed modules without additional patterning processing have been reported[15–17].

Here we demonstrate a new module architecture for manufacturing large-area printed OSC modules without the aid of additional and complicated post-patterning processing. By introducing an innovative series connection concept based on the charge recombination characteristic that occurs at the contacts between charge transport layers (CTLs), we design a monolithic interconnection that enables facile and efficient module fabrication without patterning the CTLs and producing the considerable aperture loss. Therefore, through consecutive printing processes using doctor-blade and slot-die machines, we successfully fabricate a large-area module that exhibits a high module PCE of 7.5% with a high geometric FF of 90%.

Results

Module architecture and fabrication. A schematic illustration of our module architecture is shown in Fig. 1a. The module has three inverted-type subcells consisting of three main component layers sandwiched between an indium tin oxide (ITO)

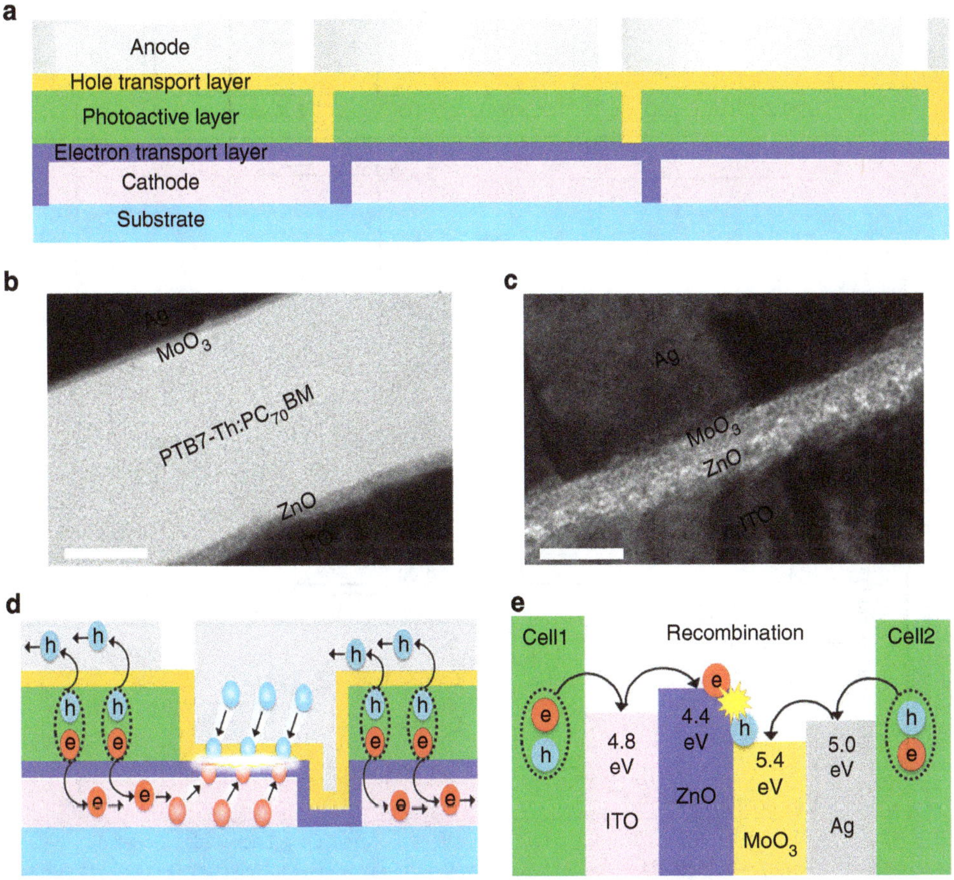

Figure 1 | Schematic illustration of the module. (**a**) Conceptual module structure consisting of patternless electron-transport and hole-transport layers and one patterned photoactive layer. (**b**,**c**) Corresponding cross-sectional TEM images of the active area (scale bar, 50 nm) (**b**) and series connection region (scale bar, 25 nm) (**c**). (**d**) A schematic image of charge recombination as it occurs in our module. (**e**) Energy level diagrams of series connection region components.

cathode and a silver (Ag) anode. For a photoactive layer, we used a bulk heterojunction composite comprising an electron-donating poly(thieno[3,4-b]thiophene-alt-benzodithiophene) derivative (PTB7-Th) and an electron-accepting [6,6]-phenyl-C_{71}-butyric acid methyl ester ($PC_{70}BM$) (ref. 18). To obtain inverted-type subcells, we introduced two CTLs, including sol-gel zinc oxide (ZnO) as an electron-transport layer and molybdenum oxide (MoO_3) as a hole-transport layer (HTL), between the photoactive layers and their respective electrodes[19–21]. In the module fabrication, the photoactive layer was patterned in stripes onto the patterned ITO cathodes ($W = 13.5$ mm) with a slight blank offset (0.5 mm), whereas the two CTLs were deposited on the surfaces of the ITO and photoactive layer in a single-layer form without any stripe patterning. Series connections between adjacent subcells were achieved by forming stripe-patterned Ag anodes ($W = 13.5$ mm) with a subtle blank offset (0.5 mm) relative to the patterned photoactive layers. By printing the component layers with a doctor-blade machine for non-patterned single-layer forms and a slot-die machine for stripe patterning, we succeeded in fabricating a monolithic printed module with a high geometric FF of 90% without the use of any post-patterning processing (Supplementary Fig. 1 and Supplementary Note 1).

Series connection mechanism in the module. The most prominent feature of our module configuration is that the CTLs were not patterned in stripe, in contrast to the conventional modules fabricated with all stripe-patterned component layers (Supplementary Fig. 2). Cross-sectional images taken with a high-resolution transmission electron microscope (TEM) clearly show not only all-component layers (ITO/ZnO/PTB7-Th:$PC_{70}BM$/MoO_3/Ag) in the active area, but also the CTLs (ZnO/MoO_3) sandwiched between the ITO and Ag electrodes in a series connection region (SCR), in which the counter electrodes of the adjacent subcells vertically overlap (Fig. 1b,c). Because of the CTLs embedded within the SCRs, our module operation will be quite different from that of typical modules in which the counter electrodes are in direct contact. We can expect the series connection mechanism of our monolithic module to be similar to that of existing multi-junction OSCs; the photogenerated charge carriers (that is, holes and electrons) from neighbouring subcells transport along the counter electrodes and are injected into the CTLs, thereby leading to series connections between subcells and voltage gains via charge recombination at the interface between the CTLs in the SCRs (Fig. 1d,e and Supplementary Fig. 3)[22,23].

Characteristics of the SCRs. To investigate the impact of the SCRs on module operation, we partitioned the module into a SCR unit cell and two subcells consisting of ITO/ZnO/MoO_3/Ag and ITO/ZnO/PTB7-TH:$PC_{70}BM$/MoO_3/Ag, respectively. Using current density–voltage (J–V) and electrochemical impedance

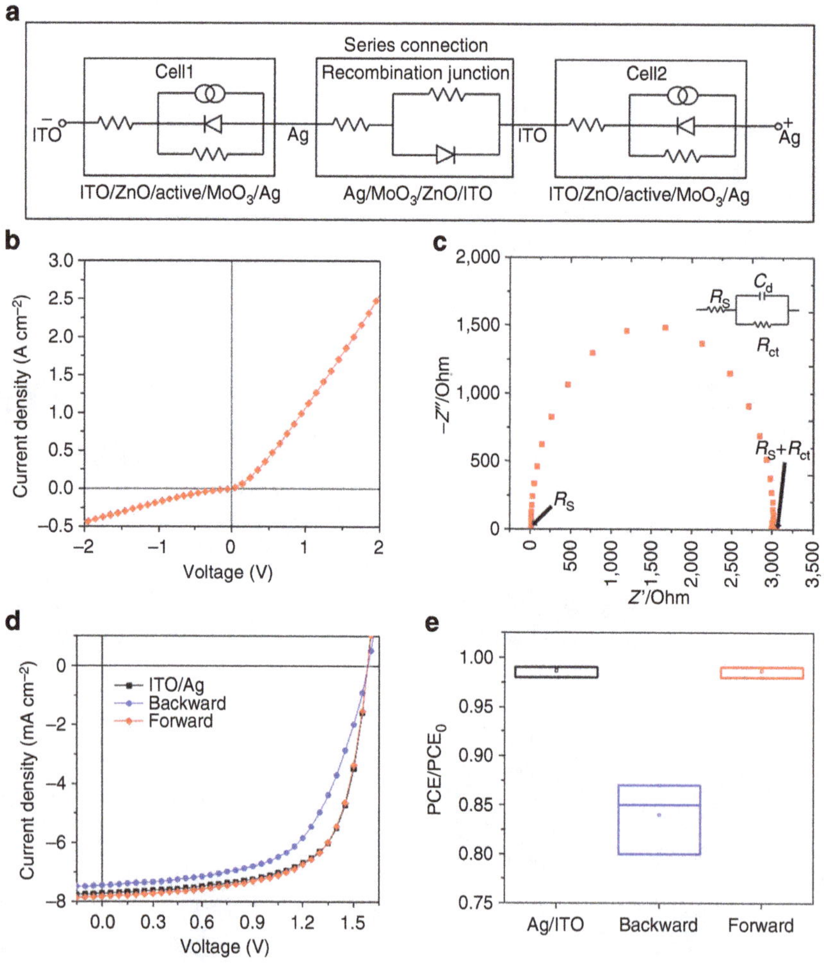

Figure 2 | Equivalent circuit of series-connected OSCs with an SCR unit cell. (**a**) The circuit comprised of two OSC unit cells with an SCR unit cell. (**b**) J–V characteristic of an SCR unit cell (Ag/MoO_3/ZnO/ITO) in the dark. (**c**) The Nyquist plot obtained from the EIS analysis of SCR (Ag/MoO_3/ZnO/ITO). (**d**,**e**) The J–V characteristics (**d**) and performance deviations (**e**) of OSCs with an SCR unit cell under AM 1.5G with 100 mW cm^{-2} (PCE_0 pertains to OSCs without any unit cell, boxes are measured values and rectangular points are average values).

spectroscopy (EIS) measurements, we formulated the equivalent circuit of the SCR unit cell (Fig. 2a). The J–V characteristic of the SCR unit cell shows the rectifying property originating from the electrical junction between the electron-transporting ZnO and hole-transporting MoO_3 layers (Fig. 2b). Meanwhile, the Nyquist plot obtained via EIS analysis exhibits a low series resistance (R_s) of 12 Ω and a relatively high shunt resistance (R_{sh}) of 3,000 Ω in the SCR unit cell (Fig. 2c and Supplementary Fig. 4). By combining these results, we can define the SCR unit cell as an electrical component composed of a diode and two resistors. On connecting this SCR unit cell between two subcells, we observed the dependence of device performance on the polarity of the SCR unit cell (Fig. 2d,e). The resulting equivalent circuit reveals that the series connection of the subcells is dominantly affected by the low R_s of the forward connection and the high R_{sh} of the backward connection. Because our module structure connects the forward SCRs with neighbouring subcells, the SCRs are expected to allow loss-free charge recombination of the adjacent subcells in the module operation.

Optimization of printing processes. To fabricate our printed module, we employed two kinds of printing techniques using doctor-blade and slot-die machines (Supplementary Fig. 5). Both printing machines have similar control factors, which depend on the viscosity of the solution, the amount of feeding solution, the coating speed and the substrate temperature. By delicately adjusting these parameters, we achieved high-quality printed films with smooth and uniform film morphologies (Supplementary Fig. 6). In particular, we designed a slot-die coating head with a positive shim mask to create a meniscus guide (Supplementary Fig. 7); the photoactive PTB7-Th:$PC_{70}BM$ solution was ejected through a narrow slot and followed the shim mask pattern, thereby forming a meniscus between the mask and substrate via capillary action (Fig. 3a). To simply control the film thickness (t), we changed the coating speed (S) while fixing other coating parameters. As shown in Fig. 3b, the thicknesses of PTB7-Th:$PC_{70}BM$ films follow power law equation of $t \approx S^{0.62}$, which has been demonstrated in meniscus coating methods using low-viscosity organic solutions[24,25]. After optimizing the film thicknesses, we obtained a high-quality printed photoactive layer with a thickness of ~125 nm that exhibited an optimal PCE of 8.5% (Fig. 3c, Supplementary Fig. 8 and Supplementary Table 1).

Performance of large-area printed OSC modules. Figure 4a displays a photograph of the complete large-area printed module with optimized film thicknesses. The current–voltage (I–V) and current density–voltage (J–V) characteristics of the module, which are measured via a large-scale calibrated solar simulator under standard illumination conditions, are shown in Fig. 4b,c and Table 1. The best performance of the new module (4.15 cm^2) yielded a remarkable PCE of 8.1% with an open-circuit voltage (V_{oc}) of 2.36 V, a short-circuit current density (J_{sc}) of 5.53 mA cm^{-2}, and a FF of 62%. The V_{oc} (2.36 V) of the module is nearly three times larger than the V_{oc} (0.79 V) of the small-area reference; the J_{sc} (5.53 mA cm^{-2}) of the module is exactly one-third of that (16.6 mA cm^{-2}) of the reference; and the FF (62%) of the module is almost comparable to that of the reference (66%). Considering its geometric FF of 90%, the module exhibits a high module PCE of 7.3%. One of the modules exhibiting the best PCE was sent to the Korea Institute of Energy Research and was returned to our laboratory with a certificated module PCE of ~7.5%, as shown in Fig. 4d. To the best of our knowledge, this PCE is the highest value in printed solar modules to date in scientific literature. These outstanding results indicate that the subcells were perfectly interconnected via effective charge recombination at the interfaces between the CTLs in the SCRs.

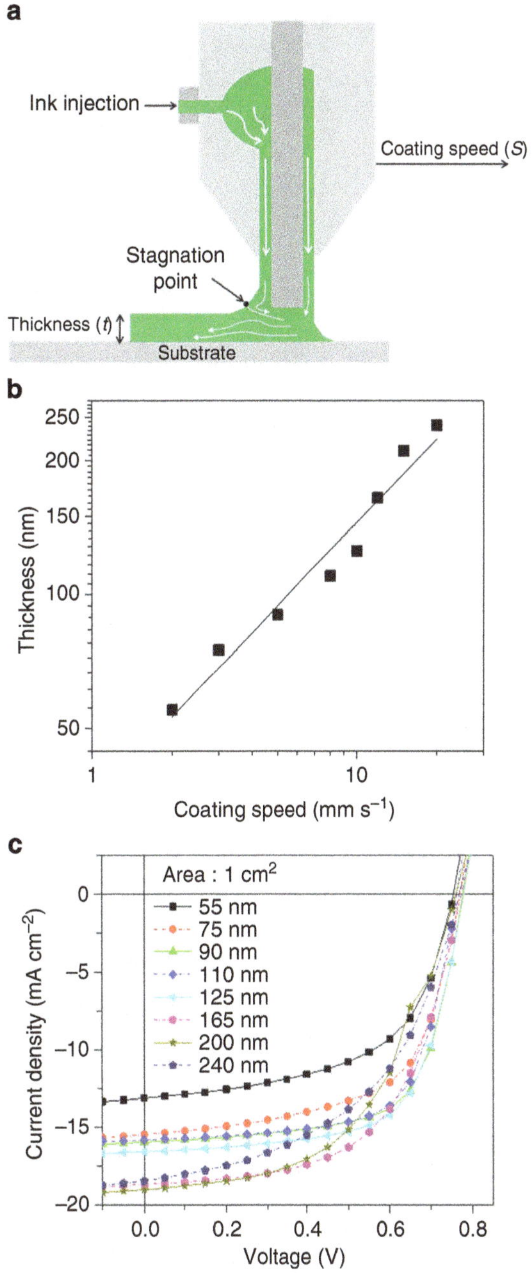

Figure 3 | OSCs fabricated using the printing method. (**a**) Schematic of meniscus formation and the streamlines near the stagnation point in the slot-die coating using a positive-shim style mask. (**b**) Thicknesses of PTB7-Th:$PC_{70}BM$ films coated via the slot-die coating method using various coating speeds from 2 to 30 mm s^{-1} (log scale). (**c**) J–V characteristics of OSCs fabricated using the slot-die coating method at various thicknesses.

Although increasing the length (size) of the module causes a slight decrease in the FF and J_{sc} values due to a few concomitant defect sites (for example, pinholes and fine dusts) within the printed films, we can overcome this problem through defect-free printing process in clean rooms, thereby enlarging the module size without suffering serious performance loss[6,26]. In addition, we introduced a solution-processed MoO_3 layer for the module fabrication[27]. By printing the MoO_3 layer, we obtained reasonable average efficiencies of 7.7% for large-area single cells and 6.9% for module, thereby demonstrating that our module can operate well even when we used all-printed CTLs (Fig. 4e,f and Supplementary Table 2).

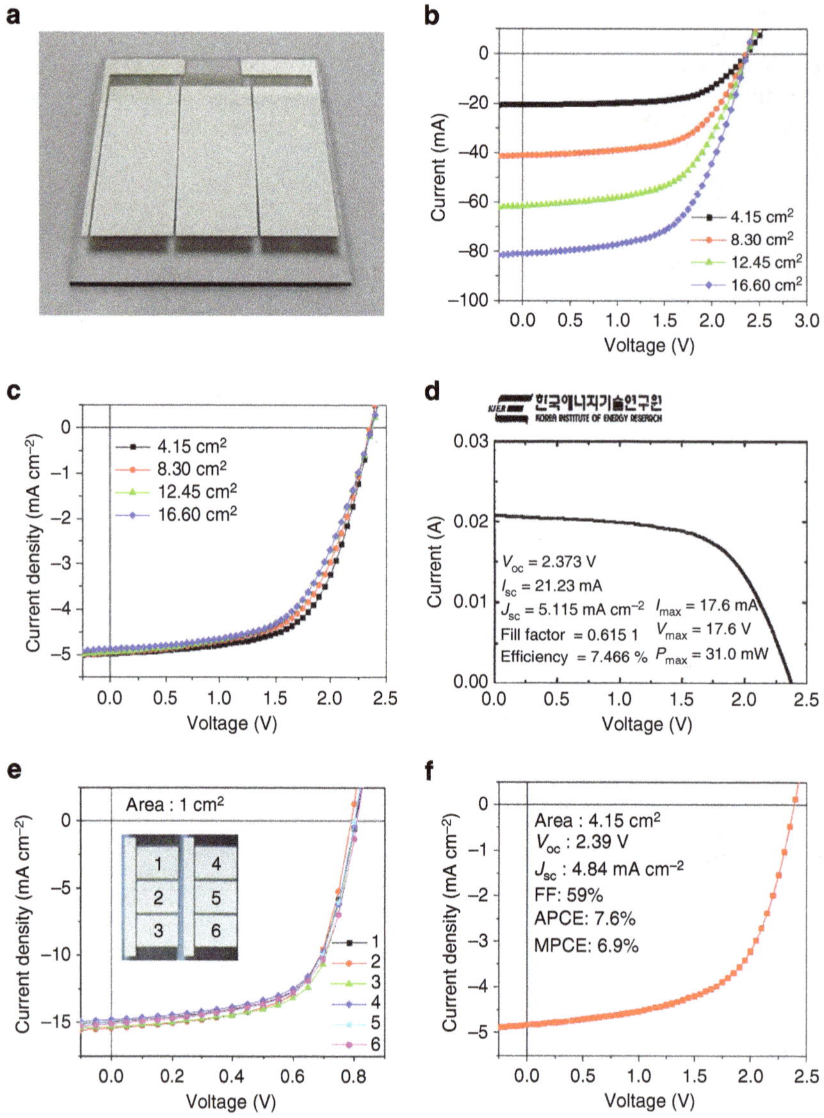

Figure 4 | OSC modules fabricated using the printing method. (**a**) A photograph image of our module (size of 60 × 44.5 mm). (**b,c**) *I–V* (**b**) and *J–V* (**c**) characteristics of OSC modules of various sizes depending on module lengths from 1 to 4 cm. (**d**) Korea Institute of Energy Research-certified *J–V* characteristics of our OSC module. (**e,f**) *J–V* characteristics of single OSCs in different positions (inset: photograph image of printed OSCs) (**e**) and OSC module (**f**) by using the printed MoO_3 layer. APCE is the PCE of the module in active area and MPCE is the PCE of the module in total area.

Discussion

Our work demonstrates a new scientific perspective, in that we have designed a simple and efficient module structure by developing a novel series connection method and a remarkable technical advance towards manufacturing printed modules for next-generation photovoltaic systems. By using a charge recombination feature arising from the electrical junctions between CTLs, we succeed in fabricating a monolithic module without the use of additional complex patterning processes. This monolithic module has achieved a certificated module PCE of 7.5% with a high geometric FF of 90%. We expect that this new approach presents a simple and useful means for transitioning from small-area laboratory-scale OSCs to large-area industrial-scale OSC modules.

Table 1 | Performance parameters of the OSC modules with increasing area.

Area (cm^2)	V_{OC} (V)	I_{SC} (mA)	J_{SC} ($mA\,cm^{-2}$)	FF (%)	APCE (%)	MPCE (%)
4.15	2.36	20.7	4.98	62	8.1	7.3
8.30	2.36	41.1	4.95	60	7.7	7.0
12.45	2.38	61.6	4.94	57	7.4	6.7
16.60	2.37	80.9	4.87	58	7.4	6.7

APCE, the PCE of the module in active area; MPCE, the PCE of the module in total area.

Methods

Material preparation. The ZnO precursor was prepared by dissolving zinc acetate dihydrate ($Zn(CH_3COO)_2 \cdot 2H_2O$, Aldrich, 99.9%, 1 g) and ethanolamine ($NH_2CH_2CH_2OH$, Aldrich, 99.5%, 0.5 g) in isopropyl alcohol ($CH_3OCH_2CH_2OH$, Aldrich, 99.8%, 50 g) via stirring for 24 h. The PTB7-Th:PC_{70}BM solution was prepared by blending PTB7-Th (1-material) and PC_{70}BM (Nano-C) at a ratio of 1:2 in chlorobenzene solvent with 1,8-diiodooctane additive (3% by volume) to obtain a total concentration of 12 mg ml^{-1}. The MoO_3 solution was prepared by dissolving bis(acetylacetonato) dioxomolybdenum (Sigma Aldrich) in a cosolvent of methanol and 1-butanol.

Single-cell fabrication. ITO/glass substrates were cleaned with detergent, after which they were sequentially washed via ultrasonic treatment in de-ionized water, acetone and IPA. The ZnO solution was coated onto the ITO/glass substrate using a doctor-blade coater (Coatmaster 509 MC, Erichsen in Germany) at 40 °C, then annealed at 150 °C for 20 min in air. The PTB7-Th:$PC_{70}BM$ composite solution was coated on top of the ZnO layer in air using a slot-die coating method (Slot-die coater, iPen in South Korea) at room temperature. The pumping rate for coating PTB7-Th:$PC_{70}BM$ solution was 0.1 ml min^{-1} when using a 50-μm-thick mask, and the film thickness was controlled by the coating speed of the slot-die header. To complete the single-cell device fabrication, MoO_3 (as a HTL) and Ag (as a top electrode) were deposited sequentially by thermal evaporation in a vacuum with a pressure of 10^{-6} torr.

Module fabrication. The ITO/glass preparation and ZnO coating process is the same as that described for single-cell fabrication. Before the module fabrication, the patterned ITO glass substrate (the stripe width of 13.5 mm and the gap of 0.5 mm) was prepared by wet-etching processing using a typical acid etchant. For the patterned PTB7-Th:$PC_{70}BM$ coating process, we used a slot-die coating machine with a 50-μm-thick three-stripe mask. The coating speed and pumping rate used to coat the PTB7-Th:$PC_{70}BM$ solution are 10 mm s^{-1} and 0.4 ml min^{-1}, respectively, and the optimized thickness of PTB7-Th:$PC_{70}BM$ is ~125 nm. The MoO_3 (as a HTL) was deposited onto the patterned photoactive layer with no patterned mask by using thermal evaporation in a vacuum with a pressure of 10^{-6} torr. The solution-processed MoO_3 layer was deposited on the photoactive layer by using a doctor-blade coating machine. Module-device fabrication was completed by thermal evaporation of the Ag metal top electrode (200 nm thickness) in a vacuum with a pressure of 10^{-6} torr. In contrast to MoO_3 deposition, Ag was deposited via a patterned mask, which is used to obtain a series connection in the module.

Characterization and analysis. The current–voltage (*I–V*) characteristics were recorded using an Iviumsoft apparatus with simulated AM 1.5 illumination (100 mW cm^{-2}) via a solar simulator (Abet Technologies Sun 3000) in normal air conditions. The thicknesses of the coated films were measured using a thickness profile metre (Surfcorder ET 3000, Kosaka Laboratory, Ltd.). Cross-sectional TEM samples of the printed OSC module were prepared using a dual beam-focused ion beam (Helios NanoLabTM). The TEM images were obtained using field emission TEM (FEI TecnaiTM G2 F30 Super-Twin) operated at 200 kV. The topographies of surface images were characterized using atomic force spectroscopy.

References

1. Zhang, Z. *et al.* A lightweight polymer solar cell textile that functions when illuminated from either side. *Angew. Chem.* **126,** 11755–11758 (2014).
2. Zhang, Z. *et al.* Integrated polymer solar cell and electrochemical supercapacitor in a flexible and stable fiber format. *Adv. Mater.* **26,** 466–470 (2014).
3. Li, G., Zhu, R. & Yang, Y. Polymer solar cells. *Nat. Photon.* **6,** 153–161 (2012).
4. Krebs, F. C., Espinosa, N., Hösel, M., Søndergaard, R. R. & Jørgensen, M. 25th anniversary article: rise to power - OPV-based solar parks. *Adv. Mater.* **26,** 29–39 (2014).
5. Espinosa, N., Hösel, M., Jørgensen, M. & Krebs, F. C. Large scale deployment of polymer solar cells on land, on sea and in the air. *Energy Environ. Sci.* **7,** 855–866 (2014).
6. Krebs, F. C., Tromholt, T. & Jørgensen, M. Upscaling of polymer solar cell fabrication using full roll-to-roll processing. *Nanoscale* **2,** 873–886 (2010).
7. Krebs, F. C. Polymer solar cell modules prepared using roll-to-roll methods: knife-over-edge coating, slot-die coating and screen printing. *Sol. Energ. Mat. Sol. C.* **93,** 394–412 (2009).
8. Søndergaard, R. R., Hösel, M. & Krebs, F. C. Roll-to-roll fabrication of large area functional organic materials. *J. Polym. Sci. B Polym. Phys.* **51,** 16–34 (2013).
9. Chen, J.-D. *et al.* Single-junction polymer solar cells exceeding 10% power conversion efficiency. *Adv. Mater.* **27,** 1035–1041 (2015).
10. He, Z. *et al.* Single-junction polymer solar cells with high efficiency and photovoltage. *Nat. Photon.* **9,** 174–179 (2015).
11. Liu, Y. *et al.* Aggregation and morphology control enables multiple cases of high-efficiency polymer solar cells. *Nat. Commun.* **5,** 5293 (2014).
12. Krebs, F. C. *Stability and Degradation of Organic and Polymer Solar Cells* (Wiley, 2012).
13. Jørgensen, M. *et al.* Stability of polymer solar cells. *Adv. Mater.* **24,** 580–612 (2012).
14. Andersen, T. R. *et al.* Scalable, ambient atmosphere roll-to-roll manufacture of encapsulated large area, flexible organic tandem solar cell modules. *Energy Environ. Sci.* **7,** 2925–2933 (2014).
15. Kang, H., Hong, S., Back, H. & Lee, K. A new architecture for printable photovoltaics overcoming conventional module limits. *Adv. Mater.* **26,** 1602–1606 (2014).
16. Spyropoulos, G. D. *et al.* Flexible organic tandem solar modules with 6% efficiency: combining roll-to-roll compatible processing with high geometric fill factors. *Energy Environ. Sci.* **7,** 3284–3290 (2014).
17. Lee, J. *et al.* Seamless polymer solar cell module architecture built upon self-aligned alternating interfacial layers. *Energy Environ. Sci.* **6,** 1152–1157 (2013).
18. Liao, S.-H., Jhuo, H.-J., Cheng, Y.-S. & Chen, S.-A. Fullerene derivative-doped zinc oxide nanofilm as the cathode of inverted polymer solar cells with low-bandgap polymer (PTB7-Th) for high performance. *Adv. Mater.* **25,** 4766–4771 (2013).
19. Sun, Y., Seo, J. H., Takacs, C. J., Seifter, J. & Heeger, A. J. Inverted polymer solar cells integrated with a low-temperature-annealed sol-gel-derived ZnO film as an electron transport layer. *Adv. Mater.* **23,** 1679–1683 (2011).
20. Sun, Y. *et al.* Efficient, air-stable bulk heterojunction polymer solar cells using MoO_x as the anode interfacial layer. *Adv. Mater.* **23,** 2226–2230 (2011).
21. He, Z. *et al.* Enhanced power-conversion efficiency in polymer solar cells using an inverted device structure. *Nat. Photon.* **6,** 591–595 (2012).
22. Kong, J. *et al.* Building mechanism for a high open-circuit voltage in an all-solution-processed tandem polymer solar cell. *Phys. Chem. Chem. Phys.* **14,** 10547–10555 (2012).
23. Chen, C.-C. *et al.* An efficient triple-junction polymer solar cell having a power conversion efficiency exceeding 11%. *Adv. Mater.* **26,** 5670–5677 (2014).
24. Hong, S., Lee, J., Kang, H. & Lee, K. Slot-die coating parameters of the low-viscosity bulk-heterojunction materials used for polymer solar cells. *Sol. Energ. Mat. Sol. C.* **112,** 27 (2013).
25. Vak, D. *et al.* 3D printer based slot-die coater as a lab-to-fab translation tool for solution-processed solar cells. *Adv. Energy Mater.* **5,** 1401539 (2015).
26. Jeong, W.-I., Lee, J., Park, S.-Y., Kang, J.-W. & Kim, J.-J. Reduction of collection efficiency of charge carriers with increasing cell size in polymer bulk heterojunction solar cells. *Adv. Func. Mater.* **21,** 343–347 (2011).
27. Murase, S. & Yang, Y. Solution processed MoO_3 interfacial layer for organic photovoltaics prepared by a facile synthesis method. *Adv. Mater.* **24,** 2459–2462 (2012).

Acknowledgements

We thank the Heeger Center for Advanced Materials (HCAM) at the Gwangju Institute of Science and Technology (GIST) of Korea for help with device fabrication and measurements. This research was supported by a grant from the National Research Foundation of Korea (NRF) funded by the Korean government (MSIP) (NRF-2014R1A2A1A09006137), the Technology Development Program to Solve Climate Changes of the NRF funded by the (MSIP) (NRF-2015M1A2A2057510), the R&D program of MSIP/COMPA (2015K000199), and the 'Basic Research Projects in High-tech Industrial Technology' Project through a grant provided by GIST in 2015. K.L. also acknowledges support provided by the Core Technology Development Program for Next-generation Solar Cells of the Research Institute for Solar and Sustainable Energies, GIST.

Author contributions

H.K. contributed a key idea. S.H., H.K. and K.L. designed the concept and the required experiments. S.H. performed the fabrication and characterization of devices. G.K. helped with the measurement of EIS. S.L. helped with the measurement of atomic force microscopy. S.K. helped with the fabrication of devices. H.B. helped with the preparation of the MoO_3 solution. S.H., H.K. and K.L. prepared the manuscript. K.L. guided and directed the research. All authors discussed the results and contributed to the writing of the paper.

Additional information

Competing financial interests: The authors declare no competing financial interests.

7

Polarized three-photon-pumped laser in a single MOF microcrystal

Huajun He[1,*], En Ma[2,*], Yuanjing Cui[1,*], Jiancan Yu[1], Yu Yang[1], Tao Song[1], Chuan-De Wu[3], Xueyuan Chen[2], Banglin Chen[1,4] & Guodong Qian[1]

Higher order multiphoton-pumped polarized lasers have fundamental technological importance. Although they can be used to *in vivo* imaging, their application has yet to be realized. Here we show the first polarized three-photon-pumped (3PP) microcavity laser in a single host-guest composite metal-organic framework (MOF) crystal, via a controllable *in situ* self-assembly strategy. The highly oriented assembly of dye molecules within the MOF provides an opportunity to achieve 3PP lasing with a low lasing threshold and a very high-quality factor on excitation. Furthermore, the 3PP lasing generated from composite MOF is perfectly polarized. These findings may eventually open up a new route to the exploitation of multiphoton-pumped solid-state laser in single MOF microcrystal (or nanocrystal) for future optoelectronic and biomedical applications.

[1] State Key Laboratory of Silicon Materials, Cyrus Tang Center for Sensor Materials and Applications, School of Materials Science and Engineering, Zhejiang University, Hangzhou 310027, China. [2] Key Laboratory of Optoelectronic Materials Chemistry and Physics, Fujian Institute of Research on the Structure of Matter, Chinese Academy of Sciences, Fuzhou, Fujian 350002, China. [3] Department of Chemistry, Zhejiang University, Hangzhou 310027, China. [4] Department of Chemistry, University of Texas at San Antonio, San Antonio, Texas 78249-0698, USA. * These authors contributed equally to this work. Correspondence and requests for materials should be addressed to X.C. (email: xchen@fjirsm.ac.cn) or to B.C. (email: banglin.chen@utsa.edu) or to G.Q. (email: gdqian@zju.edu.cn).

Polarization has been used in various fields, particularly in the field of biophotonics due to its ability to reduce multiple scattering, while to enhance the contrast and to improve tissue imaging resolution[1–3]. Through the measurement of the polarization state of the scattered light, a wealth of structural information of scatters (for example, lesions information in the tissue) can be collected given the fact that the microscopic structure of a scattering media is closely related to changes in the polarization state of the photon during a scattering process[1,3]. On the other hand, high-order multiphoton excitation can offer stronger spatial confinement, deeper tissue penetration and less Rayleigh scattering, which are significantly beneficial to the biological imaging[4–8]. To make use of the uniqueness of both polarization and high-order multiphoton excitation, the polarized three-photon and/or higher order pumped laser in single solid-state microcrystal is potentially useful for a new kind of biological imaging, so called multiphoton pumped (MPP) polarized emission biological imaging (Supplementary Fig. 1), but has never been realized. In order to produce such a unique laser, the gain medium not only needs to have a high multiphoton absorption (MPA) cross-section and lasing efficiency[4,5], but more importantly needs to be assembled into a suitable microcavity of high concentration and orientation (especially in the case that the absorption transition moment of gain medium is anisotropic) without significant luminescent quenching to enforce the high optical gain and to generate controllable and directional laser. This is really a daunting challenge. In fact, although extensive research endeavours have been pursued to target such a goal, progress has been very slow. The initial effort to diminish the significant quenching effects on the solid state was to homogeneously disperse the gain medium such as the dye molecules with high multiphoton absorption cross-section into its solution[6,9]. By employing such a strategy, it still remains extremely difficult to provide with a sufficiently high quenching concentration, which limits the realization of necessary optical gain for compensating the losses. The quenching concentration means that the aggregation-caused quenching (ACQ) gradually becomes dominant when the concentration of gain medium is higher than the quenching concentration. Furthermore, the molecules in the solution are randomly oriented, which would limit their capacities to maximize the optical gain. So far, this dispersed solution methodology can only lead to the amplified spontaneous emission instead of generating three-photon or higher order pumped laser[4,5]. Although the luminescent properties of quantum dots are intriguing, they only have generated three-photon-pumped (3PP) random lasing in which the emission direction, position and numbers of mode frequency, and the uniformity of light-emitting region of such lasers are very difficult to control[7,10]. Recently, the 3PP lasing from colloidal nanoplatelets in solution has been demonstrated by Li *et al.*[11]; however, no polarization property of the 3PP lasing has been realized. Furthermore, its liquid nature has limited practical applications. To take advantage of the pore confinement of porous materials, zeolites and nanoporous silica have been explored to incorporate dye molecules and semiconducting polymers into the corresponding crystals and thin films to develop solid-state lasing[12,13]. However, zeolite/dye composites can only generate single-photon pumped lasing, mainly due to the incompatibility between the inorganic framework and organic guest, leading to the low loading concentration (0.005 ~ 0.0005 M), uneven distribution of dye molecules and poor crystal morphology; while nanoporous silica/semiconducting polymer matrix basically leads to the single-photon pumped polarized amplified spontaneous emission.

Previously, we have used a porous metal–organic framework (MOF) for its pore confinement of a dye molecule bearing moderately high two-photon absorption cross-section, and realized the two-photon pumped lasing from a composite crystal bio-MOF-1 ⊃ DMASM (DMASM = 4-[*p*-(dimethylamino)styryl]-1-methylpyridinium) at room temperature[14]. However, the pores (two types of channels along the *c*-axis of about 7.0 and 10.0 Å, respectively) within bio-MOF-1 are still too large to exactly match the dye molecules of DMASM, thus the orientation of the dye molecules inside the pore cavities is not of a high order, particularly when the high concentration of the dye molecules are applied. Such a moderate pore confinement of bio-MOF-1 apparently has limited us to further realize the higher order multiphoton pumped laser in this solid-state crystal. In order to enhance the pore confinement efficiency of a porous MOF crystal, the pore sizes within a porous MOF need to be tuned to match the size of the dye molecule better. But the dilemma is that when the pore sizes of a porous MOF can exactly match the size of the dye molecule, the dye molecules cannot diffuse into the pore channels through the simple post-synthetic exchange process. To overcome this problem, we have developed an *in situ* self-assembly strategy[15,16]: the components for building a MOF crystal (metal ion and organic linker) and the organic dye molecule are simultaneously assembled together to form the MOF/dye single crystals. Such a methodology has enabled us to tightly incorporate the dye molecules into the porous MOF crystals, and thus the dye molecules are highly ordered and oriented. We have also managed to immobilize high concentration of the dye molecules into the MOF crystal ZJU-68 ⊃ DMASM ($(DMASM)_{0.33}H_{1.67}[Zn_3O(CPQC)_3]$, CPQC, 7-(4-carboxyphenyl)quinoline-3-carboxylate) (the average pore size of the one-dimensional channel along the *c*-axis is 6.0 Å) with the dye content over 0.4 M. Furthermore, the suitable refraction index and well-faceted MOF composite crystals of certain morphology symmetries can be naturally and efficiently utilized as the laser resonant cavities without any other fabrications. The powerful *in situ* self-assembly strategy, highly efficient pore confinement of ZJU-68 for DMASM dye molecule, and suitable refraction index as well as perfect crystal morphology have enabled us to target the first example of polarized three-photon-pumped laser in single solid-state microcrystal.

Results

Synthesis and characterization. Reaction of a new organic linker 7-(4-carboxyphenyl)quinoline-3-carboxylic acid (H_2CPQC) containing quinolone group and $Zn(NO_3)_2 \cdot 6H_2O$ in *N,N*-dimethylformamide/acetonitrile/H_2O/HBF_3 at 100 °C affords colourless hexagonal prism crystals of $H_2[Zn_3O(C_{17}H_9NO_4)_3] \cdot 2.5H_2O \cdot 0.5DMF \cdot MeCN$ (ZJU-68, Fig. 1a). Single crystal X-ray diffraction studies reveal that ZJU-68 crystallizes in the $P\bar{3}$ space group (see Supplementary Table 1 for detailed crystallographic data). As shown in Fig. 2a, trinuclear secondary building units (SBUs) of $[Zn_3O]^{4+}$ are linked by the ligands $CPQC^{2-}$ to form an anionic framework of $[Zn_3O(C_{17}H_9NO_4)_3]^{2-}$. In this structure, nine coordination sites of $[Zn_3O]^{4+}$ are completely occupied by six carboxylates and three of nitrogen atoms from the quinoline moieties, which are different from most of metal–organic frameworks with $[M_3O]^{3n-2}$ ($n = 3$ for $M = Cr^{3+}$, Fe^{3+} and so on or $n = 2$ for $M = Zn^{2+}$, Cu^{2+}) SBUs in which three sites are occupied by small capping ligands such as water and hydroxide[17,18].

The crystal has one-dimensional (1D) sub-nano channels along the *c*-axis with a hexagonal cross-section (Fig. 2b; Supplementary Fig. 2). The edge of the hexagon is about 3.0 Å. For the synthesis of laser dye functionalized crystals, we tried to introduce linear-shaped laser dye cations DMASM via an ion-exchange process, as described in our previous work[14], but failed. This is because the DMASM molecule (about 6.3 Å in the width, Supplementary

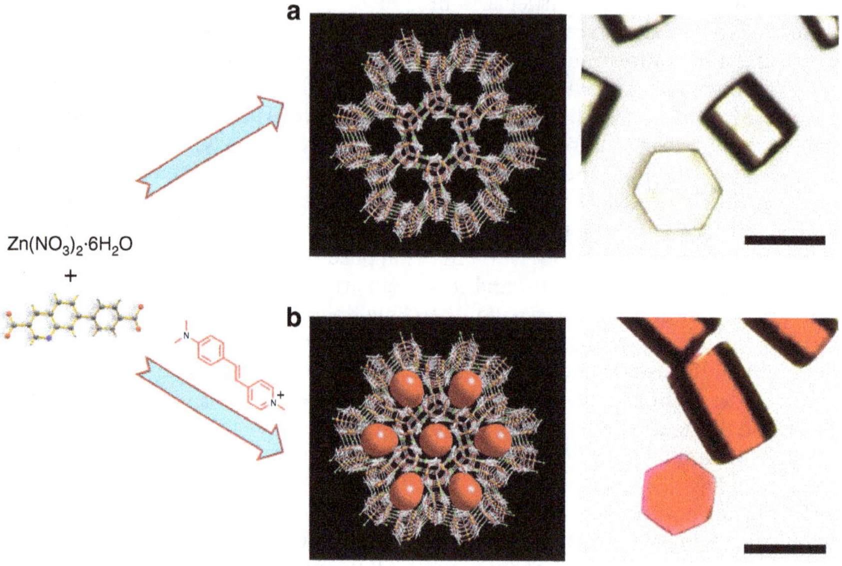

Figure 1 | Schematic synthesis of ZJU-68 and ZJU-68 ⊃ DMASM. (**a**) The synthesis and micrograph of a novel metal-organic framework ZJU-68. (**b**) *In situ* synthesis of laser dye incorporated metal-organic framework crystals ZJU-68 ⊃ DMASM. The inclusion of the red dye DMASM molecules leads to the color change from the original colourless ZJU-68 to red ZJU-68 ⊃ DMASM. Scale bar, 50 μm.

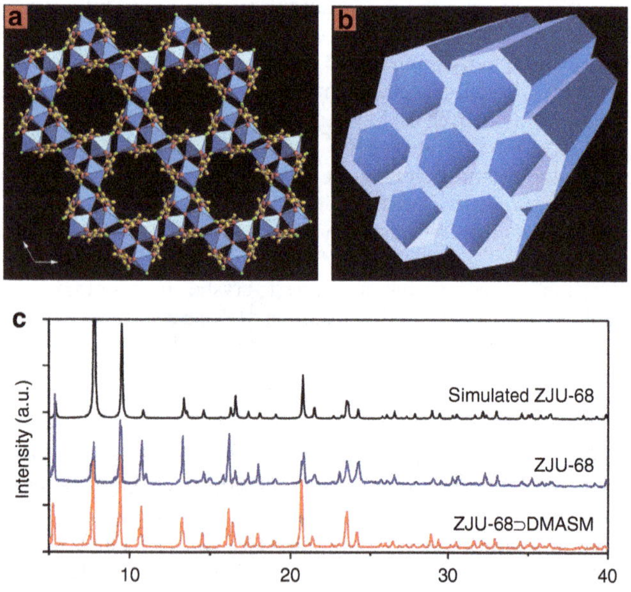

Figure 2 | The structure of a novel metal-organic framework crystal ZJU-68. (**a**) Crystal structure of ZJU-68 viewed along the crystallographic c direction (C, orange; N, green; O, red; Zn, blue polyhedra). H atoms and solvent molecules are omitted for clarity. In this structure, nine coordination sites of a trinuclear SBU $[Zn_3O]^{4+}$ are completely occupied by six carboxylates and three of nitrogen atoms from the quinoline moieties, which may play a crucial role in the stabilization of the resulting MOF, ZJU-68. (**b**) The simplified network structure of ZJU-68, displaying 1D channels along the *c*-axis. Different objects are not drawn to scale. (**c**) PXRD patterns of ZJU-68 and ZJU-68 ⊃ DMASM, which indicate that the ZJU-68 ⊃ DMASM has the identical framework structure with ZJU-68.

Fig. 2e) is too large to diffuse into the channels of ZJU-68 (ref. 19). We thus developed the *in situ* self-assembly synthetic approach in which the dye DMASM molecules were simultaneously incorporated into framework during the solvothermal synthesis of ZJU-68 by simply adding the dye molecules into the reaction solution (Fig. 1b). The resulting dye DMASM included ZJU-68 ⊃ DMASM has the same hexagonal prism crystal morphology. The inclusion of the red dye DMASM molecules leads to the colour change from the original colourless ZJU-68 to red ZJU-68 ⊃ DMASM. Both single crystal and the powder X-ray diffraction studies (Fig. 2c) confirmed that the ZJU-68 ⊃ DMASM has the identical framework structure with ZJU-68. Furthermore, both ZJU-68 and ZJU-68 ⊃ DMASM demonstrate excellent stability in the air and in the common solvents such as water, ethanol and dimethylformamide (Supplementary Fig. 3). Of course, most of the 1D hexagonal channel spaces have been occupied by DMASM molecules in ZJU-68 ⊃ DMASM. Supplementary Fig. 4 shows the fluorescence micrographs of ZJU-68 ⊃ DMASM, taken by confocal laser scanning microscope. The flat and uniform intensity profiles suggest that the DMASM dyes are homogeneously distributed inside the ZJU-68 ⊃ DMASM composite crystals. The dye contents in this composite can be finely tuned by the addition of different amount of DMASM dyes during the *in situ* self-assembly solvothermal synthesis. Generally speaking, the relatively weak MPA responses require high dye content for MPA lasing measurements[5,6], it is thus necessary to encapsulate as much dye molecules as possible into the pore space of ZJU-68. However, the high dye contents (ingredient mole ratio of $n_{DMASM}/n_{H_2CPQC} \geq 70\%$) in the reaction mixtures not only affect the *in situ* self-assembly process (formation of other MOF phases) but also lead to the formation of poor crystalline ZJU-68 ⊃ DMASM. As such, we adjusted the dye concentration in the reaction solution, which produced the optimized ZJU-68 ⊃ DMASM crystals when the ingredient mole ratio of n_{DMASM}/n_{H_2CPQC} is 35%. Accordingly, per gram of resulting ZJU-68 ⊃ DMASM crystals contain 67.7 mg dye molecules corresponding to the concentration of 0.46 M (the molar amount of dye in per unit volume of solid composite; Supplementary Fig. 5). The optimized ingredient mole ratio (35%) is determined by the measurement of fluorescence quantum yield. Among the ZJU-68 ⊃ DMASM samples with different dye loading concentrations, the ZJU-68 ⊃ DMASM composite crystals with ingredient mole ratio of 35% exhibit the strongest emission at around 635 nm with the highest quantum yield φ of 24.28 ± 5% on excitation at 450 nm (Supplementary Fig. 6). This is much higher than the quantum yield of 0.45% in dye solutions and of 1.48% solid powder[14]. These results demonstrate that the good confinement of the DMASM molecules within the size-matched channels of ZJU-68 can effectively restrain the intramolecular torsional motion and increase the conformational rigidity of the dye, thus diminishing the ACQ and populating its radiative decay pathway[20].

Multiphoton-excited fluorescence in ZJU-68 ⊃ DMASM. Figure 3a compares the single-photon-, two-photon- and three-photon-excited fluorescence spectra of a single ZJU-68 ⊃ DMASM crystal with the dye concentration of 0.46 M under the excitation of a femtosecond laser at different wavelengths. The ZJU-68 ⊃ DMASM shows a strong emission peaked at 627 nm on excitation at 532 nm, whereas the emission peak is red-shifted by 11 to 638 nm when excited at 1,064 nm. The full-width at half-maximum (FWHM) is 53.6 and 42.5 nm, respectively, in the single-photon-, two-photon-excited fluorescence spectra of ZJU-68 ⊃ DMASM. The emission spectrum on excitation at 1,380 nm is basically similar to that excited at 1,064 nm except one additional small peak at around 690 nm attributed to the second harmonic generation response. The red shift of 11 nm under multiphoton excitation can be ascribed to the reabsorption effect[14]. The diffuse reflectance ultraviolet-visible (vis) spectrum of ZJU-68 ⊃ DMASM was shown in Supplementary Fig. 7. There exists overlap between the long wavelength side of the absorption band and the short wavelength side of the fluorescence band (Fig. 3a). Furthermore, all MPP fluorescence bands in Fig. 3a are asymmetric with their left part seeming to be cut off[21]. In addition, the emission peaked at 627 nm from a single ZJU-68 ⊃ DMASM crystal on excitation at 532 nm is blue shifted relative to the spontaneous emission (maximum at 635 nm, see Supplementary Fig. 14) from multiple ZJU-68 ⊃ DMASM crystals, which also suggests the presence of reabsorption effect in ZJU-68 ⊃ DMASM (Supplementary Fig. 8). Fig. 3b shows the fluorescence intensity of the crystal with respect to the pump polarization direction when excited at 1,380 nm. The ZJU-68 ⊃ DMASM exhibits a strong emission when the pump polarization direction is parallel to the crystal channels (along the *c*-axis, denoted as 0°), but hardly emits any light when the pump polarization direction is perpendicular to the excitation direction (90°). Such significant directional fluorescence (dicroic ratio ~ 365 (ref. 15)) behaviours indicate that the absorption transition moments (approximately along the dye molecule axis[22]) are highly oriented along the crystal channels.

3PP lasing in ZJU-68 ⊃ DMASM. 3PP lasing properties were investigated on an isolated single crystal of ZJU-68 ⊃ DMASM with the dye concentration of 0.46 M under a microscope. A femtosecond laser at 1,380 nm was used to pump the crystal at room temperature. This laser beam was directed from a femtosecond optical parametric amplifier (OPA), and then was coupled to the microscope. The emission beam from the crystal was focused and collected with a fibre optic spectrometer. Representative emission spectra near the lasing threshold are shown in Fig. 4a. Under the low-pump energy (E) of 113 nJ, the emission spectrum shows a broad peak centred at ~ 649 nm with a FWHM of 57.6 nm, which corresponds to the spontaneous emission. The pump energy is defined as the laser energy directly received by the MOF crystal (after going through the objective lens and before being incident on the MOF crystal). As the pump energy increases to ≥230 nJ, a highly progressional emission pattern centred at 642.7 nm appears and grows rapidly with increasing pump energy, while the intensity of the broad spontaneous emission remains almost constant. The visible stimulated emission spectrum centred at 642.7 nm is between one half and one-third of the pumped wavelength of 1,380 nm, which means that the sum energy of two photons at 1,380 nm is not large enough to overcome the bandgap between the ground state (S_0) and excited state (S_1) of ZJU-68 ⊃ DMASM. The stimulated emission of ZJU-68 ⊃ DMASM is therefore induced by the simultaneous absorption of more than two near-infrared photons. To unravel how many photons involved in such a simultaneous absorption process, we further measured the dependence of the stimulated emission intensity on the pump energy. The right inset in Fig. 4a illustrates the pump energy dependence of the fluorescence intensity and FWHM plot as a function of pump energy, giving rise to a linear relationship with cubic pump energy and a low lasing threshold of E_{th} ~ 224 nJ as compared with other 3PP stimulated emission[4,6,7]. The FWHM plot shows a constant value below E_{th} and a sudden drop by more than two orders of magnitude when above E_{th}. The presence of a significant spectral narrowing and a threshold energy coupled with the linearly rapid increase in intensity with cubic pump energy suggest that the 3PP lasing has occurred in the ZJU-68 ⊃ DMASM crystal. The quality factor (Q) is given by $Q = \lambda/\delta\lambda$, where λ and $\delta\lambda$ are the peak wavelength and its FWHM, respectively. At pump energy of 369 nJ, the FWHM of lasing peaked at 642.7 nm is ~0.38 nm. This records a high-quality factor Q ~ 1,691 for 3PP lasing, which indicates the high crystal quality supported by our simple chemical approach without etching and coating.

For hexagonal ZJU-68 ⊃ DMASM crystal, the opposing facets can act as the mirrors of a Fabry–Pérot (F–P) cavity, or the six

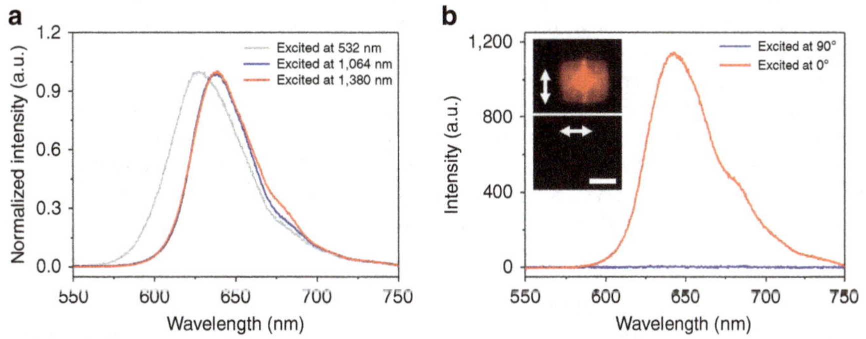

Figure 3 | Multiphoton-pumped fluorescence performance of a ZJU-68 ⊃ DMASM single crystal. (**a**) Single-photon-(532 nm), two-photon-(1,064 nm) and three-photon-(1,380 nm) excited emission spectra of ZJU-68 ⊃ DMASM. (**b**) The emission intensity versus pump polarization at two angles $\theta = 0°$ (parallel to the crystal channels) and $\theta = 90°$ (perpendicular to the crystal channels), excited at 1,380 nm. Insets: micrographs of a ZJU-68 ⊃ DMASM single crystal ($R = 36.5$ μm) with different pump polarizations excited at 1,380 nm, Scale bar, 50 μm. The high intensity ratio between the two angles indicates the high orientation of dye molecules within the channels of ZJU-68.

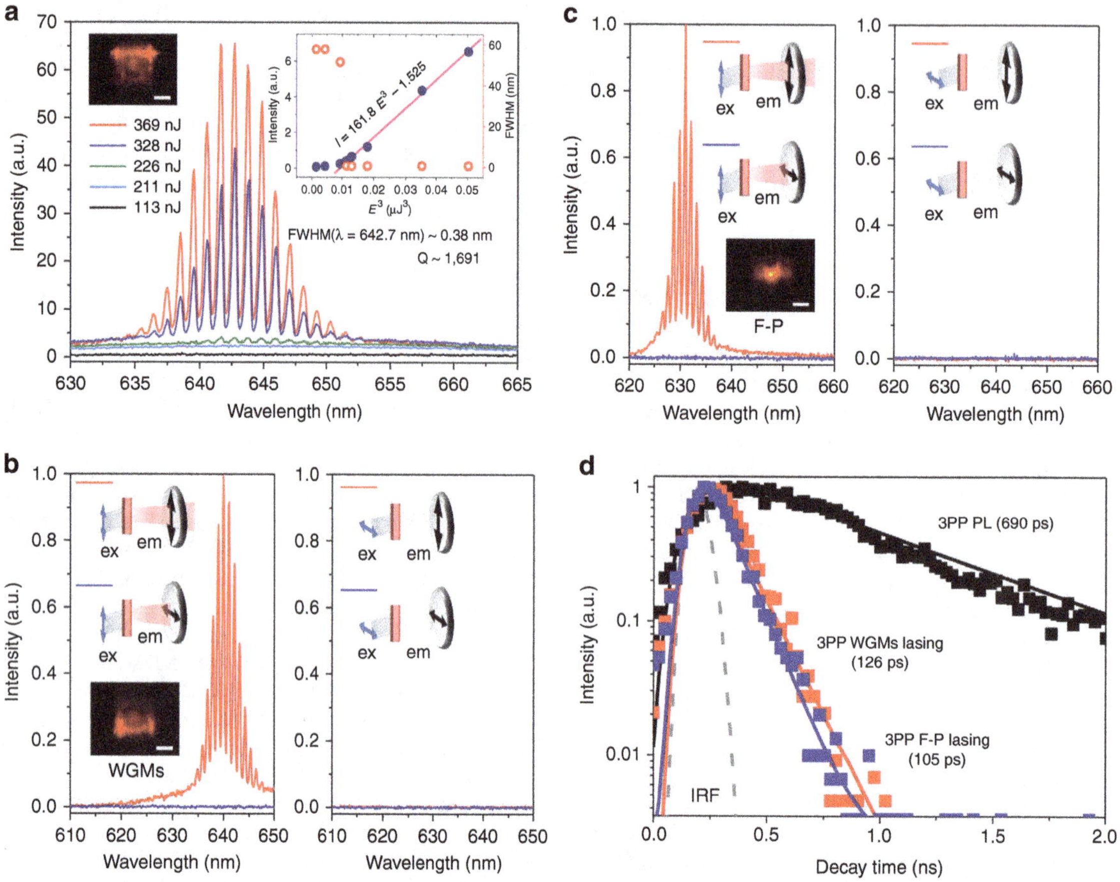

Figure 4 | Three-photon-pumped lasing performance of ZJU-68 ⊃ DMASM. (**a**) 1,380 nm pumped emission spectra of ZJU-68 ⊃ DMASM around the lasing threshold. Insets: the micrograph of a ZJU-68 ⊃ DMASM single crystal ($R = 26.2$ μm) excited at 1,380 nm (left) and emission intensity and FWHM as a function of pump energy showing the lasing threshold at ~ 224 nJ (right). At pump energy of 369 nJ, the FWHM of lasing peaked at 642.7 nm is ~ 0.38 nm, corresponding to a Q factor ~ 1,700. (**b**,**c**) Intensity-dependent emission spectra of 3PP WGMs (**b**, pump energy at 433 nJ) and F-P lasing (**c**, pump energy at 837 nJ) from two isolated crystals ($R = 26.1$ and 27.6 μm, respectively) with pump/emission-detected polarization combinations at two angles $\theta = 0°$ (parallel to the crystal channels) and $\theta = 90°$ (perpendicular to the crystal channels), excited at 1,380 nm. Insets: schematic diagrams of the measurement geometry for an individual crystal and micrographs of two kinds of lasing spot patterns due to the F-P (one-spot) and WGMs (two-spot) mechanisms, Scale bar, 20 μm. Both 3PP WGMs and F-P lasing exhibit perfectly polarized emission with DOP > 99.9%, which are attributed to the highly oriented assembly of dye molecules within the host-guest composite ZJU-68 ⊃ DMASM microcrystal. (**d**) TRPL decay kinetics measurements of ZJU-68 ⊃ DMASM under photoluminescence (PL), F-P and WGMs lasing excited at 1,380 nm. TRPL, time-resolved photoluminescence

facets can form a whispering gallery modes (WGMs) or other quasi-WGMs cavities[23] (Supplementary Fig. 9). We observed two kinds of lasing spot pattern on isolated ZJU-68 ⊃ DMASM crystals when excited at 1,380 nm: (1) the strong emission with spatial interference from two-side facets of a hexagonal prism crystal (two-spot pattern)[24], as shown in the insets of microscopy image in Figs. 4a and 4b; the strong emission from a round bright spot at the central facet of the crystal (one-spot pattern)[14], as shown in the inset of Fig. 4c. Such two lasing patterns can be attributed to the WGMs and F-P feedback mechanisms, respectively, as confirmed later. Fig. 4b,c show the anisotropic study of 3PP WGMs and F-P lasing from two crystals with side lengths of 26.1 and 27.9 μm, respectively. The red emission light from the crystal passed through a polarizer first and then was focused and collected with a fibre optic spectrometer. The schematic diagrams of the measurement geometry for an individual crystal are shown in Fig. 4b,c, where the polarization directions of the pump light and the polarizer are parallel (0°)/perpendicular (90°) to the crystal channels (along *c*-axis). We can see that both 3PP WGMs and F-P lasing with highly structured spectra occur when excited at 0° and emission polarization detected at 0°, while hardly any emission intensity can be detected in all other configurations. The corresponding pump energy is 433 nJ for WGMs lasing and 837 nJ for F-P lasing. It should be noted that the pump energy at almost 3.2 E_{th} of 3PP WGMs lasing can realize the 3PP F-P lasing in our experiments, indicating that the WGMs mechanism is more conducive to the realization of 3PP lasing due to the total internal reflection for less loss of light in such size of the crystal. The degree of polarization can be defined as $DOP = (I_{max} - I_{min})/(I_{max} + I_{min})$ in our experiments[25], and we calculated that both 3PP WGMs and F-P lasing exhibited DOP > 99.9% (limited by the spectral intensity sensitivity of our measurement system) when the excitation polarization is fixed parallel to the crystal channels, indicating a perfectly polarized 3PP lasing operation. Compared with the WGMs lasing, the parallel mirrors in a F-P cavity cannot be utilized as Brewster windows for the polarization selectivity. Therefore, the perfectly polarized 3PP F-P lasing is attributed to the highly oriented assembly of dye molecules within the host-guest composite ZJU-68 ⊃ DMASM microcrystal given the fact that the angle between the absorption transition moment and emission transition moment is close to zero in the dye molecule DMASM[26]. These anisotropic results indicate that ZJU-68 ⊃ DMASM can only be excited at the polarization direction parallel to the crystal channels, and can produce lasing perfectly polarized along the crystal channels, which exhibit a great potential for bioimaging, optical sensing and future optoelectronic integration.

To further confirm the optical-feedback mechanisms for 3PP lasing in these hexagonal ZJU-68 ⊃ DMASM crystals, several single crystals with different side lengths R were chosen for the further measurements (Supplementary Fig. 10a and b). For both feedback mechanisms, the spectra of lasing exhibit an increased mode spacing with the decrease of side length of the MOF crystals. For possible resonant modes, the mode spacing $\Delta\lambda_s$ is defined as[27]

$$\Delta\lambda_s = \frac{\lambda^2}{Ln_g} \quad (1)$$

where λ is the resonant wavelength, L is the cavity path length ($3\sqrt{3}R$ for WGMs and $2\sqrt{3}R$ for F–P cavity), and n_g is the group index of refraction. The measured mode spacing, $\Delta\lambda_s$, around 635 nm, demonstrates a linear relationship with $1/R$ for each feedback mechanism, which agrees well with Equation (1) (see Supplementary Fig. 10c,d). This result indicates that the lasing with the similar output spot pattern (one-spot or two-spot) can be attributed to the same feedback mechanism. According to the fitting formula, we calculated the ratio of slopes ($S_{one\text{-}spot}/S_{two\text{-}spot}$) to be 1.47, which is very close to the ratio of the cavity path lengths ($L_{WGMs}/L_{F\text{–}P} = 1.5$), verifying that the lasing with these two kinds of output pattern (one-spot and two-spot) should be attributed to the F–P cavity and WGMs, respectively. On the basis of Equation (1), we also derived $n_g \sim 3.27$ for F–P cavity, and $n_g \sim 3.21$ for WGMs at the wavelength of 635 nm. The relatively high group index n_g value may result from the unusual dispersion relation near the absorption band or the strong exciton–photon coupling in organic materials[28]. Further insight into the 3PP emission performance of the single crystal ZJU-68 ⊃ DMASM arises from time-resolved photoluminescence measurements (Fig. 4d). The pulse durations of the 3PP F–P lasing and WGMs lasing were determined to be 105 and 126 ps, respectively, which are much shorter than the corresponding 3PP fluorescence (below the E_{th}) decay time of 690 ps. Such temporal narrowing can be ascribed to the depletion in the population inversion of the gain medium with photon-stimulated amplification[6]. The lasing pulse durations from WGMs and F–P are almost in the same order of magnitude, indicating that the optical-feedback mechanism may have little effect on the pulse duration. Subtle differences in our measured decay times of lasing may be ascribed to the proportion of stimulated emission and spontaneous emission in the emitted light, which depends on multiple factors, for example, pump energy, crystal size and crystalline quality[29,30].

Discussion

In summary, we have achieved an unprecedented solid-state polarized frequency-upconversion lasing in a novel composite single microcrystal ZJU-68 ⊃ DMASM by simultaneous three-photon absorption in the near-infrared region. The tightly confined and highly oriented cationic DMASM dye molecules in anionic ZJU-68 nano-channels through an *in situ* assembly process efficiently increase the loaded concentration, minimize the aggregation and optimize the orientation of dye molecules within the framework, which fulfilled the high-gain lasing with highly polarized excitation response and perfectly polarized emission in a micro-sized laser cavity. Particularly, the 3PP lasing, with a low lasing threshold of ~ 224 nJ centred at 642.7 nm on excitation at 1,380 nm, has been successfully achieved with a record high-quality factor of ~ 1,700. Both F–P and WGMs optical-feedback mechanisms have been confirmed to be responsible for 3PP lasing in ZJU-68 ⊃ DMASM microcrystals. Owing to the highly oriented assembly of dye molecules within ZJU-68 ⊃ DMASM, the 3PP WGMs and especially F–P lasing show a perfect emission polarization with DOP > 99.9%. The observed solid-state frequency-upconversion polarized lasing induced by 3PP may find great potentials in practical applications such as photonics, information storage and biomedicine, to name a few. For instance, the wavelength of 1,380 nm belongs to the near-infrared-IIa window (1,300-1,400 nm), which is very promising in biological applications (especially for *in vivo* imaging) because such wavelength region not only can reach deeper penetration depths and minimize the scattering/auto-fluorescence of biological tissues, but also avoid an increased light absorption from water above 1,400 nm (ref. 31). Because the MOF strategy and design can provide us with rich structures of the systematically tuned pore/channel sizes to encapsulate various chromophores with controlled concentration and orientation[32–34], we anticipate that higher order multiphoton-pumped lasing in solid state can also be realized given that the chromophores (or other nano-sized materials) with great multiphoton absorption properties are well incorporated into the structurally matched MOFs. These findings may eventually open up a new route to the exploitation of multiphoton-pumped solid-state laser in single MOF microcrystal (or nanocrystal) for future optoelectronic and biomedical applications.

Methods

Synthesis of ZJU-68 ⊃ DMASM. A mixture of $Zn(NO_3)_2 \cdot 6H_2O$ (0.34 mmol, 149 mg), H_2CPQC (0.17 mmol, 50 mg), DMF (10 ml), MeCN (2 ml), H_2O (0.05 ml), HBF_3 (0.05 ml) and DMASM iodide (0.03 mmol, 11 mg) were sealed in a 15 ml Teflon-lined stainless-steel bomb at 100 °C for 24 h, which was then slowly cooled to room temperature. After decanting the mother liquor, the fine red hexagonal crystalline product was rinsed three times with fresh DMF (5 ml × 3) and dried in air. The synthesis of the new organic linker H_2CPQC can be found in Supplementary Fig. 11 and Supplementary Methods.

Measurements. For MPP, an optical parametric amplifier (TOPAS-F-UV2, Spectra-Physics) pumped by a regeneratively amplified femtosecond Ti:sapphire laser system (800 nm, 1 kHz, pulse energy of 4 mJ, pulse width < 120 fs, Spitfire Pro-FIKXP, Spectra-Physics), which was seeded by a femtosecond Ti-sapphire oscillator (80 MHz, pulse width < 70 fs, 710-920 nm, Mai Tai XF-1, Spectra-Physics) was used for generating the excitation pulse (1 kHz, 240–2,600 nm, pulse width < 120 fs). The incident laser was coupled to the microscope (Ti-U, Nikon), focusing on crystals through an objective lens (CFI TU Plan Epi ELWD 50 ×, numerical aperture = 0.60, work distance = 11.0 mm) with an exposure region of diameter around 30 μm (supplementary Fig. 12). The excited red light was then focused and collected by the fibre optic spectrometer (QE65Pro, Ocean Optics).

The decay curves of multiphoton-pumped emissions were measured by a picosecond lifetime spectrometer (Lifespec-ps, Edinburgh Instruments). For the lifetime measurement of upconverted fluorescence, the pump energy was under the lasing threshold to ensure that no stimulated emission was generated. To measure the decay of the multiphoton-pumped lasing, the pump energy was enhanced over the threshold so that the ultra-strong lasing could be achieved.

Contents of well-dried dye-included ZJU-68 ⊃ DMASM crystals were determined by ^1H NMR. As shown in Supplementary Figs 5 and 13c, we calibrated and obtained peak area values of peaks that belong to H_2CPQC and DMASM, respectively. The ratio (R_a) of their peak area values represents the ratio of their contents in the crystal. The dye concentration of the ZJU-68 ⊃ DMASM composite is calculated from $c = 2R_a/N_AV$, where $V = 2403.91$ Å^3 and $N_A = 6.02 \times 10^{23}\,mol^{-1}$ is Avogadro's constant.

References

1. Jameson, D. M. & Ross, J. A. Fluorescence polarization/anisotropy in diagnostics and imaging. *Chem. Rev.* **110**, 2685–2708 (2010).
2. Ghosh, N. & Vitkin, I. A. Tissue polarimetry: concepts, challenges, applications, and outlook. *J. Biomed. Opt.* **16**, 110801 (2011).
3. Gurjar, R. S. *et al.* Imaging human epithelial properties with polarized light-scattering spectroscopy. *Nat. Med.* **7**, 1245–1248 (2001).
4. He, G. S., Tan, L. S., Zheng, Q. & Prasad, P. N. Multiphoton absorbing materials: molecular designs, characterizations, and applications. *Chem. Rev.* **108**, 1245–1330 (2008).

5. Guo, L. & Wong, M. S. Multiphoton excited fluorescent materials for frequency upconversion emission and fluorescent probes. *Adv. Mater.* **26,** 5400–5428 (2014).
6. Zheng, Q. D. *et al.* Frequency-upconverted stimulated emission by simultaneous five-photon absorption. *Nat. Photon.* **7,** 234–239 (2013).
7. Wang, Y. *et al.* Stimulated emission and lasing from CdSe/CdS/ZnS core-multi-shell quantum dots by simultaneous three-photon absorption. *Adv. Mater.* **26,** 2954–2961 (2014).
8. Hoover, E. E. & Squier, J. A. Advances in multiphoton microscopy technology. *Nat. Photon.* **7,** 93–101 (2013).
9. He, G. S., Markowicz, P. P., Lin, T. C. & Prasad, P. N. Observation of stimulated emission by direct three-photon excitation. *Nature* **415,** 767–770 (2002).
10. Gomes, A. S., Carvalho, M. T., Dominguez, C. T., de Araujo, C. B. & Prasad, P. N. Direct three-photon excitation of upconversion random laser emission in a weakly scattering organic colloidal system. *Opt. Express* **22,** 14305–14310 (2014).
11. Li, M. *et al.* Ultralow-threshold multiphoton-pumped lasing from colloidal nanoplatelets in solution. *Nat. Commun.* **6,** 8513 (2015).
12. Vietze, U. *et al.* Zeolite-dye microlasers. *Phys. Rev. Lett.* **81,** 4628–4631 (1998).
13. Martini, I. B. *et al.* Controlling optical gain in semiconducting polymers with nanoscale chain positioning and alignment. *Nat. Nanotechnol.* **2,** 647–652 (2007).
14. Yu, J. *et al.* Confinement of pyridinium hemicyanine dye within an anionic metal-organic framework for two-photon-pumped lasing. *Nat. Commun.* **4,** 2719 (2013).
15. Martinez-Martinez, V., Garcia, R., Gomez-Hortiguela, L., Perez-Pariente, J. & Lopez-Arbeloa, I. Modulating dye aggregation by incorporation into 1D-MgAPO nanochannels. *Chemistry* **19,** 9859–9865 (2013).
16. Martínez-Martínez, V. *et al.* Highly luminescent and optically switchable hybrid material by one-pot encapsulation of dyes into MgAPO-11 unidirectional nanopores. *ACS Photon.* **1,** 205–211 (2014).
17. Mao, C. *et al.* Anion stripping as a general method to create cationic porous framework with mobile anions. *J. Am. Chem. Soc.* **136,** 7579–7582 (2014).
18. Ferey, G. *et al.* A chromium terephthalate-based solid with unusually large pore volumes and surface area. *Science* **309,** 2040–2042 (2005).
19. Zhao, C. F., He, G. S., Bhawalkar, J. D., Park, C. K. & Prasad, P. N. Newly synthesized dyes and their polymer/glass composites for one-photon and 2-photon pumped solid-state cavity lasing. *Chem. Mater.* **7,** 1979–1983 (1995).
20. Cui, Y. J. *et al.* Dye encapsulated metal-organic framework for warm-white LED with high color-rendering index. *Adv. Funct. Mater.* **25,** 4796–4802 (2015).
21. Ren, Y. *et al.* Synthesis, structures and two-photon pumped up-conversion lasing properties of two new organic salts. *J. Mater. Chem.* **10,** 2025–2030 (2000).
22. Weiß, Ö. *et al.* in *Host-Guest-Systems Based on Nanoporous Crystals* 544–557 (Wiley-VCH Verlag GmbH & Co., 2005).
23. Wang, X. *et al.* Whispering-gallery-mode microlaser based on self-assembled organic single-crystalline hexagonal microdisks. *Angew. Chem. Int. Ed.* **53,** 5863–5867 (2014).
24. Braun, I. *et al.* Hexagonal microlasers based on organic dyes in nanoporous crystals. *Appl. Phys. B* **70,** 335–343 (2000).
25. Zhu, H. *et al.* Lead halide perovskite nanowire lasers with low lasing thresholds and high quality factors. *Nat. Mater.* **14,** 636–642 (2015).
26. Gozhyk, I. *et al.* Polarization properties of solid-state organic lasers. *Phys. Rev. A* **86,** 043817 (2012).
27. Choi, S., Ton-That, C., Phillips, M. R. & Aharonovich, I. Observation of whispering gallery modes from hexagonal ZnO microdisks using cathodoluminescence spectroscopy. *Appl. Phys. Lett.* **103,** 171102 (2013).
28. Takazawa, K., Inoue, J., Mitsuishi, K. & Takamasu, T. Fraction of a millimeter propagation of exciton polaritons in photoexcited nanofibers of organic dye. *Phys. Rev. Lett.* **105,** 067401 (2010).
29. Zhang, C. *et al.* Two-photon pumped lasing in single-crystal organic nanowire exciton polariton resonators. *J. Am. Chem. Soc.* **133,** 7276–7279 (2011).
30. Liu, X. *et al.* Whispering gallery mode lasing from hexagonal shaped layered lead iodide crystals. *ACS Nano* **9,** 687–695 (2015).
31. Hong, G. S. *et al.* Through-skull fluorescence imaging of the brain in a new near-infrared window. *Nat. Photon.* **8,** 723–730 (2014).
32. Furukawa, H., Cordova, K. E., O'Keeffe, M. & Yaghi, O. M. The chemistry and applications of metal-organic frameworks. *Science* **341,** 1230444 (2013).
33. Kitagawa, S., Kitaura, R. & Noro, S. Functional porous coordination polymers. *Angew. Chem. Int. Ed.* **43,** 2334–2375 (2004).
34. Chen, B., Xiang, S. & Qian, G. Metal-organic frameworks with functional pores for recognition of small molecules. *Acc. Chem. Res.* **43,** 1115–1124 (2010).

Acknowledgements

We acknowledge the financial support from the National Natural Science Foundation of China (Nos. 51229201, 51272229, 51272231, 51402259, 51472217, 51432001, U1305244 and 21325104) and Zhejiang Provincial Natural Science Foundation of China (Nos. LR13E020001 and LZ15E020001). This work is also partially supported by Welch Foundation (AX-1730) and National Science Foundation of United States (ECCS-1407443). X.C. and E.M. acknowledge the support from Special Project of National Major Scientific Equipment Development of China (No. 2012YQ120060) and the CAS/SAFEA International Partnership Program for Creative Research Teams. We also thank Dr Ghezai Musie for proof-reading the manuscript.

Author contributions

H.H., E.M., Y.C., J.Y. and G.Q. conceived and designed the experiments. H.H. synthesized the materials. J.Y. and C.W. Analysed the crystal structure. H.H. and E.M. performed the multiphoton experiments. Y.C., J.Y., Y.Y. and T.S. assisted with the linear optical property measurements and characterization of the material. H.H., E.M., Y.C., X.C., B.C. and G.Q. analysed the data and co-wrote the manuscript. All authors discussed the results and commented on the manuscript.

Additional information

Accession codes: The X-ray crystallographic coordinates for structure reported in this study has been deposited at the Cambridge Crystallographic Data Centre (CCDC), under deposition number 1046524. These data can be obtained free of charge from the Cambridge Crystallographic Data Centre via www.ccdc.cam.ac.uk/data_request/cif.

Competing financial interests: The authors declare no competing financial interests.

8

Damage-free vibrational spectroscopy of biological materials in the electron microscope

Peter Rez[1], Toshihiro Aoki[2], Katia March[3], Dvir Gur[4], Ondrej L. Krivanek[1,5], Niklas Dellby[5], Tracy C. Lovejoy[5], Sharon G. Wolf[6] & Hagai Cohen[6]

Vibrational spectroscopy in the electron microscope would be transformative in the study of biological samples, provided that radiation damage could be prevented. However, electron beams typically create high-energy excitations that severely accelerate sample degradation. Here this major difficulty is overcome using an 'aloof' electron beam, positioned tens of nanometres away from the sample: high-energy excitations are suppressed, while vibrational modes of energies <1 eV can be 'safely' investigated. To demonstrate the potential of aloof spectroscopy, we record electron energy loss spectra from biogenic guanine crystals in their native state, resolving their characteristic C-H, N-H and C=O vibrational signatures with no observable radiation damage. The technique opens up the possibility of non-damaging compositional analyses of organic functional groups, including non-crystalline biological materials, at a spatial resolution of ~10 nm, simultaneously combined with imaging in the electron microscope.

[1] Department of Physics, Arizona State University, Tempe, Arizona 85287, USA. [2] LeRoy Eyring Center for Solid State Science, Arizona State University, Tempe, Arizona 85287, USA. [3] Laboratoire de Physique des Solides, Université Paris-Sud, CNRS, UMR8502, Orsay 91405, France. [4] Department of Structural Biology, Weizmann Institute of Science, Rehovot 76100, Israel. [5] Nion Co., 11511 NE 118th St., Kirkland, Washington 98034, USA. [6] Department of Chemical Research Support, Weizmann Institute of Science, Rehovot 76100, Israel. Correspondence and requests for materials should be addressed to P.R. (email: peter.rez@asu.edu).

A central paradigm in biology is the inseparable connection between structure and function of biogenic molecules. Imaging of vitrified biological cells, tissues and macromolecules by transmission electron microscopy (TEM) has progressed rapidly in extracting complex structural information[1–3]. These techniques provide structural information, but do not probe the sample properties. They are also insensitive to the presence of hydrogen, the most common element in organic molecules, and to how hydrogen is bonded to other atoms. Although it is possible to acquire infrared spectra at a spatial resolution of better than 100 nm using tip-enhanced infrared absorption in an atomic force microscopy[4,5], the technique is limited to films that are flat over an extended region. Selecting the amide I absorption peak, Berweger *et al.*[6] showed that it was possible to map the distribution of bacteriorhodopsin in phospholipid bilayer. If spectroscopy carried out in the electron microscope could be extended to the visible and infrared region, then it should be possible to use electron microscopy to not only determine overall morphology but also to find where functionally significant biomolecules, such as chromophores and proteins, are located within the overall structure at high spatial resolution.

The damage an energetic electron beam causes to the biological sample is another key limitation of present-day electron microscope imaging and analysis. Avoiding the damage requires that the electron fluence be kept at a very low level, typically of the order of $<10\,e\,Å^{-2}$ (refs 7,8), and this severely limits the spatial resolution at which biological samples can be imaged and analysed[9]. Methods for minimizing the radiation damage include averaging the information over many unit cells of a periodic sample, averaging over many identical structural units in a non-periodic sample[10] examining cryogenically preserved specimens at liquid nitrogen temperatures (cryo-TEM)[11] and using direct electron detectors to fractionate dose among many image frames[3,12,13].

Recent developments have demonstrated electron microscope resolution of infrared features, by electron energy loss spectroscopy (EELS) carried out with an energy resolution of ~10 meV (ref. 14). This has made it possible to record spectra showing vibrational features of hydrogen-containing inorganic materials such as metal hydrides, but the approach has not been applied to biological samples. Aloof beam spectroscopy[15,16] was proposed by Cohen *et al.*[17] as a near-field spectroscopy probe and was used by Krivanek *et al.*[14] to acquire vibrational spectra from titanium hydride. The theoretical advantage of using aloof beams to avoid radiation damage has recently been explored by Egerton[18]. The key advantage of aloof spectroscopy is that for a tightly focused beam passing a distance *d* outside the sample, damage-causing high-energy interactions are suppressed relative to information-providing low-energy interactions. In other words, an ability to accurately control the beam–sample distance has an important advantage: one can choose the effective range of energy transferred to the sample, similar to what can be achieved with photon techniques by adjusting the beam energy range. Thus, it is not just the beam intensity, or scattering cross-section that is varied, but the fundamental characteristics of beam–sample interaction. Specifically, by selecting *d*-values that do not allow energy transfers above the infrared regime, one can perform measurements that are, in principle, similar to probing with infrared irradiation, hence with minimal damage.

Here we show that aloof electron beam vibrational spectroscopy can be carried out at an energy resolution sufficient to probe different types of bonds in biological samples, including those of hydrogen, carbon, nitrogen and oxygen, with no significant radiation damage.

Results

Vibrational spectroscopy. To demonstrate control over beam–sample interaction, we apply the aloof beam EELS configuration (Fig. 1b) for the detection of electronic and vibrational peaks in guanine, one of the DNA bases. Anhydrous guanine in its crystalline form has a strong anisotropic dielectric response that is used in nature by organisms such as fish[19,20], arthropods[21] and reptiles[22,23] for manipulating light. Crystals extracted from the scales of the Japanese Koi fish (*Cyprinus carpio*) were purified and placed on Cu holey carbon-coated TEM grid (see Methods for details). They are in the form of rhombohedral plates ~10-μm long and 30-nm thick. A typical guanine crystal is shown in the TEM image (Fig. 1a). It is lying on a holey carbon film, such that the e-beam of a Nion UltraSTEM can be directed through vacuum at a distance *d* from the side of the crystal that can be accurately controlled in a range of 10–250 nm, as shown schematically in Fig. 1b. The scattering wavevectors perpendicular to the beam direction q_r in the plane of the specimen are much greater than q_z under our experimental conditions (Fig. 1c). Also, under our experimental conditions, dipole scattering dominates[24]. The planar guanine molecules (inset in Fig. 1a) with 1 C=O bond, 1 NH_2 group, 1 CH bond, 2 NH bonds and 7 C–N bonds all in the same plane are stacked parallel to the surface of the crystal (Fig. 1b). This means that we selectively excite vibrations in the plane of the guanine molecule.

Figure 2a presents an EEL spectrum in the infrared range, as recorded at a distance $d=30$ nm from the crystal edge with beam energy of 60 keV. Close agreement with the infrared absorption spectrum, Fig. 2a, of anhydrous guanine crystals (acquired *ex situ*) is found, including the detailed peak positions given in

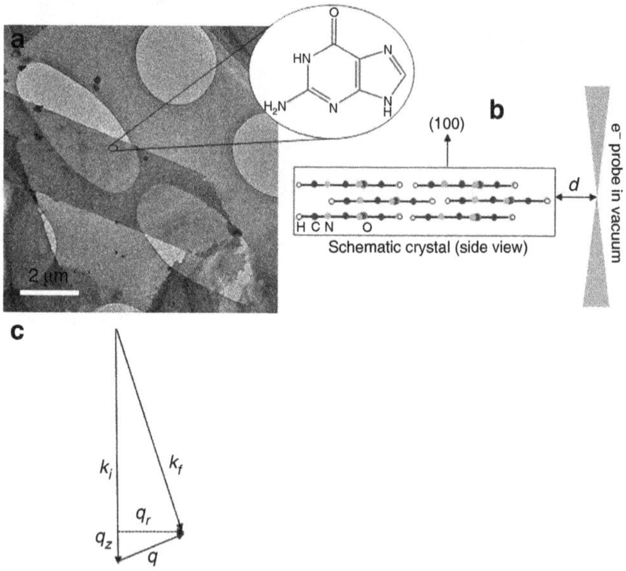

Figure 1 | Position of the electron beam with respect to the guanine crystal. (**a**) Low-magnification 'top-view' image of guanine crystals laying on holey carbon film. The long edges of the crystal are in the (010) direction using the unit cell of Hirsch *et al.*[37] (**a**, inset) one schematic guanine molecule—the molecules are arranged in layers parallel to specimen surface to form the crystal as shown schematically from the side in **b**. Also shown in **b** is the position of the electron probe a distance *d* outside the crystal in the vacuum. (**c**) Schematic diagram showing the scattering wavevector for the electrons (side view). The energy loss is small compared with the energy of the incident electrons and the scattering angles are also small. The scattering wavevector can be decomposed into a component q_z along the incident beam direction related to the energy loss (equation 1) and q_r perpendicular to the incident beam direction. Under our experimental conditions, q_r is much larger than q_z.

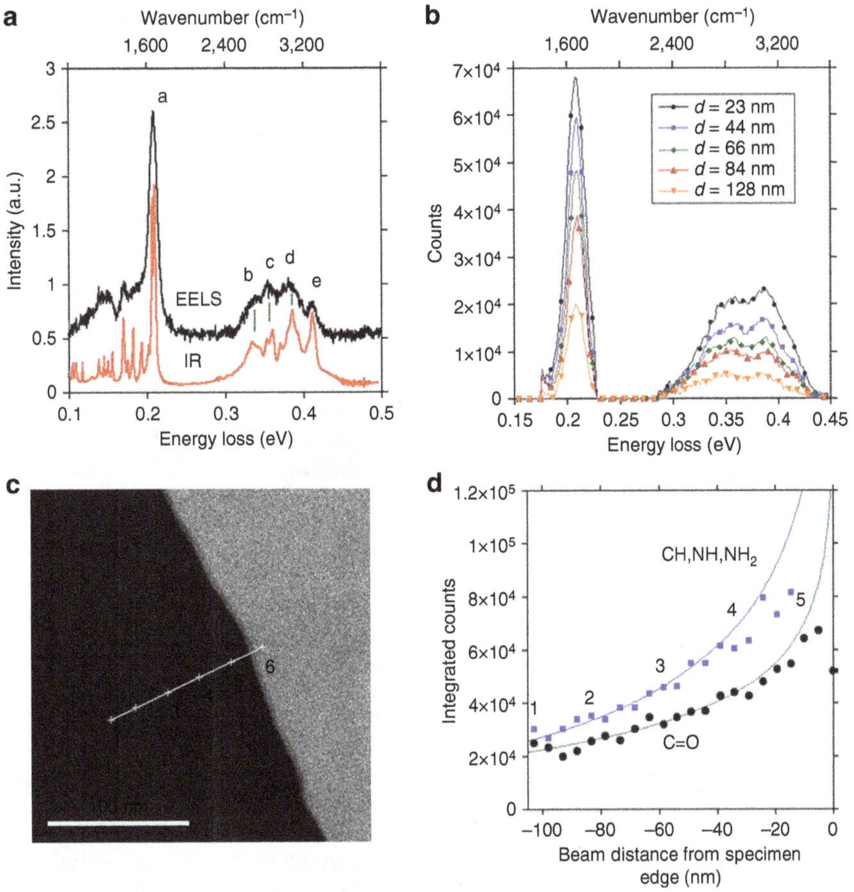

Figure 2 | Variation of infrared region spectra with electron beam position. (**a**) An EEL spectrum at the infrared region, collected when the electron probe is 30 nm from the edge of the crystal, compared with an *ex situ* FTIR spectrum. Peaks corresponding to C=O, NH, CH , NH_2 symmetric and NH_2 antisymmetric stretches can be seen in the EEL spectrum, matching corresponding features in the FTIR (Table 1). (**b**) A set of spectra showing how the peaks in the infrared region increase in height as the probe is moved closer to the specimen. (**c**) Dark-field image showing guanine crystal-vacuum interface and line along which spectra were taken. (**d**) Variation of total intensity of the C=O peak (circles) and the combination of CH, and NH and NH_2 peaks (squares, peaks b-e from (**a**)) from the EELS signal as the probe approaches the sample compared with the theoretical variation derived from classical dielectric theory given as equation (2) (solid lines). The acquisition time was 1.6 s a point.

Table 1. The most prominent feature is the peak from the C=O stretch at 209 meV. All the peaks that can be attributed to stretching of hydrogen covalently bonded to nitrogen or carbon have been identified in the spectrum as shown in Table 1. Thus, this spectrum demonstrates not only the detection of hydrogen but also the capability of resolving different hydrogen-related vibrations. The peak intensities increase as the electron probe is moved closer to the specimen edge, as shown in Fig. 2b. Supplementary Fig. 1 is a dark-field image showing the positions of the electron probe. It is interesting to note that vibrational signals can still be detected with a probe 100-nm away from the specimen. Figure 2d presents details of the intensity variations of the C=O intensity, along a line scan shown in the annular dark field image Fig. 2c, with points every 0.5 nm. Also shown in Fig. 2d is the integrated intensity encompassing the C-H, N-H and NH_2 stretches.

Energy and momentum conservation leads to a scattering wavevector along the beam direction, q_z, that depends on the energy loss, ΔE such that:

$$q_z = k_i \frac{\Delta E}{mv^2} = \frac{\omega}{v} \qquad (1)$$

where v is the electron velocity, m is the electron mass, k_i is the initial wavevector for the fast electron and $\omega = \Delta E/\hbar$ is the frequency of vibration.

Table 1 | Assignment of peaks observed by aloof EELS in the infrared region.

Peak	Energy (meV)	Frequency (cm^{-1})	Assignment
a	209	1,666	C=O stretch
b	334	2,663	C-H stretch
c	357	2,846	N-H stretch
d	386	3,078	Symmetric NH_2
e	411	3,277	Antisymmetric NH_2

EELS, electron energy loss spectroscopy.

Together with the condition in equation (1), the classical dielectric response theory for an electron beam running parallel to a slab a distance d from the edge gives a functional dependence of

$$\sigma(d) \propto K_0\left(\frac{2\omega}{\gamma v} d\right) \qquad (2)$$

for the scattering cross-section $\sigma(d)$, where K_0 is a Bessel function of the second kind[25,26] and γ is the relativistic Lorentz factor.

Beyond 30 nm from the edge the intensity variation is very well described by equation (2), based on classical dielectric theory. The deviation from the theoretical model near the sample, especially in the integrated counts of the peak due to vibrations associated with hydrogen, is due to beam damage.

The ultraviolet energy range. In contrast to most optical probes, EELS covers a broad spectral range, and spectral characteristics of both the infrared and ultraviolet-visible regions can in principle be recorded simultaneously. Figure 3 focuses on the ultraviolet range, comparing spectra with the probe at positions 12 nm inside the sample and 12-nm away from the edge. Strong peaks at 4.04 and 6 eV are observed, with a weaker peak at 4.94 eV, superimposed on a broad spectral band. These features have been observed in the dielectric response derived from ultraviolet ellipsometry of thin films of guanine on Si(111) substrates[27], and polarized reflection spectra of 9-ethylguanine and guanine hydrochlodirde dihydrate[28]. The 4.04 eV peak is due to π–π^* transitions from the highest occupied molecular orbital to the lowest unoccupied molecular orbital as supported by both self consistent field[29] and density functional theory[30] calculations. The 6 eV peak is due to transitions from the highest occupied molecular orbital to a σ^* level. The broad band is associated with the tails of a π-collective plasmonic excitation. The 4–6 eV peaks become visible when the probe is within 15 nm of the edge, but as shown in Supplementary Fig. 2 the intensity of higher-energy loss features increases significantly only when the electron probe is very close to the specimen edge.

Damage-free vibrational spectroscopy. Figure 4a,c shows how the intensity of vibrational EELS signals varies with time for a beam situated at a fixed position. A slow decay is observed for $d = 10$ nm, taking place on a timescale of 30 min. The time for the intensity to decay by a factor of $1/e$ is 53 min for the bonds of carbon and nitrogen with hydrogen and 68 min for the carbon oxygen bond. This would imply that hydrogen is broken off in the first stage of radiation damage. Visible signs of damage are also observed in the image taken immediately after this data was acquired (Fig. 4b). These patches are similar in appearance to what is observed when the electron probe is stationary on the specimen for 20 s as shown in Supplementary Fig. 3. Even with a low electron flux of 0.02 electrons per nm^2 per s, as would be typical in cryo-electron microscopy, there is little high-resolution structural information remaining after exposure for just over 1 min, as shown in Supplementary Fig. 4.

It would be expected that the lateral distribution of damage would be confined to $\sim$10 nm, which is the distance of the probe from the sample edge and hence the scale of dominant interaction. However, damage is also observed further away in this case. In a discussion on damage caused by fast secondary electrons[31–33], it has been proposed that it mainly originates from K shell excitations. For a beam 10 nm outside the specimen, it is not possible to excite K edges; hence the electrons causing this damage very likely originate from lower-energy excitations. Remarkably, and in contrast to the measurement at 10 nm distance, no damage is observed when the beam is 30 nm from the edge, see Fig. 4 c,d. The reason for this striking difference between the two beam positions is that a critical energy is crossed at this d-range: the ionization energy of guanine is $\sim$7 eV, which translates into $\sim$14 nm (refs 30,34).

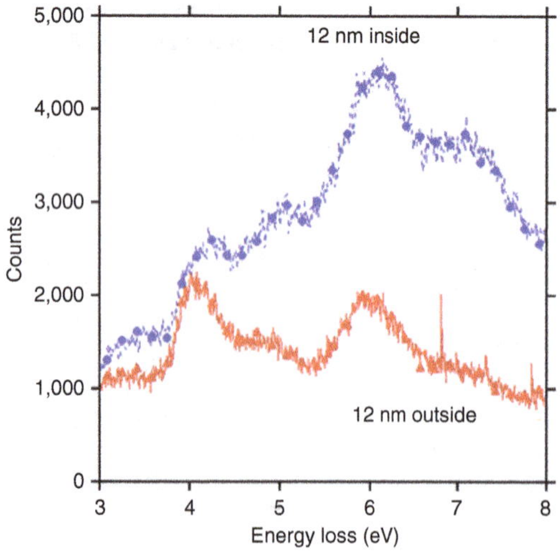

Figure 3 | Ultraviolet region spectra with the beam inside and outside the sample. Comparison of EEL spectra of the ultraviolet region, taken with probe 12 nm inside and 12 nm outside the sample in the vacuum region. The spectrum with the probe outside the specimen lacks the high-energy features at $\sim$7 eV that are believed to be responsible for ionization. The ratio of the peaks at 4 and 6 eV changes when moving from outside to inside the specimen, since lower-energy peaks can be excited at greater distance from the specimen edge.

Discussion

The lateral spatial resolution is approximately given by the distance of the electron beam from the specimen edge. In the electron microscopy of beam-sensitive specimens, spatial resolution is limited by the need to restrict the electron beam fluence to avoid damage[9]. In this sense, aloof beam spectroscopy is similar to other forms of electron microscopy and spectroscopy in that there is a tradeoff between damage and spatial resolution. An advantage of aloof beam spectroscopy is that vibrational spectral features can be safely detected before bringing the beam closer to the sample and before damage becomes apparent. Furthermore, it is possible to directly monitor which bonds are being broken given that radiation damage often involves ejection of hydrogen, either by a knockon impact process or by radiolysis. By resolving the infrared signatures of hydrogen attached to different atoms as the electron probe gets closer to the specimen within a range of 10 nm, the sensitivity of radiation damage to ionization events of different energies can be explored.

In practical applications aloof beam spectroscopy will be complementary to tip-enhanced infrared absorption. The main difference is the nature of the specimen. Flat films are better studied using scanning probe infrared absorption, while aloof beam EELS is more appropriate for TEM specimens where the features of interest lie close to an edge. The spatial resolution of electron beam techniques applied to biological specimens is limited by the ability to tolerate damage. If high resolution is required, the electron beam can be moved very close to, or even on to, the specimen for brief periods to achieve nanometre or better resolution. The spatial resolution of tip-enhanced infrared absorption is approximately the tip radius of $\sim$20 nm, as demonstrated by Huth *et al.*[5]. At present, tip-enhanced infrared spectroscopy has superior energy resolution, but the spectral range is limited by the laser power to 600 cm^{-1} (75 meV). In electron spectroscopy (within the electron microscope), excitations of all energies take place simultaneously, from vibrational modes to plasmons and single-electron excitations with energies equivalent to ultraviolet and soft X-ray photons. We have shown that by controlling the distance of an external narrow electron probe from the edge of a specimen, we can selectively probe vibrational modes without exciting energies above the energy thresholds that potentially lead to radiation damage. This can be seen graphically in Supplementary Fig. 5, which shows strong infrared peaks when the electron probe is outside the specimen while the features corresponding to ultraviolet

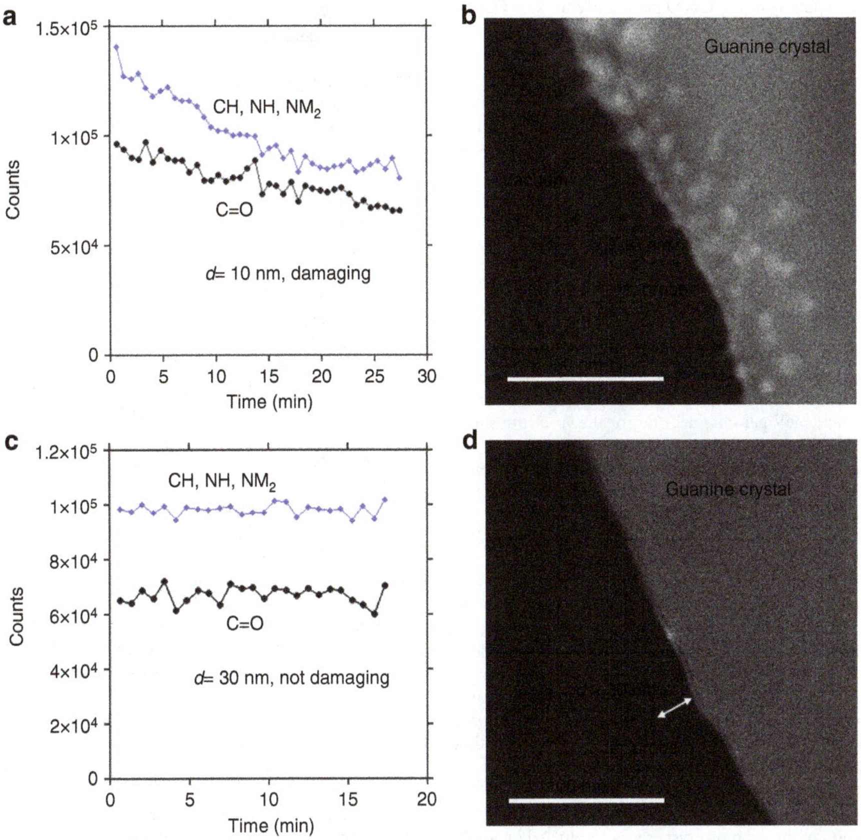

Figure 4 | Time dependence of infrared region peaks. The time dependence of the C = O peak height and the integrated peak area from the hydrogen-stretching modes for an electron probe placed (**a**) 10 nm from the edge of the specimen and (**c**) 30 nm from the edge of the specimen. Images taken after acquiring the data in **a**,**c** are shown in **b**,**d**, respectively. Note the visible signs of damage when the probe was positioned 10 nm from the edge of the specimen, in contrast to the absence of damage when the probe was at a distance of 30 nm. Also note that damage is present at distances greater than would be expected from the 'delocalization' of the probe $\frac{\gamma v}{\omega}$.

excitations are hardly detectable. Moreover, by stepping outside the sample, we avoid high-momentum transfer impact processes (knock on damage) that eject or displace light atoms. This is especially significant for biological samples that are predominately composed of light elements. Notably, in contrast to optical spectroscopy, we still retain all the advantages of electron microscopy, with fast and easy control of the electron beam, and with the ability to easily change *d*-values or move to a different area on the sample.

The superior high-energy resolution provided by the present monochromated electron microscope[35] is an important ingredient of the proposed approach. For example, we have successfully distinguished the vibrational modes of hydrogen covalently bonded to carbon versus nitrogen and clearly resolved the peak due to stretching of the C = O bond. Further improvements in the energy resolution will make it possible to record all peaks corresponding to modes with frequencies greater than ~50 meV (400 cm^{-1}), encompassing much of the useful range in infrared absorption spectroscopy.

To realize its full potential for characterizing biological specimens damage free, aloof beam EELS should be combined with cryo-TEM. Complementary information can then be obtained from low-dose imaging and aloof beam electron spectroscopy. We have already resolved the C = O stretch, one of the amide peaks characteristic of proteins. An improvement in resolution to 5 meV would make it possible to clearly resolve the other amide peaks[36] and analyse the secondary structure of proteins in the electron microscope in a damage-free mode, while determining overall morphology using standard low-dose imaging techniques in the same instrument. Using the NH stretch vibrational mode, it should be possible to localize proteins in membranes at nanometre resolution. Exciting possibilities of distinguishing other significant biological molecules from their vibrational or optical signature are expected to be realized by our approach, enabling nanometre-level resolution of functional groups in biological structures.

Methods

Specimen preparation. Japanese Koi fish (*Cyprinus capio*) were purchased from Peka (Rehovot, Israel) and were maintained in an aquarium at the animal care facility in the Weizmann Institute of Science.

Guanine crystals extraction. Fish scales were mechanically removed and washed with double deionized water (DDW). The guanine crystals were extracted by detaching the skin from underneath the scale in a small volume of DDW, dispersing the crystals into the water. The crystal suspension was collected and centrifuged (10 min, 10,600 r.p.m., Eppendorf centrifuge 5417C). The supernatant was replaced with DDW, the crystal pellet was resuspended and the procedure was repeated twice more.

TEM grids. The crystal suspension was dispersed on a Cu holy carbon-coated TEM grid (MultiA, Quantifoil), and was left in vacuum until completely dehydrated.

FTIR. Synthetic anhydrous guanine crystals (Sigma Aldrich) was transferred to an agate mortar and was lightly crushed with KBr and a pellet was prepared. The spectra were recorded on a NICOLET is5 Fourier transform infrared (FTIR) spectrometer. The FTIR shows a splitting of the C = O peak, since the C = O are in two different environments.

EELS. Spectra, with the exception of the spectrum shown in Fig. 2a, were acquired at Arizona State University with a Nion High-Resolution Monochromated EELS STEM (HERMES) system consisting of a Nion STEM100 scanning transmission electron microscope (STEM) equipped with a Nion high-energy resolution monochromator and a modified Gatan Enfinium EEL spectrometer. The energy resolution, measured from the half width of the zero loss peak (ZLP), was 15 meV. The accelerating voltage was 60 kV, the probe current ~1 pA, the beam convergence was 12 mrad and the electron probe diameter was ~2 nm. A 1-mm spectrometer entrance aperture was used, defining a collection angle of 15 mrad.

As a first stage, the raw spectra were corrected for energy drift, which was a major problem in all acquisitions with high dispersion. Since the ZLP was saturated, the centroid was found by assessing where the intensity was 5% lower than its peak value and taking the mean of these two energies. This procedure corrected drift to within one or two channels. Acquisition parameters for the spectrum in Fig. 2b were 40 s and 2 meV per channel, and for the spectrum shown in Fig. 3 were 7 s and 5 meV per channel.

The spectrum shown in Fig. 2a was acquired using the Nion HERMES system at Rutgers University. It was equipped with a Nion prototype spectrometer with first- to fourth-order aberration correction and a 2k × 2k pixel lens optically coupled with CMOS camera. The scintillator is constructed such that the ZLP completely missed the scintillator, and hence the ZLP tail due to light spread in the scintillator was greatly suppressed. Supplementary Fig. 6 shows how this significantly reduces the background intensity. A power-law background was fitted between 66.7 and 96.13 meV. The EELS dispersion was 0.46 meV per channel, and the energy resolution (half width of the ZLP, measured in a spectrum recorded separately) was 8 meV. The accelerating voltage was 60 kV, the probe diameter was ~2 nm and the probe current ~2 pA. The probe convergence semi-angle was 15 mrad, and the EELS collection semi-angle was 13 mrad. The spectrum shown in Fig. 2a is an average of 19 energy-aligned spectra, each with 10 s acquisition time, taken from a point 30 nm from the edge of the guanine crystal.

TEM diffraction. TEM diffraction of the guanine crystals was conducted on an FEI Tecnai F20 microscope with the beam spread out over micron-sized regions at room temperature. Electron diffraction patterns were recorded on a Gatan US4000 charge-coupled device camera.

Spectrum processing. With the exception of the spectrum shown in Fig. 2a, separate backgrounds were fit under the C = O peak, and the CH, NH and NH_2 peaks. The fitting regions were 0.16–0.18 and 0.23–0.25 eV for the C = O peak, and 0.25–0.29 and 0.45 eV–0.52 eV for the stretching modes involving hydrogen. The C = O peak was the intensity remaining from 0.18 to 0.23 eV after background subtraction, and the peak corresponding the CH, NH and NH_2 stretches was the intensity remaining from 0.29 to 0.45 eV after background subtraction. These energy windows were used for the integrated intensities plotted in Figs 2d and 4a,b, though the window for fitting the power-law background under the compound CH, NH and NH_2 stretch peaks was modified to stop at 0.48 eV and the integration window ended at 0.43 eV, due to problems with a defective pixels in the detector array in the spectra used for Fig. 2d. Supplementary Fig. 7 shows the raw and background subtracted spectra.

References

1. Kuhlbrandt, W. The resolution revolution. *Science* **343,** 1443–1444 (2014).
2. Nogales, E. & Scheres, S. H. W. Cryo-EM: a unique tool for the visualization of macromolecular complexity. *Mol. Cell* **58,** 677–689 (2015).
3. Schroder, R. R. Advances in electron microscopy: a qualitative view of instrumentation development for macromolecular imaging and tomography. *Arch. Biochem. Biophys.* **581,** 25–38 (2015).
4. Knoll, B. & Keilmann, F. Near-field probing of vibrational absorption for chemical microscopy. *Nature* **399,** 134–137 (1999).
5. Huth, F. *et al.* Nano-FTIR absorption spectroscopy of molecular fingerprints at 20nm spatial resolution. *Nano Lett.* **12,** 3973–3978 (2012).
6. Berweger, S. *et al.* Nano-chemical infrared imaging of membrane proteins in lipid bilayers. *J. Am. Chem. Soc.* **135,** 18292–18295 (2013).
7. Baker, L. A. & Rubinstein, J. L. Radiation damage in electron cryomicroscopy. *Methods Enzymol.* **481,** 371–388 (2010).
8. Karuppasamy, M., Karimi Nejadasi, F., Vulovic, M., Koster, A. J. & Ravelli, R. B. G. Radiation damage in single particle cryo-electron microscopy: effects of dose and dose rate. *J. Synchrotron Radiat.* **18,** 398–412 (2011).
9. Egerton, R. F. Limit of the spatial, energy and momentum resolution of electron energy-loss spectroscopy. *Ultramicroscopy* **107,** 575–586 (2007).
10. Frank, J. *Three-Dimensional Electron Microscopy of Macromolecular Assemblies: Visualization of Biological Molecules in their Native State* (Oxford Univ. Press, 2006).
11. Bammes, B. E., Jakana, J., Schmid, M. F. & Chiu, W. Radiation damage effects at four specimen temperatures from 4 to 100° K. *J. Struct. Biol.* **169,** 331–341 (2010).
12. Scheres, S. H. W. Beam-induced motion correction for sub-megadalton cryo-EM particles. *eLife* **3,** e03665 (2014).
13. Li, X. *et al.* Electron counting and beam-induced motion correction enable near-atomic-resolution single-particle cryo-EM. *Nat. Methods* **10,** 584–590 (2013).
14. Krivanek, O. L. *et al.* Vibrational spectroscopy in the electron microscope. *Nature* **514,** 209–212 (2014).
15. Echenique, P. M. & Howie, A. Image force effects in electron microscopy. *Ultramicroscopy* **16,** 269–272 (1985).
16. Howie, A. & Milne, R. H. Excitations at interfaces and small particles. *Ultramicroscopy* **18,** 427–433 (1985).
17. Cohen, H. *et al.* Near-field electron energy loss spectroscopy of nanoparticles. *Phys. Rev. Lett.* **80,** 782–785 (1998).
18. Egerton, R. F. Vibrational-loss EELS and the avoidance of radiation damage. *Ultramicroscopy* **159,** 95–100 (2015).
19. Denton, E. & Land, M. F. Mechanism of reflection in silvery layers of fish and cephalophods. *Proc. Roy. Soc. Lond. B* **178,** 43–61 (1971).
20. Gur, D. *et al.* Guanine-based photonic crystals in fish scales form from an amorphous precursor. *Angew. Chem. Int. Ed.* **52,** 388–391 (2013).
21. Mueller, K. P. & Labhart, T. Polarizing optics in a spider eye. *J. Comp. Physiol. A* **196,** 335–348 (2010).
22. Morrsion, R. L., Sherbrooke, W. C. & Frost-Mason, S. K. Temperature sensitive, physiologically active iridophores in the lizard Urosurus ornatus: an ultrastructural analysis of color change. *Copeia* **1996,** 804–812 (1996).
23. Teyssier, J., Saenko, S. V., van der Marel, D. & Milinkovitch, M. C. Photonic crystals cause active colour change in chameleons. *Nat. Commun.* **6,** 6368 (2015).
24. Ibach, H. & Mills, D. L. *Electron Energy-Loss Spectroscopy and Surface Vibrations* (Academic Press, 1982).
25. Garcia de Abajo, F. J. Optical excitations in electron microscopy. *Rev. Mod. Phys.* **82,** 109–175 (2010).
26. Wang, Z. L. Valence electron excitations and plasmon oscillations in thin films, surfaces, interfaces and small particles. *Micron* **27,** 265–299 (1996).
27. Silaghi, S. D. *et al.* Dielectric functions of DNA base films from near-infrared to ultraviolet. *Phys. Stat. Sol. (b)* **242,** 3047–3052 (2005).
28. Clark, L. B. Electronic spectra of crystalline 9-ethylguanine and guanine hydrochloride. *J. Am. Chem. Soc.* **99,** 3934–3938 (1977).
29. Fluscher, M. P., Serrano-Andres, L. & Roos, B. O. A theoretical study of the electronic spectra of adenine and guanine. *J. Am. Chem. Soc.* **119,** 6168–6176 (1997).
30. Preuss, M., Schmidt, W. G., Seino, K., Furthmuller, J. & Bechstedt, F. Ground and excited state properties of DNA base molecules from plane-wave calculations using ultrasoft pseudopotentials. *J. Comput. Chem.* **25,** 112–122 (2003).
31. Egerton, R. F., Lazar, S. & Libera, M. Delocalized radiation damage in polymers. *Micron* **43,** 2–7 (2012).
32. Howie, A., Rocca, F. J. & Valdre, U. Electron beam ionization damage processes in p-terphenyl. *Philos. Mag. B* **52,** 751–757 (1985).
33. Siangchaew, K. & Libera, M. The influence of fast secondary electrons on the aromatic structure of polystyrene. *Philos. Mag. A* **80,** 1001–1016 (2000).
34. Orlov, V. M., Smirnov, A. N. & Varshavsky, Y. M. Ionizarion potentials and electron donor ability of nucleic acid bases and their analogues. *Tetrahedron Lett.* **48,** 4377–4378 (1976).
35. Krivanek, O. L. *et al.* High-energy-resolution monochromator for aberration-corrected scanning transmission electron microscopy/electron energy-loss spectroscopy. *Phil. Trans. R. Soc. A* **367,** 3683–3697 (2009).
36. Byler, D. M. & Susi, H. Examination of the secondary structure of proteins by deconvolved FTIR spectra. *Biopolymers* **25,** 469–487 (1986).
37. Hirsch, A. *et al.* 'Guanigma': the revised structure of biogenic anhydrous guanine. *Chem. Mater.* **27,** 8289–8297 (2015).

Acknowledgements

We gratefully acknowledge the use of facilities within the Leroy Eyring Center for Solid State Science at Arizona State University. We also would like to thank Professor P.E. Batson for the use of experimental facilities at Rutgers University, supported by the Department of Energy grant DE-SC0005132. We acknowledge partial financial support from the National Science Foundation grant CHE-1508667. Financial support for the purchase of the microscopes was provided by the National Science Foundation grant DMR MRI #0821796 (Arizona State University) and National Science Foundation grant DMR MRI-R2 #959905 (Rutgers University). Partial funding for the work at the Weizmann Institute was provided by the Kimmel Center for Nanoscale Science and the Irving and Cherna Moskowitz Center for Nano and Bio-Nano Imaging. P.R. would like to acknowledge support as the Erna and Jakob Michael Visiting Professor while at the Weizmann Institute of Science.

Author contributions

H.C. and P.R. initiated the project; T.A., K.M., T.C.L., N.D. and O.L.K. acquired spectra or images; H.C. and S.G.W. acquired diffraction patterns; D.G. prepared specimens and acquired infrared and ultraviolet spectra; P.R. processed spectra; K.M. processed images; P.R., K.M. and T.C.L. prepared figures; H.C., P.R., O.L.K. and S.G.W. all made significant contributions to writing the paper.

Additional information

Competing financial interests: O.L.K., N.D. and T.C.L. have a financial interest in Nion Co.

9

Chirality of nanophotonic waveguide with embedded quantum emitter for unidirectional spin transfer

R.J. Coles[1], D.M. Price[1], J.E. Dixon[1], B. Royall[1], E. Clarke[2], P. Kok[1], M.S. Skolnick[1], A.M. Fox[1] & M.N. Makhonin[1]

Scalable quantum technologies may be achieved by faithful conversion between matter qubits and photonic qubits in integrated circuit geometries. Within this context, quantum dots possess well-defined spin states (matter qubits), which couple efficiently to photons. By embedding them in nanophotonic waveguides, they provide a promising platform for quantum technology implementations. In this paper, we demonstrate that the naturally occurring electromagnetic field chirality that arises in nanobeam waveguides leads to unidirectional photon emission from quantum dot spin states, with resultant in-plane transfer of matter-qubit information. The chiral behaviour occurs despite the non-chiral geometry and material of the waveguides. Using dot registration techniques, we achieve a quantum emitter deterministically positioned at a chiral point and realize spin-path conversion by design. We further show that the chiral phenomena are much more tolerant to dot position than in standard photonic crystal waveguides, exhibit spin-path readout up to $95 \pm 5\%$ and have potential to serve as the basis of spin-logic and network implementations.

[1]Department of Physics and Astronomy, University of Sheffield, Hicks Building, Sheffield S3 7RH, UK. [2]EPSRC National Centre for III-V Technologies, Department of Electronic and Electrical Engineering, University of Sheffield, Sheffield S1 3JD, UK. Correspondence and requests for materials should be addressed to M.N.M. (email: m.makhonin@sheffield.ac.uk).

Quantum information processing promises a dramatically increased performance in computing, secure communications and simulations of quantum systems. In addition, a network of quantum nodes enables distributed quantum computing[1], and may facilitate the quantum internet[2]. Such a distributed architecture requires the conversion between matter qubits for local quantum memories and photonic qubits for quantum communication between nodes[3–5]. Furthermore, any scalable quantum technology must solve the miniaturization and fabrication problem, which likely demands that the quantum nodes must be implemented on integrated circuits and waveguides[6]. Recent developments of passive components[7] with embedded quantum dots (QDs) using advanced semiconductor technologies and enhanced-coherence using resonant techniques on-chip[8] contribute to potential solutions. The spin of an electron or hole in a QD is a promising candidate to serve as the matter-qubit in a quantum network, but its implementation requires an efficient on-chip spin–photon interface.

The coupling of spin to the direction of photon emission in nanophotonic waveguides provides a potential solution. The subject of unidirectional light propagation in nanophotonic structures under circularly polarized laser excitation has been advancing rapidly since the first reports in 2013 for surface plasmons and atomic dipoles[9,10]. The chiral effect is understood to arise from the longitudinal component of evanescent fields in nanophotonic structures, which enables circularly polarized states to propagate with the field rotating within the longitudinal plane defined by the geometry of the waveguide[9–16]. An important step forward was made very recently in which directional emission was demonstrated for a quantum-dot emitter embedded within a specially engineered glide-plane waveguide[14], confirming theoretical predictions for chiral emission in photonic crystal waveguides (PhC WGs)[14,15].

In this work, we demonstrate the internal intrinsic chirality of the electromagnetic field in a system with no specially engineered chirality using the QD as an internal probe. We demonstrate the importance of the position of the quantum emitter, in such a way that spin-dependent directional emission is achieved. We achieve efficient coupling of QD exciton spin to the direction of photon emission in nanophotonic waveguides. This chiral behaviour occurs even though the waveguide geometry is completely symmetric, and the dielectric material is non-chiral. We use numerical simulations to demonstrate that the chiral emission originates from the chirality of the electromagnetic field inside the waveguide and exploit it for efficient coupling of the QD emitter by lateral positioning within the waveguide. The dots located at the centre of the waveguide exhibit non-chiral emission, but those displaced from the centre show a varying degree of chiral emission with distance from the centre. The maximal spin-dependent directionality occurs for QDs located at chiral points shifted from the waveguide centre by 32% of the waveguide width. We first confirm this hypothesis by measuring the variation of the spin-dependent directionality of a large number of dots, with good agreement found between the experimental data and the predictions of simulations for randomly positioned emitters within the waveguide. We compare results for nanobeam and PhC WGs, and find that the simpler nanobeam structure provides substantially better spin-dependent directionality because of the lower sensitivity of the chirality to the exact dot position. We then demonstrate designable spin-path coupling by using enhanced registration techniques[17,18] (see the Methods for details) to achieve highly directional emission for a QD deterministically positioned at a chiral point. We conclude that directionality factors approaching 100% ($95 \pm 5\%$) are achievable for the exciton spin of an embedded dot to a single photon state propagating in a nanophotonic waveguide. We also confirm the expected bi-directional emission for a dot deterministically positioned at the centre of the waveguide. We further note that our results provide a natural explanation for the bi-directional emission reported but not explained in cross-waveguide structures[19]. Taken together, the results establish a route towards the transfer and entanglement of the spin state of an emitter to a photonic qubit in an integrated network.

Results

Directional emission in dielectric nanobeam waveguides. The system we investigate is a QD in a nanophotonic waveguide. The dot serves as an integrated quantum emitter with addressable spin-exciton eigenstates, generating spin- and position-dependent directional emission of single photons in the waveguide by exciton recombination. Nanophotonic structures support both longitudinal and transverse field components (E_x and E_y) and, as we show, permit the transfer of in-plane circular polarization ($E_x \pm iE_y$; see Fig. 1a,b for definition of axes). Despite its non-chiral geometry and material, the waveguide exhibits a chiral electromagnetic field mode. As a result, a well-positioned QD will emit in-plane circularly polarized single photons unidirectionally, with a high correlation between the spin state of the QD and the which-path information of the photon.

A schematic of the experimental geometry is given in Fig. 1a. A cross-section of the device including the layer of QDs is shown schematically in Fig. 1b. The left- and right-circularly polarized photons arising from recombination of up and down exciton spin states from a QD embedded within a suspended nanobeam waveguide (NWG) are coupled to NWG modes, and then diffracted by out-coupler gratings at opposite ends towards external detectors. As shown in Fig. 1c–e, for all three charge states of the dots, exciton recombination leads to circularly polarized photon emission. In a non-chiral structure, the emission probability for a circularly polarized dipole is identical in both directions, and this behaviour is indeed observed when the dot is located at the centre of the waveguide (see Fig. 1f obtained from finite difference time domain (FDTD) simulations, see the Methods). By contrast, when the dot is displaced from the centre of the waveguide, the emission direction depends on the spin, with σ^+ photons emitted in one direction and σ^- in the other. The unidirectional emission is shown in Fig. 1g,h (the results of FDTD simulations), and occurs through the coupling of the σ^+/σ^- dipole emitters to direction-dependent modes at the chiral points of the NWG.

Electric field distributions in NWG. The origin of the unidirectional emission can be understood by consideration of the electromagnetic field distribution within the laterally confined nanophotonic geometry. Figure 2a shows a schematic representation of an unterminated NWG with the field distribution in the xy-plane at $z=0$ (position of the QDs layer) calculated by solving Maxwell's equations (Supplementary Note 1). The amplitudes of the E_y and E_x field components, together with their relative phases, are shown in Fig. 2b. The $|E_y|$ component has a maximum at the centre of the NWG, whereas $|E_x|$ has two maxima closer to the edges (Supplementary Note 2). The relative phase between E_x and E_y is constant at either $\pm\pi/2$ and changes sign when crossing $y=0$. It is this asymmetry in the phase that enables chiral behaviour: the field is right- or left-circularly polarized (that is, $\sigma^\pm$ corresponding to $E_x \pm iE_y$ fields) at points where $|E_y|=|E_x|$ and the phase is $\pm\pi/2$. This is shown more clearly in Fig. 2c, where the electric field is plotted vectorially at an instant in time. The chiral points are positions where the electric field rotates in time during propagation of the waveguide mode. The rotation direction depends on the propagation

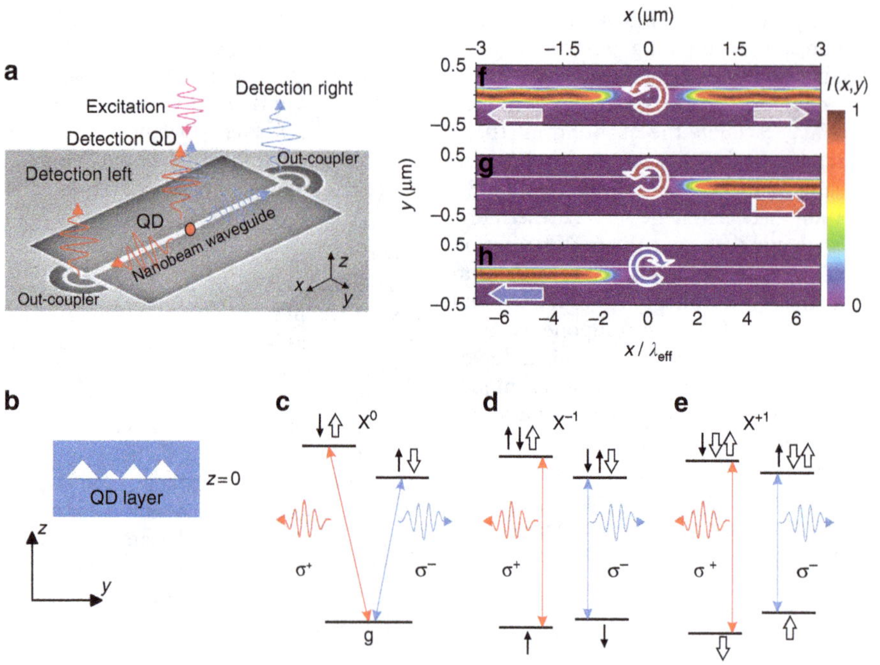

Figure 1 | Directional emission in dielectric nanobeam waveguides. (**a**) Schematic representation of the experiments on directional emission from quantum dots embedded inside a single-mode nanobeam waveguide (NWG) with out-couplers for photon collection. Red (blue) arrows correspond to photons originating from right (left) circularly polarized dipoles in the quantum dot. Scheme labels are plotted on top of an SEM image of the real device. (**b**) A schematic of *yz* cross-section of the NWG with the QD layer at $z=0$. Level diagrams and optical transitions for (**c**) neutral (X^0) and (**d**,**e**) charged (X^{-1}) and (X^{+1}) exciton QD states in $\mathbf{B}\neq 0$. (**f**) Time averaged intensity distribution of the emission $I(x,y)$ (colour scale) in the *xy*-plane at $z=0$ from circularly polarized σ^+ dipole at the centre of the NWG ($y=0$). (**g**,**h**) Emission for a circularly polarized dipole at the chiral point, displaced from the centre of the waveguide by 32% of the waveguide width: σ^+ dipole; σ^- dipole. The undulation arises from periodic coupling with time of the rotating circular dipole at the non-chiral point. The intensity distributions in (**f-h**) are calculated after the dipole source is switched off. The *x* axis is normalized to the NWG mode effective wavelength λ_{eff}. The white horizontal lines show the nanostructure boundaries. Fully animated versions of the simulations may be found in the Supplementary Notes 1-6.

direction as indicated in Fig. 2d, so that $\sigma^{\pm}$ emitters couple preferentially to modes propagating in opposite directions. By contrast, the point at the centre of the waveguide ($y=0$) has no longitudinal component ($E_x=0$) and is therefore linearly polarized (Fig. 2c,d). Translation along *y* from the centre to the chiral points thus transforms the field polarization from linear, through elliptical, to circular.

Spin readout in a NWG and PhC WG. To demonstrate the chiral effects experimentally, we use a single-mode NWG with randomly distributed self-assembled QDs, positioned by growth in the *xy*-plane at $z=0$. The guided photoluminescence (PL) of single QDs is detected from out-couplers at opposite ends of the NWG (Fig. 1a). The inset in Fig. 3c shows a typical auto-correlation function $g^{(2)}(\tau)$ confirming the single-photon nature of the QD emission. The sample was placed in an out-of-*xy*-plane magnetic field B_z to quantize the QD spins into $S_z=\pm 1$ states (spin up/down). This enables identification of exciton Zeeman spin components from the energy of the circularly polarized (σ^+/σ^-) photons emitted: $h\nu_x(S_z)=h\nu_0+\mu_B g_x B_z S_z$, where g_X is an excitonic g-factor that varies from dot to dot with a typical value of ~ 1.2. $B_Z=1$ T for the data in Figs 3 and 4. Self-assembled growth leads to random positions of QDs, enabling the observation of emission from both chiral and non-chiral areas of the waveguide. Figure 3 contrasts the behaviour for two dots selected from the random distribution, with one showing chiral behaviour (Fig. 3a), and the other not (Fig. 3b). For the QD in the chiral point in Fig. 3a, both Zeeman-split components are seen when detecting directly from the dot ('det QD'), but only one component is observed for detection from the left and right out-couplers ('det L' and 'det R'). The two Zeeman components emit in opposite directions, implying spin-dependent directional emission. By contrast, non-directional emission is observed for the dot in centre of the waveguide (Fig. 3b), with both Zeeman components detected from both out-couplers. The degree of contrast in spin-readout (the directionality factor) is defined as:

$$C_{\text{LEFT/RIGHT}} = \frac{I_{\sigma+}^{L/R} - I_{\sigma-}^{L/R}}{I_{\sigma+}^{L/R} + I_{\sigma-}^{L/R}} \tag{1}$$

where the superscripts L and R refer to the left and right out-couplers. The results for 50 QDs are shown in Fig. 3c. The diagonal line across the figure is the expectation for a random distribution of dots in the *xy*-plane at $z=0$ whose circular dipoles couple to the confined electromagnetic fields. QDs at the linear point at the centre of the waveguide correspond to zero contrast, whereas dots at the chiral points give rise to contrasts of ± 1. Agreement between the trends in contrast between experiment and simulation is seen (the experimental scatter around the diagonal is discussed below).

The data from Fig. 3c are plotted as a histogram of numbers of dots of absolute contrast $C=(|C_{\text{LEFT}}|+|C_{\text{RIGHT}}|)/2$ in Fig. 3d and compared with binned data of the FDTD simulations of contrast (see Supplementary Fig. 7 for details) for out-coupler-terminated waveguide in Fig. 3e. Details of the comparison between infinite and terminated waveguides are given in the Supplementary Notes 2 and 3. The out-coupler gratings introduce back-reflections that lead to modulation of the field distributions; they do not change the mechanisms responsible for unidirectionality, but result in reduction in the probability for dots with high

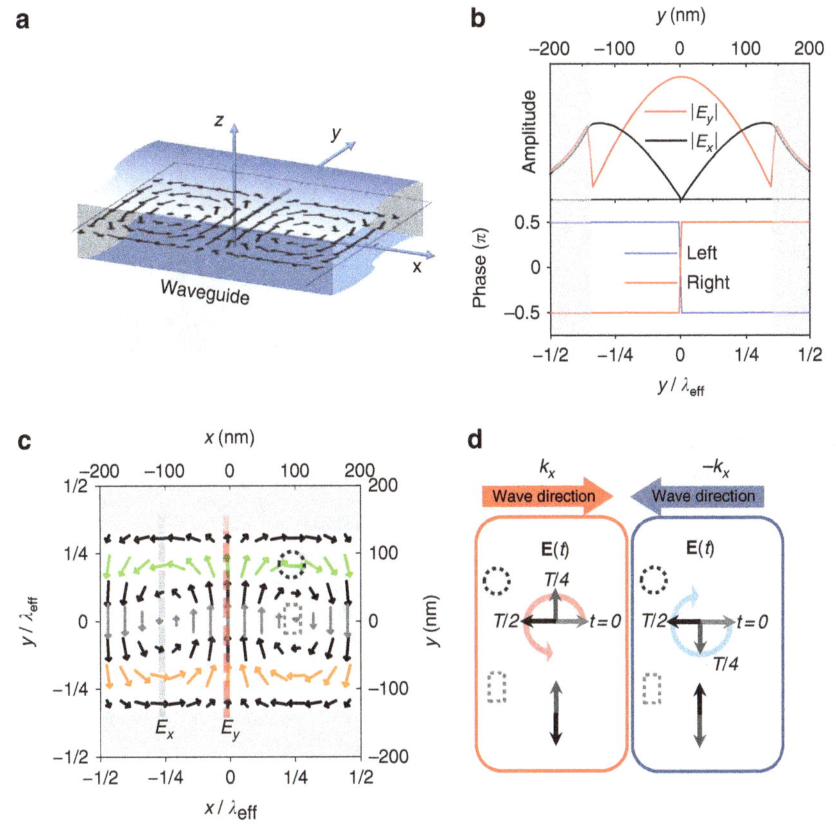

Figure 2 | Electric field distributions in nanobeam waveguide. (**a**) Schematic of an infinite length NWG showing the orientation of the axes and the field distribution in the *xy*-plane at $z=0$, as defined by the central layer of the waveguide containing QDs. (**b**) Electric field amplitudes (top) for the *x* and *y* components and their relative phase (bottom) for left/right propagating modes. (**c**) Simulated distribution of the electric field vector for the first fundamental mode at a fixed moment in time. Arrows show the direction of the field and the length of the magnitude. Colour is used as a guide to the eye: green/orange for opposite-helicity chiral points, and grey for central points with linear polarization. The left/bottom scales are normalized to the NWG mode effective wavelength λ_{eff}. (**d**) Electric field time evolution in space from **c** for propagating modes at chiral (circular dotted area) and non-chiral (rectangular dotted area) points in the NWG. The red and blue arrows show the k_x and $-k_x$ mode propagation direction, respectively. The rotation of the electric field vectors exhibiting circular right/left polarization is shown by red/blue circular arrows. The grey areas in **b** and **c** correspond to air cladding regions at the NWG edges.

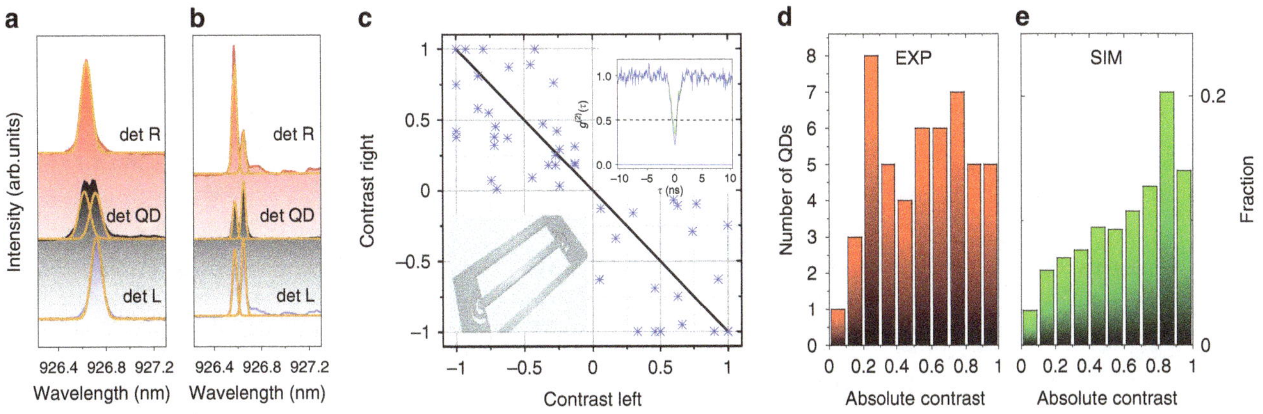

Figure 3 | Experimental demonstration of spin readout in nanobeam waveguides. (**a**) Photoluminescence spectra of QDs with high contrast in NWGs. (**b**) Experimental photoluminescence spectra of QDs with low contrast in NWGs. Fitted peaks in **a** and **b** are shown with orange curves. (**c**) Readout contrast $C=(I_{\sigma+}-I_{\sigma-})/(I_{\sigma+}+I_{\sigma-})$ for left and right out-couplers for randomly distributed self-assembled QDs in NWGs (50 QDs). The points correspond to experimental data, and the black diagonal line shows the expected distribution for ideal circularly polarized dipoles. (**d**) Statistical distributions of the absolute average contrasts $C=(|C_{\mathrm{LEFT}}|+|C_{\mathrm{RIGHT}}|)/2$ of QD spin readout taken from the experimental data in **c** for NWGs. All experimental data are recorded at $B_Z=1$ T. (**e**) Simulated statistical distributions of absolute contrasts calculated from the coupling of circularly polarized dipoles distributed across the WGs in the *xy*-plane ($z=0$) taking into account the electric field distributions (Supplementary Note 3) due to the out-couplers for NWG. The top inset in **c** shows a typical QD autocorrelation function $g^{(2)}(\tau)$ (blue curve with background subtraction $g^{(2)}(0)=0.22$, green without $g^{(2)}(0)=0.34$). The inset in **c** shows schematic of the NWG.

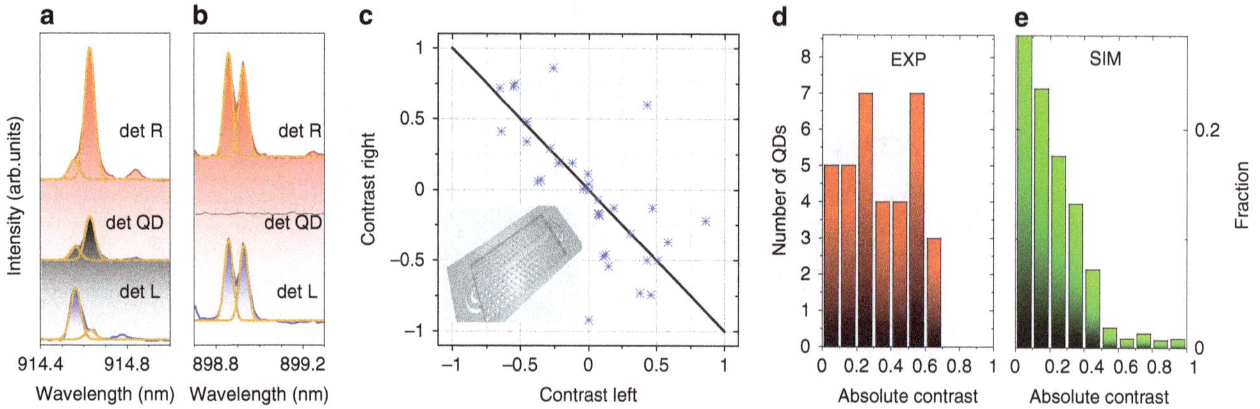

Figure 4 | Experimental demonstration of spin readout in photonic crystal waveguides. (**a**) Photoluminescence spectra of QDs with high contrast in PhC WGs. (**b**) Experimental photoluminescence spectra of QDs with low contrast in PhC WGs. Fitted peaks in **a** and **b** are shown with orange curves. (**c**) Readout contrast $C=(I_{\sigma+}-I_{\sigma-})/(I_{\sigma+}+I_{\sigma-})$ for left and right out-couplers for randomly distributed self-assembled QDs in PhC WGs (35 QDs). The points correspond to experimental data, and the black diagonal line shows the expected distribution for ideal circularly polarized dipoles. (**d**) Statistical distributions of the absolute average contrasts $C=(|C_{LEFT}|+|C_{RIGHT}|)/2$ of QD spin readout taken from the experimental data in **c** for PhC WGs. All experimental data are recorded at $B_Z=1$ T. (**e**) Simulated statistical distributions of absolute contrasts calculated from the coupling of circularly polarized dipoles distributed across the WGs in the *xy*-plane ($z=0$) taking into account the electric field distributions (Supplementary Note 4) due to the out-couplers for PhC WGs. The inset in **c** shows schematic of the PhC WGs.

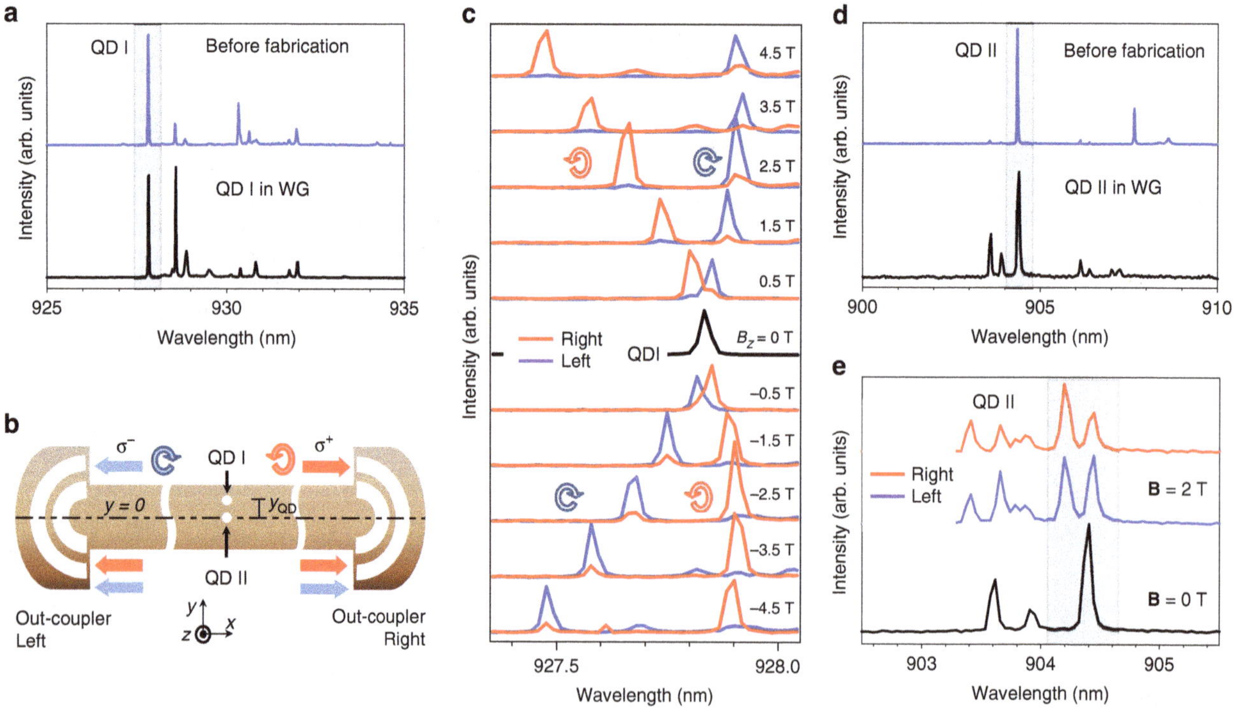

Figure 5 | Quantum dot registration and spin readout. (**a**) Spectra of QD I before and after NWG fabrication. (**b**) Schematic of spin readout for registered QD I displaced vertically from the centre by 32% of the waveguide width (the Sample details in the Methods) and the absence of spin-readout for registered QD II located at the centre of the waveguide. (**c**) Magnetic field dependence of emission from the registered QD I collected from the left out-coupler (blue) and right out-coupler (red). (**d**) Spectra of QD II before and after NWG fabrication. (**e**) Spectra of QD II in magnetic field **B** = 2 T collected from left and right out-couplers and spectrum of QD II at **B** = 0 T for reference.

(>90%) spin readout conversion by ~52%. Qualitative agreement between experiment (Fig. 3d) and simulation (Fig. 3e) is observed, with an increasing number of QDs to high contrast clearly seen, and levelling-off around a contrast of 80%. It is also notable that QDs with extremely high directionality exceeding 90% are observed experimentally (Fig. 3c,d). These results support the findings in the simulations which show chiral areas (Supplementary Fig. 6d) of 14% of the waveguide where the contrast exceeds 90% (and 34% exceeding 80%).

The circular dipole unidirectional coupling efficiency, defined as the fraction of power emitted into a waveguide mode propagating in one direction, is calculated to be 68% for a dipole positioned at a chiral point of the NWG (simulations in the Methods). This is an increase of more than ~2.8 compared with the coupling of a circular-dipole at the centre of the waveguide into one spatial direction. It is also ~1.4 times greater than for a linear-dipole co-polarized with the field at the centre of the NWG ($y=0$)[8]. Both comparisons show the favourable degrees of coupling efficiency obtained for spin readout at the chiral points.

In order to compare to PhC WGs, we perform similar experiments and simulations for them. QD spectra with high/low contrasts are shown in Fig. 4a,b. The contrasts for the 35 QDs studied are shown in Fig. 4c. The overall trends in the behaviour are similar to those seen in Fig. 3c for the NWGs, but with the very marked difference that no QDs exhibit high directionality values above 70% (Fig. 4c,d). This is expected as the chiral areas (Supplementary Fig. 9d) inside the PhC WG, where the contrast exceeds 90%, only constitute $\sim 0.8\%$ of the waveguide area ($\sim 1.5\%$ of area for $>80\%$ contrast) as shown in Supplementary Note 4. They are furthermore located in areas of low field intensity (Supplementary Fig. 9a), leading to poor dipole coupling. Both the high-fidelity chiral areas (~ 20 times smaller than for the NWGs) and field intensities for the PhCs indicate the advantageous characteristics of the NWGs for spin readout, with their near continuous translational symmetry, relative to PhC WGs. This situation can likely be improved for PhC WGs by the use of glide plane techniques, as shown in ref. 14.

QD registration and spin readout. Having identified the location of chiral points theoretically, and observed high contrast unidirectional emission for randomly distributed dots, we now demonstrate control of the emission direction, using QD registration techniques[17,18] (see the Methods for details). First, QD spectra were compared before and after the registration and NWG fabrication to confirm that the QD (we term this dot QD I) is successfully integrated inside the NWG (Fig. 5a). The waveguide is fabricated (see the Methods for details) such that QD I is displaced from the centre of the waveguide at the chiral point, with y_{QD} $\sim 30\%$ of the waveguide width (see Fig. 5b schematic for details). For this position, we expect the σ^+ dipole to couple to the right propagating mode and the σ^- dipole to the left (Fig. 5b). The spectra in magnetic field reveal high directionality ($|C_{LEFT}| = 92 \pm 3\%$, $|C_{RIGHT}| = 80 \pm 3\%$) of emission with absolute contrast independent of magnetic field to within 10%. Furthermore, the expected signs of the contrast correlate with the QD I position, with the σ^+ line appearing at the right out-coupler with high energy when the field is positive, whereas the σ^- line is collected from the left out-coupler. Moreover, when the direction of the magnetic field is changed, the order of the lines is reversed, but the dipole coupling direction remains the same (Fig. 5c). Thus, we not only achieve control of spin-directionality by registration but also the reversal of the emission direction at a given energy by magnetic field control[14]. We further demonstrate the effectiveness of the registration techniques by fabricating a control sample with a QD (QD II) positioned at the centre of the waveguide (see Fig. 5b schematic for details). Spectra before and after NWG fabrication are compared to prove the integration of the QD within the waveguide (Fig. 5d). Low spin-readout contrast of ($|C_{LEFT}|$ $\sim 3 \pm 6\%$, $|C_{RIGHT}| = 24 \pm 4\%$, Fig. 5e) is found, as expected for a dot at the centre of the waveguide. We note that the approach with dot registration is not limited to one QD and can be scaled up to create more complex circuits with deterministically coupled QDs to realize spin–photon and spin–spin entanglement on chip.

Discussion

As noted above, the experimental points in Figs 3c and 4c show scatter around the diagonal, compared with theoretical expectations. This behaviour is not fully understood. It may arise due to back reflections in the finite length waveguides, in combination with elliptically polarized QDs. Fine structure splitting[20] of the QDs can be excluded as we observe no magnetic field dependence for asymmetrically coupled QDs. The circular polarization for the QDs may also be affected in photonic structures by the nanostructure and surface proximity[21]. Moreover, the finite size of the dot, which is known to cause a breakdown in the point dipole approximation[22] used for our simulations in nanophotonic structures may also affect the contrast.

To conclude, we demonstrate experimentally and theoretically chiral effects in a simple nanophotonic system consisting of waveguides containing embedded quantum emitters. The unidirectional phenomena we report may be used for spin read out and to transfer spin information from localized emitters in waveguide geometries. Deterministic positioning of a QD at a chiral point of the waveguide is achieved, with accompanying spin readout, a possible route to scalability. Larger areas for chiral behaviour found in the NWGs because of their continuous translational symmetry together with simplified fabrication techniques and low loss may provide significant advantages over PhC WGs. The findings and techniques presented could contribute to the creation of spin-optical on-chip networks and processing devices based on nanophotonic waveguides.

Methods

Sample details. The sample was grown by molecular beam epitaxy on an undoped [100] GaAs substrate, and consisted of a 140-nm-thick GaAs membrane containing a single layer of self-assembled InGaAs QDs at the centre grown on a 1-μm sacrificial $Al_{0.6}Ga_{0.4}As$ layer. The photonic structures were patterned by electron beam lithography (EBL) and the pattern transferred to the membrane using inductively coupled plasma etching. To release the membranes, the sacrificial AlGaAs layer was removed by selective wet etch using a buffered hydrogen fluoride solution. The suspended NWGs were 15 μm long, 280 nm wide and 140 nm height. An scanning electron microscopy (SEM) image of the suspended NWG is shown in Supplementary Fig. 1a. The W1 PhC WGs were fabricated with lattice constant $a = 254$ nm and hole radius $r = 0.31\ a$. An SEM image of the PhC WG is presented in Supplementary Fig. 1b. Both waveguides were terminated with semi-circular $\lambda/2n$ air-GaAs out-coupler gratings designed for optimum operation at $\lambda = 950$ nm at the centre of the QD ensemble PL emission[23].

Experimental set-up. All measurements were performed in a helium bath cryostat at $T = 4.2$ K within which a superconducting magnet provided magnetic field of up to 5 T normal to the sample plane. Ultra-stable positioning of the sample within the system is provided by X, Y, Z home-made piezo stages. The cryostat insert had optical access to the sample in a confocal scanning microscope arrangement[24]. The excitation and collection spots were below 1.5 μm in diameter and could be separately moved by more than 15 μm by scanning mirrors to obtain the exact geometry required for each experiment. Optical excitation was provided by an 808 nm diode laser coupled to an optical fibre. The collected PL signal was fibre coupled and spectral measurements were performed using a spectrometer with a LN2 cooled charged-coupled device (CCD). Using a second exit port on the spectrometer, spectrally filtered PL signals were sent to two avalanche photodiodes (APDs) for QD autocorrelation $g^{(2)}(\tau)$ measurements using a single-photon counting module.

QD registration method. Dot registration[17,18] is carried out in a scanning micro-PL set-up with two collection paths, as illustrated in Supplementary Fig. 2. The dot registration process involves three stages: pre-registration markers are fabricated, individual QDs are registered and photonic structures are deterministically fabricated around a QD. Registration markers are patterned using EBL and a positive resist. The sample is developed in xylene leaving the wafer surface exposed in the desired pattern, as shown in Supplementary Fig. 3a. The inner double markers are used to register the relative position of a QD, whereas the outer angular markers are used to re-align the EBL during the final fabrication stage. A layer of 5 nm of titanium followed by 20 nm of gold are then evaporated onto the wafer. The sample is then placed in a solvent bath to remove the remaining layer of resist and any unwanted titanium and gold.

A 1-mm diameter cubic zirconia (ZrO_2) Weierstrass solid immersion lens is placed directly above a registration grid, with an aberration free image size of ~ 50 μm in diameter. A low QD density of $\sim 1.6 \cdot 10^7$ cm^{-2} and a solid immersion lens-enhanced excitation spot size ~ 350 nm enables individual QDs to be measured. The closed-loop three axes scanning piezo stage, with a resolution of 1 nm, is used to scan the objective lens, which simultaneously moves the excitation and collection spots. One collection path is spectrally filtered through a monochromator to isolate a single exciton line before it is measured with an APD. The second collection path is used to measure the reflected laser signal from the gold markers to provide reference points in the scan from which the QD position is measured. No additional spectral filtering is used on the second collection path before it is detected with an APD, although it is heavily attenuated using neutral density filters. Horizontal and vertical line scans are used to determine the relative QD position. The signals from both APDs were recorded during the scans as a function of the stage position (Supplementary Fig. 3b).

The spacing between the gold markers is designed to be 15 μm. The relative position of the QD is then determined by repeating the line scans 100 times to provide a statistical average (Supplementary Fig. 3c). A Gaussian curve is fitted to the distribution to determine the centre. The minimum standard deviation observed is $\sigma < 5$ nm, which is exceeding the accuracy of the best position uncertainty previously achieved[17,18]. The position of the suspended NWGs to be fabricated is placed within the original EBL design file at a suitable location, such that the QD are in the centre or laterally displaced. To align the EBL writing field to the pattern previously fabricated, the electron beam is scanned over the outer (blue) registration markers. The reflected electron flux is recorded and used to determine the central position of each marker. The suspended nanobeam is then fabricated (SEM image in Supplementary Fig. 3d) as described above in the Sample details section.

Calculations of the fields in the waveguides. FDTD simulations were performed using freely available software package[25] and commercial-grade simulator[26]. The GaAs waveguide devices simulated had the dimensions described in the Sample details section above. Details of the simulations are available in Supplementary Notes 1–4.

References

1. Barz, S. *et al.* Demonstration of blind quantum computing. *Science* **335,** 303–308 (2012).
2. Kimble, H. J. The quantum Internet. *Nature* **453,** 1023–1030 (2008).
3. Kok, P. & Lovett, B. W. *Introduction to Optical Quantum Information Processing* (Cambridge Univ., 2010).
4. DiVincenzo, D. P. The physical implementation of quantum computation. *Fortschr. Phys.* **48,** 771–783 (2000).
5. Barrett, S. D. & Kok, P. Efficient high-fidelity quantum computation using matter qubits and linear optics. *Phys. Rev. A* **71,** 060310 (2005).
6. O'Brien, J. L. Optical quantum computing. *Science* **318,** 1567–1570 (2007).
7. Prtljaga, N. *et al.* Monolithic integration of a quantum emitter with a compact on-chip beam-splitter. *Appl. Phys. Lett.* **104,** 231107 (2014).
8. Makhonin, M. N. *et al.* Waveguide coupled resonance fluorescence from on-chip quantum emitter. *Nano Lett.* **14,** 6997–7002 (2014).
9. Rodriguez-Fortuño, F. J. *et al.* Near-field interference for the unidirectional excitation of electromagnetic guided modes. *Science* **340,** 328–330 (2013).
10. Junge, C., O'Shea, D., Volz, J. & Rauschenbeutel, A. Strong coupling between single atoms and nontransversal photons. *Phys. Rev. Lett.* **110,** 213604 (2013).
11. Mitsch, R., Sayrin, C., Albrecht, B., Schneeweiss, P. & Rauschenbeutel, A. Quantum state-controlled directional spontaneous emission of photons into a nanophotonic waveguide. *Nat. Commun.* **5,** 5713 (2014).
12. Petersen, J., Volz, J. & Rauschenbeutel, A. Chiral nanophotonic waveguide interface based on spin-orbit coupling of light. *Science* **346,** 67–71 (2014).
13. le Feber, B., Rotenberg, N. & Kuipers, L. Nanophotonic control of circular dipole emission. *Nat. Commun.* **6,** 6695 (2015).
14. Söllner, I. *et al.* Deterministic photon–emitter coupling in chiral photonic circuits. *Nat. Nano* **10,** 775–778 (2015).
15. Young, A. B. *et al.* Polarization engineering in photonic crystal waveguides for spin-photon entanglers. *Phys. Rev. Lett.* **115,** 153901 (2015).
16. Bliokh, K. Y., Rodríguez-Fortuño, F. J., Nori, F. & Zayats, A. V. Spin–orbit interactions of light. *Nat. Photon* **9,** 796–808 (2015).
17. Thon, S. M. *et al.* Strong coupling through optical positioning of a quantum dot in a photonic crystal cavity. *Appl. Phys. Lett.* **94,** 111115 (2009).
18. Dousse, A. *et al.* Controlled light-matter coupling for a single quantum dot embedded in a pillar microcavity using far-field optical lithography. *Phys. Rev. Lett.* **101,** 267404 (2008).
19. Luxmoore, I. J. *et al.* Optical control of the emission direction of a quantum dot. *Appl. Phys. Lett.* **103,** 241102 (2013).
20. Gammon, D. *et al.* Fine structure splitting in the optical spectra of single GaAs quantum dots. *Phys. Rev. Lett.* **76,** 3005–3008 (1996).
21. Stepanov, P. *et al.* Quantum dot spontaneous emission control in a ridge waveguide. *Appl. Phys. Lett.* **106,** 041112 (2015).
22. Tighineanu, P., Sørensen, A. S., Stobbe, S. & Lodahl, P. Unraveling the mesoscopic character of quantum dots in nanophotonics. *Phys. Rev. Lett.* **114,** 247401 (2015).
23. Faraon, A. *et al.* Dipole induced transparency in waveguide coupled photonic crystal cavities. *Opt. Express* **16,** 12154–12162 (2008).
24. Grazioso, F., Patton, B. R. & Smith, J. M. A high stability beam-scanning confocal optical microscope for low temperature operation. *Rev. Sci. Instrum* **81,** 093705 (2010).
25. Johnson, S. S. & Joannopoulos, J. J. Block-iterative frequency-domain methods for Maxwell's equations in a planewave basis. *Opt. Express* **8,** 173–190 (2001).
26. Lumerical FDTD Solutions. Lumerical Solutions, Inc. http://www.lumerical.com/tcad-products/fdtd/ (2014).

Acknowledgements

This work has been supported by the Engineering and Physical Sciences Research Council (Programme Grant EP/J007544/1). We thank D.M. Whittaker, D.N. Krizhanovksii, E. Cancellieri and F. Li for fruitful discussions.

Author contributions

D.M.P., J.E.D. and M.N.M. carried out the experiments, J.E.D. and B.R. performed the dot registration and sample fabrication, R.J.C. carried out the simulations, E.C. grew the samples. All authors analysed the data and contributed to the writing of the manuscript, with P.K. providing quantum technology context. A.M.F., M.S.S. and M.N.M. conceived the project and provided overall guidance and management.

Additional information

Competing financial interests: The authors declare no competing financial interests.

10

Pure-quartic solitons

Andrea Blanco-Redondo[1], C. Martijn de Sterke[1], J.E. Sipe[2], Thomas F. Krauss[3], Benjamin J. Eggleton[1] & Chad Husko[1]

Temporal optical solitons have been the subject of intense research due to their intriguing physics and applications in ultrafast optics and supercontinuum generation. Conventional bright optical solitons result from the interaction of anomalous group-velocity dispersion and self-phase modulation. Here we experimentally demonstrate a class of bright soliton arising purely from the interaction of negative fourth-order dispersion and self-phase modulation, which can occur even for normal group-velocity dispersion. We provide experimental and numerical evidence of shape-preserving propagation and flat temporal phase for the fundamental pure-quartic soliton and periodically modulated propagation for the higher-order pure-quartic solitons. We derive the approximate shape of the fundamental pure-quartic soliton and discover that is surprisingly Gaussian, exhibiting excellent agreement with our experimental observations. Our discovery, enabled by precise dispersion engineering, could find applications in communications, frequency combs and ultrafast lasers.

[1] Centre for Ultrahigh bandwidth Devices for Optical Systems (CUDOS), Institute of Photonics and Optical Science (IPOS), School of Physics, The University of Sydney, Sydney, New South Wales 2006, Australia. [2] Department of Physics, University of Toronto, 60 Street George Street, Toronto, Ontario, Canada M5S 1A7. [3] Department of Physics, University of York, York, YO10 5DD, UK. Correspondence and requests for materials should be addressed to A.B-R. (email: andrea.blancoredondo@sydney.edu.au).

The fascinating phenomenon of optical solitons, solitary optical waves that propagate in a particle-like fashion over long distances[1], has been the subject of intense research during the last decades due to its major role in breakthrough applications such as mode locking[2], frequency combs[3,4] and supercontinuum generation[5,6] among others[7–9]. Temporal solitons in optical media[10,11], as studied to date, arise from the balance of the phase shift due to anomalous quadratic group-velocity dispersion (GVD), that is, negative GVD parameter $\beta_2 = (\partial^2 k/\partial\omega^2) < 0$, and the self-phase modulation (SPM) due to the nonlinear Kerr effect.

In practice, higher-order nonlinear and dispersive effects often perturb this behaviour. In silicon (semiconductor) waveguides the most significant higher-order nonlinearities are associated with free carriers (FCs) generated by two-photon absorption (TPA)[12,13], which have hampered the observation of soliton-based effects in this material. Recently, some of us achieved higher-order soliton compression of picosecond pulses in silicon[14] by using a dispersion engineered photonic crystal waveguide (PhC-wg). Turning to higher-order dispersive effects, the presence of third order dispersion (TOD; $\beta_3 = \partial^3 k/\partial\omega^3$) leads to soliton instability[15], whereas positive fourth-order dispersion (FOD; $\beta_4 = (\partial^4 k/\partial\omega^4) > 0$), can give rise to radiation at specific frequencies[16]. In the presence of negative FOD ($\beta_4 < 0$), as was shown by a series of theoretical works in optical fibres[17–21], solitons can be stable. These studies[17–21] led to the concept of quartic solitons[22], solitary pulses resulting from the interaction of anomalous GVD and SPM but modified by the presence of FOD.

In the following we report the experimental discovery and physical description of an entirely new class of solitons originating purely from the interaction of negative FOD and SPM, as conceptually depicted in Fig. 1a, which can occur even when the GVD vanishes or is normal. Since they arise just from quartic dispersion and SPM, and to distinguish them from the solitary waves studied earlier[17–22], we propose the name of pure-quartic soliton for this new class of solitary wave. This experimental discovery is enabled by the unique dispersion properties of PhC-wgs, which allow us to combine very large negative β_4 with small positive β_2 and negligible β_3 for the wavelength under study. Though our work directly pertains to pulse propagation in guided wave structures, the same ideas apply to spatial solitons, particularly to subdiffractive matter-wave solitons in regimes where the fourth-order diffraction is the dominant diffractive effect[23,24]. Furthermore, it was shown that in Ti:sapphire lasers FOD ultimately limits the minimum pulse duration in cavities with near-zero GVD and TOD[25–27], hinting that the pulse shaping behaviour in the laser cavity for ultrashort pulses (below 10 fs) arises from the balance of SPM and FOD[28].

Results

Experimental signatures of pure-quartic solitons. In our experiments we performed time- and phase-resolved propagation measurements on the sample using a frequency-resolved electrical gating (FREG) apparatus, depicted in Fig. 1b, which can be modelled using a generalized nonlinear Schrodinger equation (GNLSE). We show shape-preserving propagation and flat temporal phase for fundamental pure-quartic solitons and temporal compression and convex nonlinear phase for higher-order pure-quartic solitons. In spite of maintaining these well-known signatures of soliton-like behaviour, we show that pure-quartic solitons present remarkably different properties than solitons studied to date[10,11,17–21]. Importantly, the energy scaling of pure-quartic solitons suggests much higher energies for ultrashort pulses, which may inspire a new wave of soliton laser developments. Finally, we derive the approximate shape of fundamental pure-quartic solitons and find that it is close to a Gaussian, which is remarkable given that the solitary wave

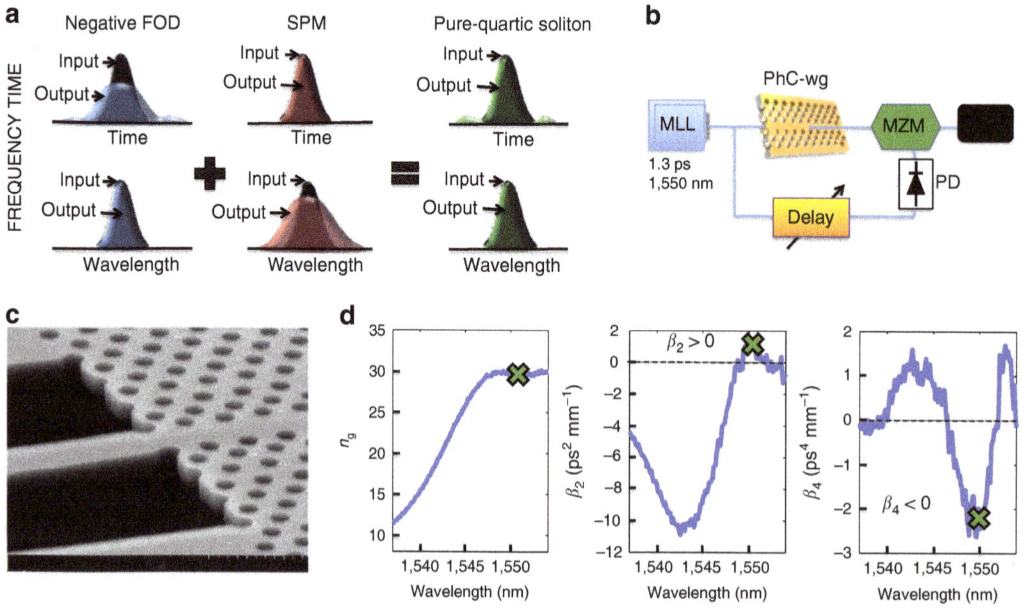

Figure 1 | Concept of pure-quartic solitons and their experimental demonstration. (**a**) Schematics of pure-quartic solitons: (Left) Fourth-order dispersion (FOD) gives rise to temporal pulse broadening (blue output pulse versus black input pulse in time) without affecting the spectrum; (Centre) self-phase modulation (SPM) generates spectral broadening (red output pulse versus black input pulse in frequency) without affecting the temporal pulse shape; (Right) the interplay of FOD and SPM can give rise to pure-quartic solitons which remain nearly unperturbed (green output pulses versus black input pulses in both frequency and time); (**b**) Frequency-resolved electrical gating set-up: mode locked laser (MLL), photonic crystal waveguide (PhC-wg), tunable delay, ultrafast photodiode (PD), Mach-Zehnder modulator (MZM), and optical spectrum analyser (OSA); (**c**) Scanning electron microscope image of the sample; (**d**) Measured dispersion of the silicon photonic crystal waveguide used in our experiments: group index (n_g), second-order dispersion parameter (β_2) and fourth-order dispersion parameter (β_4).

solutions found to date are always a function of hyperbolic secant[10,18,20].

For demonstrating the existence of pure-quartic solitons we used the 396-μm-long dispersion engineered slow-light silicon PhC-wg[29] shown in Fig. 1c (see the Methods section). Figure 1d shows the waveguide dispersion measured using an interferometric technique[30]. At the pulse central wavelength, 1,550 nm, the PhC-wg has a measured group index of $n_g = 30$, a GVD of $\beta_2 = +1\,\mathrm{ps^2\,mm^{-1}}$, corresponding to normal dispersion, a TOD of $\beta_3 = +0.02\,\mathrm{ps^3\,mm^{-1}}$, and a FOD of $\beta_4 = -2.2\,\mathrm{ps^4\,mm^{-1}}$. Note that β_4 is negative in a 6-nm wavelength range.

To perform a complete temporal and spectral characterization of the sub-picojoule ultrafast nonlinear dynamics in the waveguide we used a FREG apparatus[31] in a cross-correlation configuration, as schematically depicted in Fig. 1b (see the Methods section). This set-up provides a series of spectrograms, that is, the gated optical power versus delay, for varying input powers. From these spectrograms we then extract the optical pulses' electric field envelope and phase using a numerical algorithm[32]. Figure 2 shows the measured intensity (red dashed lines) and phase (black dashed) at the output of the PhC-wg, when injecting 1.3 ps Gaussian pulses (full-width at half-maximum, FWHM) at 1,550 nm with different input peak powers, P_0. Figure 2a shows the frequency domain and Fig. 2b shows the temporal domain. The physical length scales of the dispersion orders for this pulse duration are: $L_{\mathrm{GVD}} = T_0^2/|\beta_2| = 0.615\,\mathrm{mm}$, $L_{\mathrm{TOD}} = T_0^3/|\beta_3| = 22.6\,\mathrm{mm}$, $L_{\mathrm{FOD}} = T_0^4/|\beta_4| = 0.168\,\mathrm{mm}$, with $T_0 = \mathrm{FWHM}/1.665$ for Gaussian pulses. These length scales indicate that FOD is dominant, with the total length of the sample being $L = 2.4.L_{\mathrm{FOD}}$ and the GVD length being $L_{\mathrm{GVD}} = 3.66.L_{\mathrm{FOD}}$. TOD is negligible in this sample for our pulses.

To understand the origin of the experimental observations in Fig. 2 we employ a GNLSE model to describe the propagation in the silicon PhC-wg:

$$\frac{\partial A}{\partial z} = -\frac{\alpha_{\mathrm{l,eff}}}{2}A - i\frac{\beta_2}{2}\frac{\partial^2 A}{\partial t^2} + \frac{\beta_3}{6}\frac{\partial^3 A}{\partial t^3} + i\frac{\beta_4}{24}\frac{\partial^4 A}{\partial t^4} + \left(i\gamma_{\mathrm{eff}} - \frac{\alpha_{\mathrm{TPA,eff}}}{2}\right)|A|^2 A + \left(ik_0 n_{\mathrm{FC,eff}} - \frac{\sigma_{\mathrm{eff}}}{2}\right)N_c A. \quad (1)$$

Here $A(z,t)$ is the slowly varying amplitude of the pulse, $\alpha_{\mathrm{l,eff}}$

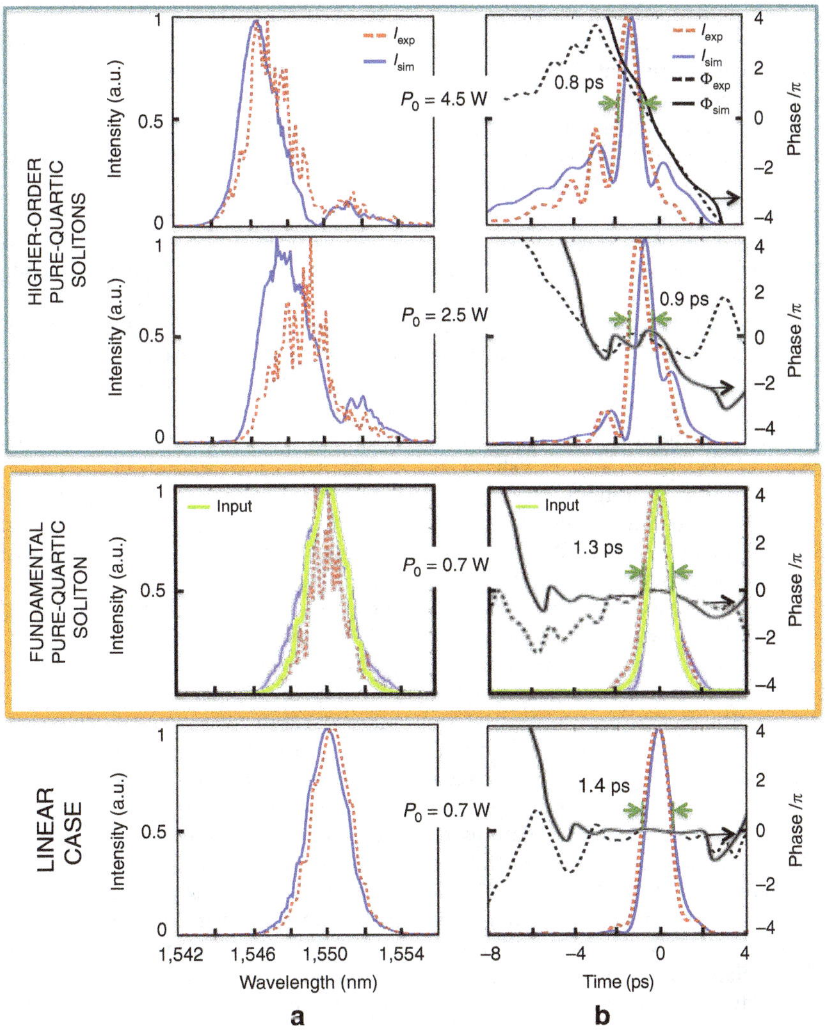

Figure 2 | Experimental and modelling results. **(a)** Frequency and **(b)** time domain results for different input powers. The dashed red lines represent the intensity measurements, the blue solid lines represent the intensity simulations, the black dashed line represents the measured phase, and the solid black line represents the simulated phase. The green solid line at 0.7 W represents the normalized input intensity. The yellow box encompasses the fundamental pure-quartic soliton, showing nearly unperturbed propagation and flat temporal phase. The turquoise box includes two cases of higher-order pure-quartic solitons, showing temporal compression and nonlinear spectral broadening. The higher-order pure-quartic solitons observed here are greatly perturbed by the presence of free carriers.

denotes the linear loss, γ_{eff} and $\alpha_{\mathrm{TPA,eff}}$ are the effective nonlinear Kerr and TPA parameters, respectively; $n_{\mathrm{FC,eff}}$ and σ_{eff} represent the free-carrier dispersion (FCD) and the free-carrier absorption effective parameters for a free-carrier concentration of N_c. Since we use a slow-light PhC-wg, the effective coefficients vary with the slow-down factor $S = n_g/n_0$ (ref. 13). We use the measured envelope amplitude of the input pulse as the input to our GNLSE model with the sample parameters detailed in the Methods section. As shown in Fig. 2, the model (solid lines) agrees well with the experimental data (dashed lines) in both frequency and time.

We first focus on the time domain results of Fig. 2b. At the low coupled power of 0.07 W, nonlinear effects can be neglected and we simply observe small temporal broadening, mainly due to quartic dispersion. The different signs of β_2 and β_4 counteract each other to some degree, leading to a modest temporal broadening at the output of this short PhC-wg (from 1.3 to 1.4 ps). We have verified, by running the GNLSE for longer lengths that the pulse width keeps increasing with the propagation distance in the linear case. Increasing the input power up to 0.7 W, where the nonlinear length $L_{\mathrm{NL}} = 1/(\gamma_{\mathrm{eff}} P_0)$ becomes comparable to L_{FOD}, the pulse preserves its initial shape and duration, as illustrated by the good matching between the measured output intensity (dashed red line) and the normalized input intensity (green solid line). Furthermore, the temporal phase across the pulse duration is nearly flat. These are two signatures of fundamental soliton behaviour[10]. Simple estimates, confirmed with GNLSE simulations, show that the loss due to TPA at this power level is quite small and that free carriers do not yet play a role. At 2.5 W, where $L_{\mathrm{NL}} << L_{\mathrm{FOD}}$, the phase becomes convex due to the stronger nonlinear Kerr effect and the main peak of the pulse narrows, temporal signatures of a higher-order soliton. At even higher powers, 4.5 W, the main peak of the pulse narrows even more, corresponding to a higher-order soliton with a higher-soliton number[11]. In addition, a long tail develops towards the leading edge of the pulse. We provide an explanation for this effect below.

Next we examine the frequency domain in Fig. 2a. At 0.07 W, since the nonlinearities are negligible and the pulse spectrum is not affected by the dispersion, the pulse spectral shape is maintained. At 0.7 W the pulse preserves its initial spectral shape, again consistent with fundamental soliton behaviour in the spectral domain. At higher powers (2.5 and 4.5 W) the pulse experiences spectral broadening and splits into two peaks, spectral signatures of higher-order solitons. The observed blue shift and asymmetry are associated mainly with FCD. Note that the oscillations in the measured spectra simply correspond to Fabry–Perot reflections at the input and output facets of the PhC-wg and disorder in the periodic media[33].

Whereas we previously reported shape-preserving fundamental solitons and higher-order soliton compression in silicon[14], such behaviour was unforeseen for the normal GVD here. By setting $\beta_2 = 0$ in our numerical model we find that the signatures of soliton behaviour are maintained: the shape is preserved and the phase is flat for the fundamental soliton at 0.7 W, and at 2.5 and 4.5 W, the higher-order solitons undergo nonlinear temporal narrowing. This demonstrates that GVD is not important in this system and, since we established that TOD is also negligible, that the soliton behaviour stems purely from the interaction of FOD and SPM. Furthermore, we have verified that the long tail at the leading edge observed at high powers (Fig. 2b), as well as the self-acceleration of the pulse, originate from the interaction of negative FOD and FCD. The FCD generates additional blue components (Fig. 2a) and the negative FOD makes them travel faster than the red components of the pulse, analogous to our earlier results with negative GVD[13,14,34].

These observations at the output of the silicon PhC-wg suggest the existence of a new type of soliton: pure-quartic solitons. We use the term soliton here to refer to solitary optical waves that propagate essentially unperturbed over long distances, not to exact localized solutions of integrable nonlinear differential equations[35].

As expected in a silicon system at 1,550 nm, pure-quartic solitons are strongly perturbed by TPA and FC as we just described, and thus the measured behaviour differs from the simple case with just SPM and FOD. Therefore, to elucidate the dynamics of pure-quartic solitons in the absence of higher-order nonlinearities we next numerically study the propagation of picosecond pulses along the PhC-wg neglecting all effects but SPM and FOD.

Propagation behaviour of pure-quartic solitons. Figure 3a,b depict the propagation dynamics of undistorted pure-quartic solitons, that is, in the presence of SPM and FOD only. Such a system is governed by the biharmonic nonlinear Schrodinger equation

$$\frac{\partial A}{\partial z} = i\frac{\beta_4}{24}\frac{\partial^4 A}{\partial t^4} + i\gamma_{\mathrm{eff}}|A|^2 A. \qquad (2)$$

In our simulations we consider two different power levels: fundamental pure-quartic solitons occur at moderate powers (Fig. 3a), whereas at high powers higher-order pure-quartic solitons result (Fig. 3b).

The simulations in Fig. 3a show shape-preserving pulse propagation in time and frequency over five quartic dispersion lengths L_{FOD} for a fundamental pure-quartic soliton. The very slight increase in the maximum intensity at $t = 0$ corresponds to the pulse adapting itself from the standard Gaussian pulse used as an input to the model, to the soliton form whose approximate shape we provide in the next section. The output, represented by the blue curve to the right of the propagation plot, shows that the pulse maintains essentially the same amplitude, shape (Gaussian), and duration (1.3 ps) as the input pulse. Importantly, the temporal phase at the output, represented by the black solid line, is flat across the duration of the pulse.

Figure 3b reveals a higher-order pure-quartic soliton, with the pulse experiencing periodic recurrent propagation. In time the pulse undergoes compression and then periodically returns to its initial shape. In frequency the pulse splits into two and then recombines to recover its initial spectral shape after the same period. At the maximum compression point, the pulse reaches the minimum duration of 0.54 ps, a compression factor of 2.4 compared with the initial pulse duration, with a peak intensity of two times that of the initial pulse. Our simulations show the compression factor of the pure-quartic soliton roughly follows the same trend as conventional solitons[11]. Specifically, larger intensities lead to higher compression factors and to compression occurring at an earlier spatial position along the waveguide. Crucially, while the trends are superficially similar to the behaviour of conventional solitons, the different scaling of SPM and FOD with pulse length suggests that the well-known definitions of soliton number and soliton period will not be appropriate for pure-quartic solitons; further studies are underway to derive the appropriate parameters and physical scaling laws for these field structures.

To understand why the experimental observations of pure-quartic solitons in Fig. 2 differ from the numerical results of the undistorted system in Fig. 3a,b, we simulate the propagation along the PhC-wg including all the effects in the real system indicated in equation 1. The outputs shown in Fig. 3c,d match

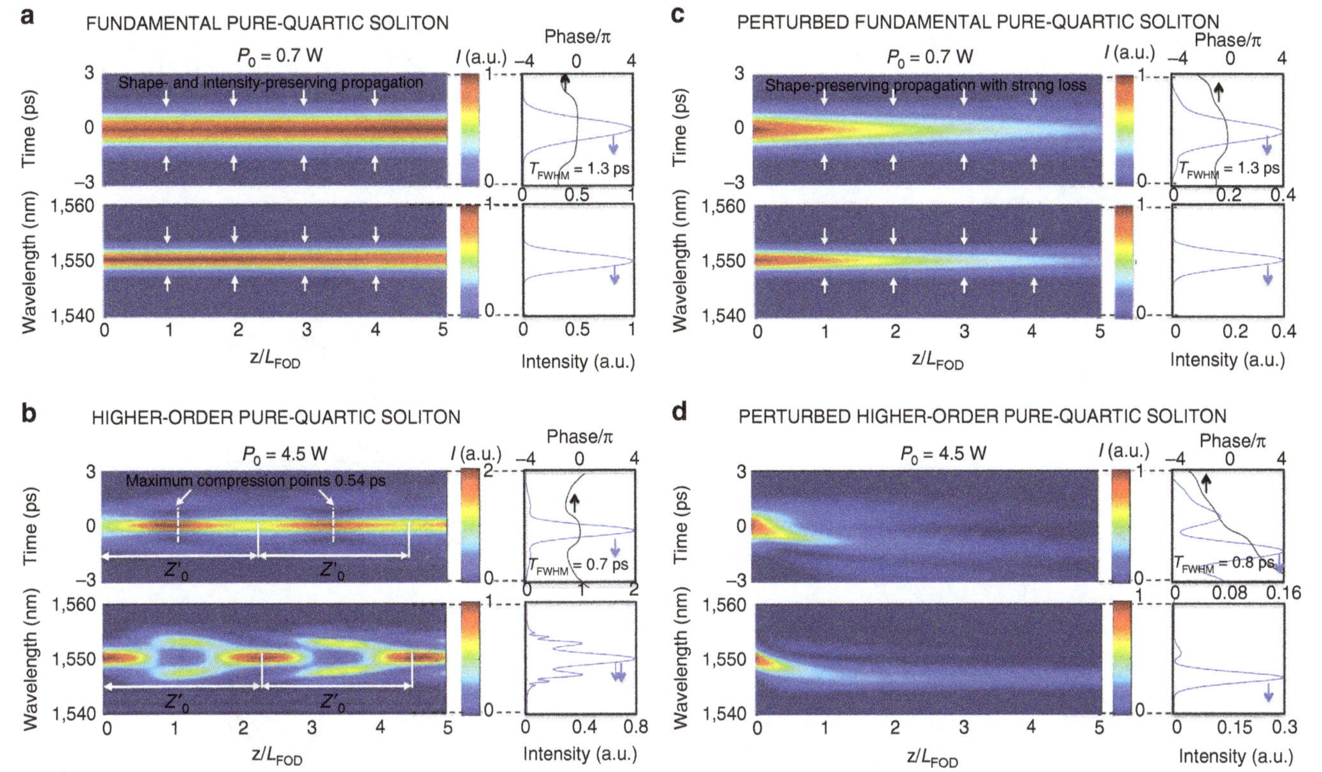

Figure 3 | Simulations of the propagation of a fundamental and a higher-order pure-quartic soliton along five quartic dispersion lengths, L_{FOD}. (**a**) Fundamental ($P_0 = 0.7$ W) pure-quartic soliton with only self-phase modulation and quartic dispersion present, and (**c**) in the more realistic scenario for our silicon waveguide with two-photon absorption and free carriers; (**b**) and (**d**) are similar but for a higher power level ($P_0 = 4.5$ W) where a higher-order pure-quartic soliton results.

our experimental measurements at $P_0 = 0.7$ W and $P_0 = 4.5$ W. Figure 3c shows that the signatures of the fundamental pure-quartic soliton remain in realistic simulations: the pulse maintains its shape and width, and the phase at the output remains almost flat. However, the intensity of the pulse decreases due, predominantly, to the linear loss in the slow-light waveguide ~70 dB cm^{-1}. Since the intensity decreases as the pulse propagates, the linear FOD will eventually dominate over the SPM for longer distances, leading to temporal broadening of the fundamental pure-quartic soliton. The higher-order pure-quartic soliton in the realistic scenario of Fig. 3d differs considerably from Fig. 3b. The TPA clamps the intensity in the waveguide from the early stages of propagation. The FCD introduces blue components that lead to the self-acceleration of the pulse and an asymmetry, and the free-carrier absorption induces absorption on the trailing edge. These effects of TPA and FCs on the propagation of higher-order pure-quartic solitons in silicon are analogous to the effects of FCs on conventional solitons[14].

Approximate solution for the fundamental pure-quartic soliton. We now derive an analytic expression for the fundamental pure-quartic soliton. The experimental observations and numerical simulations indicate that the central part of fundamental pure-quartic solitons appears to be Gaussian. Assuming this shape, we look for an approximate solution to equation (2) in two separate ways for verification purposes: using the variational principle and looking for a local approximation near the centre. The complete derivations for the variational and local approximate solutions are described in Supplementary Note 1 and Supplementary Note 2, respectively.

In both cases, after a simple dimensional analysis, we take the central part of the pure-quartic soliton to be of the form

$$A(z,t) = A_0 e^{i\mu\gamma_{\text{eff}} A_0^2 z} e^{-\left(A_0\sqrt{\nu\frac{\gamma_{\text{eff}}}{|\beta_4|}}\right)t^2}, \qquad (3)$$

where, since $\beta_4 < 0$, we have written $\beta_4 = -|\beta_4|$for convenience, and μ and ν are free parameters. The variational approach gives $\mu = 7/(8\sqrt{2}) \approx 0.62$, $\nu = 1/\sqrt{2}$,whereas the local approximation gives $\mu = \frac{1}{2}$, $\nu = 1$. Thus, these approaches predict pulse widths which differ only by a factor $2^{\frac{1}{8}}$ or by <10%. The fact that these different approximations give very similar results reinforces our confidence in them. Based on these results, the argument of Akhmediev and Karlsson[36] suggests that pure-quartic solitons do not lose energy due to linear radiation. []ally, taking the Fourier transform of the right-hand side equation (3) in time and position leads to a straight line in an ω-k diagram. Since this straight line does not intersect the linear dispersion relation of the medium, the soliton cannot lose energy to dispersive waves.

To test the validity of this analytic approximate solution we numerically solve equation (2) with the β_4 and γ_{eff} of our sample (see Methods section) and obtain the output of the system at the power level corresponding to a fundamental pure-quartic soliton for a propagation length $L = 30 \cdot L_{FOD}$. This long propagation distance ensures convergence of the pulse evolution. The results of this numerical experiment for three different pulse shapes: a Gaussian, a hyperbolic secant (sech), and a super Gaussian of order four, are depicted by a solid blue curve in Fig. 4a–c, respectively. Importantly, the hyperbolic secant and super Gaussian inputs (black solid curve in Fig. 4b,c, respectively) evolve into the solitary wave Gaussian shape, constituting an additional signature of soliton-like behaviour and proving that

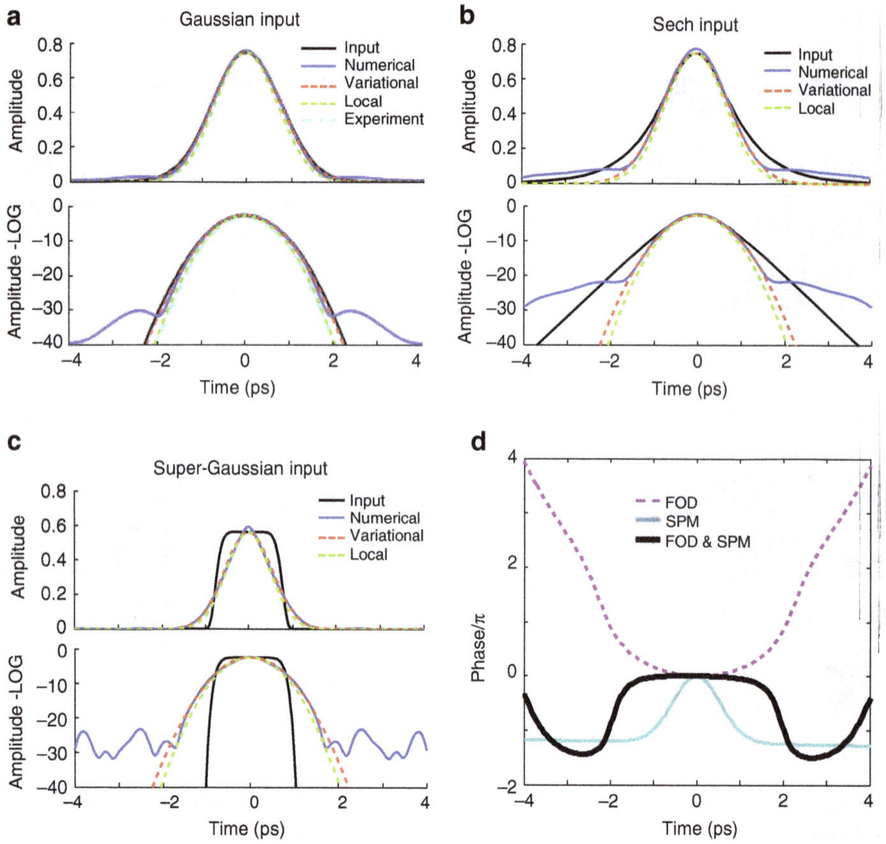

Figure 4 | Approximate solutions to a fundamental pure-quartic soliton and phase diagram. Comparison between the variational and local approximate solutions for the fundamental pure-quartic soliton and the numerical output after propagating over thirty quartic dispersion lengths for (**a**) a Gaussian input, (**b**) a hyperbolic secant input, and (**c**) a super Gaussian input of order 4. In the Gaussian case the measured output pulse at 0.7 W is shown in the background (dot-dash cyan curve). (**d**) Phase shift induced by the fourth-order dispersion (dashed purple) and the self-phase modulation (solid turquoise) independently and its combined phase shift (black).

this type of soliton acts as an attractor. These results are overlapped with the variational approximation (dashed red curve) and the local approximation (green dashed curve), using the same parameters. The agreement between the numerical and the variational solution in the central part of the pulse is remarkable. The local approximation deviates only slightly. In addition we numerically find that $\mu \approx 0.63$, again very close to the variational result. In the background of Fig. 4a, we show the measured pulse at the output of the chip at $P_0 = 0.7$ W (cyan dot-dashed curve), which matches variational solution perfectly. The wings observed in the numerical solution relate to the fact that the phase shift profile of the FOD has a quartic dependence with time, whereas the SPM varies quadratically, as illustrated in Fig. 4d. This allows both phase shift profiles to perfectly counterbalance each other close to the centre of the pulse, but deviate from each other at the edges. This fact is not captured in the approximate analytic solution since a perfect balance between FOD and SPM is assumed.

Since temporal optical solitons studied to date have some kind of hyperbolic secant shape[10,18,20], it is surprising that pure-quartic solitons are approximately Gaussian. To understand this better, we apply the argument of Dudley *et al.*[37] to our system, according to which the dispersive and nonlinear phase components developed during short propagation distances must cancel each other across the pulse duration to lead to the formation of a soliton. Figure 5a shows the FOD- (red) and the SPM-induced chirp (blue) after propagating the Gaussian variational solution in equation 3 (dashed black curve) for a distance $L_{FOD}/10$ and demonstrates how this solution leads to the mentioned cancellation across the central part of the pulse. To highlight the different nature of the solutions found here with respect to the previously studied NLS solitons with FOD, we apply the same verification to the solution found in ref. 18, in the limit $\beta_2 = 0$. Figure 5b shows how the sech^2 solution obtained from ref. 18 (dashed black curve) does not provide the necessary cancellation of the SPM- and FOD-induced chirp required for the formation of a stable solitary wave in the presence of just SPM and FOD.

Discussion

The experimental results and the numerical simulations presented here have established the existence of a new class of solitons: pure-quartic solitons, arising from the interaction of SPM and FOD only. In particular, we experimentally demonstrated shape-preservation and flat-phase behaviour for the fundamental pure-quartic soliton, and temporal compression for the higher-order pure-quartic solitons. We numerically demonstrated that the higher-order pure-quartic soliton would undergo recurrent periodic propagation in the absence of loss and higher-order nonlinearities. Although we have verified that these signatures of soliton propagation are preserved for long propagation distances in the presence of just FOD and SPM, the disparity between the quartic profile of the FOD-induced phase shift and the quadratic profile of the SPM-induced phase shift affecting the edges of the pulse could lead to stability issues that should be studied. Establishing appropriate definitions of concepts such as soliton number and

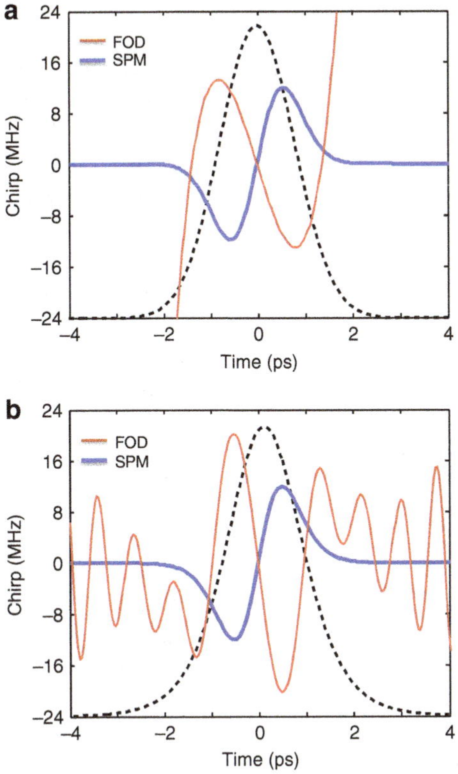

Figure 5 | The cancellation of nonlinear and dispersive phase components on the fundamental pure-quartic soliton. (**a**) FOD-induced (red) and SPM-induced (blue) frequency chirps after a propagation of $L_{FOD}/10$ for the Gaussian pure-quartic soliton of equation (3); (**b**) similar, but for the sech^2 type solutions obtained taking ref. 18 with $\beta_2 = 0$. The dashed black curves in the background of (**a**) and (**b**) represent the input pulse intensity: the Gaussian solution of equation 3 and the sech^2 solution in ref. 18 respectively.

soliton period for pure-quartic solitons is an open theoretical challenge.

Our discovery was facilitated by the unique dispersion properties of PhC-wgs that provide the design freedom to achieve a wide variety of dispersion profiles. However, other guided wave systems such as highly nonlinear fibres[38], photonic crystal fibres[8] or specially designed silicon waveguides[22,39], could also be engineered to observe pure-quartic solitons. The main condition to fulfil is $L_{FOD} << L_{GVD}$, L_{TOD} (with $\beta_4 < 0$), and, in practice, we have verified that $L_{FOD} << L_{GVD}/3$ across most of the pulse bandwidth is enough for a robust observation. The pure-quartic soliton behaviour starts to become observable when L_{FOD} becomes comparable to the sample length. However, our initial simulations show that the pure-quartic soliton does not reach a steady state until it propagates for several quartic dispersion lengths, similar to conventional solitons[40]. Analogous regimes of evolution have been demonstrated in Ti:sapphire laser cavities[25–28], taking advantage of the rich variety of physical regimes offered by the discrete structure of the laser cavity. For example, Zhou *et al.* demonstrated in ref. 26 8.5 fs pulses from a Ti:sapphire laser operating near zero GVD with the minimum pulse duration limited by FOD, and later Christov *et al.*[28] hinted that the 'soliton-like pulse' inside such a laser was 'fourth-order dispersion limited'. Here we experimentally demonstrate that the balance between SPM and FOD gives rise to robust soliton-like behaviour. Hence, the scope of our findings is not just limited to nonlinear guided wave optics, but may provide novel insights into extreme regimes of ultrafast lasers operation.

The Gaussian variational solution provided here constitutes a good approximation to the central form of the fundamental pure-quartic soliton. The results of our study on the cancellation of the nonlinear and quartic dispersion phase components in short propagation distances proved that no previously found solitary wave solution[10,18,20] can describe the behaviour of pure-quartic solitons. This approximate solution, valid only for $\beta_4 < 0$, could stimulate new efforts in finding solutions to the biharmonic nonlinear Schrodinger equation[41,42] also of interest in the field of spatial solitons[23,24,43,44]. Recent interest in temporal cavity solitons in both microresonators[3] and optical fibres[7] with applications in Kerr frequency combs[45] and low-noise microwave generation[46] could also benefit from exploring pure-quartic solitons in their systems. Furthermore, it would be interesting to investigate analytic solutions supported by the pure-quartic soliton system including the effects of linear loss, TPA and FCs.

Aside from their different physical origin, pure-quartic solitons present significant potential advantages with respect to conventional solitons. As mentioned, pure-quartic solitons open the door to soliton functionality in the normal GVD regime of optical media. More importantly, perhaps, the energy of conventional solitons scales like $(T_0)^{-1}$, whereas the energy of pure-quartic solitons scales like $(T_0)^{-3}$, which suggests that they are more energetic for ultrashort pulses. We expect that the understanding of pure-quartic solitons provided in this paper, combined with the previous advances in the laser literature[25–28], will inspire a new wave of ultrafast laser development.

Methods

Device and linear characterization. The present experiment was performed using a silicon photonic crystal air-suspended structure with a hexagonal lattice (p6m symmetry group) constant $a = 404$ nm, a hole radius $r = 116$ nm, and a thickness $t = 220$ nm. A 396-µm-long dispersion engineered PhC-wg was created by removing a row of holes and shifting the two innermost adjacent rows 50 nm away from the line defect. The air-clad devices were fabricated with a combination of electron beam lithography, reactive ion and chemical wet etching. The measured linear propagation loss in this slow-light region is ~70 dB cm^{-1}, with a total linear insertion loss of ~13 dB (5 dB per facet). Light was coupled in with tapered lensed fibres to SU8 polymer waveguides with inverse tapers.

Phase-resolved characterization method. For the nonlinear experiments, we used a mode locked laser (Alnair) fed into a pulse shaper (Finisar) generating near transform-limited 1.3 ps pulses at 1,550 nm at a 30 MHz repetition rate. These pulses were then input into the FREG apparatus. The pulses were split into two branches by a fiber-coupler, with the majority of the energy coupled into the PhC-wg. The remaining fraction was sent to a reference branch with a variable delay, before being detected by a fast photodiode and transferred to the electronic domain. This electronic signal drove a Mach–Zender modulator that gated the optical pulse output from the PhC-wg. Using an optical spectrum analyser, we measured the spectra as a function of delay to generate a series of optical spectrograms. We de-convolved the spectrograms with a numerical algorithm (256 × 256 grid-retrieval errors $G < 0.005$), to retrieve the pulse intensity and the phase in both the temporal and spectral domain[32].

Generalized nonlinear Schodinger equation model. The parameters used in our GNLSE model for the slow-light dispersion engineered PhC-wg are: slow-down factor $S = n_g/n_0 = 8.64$, effective linear absorption $\alpha_{l,eff} = 13.9\,\mathrm{cm}^{-1}$; $\beta_2 = +1\,\mathrm{ps}^2\,\mathrm{mm}^{-1}$, a TOD parameter of $\beta_3 = 0.02\,\mathrm{ps}^3\,\mathrm{mm}^{-1}$, and a FOD parameter of $\beta_4 = -2.2\,\mathrm{ps}^4\,\mathrm{mm}^{-1}$; effective nonlinear parameter $\gamma_{eff} = \frac{2\pi n_2}{\lambda_0 A_{eff}} S^2 = 4{,}072(\mathrm{Wm})^{-1}$, with $n_2 = 6 \times 10^{-18}\,\mathrm{m}^2\,\mathrm{W}^{-1}$ and $A_{eff} = 0.44\mu\mathrm{m}^2$; effective TPA parameter $\alpha_{TPA,eff} = \frac{\beta_{TPA}}{A_{eff}} S^2 = 1{,}674(\mathrm{Wm})^{-1}$, with $\beta_{TPA} = 10 \times 10^{-12}\,\mathrm{m}\,\mathrm{W}^{-1}$; effective free-carrier dispersion parameter $n_{FC,eff} = -6 \times 10^{-27}\,\mathrm{S}\,\mathrm{m}^3$; effective free-carrier absorption parameter $\sigma_{eff} = 1.45 \times 10^{-21}\,\mathrm{S}\,\mathrm{m}^2$. The simulation results in Fig. 2 were obtained by using the measured input pulse as the input to the model. The simulation results in Fig. 4 were obtained using a perfect Gaussian, hyperbolic secant, and a super Gaussian (order 4) pulse of the same width as the experimental pulse, 1.3 ps. The linear loss in the nanowires that couple light into and out of the PhC-wg was negligible. Nonlinear absorption in the coupling nanowire (effective area, ~0.2 µm^2) was taken into account in the NLSE model.

References

1. Zabulsky, N. J. & Kruskal, M. D. Interactions of "solitons" in a collisionless plasma and the recurrence of initial states. *Phys. Rev. Lett.* **15,** 240–243 (1965).
2. Grelu, P. & Akhmediev, N. Dissipative solitons for mode-locked lasers. *Nat. Photon.* **6,** 84–92 (2012).
3. Herr, T. *et al.* Temporal solitons in optical microresonators. *Nat. Photon.* **8,** 145–152 (2014).
4. Cundiff, S. T. & Ye, J. Colloquium: femtosecond optical frequency combs. *Rev. Mod. Phys.* **75,** 325–342 (2003).
5. Husakou, A. V. & Herrmann, J. Supercontinuum generation of higher-order solitons by fission in photonic crystal fibers. *Phys. Rev. Lett.* **87,** 20390 (2001).
6. Dudley, J. M., Genty, G. & Coen, S. Supercontinuum generation in photonic crystal fiber. *Rev. Mod. Phys.* **78,** 1135–1185 (2006).
7. Leo, F. *et al.* Temporal cavity solitons in one-dimensional Kerr media as bits in an all-optical buffer. *Nat. Photon.* **4,** 471–476 (2010).
8. Reeves, W. H. *et al.* Transformation and control of ultra-short pulses in dispersion-engineered photonic crystal fibres. *Nature* **424,** 511–515 (2003).
9. Foster, M., Gaeta, A., Cao, Q. & Trebino, R. Soliton-effect compression of supercontinuum to few-cycle durations in photonic nanowires. *Opt. Express* **13,** 6848–6855 (2005).
10. Hasegawa, A. & Tappert, F. Transmission of stationary nonlinear optical physics in dispersive dielectric fibers i: anomalous dispersion. *Appl. Phys. Lett.* **23,** 142–144 (1973).
11. Mollenauer, L. F., Stolen, R. H., Gordon, J. P. & Tomlinson, W. J. Extreme picosecond pulse narrowing by means of soliton effect in single-mode optical fibers. *Opt. Lett.* **8,** 289–291 (1983).
12. Yin, L. & Agrawal, G.P. Impact of two-photon absorption on self-phase modulation silicon waveguides. *Opt. Lett.* **32,** 2031–2033 (2007).
13. Blanco-Redondo, A. *et al.* Controlling free-carrier temporal effects in silicon by dispersion engineering. *Optica* **1,** 299–306 (2014).
14. Blanco-Redondo, A. *et al.* Observation of soliton compression in silicon photonic crystals. *Nat. Commun.* **5,** 3160 (2014).
15. Wai, P.-K. A., Chen, H. H. & Lee, T. C. Radiations by "solitons" at the zero group-dispersion wavelength of single-mode optical fibers. *Phys. Rev. A* **41,** 426–439 (1990).
16. Höök, A. & Karlsson, M. Ultrashort solitons at the minimum-dispersion wavelength: effects of fourth-order dispersion. *Opt. Lett.* **18,** 1388–1390 (1993).
17. Blow, K. J., Doran, N. J. & Wood, D. Generation and stabilization of short soliton pulses in the amplified nonlinear Schrödinger equation. *J. Opt. Soc. Am. B* **5,** 381–389 (1988).
18. Karlsson, M. & Höök, A. Soliton-like pulses governed by fourth-order dispersion in optical fibers. *Opt. Commun.* **104,** 303–307 (1994).
19. Akhmediev, N., Buryak, A. V. & Karlsson, M. Radiationless optical solitons with oscillating tails. *Opt. Commun.* **110,** 540–544 (1994).
20. Piché, M., Cormier, J.-F & Zhu, X. Bright optical soliton in the presence of fourth-order dispersion. *Opt. Lett.* **21,** 845–847 (1996).
21. Zhakarov, V. E. & Kuznetsov, E. A. Optical solitons and quasisolitons. *J. Exp. Theor. Phys.* **86,** 1035–1045 (1998).
22. Roy, S. & Biancalana, F. Formation of quartic solitons and a localized continuum in silicon-based waveguides. *Phys. Rev. A* **87,** 025801 (2013).
23. Staliunas, K., Herrero, R. & De Valcárcel, G. J. Subdiffractive bad-edge solitons in Bose-Einstein condensates in periodic potentials. *Phys. Rev. E* **73,** 065603 (2006).
24. Staliunas, K., Herrero, R. & De Valcárcel, G. J. Arresting soliton collapse in two-dimensional nonlinear Schrödinger systems via spatiotemporal modulation of the external potential. *Phys. Rev. A* **75,** 011604 (2007).
25. Lemoff, B. E. & Barty, C. P. J. Quintic-phase-limited, spatially uniform expansion and recompression of ultrashort optical pulses. *Opt. Lett.* **18,** 1651–1653 (1993).
26. Zhou, J. *et al.* Pulse evolution in a broad-bandwidth Ti:sapphire laser. *Opt. Lett.* **19,** 1149–1151 (1994).
27. Stingl, A., Lenzner, M., Spielman, Ch. & Krausz, F. Sub-10-fs mirror-dispersion-controlled Ti:sapphire laser. *Opt. Lett.* **20,** 602–6014 (1995).
28. Christov, I., Murnane, M. N., Kapteyn, H. C., Zhou, J. & Huang, C.-P. Fourth-order dispersion-limited solitary pulses. *Opt. Lett.* **19,** 1465–1467 (1994).
29. Li, J., O'Faolain, L., Rey, I. H. & Krauss, T. F. Four-wave mixing in photonic crystals waveguides: slow light enhancement and limitations. *Opt. Express* **19,** 4460–4463 (2011).
30. Soller, B. J., Gifford, D. K., Wolfe, M. S. & Froggatt, M. E. High resolution optical frequency domain reflectometry for characterization of components and assemblies. *Opt. Express* **13,** 666–674 (2005).
31. Dorrer, C. Simultaneous temporal characterization of telecommunication optical pulses and modulators by use of spectrograms. *Opt. Lett.* **27,** 1315–1317 (2002).
32. Thomsen, B. C., Roelens, M. A. F., Watts, R. T. & Richardson, D. J. Comparison between nonlinear and linear spectrographic techniques for the complete characterization of high bit-rate pulses used in optical communications. *IEEE Phot. Tech. Lett.* **17,** 1914–1916 (2005).
33. Patterson, M. *et al.* Disorder-Induced Coherent Scattering in Slow-Light Photonic Crystal Waveguides. *Phys. Rev. Lett.* **102,** 253903 (2009).
34. Lefrancois, S., Husko, C., Blanco-Redondo, A. & Eggleton, B. J. Nonlinear silicon photonics analyzed with the moment method. *JOSA B* **32,** 218–226 (2015).
35. Ablowitz, M. J., Kaup, D. J., Newell, A. C. & Segur, H. The inverse scattering transform - Fourier analysis for nonlinear problems. *Stud. Appl. Math.* **53,** 249–315 (1974).
36. Akhmediev, N. & Karlsson, M. Cherenkov radiation emitted by solitons in optical fibers. *Phys. Rev. A.* **51,** 2602–2607 (1995).
37. Dudley, J. M, Peacock, A. C. & Millot, G. The cancellation of nonlinear and dispersive phase components on the fundamental optical fiber soliton: a pedagogical note. *Opt. Commun.* **193,** 253–259 (2001).
38. Hirano, M., Nakanishi, T., Okuno, T. & Onishi, M. Silica-based highly nonlinear fibers and their application. *IEEE J. Quant. Electron.* **15,** 103–113 (2009).
39. Castelló-Lurbe, D., Torres-Company, V. & Silvestre, E. Inverse dispersion engineering in silicon waveguides. *J. Opt. Soc. Am. B* **31,** 1829–1835 (2014).
40. Agrawal, G. P. *Nonlinear Fiber Optics* (Academic, 2007).
41. Karpman, V. I. & Shagalov, A. G. Stability of soliton described by nonlinear Schrödinger-type equations with higher-order dispersion. *Physica D* **144,** 194–210 (2000).
42. Baruch, G., Fibich, G. & Mandelbaum, E. Singular solutions of the biharmonic nonlinear Schrödinger equation. *SIAM J. Appl. Math.* **70,** 3319–3341 (2010).
43. Turitsyn, S. K. Three-dimensional dispersion nonlinearity and stability of multidimensional solitons. *Teoret. Mat. Fiz.* **64,** 226–232 (1985).
44. Cole, J. T. & Musslimani, Z. H. Band gaps and lattice solitons for the higher-order nonlinear Schrödinger equation with a periodic potential. *Phys. Rev. A* **90,** 013815 (2014).
45. Xue, X. *et al.* Mode-locked dark pulse Kerr combs in normal dispersion microresonators. *Nat. Photon.* **9,** 594–600 (2015).
46. Savchenkov, A. A. *et al.* Tunable optical frequency comb with a crystalline whispering gallery mode resonator. *Phys. Rev. Lett.* **101,** 93902 (2008).

Acknowledgements

This work was supported in part by the Center of Excellence CUDOS (CE110001018), Laureate Fellowship (FL120100029) schemes of the Australian Research Council (ARC) and by The University of Sydney and the Technion collaborative photonics research project funded by The Department of Trade and Investment, Regional Infrastructure and Services of the New South Wales Government and The Technion Society of Australia NSW. T.F.K. was supported by EPSRC UK Silicon Photonics (Grant reference EP/F001428/1). C.H. was supported by the ARC Discovery Early Career Researcher award (DECRA—DE120102069).

Author contributions

A.B.-R. and C.H. conducted the experiments. A.B.-R. performed the retrievals, developed the GNLSE model, analysed the results, contributed with physical insight to the derivation of the approximate analytic solutions, and wrote the paper. C.M.d.S. obtained the local approximate analytic solution, contributed to the analysis of the results, and proposed numerous verification tests. J.E.S. obtained the variational approximate analytic solution. T.F.K. fabricated the sample. B.J.E. provided guidance to the project and proposed the verification test for the Gaussian solution. C.H. built the set-up and contributed to analysing the results. All authors contributed in editing the paper.

Additional information

Competing financial interests: The authors declare no competing financial interests.

11

Confining energy migration in upconversion nanoparticles towards deep ultraviolet lasing

Xian Chen[1,*], Limin Jin[2,*], Wei Kong[1], Tianying Sun[1], Wenfei Zhang[2], Xinhong Liu[3], Jun Fan[1,4], Siu Fung Yu[2] & Feng Wang[1,4]

Manipulating particle size is a powerful means of creating unprecedented optical properties in metals and semiconductors. Here we report an insulator system composed of $NaYbF_4$:Tm in which size effect can be harnessed to enhance multiphoton upconversion. Our mechanistic investigations suggest that the phenomenon stems from spatial confinement of energy migration in nanosized structures. We show that confining energy migration constitutes a general and versatile strategy to manipulating multiphoton upconversion, demonstrating an efficient five-photon upconversion emission of Tm^{3+} in a stoichiometric Yb lattice without suffering from concentration quenching. The high emission intensity is unambiguously substantiated by realizing room-temperature lasing emission at around 311 nm after 980-nm pumping, recording an optical gain two orders of magnitude larger than that of a conventional Yb/Tm-based system operating at 650 nm. Our findings thus highlight the viability of realizing diode-pumped lasing in deep ultraviolet regime for various practical applications.

[1] Department of Physics and Materials Science, City University of Hong Kong, 83 Tat Chee Avenue, Hong Kong SAR, China. [2] Department of Applied Physics, The Hong Kong Polytechnic University, Hung Hom, Hong Kong SAR, China. [3] Department of Electronic Engineering, City University of Hong Kong, 83 Tat Chee Avenue, Hong Kong SAR, China. [4] City University of Hong Kong Shenzhen Research Institute, Shenzhen 518057, China. * These authors contributed equally to this work. Correspondence and requests for materials should be addressed to S.F.Y. (email: sfyu21@hotmail.com) or to F.W. (email: fwang24@cityu.edu.hk).

The construction of functional materials with designable optical properties is fundamentally important for scientific research and technological applications in diverse fields encompassing energy, environment and biomedicine[1–11]. Given the constraints in designing materials using different combinations of elements, nanoscale manipulation of matters has become a promising alternative to the creation of novel functional materials[12–19]. Particularly, by taking the advantage of size confinement effects, the energy band structure in semiconductors can be precisely modified to offer size-tunable emission wavelengths[20,21]. Despite the attractions, the size effect is largely unexplored in lanthanide-doped upconversion nanoparticles, which represents an important family of optical materials characterized by large anti-Stokes shift, narrow emission bandwidths and long excited-state lifetimes.

Photon absorption and emission in upconversion nanoparticles are due to the lanthanide dopants localized on the lattice sites[22,23]. In principle, a high concentration of lanthanide dopants enhances upconversion processes as a result of an elevated capacity to sustain excitation energy[24–27]. However, a high lanthanide content also enhances energy migration through the crystal lattice, which usually leads to a depletion of the excitation energy[24,28]. To minimize nonradiative energy losses, energy migration is typically inhibited by doping low concentrations of lanthanide ions[27,28] or by using special host lattices[24]. Currently, there lacks a general approach to maximize upconversion luminescence in stoichiometric lanthanide lattices.

In this work, we describe an investigation of energy migration in a nanosized $NaYbF_4$ lattice. We demonstrate fine tuning of energy migration through controlling the dimensions of the crystal lattice. Our mechanistic investigation reveals a spatial confinement of energy migration that prevents energy loss to the crystal lattice and increases the local density of excitation energy. Through the use of Tm^{3+} ion as an energy accumulator, the excitation energy can be maximally amassed to generate intense ultraviolet emissions on near-infrared excitation. We show that the technological advancement may revolutionize the fabrication of cost effective and compact diode-pumped solid-state deep ultraviolet lasers that are useful for environmental, life science and industrial applications[29].

Results

Synthesis and characterization. As a proof-of-concept experiment, we confined Yb^{3+} ions in the inner shell layer of a hexagonal phase $NaYF_4@NaYbF_4:Tm@NaYF_4$ host (Fig. 1a), which is known to render high upconversion efficiencies[3]. In our study, the concentration of Tm^{3+} was fixed at 1 mol% to maximize upconversion emission in the ultraviolet region (Supplementary Fig. 1). We did not employ a $NaYbF_4:Tm@NaYF_4$ core–shell structure because existing synthetic protocols give essentially no access to sub-10 nm β-$NaYbF_4$ nanoparticles of a tunable particle size, which is critical prerequisite for assessing the effect of confining energy migration on upconversion. Although Yb-doped β-$NaYF_4$ nanoparticles with small feature size can be synthesized by several complimentary methods[23,30], pure β-$NaYbF_4$ tends to form big particles (Supplementary Fig. 2a) due to rapid growth of the crystal[31]. Through the use of preformed $NaYF_4$ core nanoparticle as a template, the growth of the $NaYbF_4$ crystal can be effectively regulated, thereby offering exquisite control over the lattice dimensions (Supplementary Note 1). Note that it is also critical to enclose the Yb sublattice in an inert protection layer (that is, $NaYF_4$) because a $NaYF_4@NaYbF_4$:Tm core–shell structure yields luminescence that is substantially weak due to surface quenching (Supplementary Fig. 2b,c).

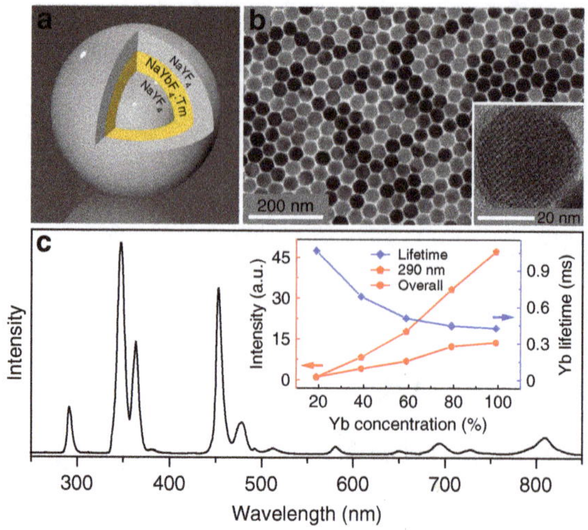

Figure 1 | Deep ultraviolet upconversion in core-shell-shell nanoparticles. (**a**) Schematic design of a $NaYF_4@NaYbF_4:Tm@NaYF_4$ core-shell-shell nanoparticle for confining the migration of excitation energy generated in the Yb^{3+} ions. (**b**) TEM image of the as-synthesized nanoparticles. Inset: high-resolution TEM image reveals single-crystalline nature of the particle. (**c**) Upconversion emission spectrum of the nanoparticles under 980 nm excitation (CW laser diode, 20 W cm^{-2}). Inset: $^2F_{5/2}$ lifetime of Yb^{3+}, emission intensity at 290 nm and integrated emission intensity over 250–850 nm range versus dopant concentration of Yb^{3+}, respectively. Note that the solid lines are intended to guide the eye.

The nanoparticles were fabricated by a layer-by-layer epitaxial growth process (Supplementary Note 1). Transmission electron microscope (TEM) images (Fig. 1b) reveal a highly uniform morphology of the nanoparticles with an average size of 38 nm. High-resolution TEM (inset of Fig. 1b) and X-ray powder diffraction (Supplementary Fig. 3) experiments confirm the single-crystalline nature of the as-synthesized nanoparticles with a hexagonal phase. To verify the formation of the multilayer structure, we intentionally doped Gd^{3+} ions in the inner shell layer to create a contrast under electron energy loss spectroscopy analysis. The difference in the elemental distribution of Y and Gd clearly indicates the presence of multiple core–shell interfaces (Supplementary Fig. 4).

Figure 1c shows a representative upconversion emission spectrum of $NaYF_4@NaYbF_4:Tm@NaYF_4$ nanoparticles on 980 nm excitation with a continuous wave (CW) laser diode at a power density of 20 W cm^{-2}. The spectrum consists of characteristic emission peaks that can be assigned to $^1I_6 \rightarrow ^3H_6$ and $^3F_4$ (290 and 350 nm), $^1D_2 \rightarrow ^3H_6$ and $^3F_4$ (360 and 450 nm), $^1G_4 \rightarrow ^3H_6$ and $^3F_4$ (475 and 650 nm) and $^3H_4 \rightarrow ^3H_6$ (800 nm) transitions of Tm^{3+}, respectively. Both the violet and overall emissions surpass that of the $NaYF_4@NaYbF_4:Tm/Y@NaYF_4$ counterparts comprising lower Yb^{3+} contents (inset of Fig. 1c and Supplementary Figs 5 and 6). Notably, Tm^{3+} emission at 290 nm originating from a five-photon upconversion declined by over 45-fold when the Yb^{3+} concentration dropped to 19 mol%, which in conventional systems typically produces the maximum emission of Tm^{3+} ions[24,28].

Confinement of energy migration. We attribute the observations to confined migration of excitation energy within the nanoshells, which prevents the excitation energy from travelling a long distance at a high Yb^{3+} concentration (99 mol%). The absence of long-distance energy migration is likely to suppress energy loss to the crystal lattice accounting for luminescence quenching. Furthermore, the localization of excitation energy raises the rate of energy transfer to a nearby Tm^{3+} activator, which facilitates the multiphoton upconversion process.

An assessment of a series of $NaYF_4$@$NaYbF_4$:Tm@$NaYF_4$ nanoparticles of varying inner shell thickness from 1 to 17 nm verified the spatial confinement of energy migration (Supplementary Fig. 7). Luminescence decay studies reveal a markedly lengthened lifetime of the Yb^{3+} by a factor of over nine with decreasing inner shell thickness from 17 to 1 nm (Supplementary Fig. 8a), confirming the suppression of energy loss to the host lattice in thin shells. In contrast, the decay time of localized Tm^{3+} transition was only increased by less than twofold for the same series of samples (Supplementary Fig. 8b), suggesting that the defect density in the host lattice were marginally modified. Therefore, the suppressed depletion of excitation energy of Yb^{3+} may be dominantly ascribed to the spatial confinement of energy migration, which reduces the quantity of defects accessible to the Yb sublattice. In line with the reduced energy loss, we observed a steady enhancement of upconversion emission, especially the part in the 280 – 356 nm range that originates from the five-photon process, accompanied by a decrease of the inner shell thickness (Fig. 2a; Supplementary Fig. 9). The slight drop in the ratio of five-photon upconversion to overall emission for substantially thin shells (that is, 1 and 2 nm) can be attributed to the reduced amount of Yb^{3+} ions in the vicinity of a Tm^{3+} activator (Fig. 2b), which limits the quantity of energy that can be captured by a Tm^{3+} activator in a photo cycle. It is noted that the spatial confinement of energy migration also plays a role in cubic phase $NaYbF_4$:Tm (1%)@$NaYF_4$ core–shell nanoparticles (Supplementary Fig. 10)[32], implying that the geometry of the host lattice is not a formative factor in restraining the migration of excitation energy.

To shed more light on energy migration in the core–shell–shell nanostructure, we calculated the probability distribution function of excitation energy as a function of space within the Yb shell. For simplification, we assumed that the excitation energy randomly hops in the inner shell layer through the Dexter energy transfer (Supplementary Note 2). As shown in Fig. 2c, the energy migrates to smaller areas with decreasing thickness of the Yb shell from 12 to 6 and 3 nm, supporting reduced coupling of excitation energy to defects. The high probability of finding the excitation energy in a thin Yb shell further validates a favourable energy transfer to an adjacent Tm^{3+} activator.

Lasing through upconversion. To facilitate the use of the upconversion nanoparticles as gain media for lasing applications, we further developed a Gd^{3+} doping method for optimizing the optical properties (Supplementary Fig. 11). We used a 3-nm inner shell for the study due to a high-intensity ratio of five-photon upconversion emission and a relatively low mass ratio of the optically inert $NaYF_4$ layers. Gd^{3+} ions are able to extract the excitation energy of Tm^{3+} ions and generate a new emission peak centred at around 311 nm, owing to the reasonably matched energy levels (that is, $^6P_{7/2}$ level of Gd^{3+} and $^1I_6$ level of Tm^{3+}) (Fig. 3a,b). Importantly, the large energy gap (32,200 cm^{-1}) in Gd^{3+} favours the preservation of the excitation energy as supported by time decay studies (Fig. 3c). The long-lived excited state contributes to high optical gains of around 150 cm^{-1} through five-photon upconversion (Fig. 3d), which is comparable to that of the GaN-based semiconductor quantum wells operating in deep ultraviolet at room temperature[33,34]. The optical gain is also two orders of magnitude higher than that of a conventional Yb/Tm-based system operating at 650 nm through three-photon upconversion[35]. Moreover, efficient emission is attained at a high Gd^{3+} concentration (30 mol%; Supplementary Fig. 12), which provides abundant

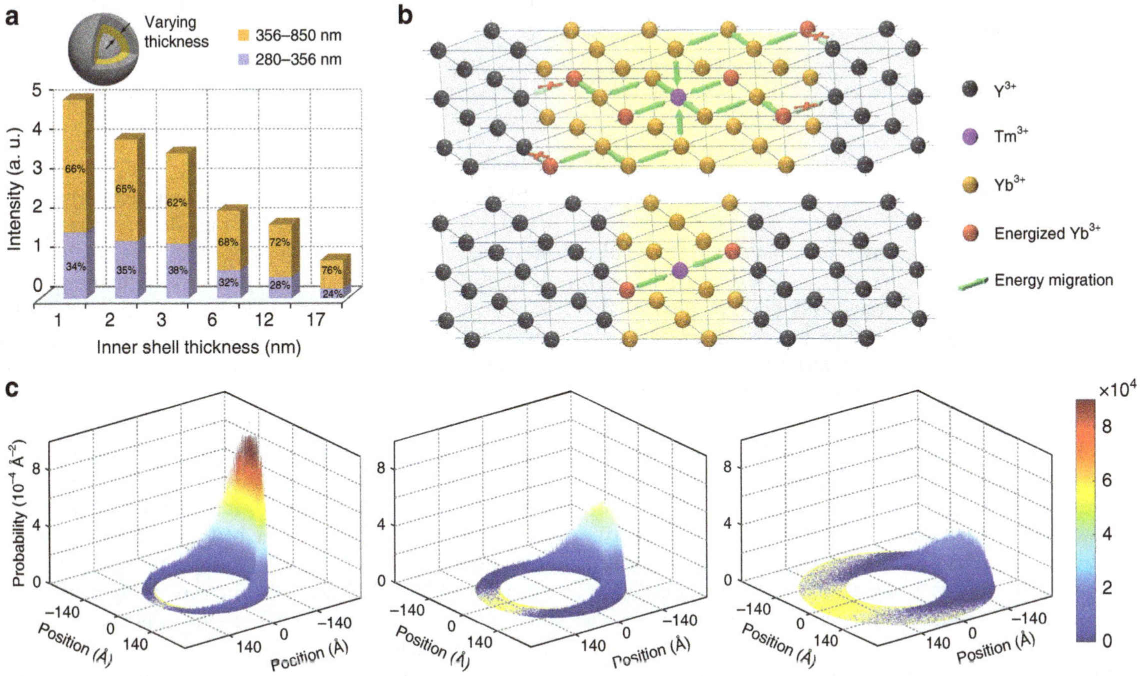

Figure 2 | The effect of inner shell thickness on upconversion. (**a**) Upconversion emission intensity versus inner shell thickness (1-17 nm). The emission intensities were calculated by integrating the spectral intensity of the emission spectra that are normalized to the absorption of Yb^{3+} at 980 nm. (**b**) Schematic illustration showing proposed energy transfer from Yb^{3+} to Tm^{3+} in Yb-sublattice of varying dimensions. Note that only partial lattice sites are shown for clarification. (**c**) The probability of finding the excitation energy on the equatorial section of core-shell-shell nanoparticles of varying inner shell thickness. With increasing inner shell thickness (from left to right panels), the energy migrates to a larger area and the probability of finding the excitation energy in the vicinity of the starting point drops significantly.

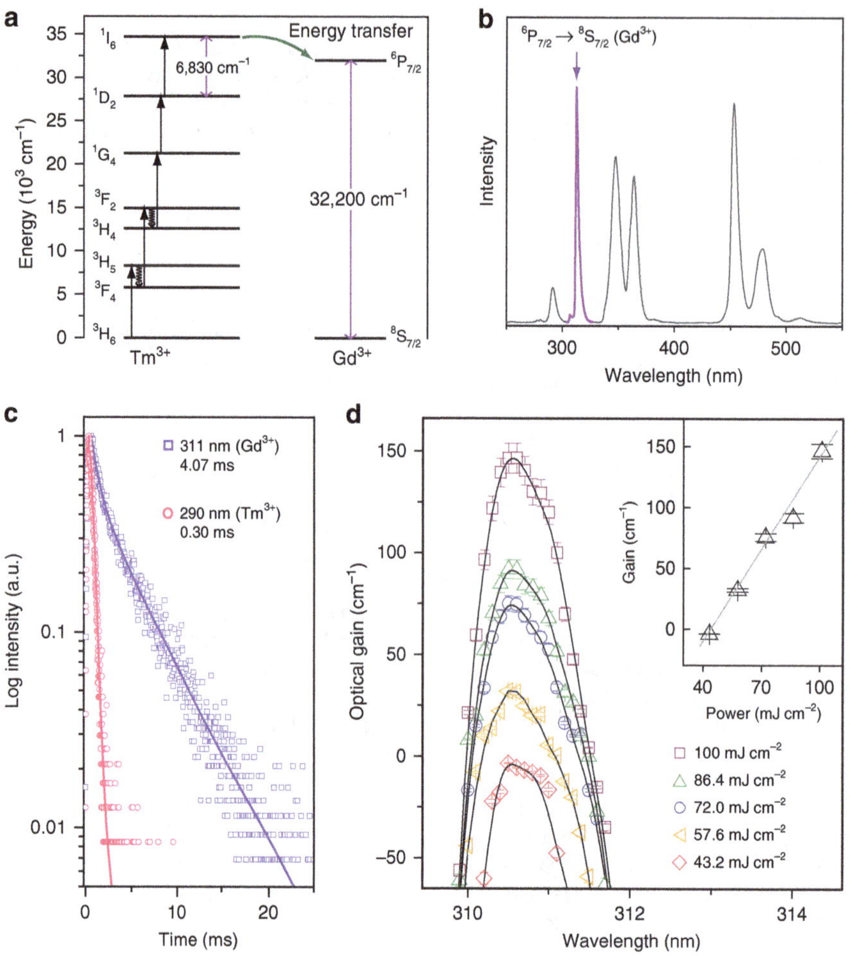

Figure 3 | The effect of Gd^{3+} doping on the deep ultraviolet upconversion. (**a**) Simplified energy level diagram showing the energy gaps in Tm^{3+} and Gd^{3+} activators, respectively. (**b**) Upconversion emission spectra of the core-shell-shell nanoparticles co-doped Gd^{3+} (30 mol%) in the inner shell layer (CW laser diode, 20 W cm^{-2}). (**c**) A comparison of the excited state lifetime between $^1I_6$ state of Tm^{3+} and $^6P_{7/2}$ state of Gd^{3+} in the $NaYF_4$@$NaYbF_4$:Tm/Gd (1/30%)@$NaYF_4$ core-shell-shell nanoparticles. (**d**) Gain spectra of the nanoparticles in **b** as a function of excitation power (pulse laser). The inset gives the corresponding optical gain versus pump power at a wavelength of 310.5 nm. The straight line is the linear regression of the measured data. Error bars shown represent the s.d.'s from five sets of repeated measurements.

carriers to sustain optical gains at high excitation powers without saturation (Fig. 3d).

To realize lasing emission, we constructed a five-pulse pumping scheme to excite the upconversion process (Fig. 4a). The pulse excitation scheme is primarily intended to alleviate the problems of catastrophic optical damage and the thermal effects associated with CW excitation, which terminates upconversion lasing actions. Furthermore, the five-pulse system is advantageous over the single-pulse system for pumping the multiphoton upconversion (Supplementary Fig. 13), as a result of improved alignment with the excitation process where the absorption of photons occurs sequentially[36].

The laser cavity was fabricated by coating a drop of silica resin containing the nanoparticles onto a standard optical fibre. Driven by surface tension, the silica resin tends to form a bottle-like microresonator, which supports whispering gallery modes at a thin equatorial ring near the surface of the microresonator (Fig. 4b). Notably, the emission features such as mode spacing and threshold pump power of the microresonator can be readily tuned by controlling the diameter (D_m) of the resonator (Fig. 4b)[37,38], which provides a general platform for assessing the optical characteristics of the upconversion nanoparticles.

We validated the lasing action by measuring the optical emission in a typical microresonator ($D_m = 75\,\mu m$) under excitation of varying powers at room temperature. The light input–output curve shown in Fig. 4c exhibits a well-defined nonlinear excitation power-dependent behaviour with three distinct regions separated by two threshold pump powers (that is, $P_a = \sim 78$ mJ cm^{-2} and $P_{th} = \sim 86$ mJ cm^{-2}). This *S*-like spectrum clearly indicates a transition from a spontaneous emission to an amplified spontaneous emission and to a lasing emission. Figure 4d shows the corresponding emission spectra under various pumping powers. At low pump power ($<P_a$), a relatively broad spontaneous band is observed. As the excitation power increases slightly above P_a, a sharp peak ascends from the emission spectrum. Through further increases in the excitation power above P_{th}, well-defined sharp peaks with a linewidth <0.11 nm emerge from the spectrum. The measured mode spacing ($0.25 - 0.27$ nm) is in good agreement with the theoretical value (~ 0.26 nm), confirming that lasing emissions have been achieved. Notably, single-mode lasing was also obtained by using a thinner microresonator ($D_m = 20\,\mu m$; Fig. 4e). The single-mode emission is a result of a relatively large mode spacing (~ 1.0 nm) with respect to the full width at half maximum of the resonant frequency. This narrow gain bandwidth is a unique signature of the upconversion nanoparticles and unlikely to be realized from semiconductor nanostructures. The *Q* factor, which is defined as the ratio of

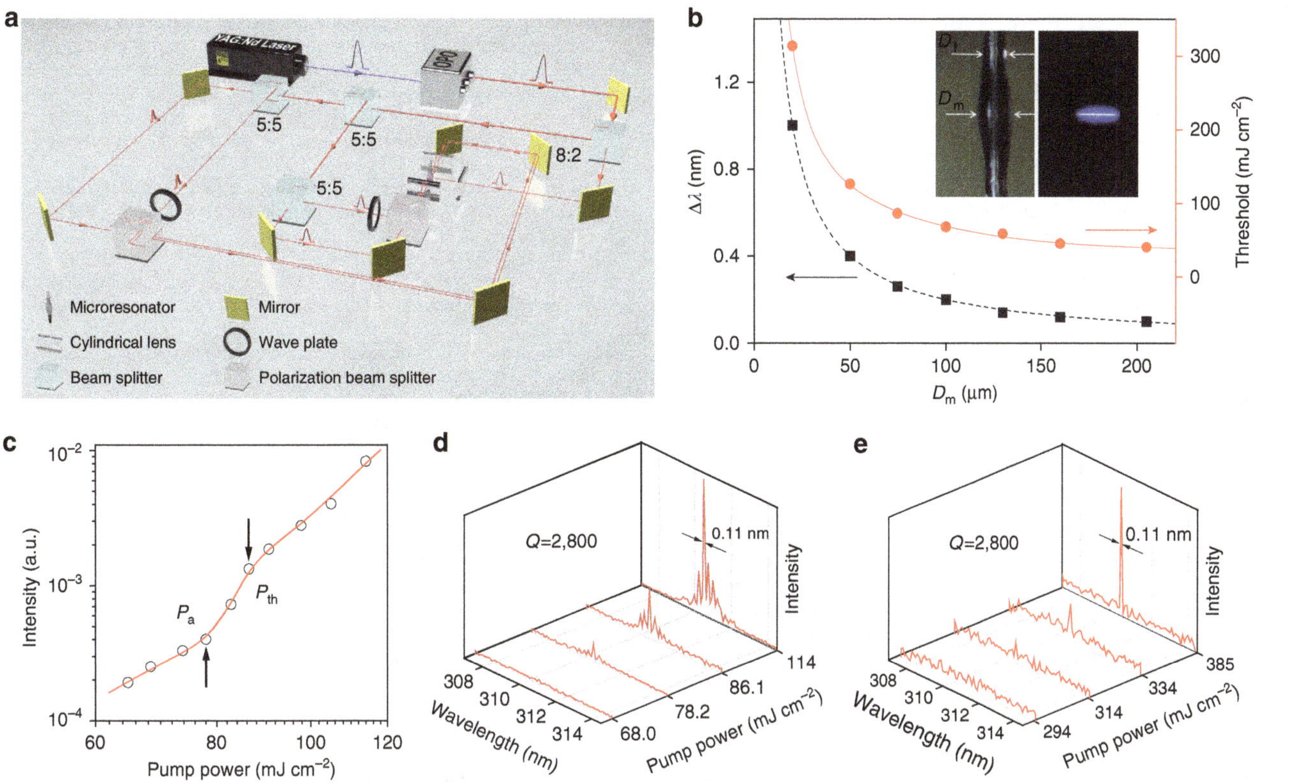

Figure 4 | Upconversion lasing characteristics of the microresonator. (**a**) Schematic diagram of the optical set-up for the five-pulse excitation scheme. (**b**) Plots of measured $\Delta\lambda$ and P_{th} of the microresonator as a function of D_m. The red and black lines are fitted and calculated curves, respectively. The insets show photographs of the microresonator with and without optical excitation. D_f and D_m denote the diameters of the fibre and microcavity, respectively. (**c**) Logarithmic plot of output intensity versus excitation power of a microresonator with $D_m = 75\,\mu m$. The red line is fitted curve. (**d**) The corresponding lasing spectra at different excitation power ($D_m = 75\,\mu m$). (**e**) Single mode lasing spectra measured from a microresonator with $D_m = 20\,\mu m$.

the resonant frequency to its full width at half maximum, was estimated to be ~2,800, revealing the high quality of the upconversion-based laser system. It is also worth noting that the lasing emission can be readily extended to violet and blue spectral regions with the same upconversion nanoparticles (Supplementary Fig. 14).

Discussion

Our investigation of energy migration in nanostructured hosts highlights an innovative strategy to manipulating optical transitions in lanthanide-doped upconversion nanoparticles. In addition, it initiates a novel tactic to obtain effective upconversion laser materials in deep ultraviolet regime with very narrow optical gain bandwidth to support single-mode excitation. Emission, in principle, can be tuned to shorter wavelengths (that is, well below 300 nm) by further refining the upconversion process. Hence, our study will lead to the development of near-infrared diode-pumped deep ultraviolet lasers which can avoid the difficulty of shifting the operating wavelength of GaN-based laser diodes below 300 nm (ref. 39), evade using nonlinear optical crystal that requires tight control in optical alignment, antireflective coating and environmental control and adopt inexpensive Q-switched near-infrared diode as the pumping source to construct compact, deep ultraviolet lasers for unexplored applications in the fields of information technology, biomedicine and biophotonics.

Methods

Nanoparticle synthesis. We synthesized the core–shell–shell nanoparticles using the method described in ref.16. Additional experimental details are provided in the Supplementary Note 1.

Optical gain measurement. Net optical gain of the nanoparticle colloid was measured using variable stripe length method[40]. The longer side of a quartz cuvette filled with nanoparticle colloid was excited by a pump stripe with width and length of ~30 µm and L, respectively in the orientation perpendicular to the length of the cuvette. Photoluminescence intensity emitted from the shorter side of the cuvette, $I_{tot}(\lambda)$, was recorded by the monochromator set-up. The net optical gain, $G(\lambda)$, was deduced by fitting $I_{tot}(L, \lambda) = I_{sp}(\lambda)\,[\exp(G(\lambda)L) - 1]/G(\lambda)$ with the measured values of $I_{tot}(\lambda)$, where λ is the wavelength and $I_{sp}(\lambda)$ is the spontaneous emission intensity.

Five-pulse excitation scheme. The optical set-up consists of a Powerlite DLS 9010 Q-switched Nd:YAG laser and a continuum Panther EX optical parametric oscillator. A 980-nm laser pulse (6 ns, 10 Hz) with a diameter of ~10 mm was generated from the optical parametric oscillator system under the excitation of the Nd:YAG laser. By splitting the 980-nm pulse into five equal-power pulses through the use of four beam splitters (that is, one 80/20 and three 50/50 beam splitters), we can obtain a five-pulse (time delay between adjacent pulses is 10 ns) laser beam. This is possible because the pulses are forced to travel in five different distances to obtain a time delay of 10 ns between the adjacent pulses. These five pulses are then combined through two polarization-dependent beam splitters and two $\lambda/2$ waveplates to form three laser beams. All the laser beams, which are spatially overlapped, are focussed onto a sample through three cylindrical lenses to form a pump stripe of width equal to ~30 µm. Photoluminescence emission from the sample was collected and analysed by an optical fibre (core diameter of 400 µm) coupled to an Oriel MS257 monochromator attached with a photomultiplier tube. The spectral resolution of the monchromator is about 0.1 nm.

Fabrication and excitation of microresonators. For the fabrication of the bottle-like microresonator, a bared standard optical fibre was coated with a tiny drop of nanoparticles and silica resin mixture. The prolate surface-tention-induced microresonator was then solidified in an arid and clean atmosphere. The surrounding temperature of the sample was kept at 23 °C to avoid deformation due to the influence of thermal effects. Whispering gallery modes can be excited by optically pumping the equatorial zone of the microresonator. Notably, the pump stripe is oriented perpendicular to the length of the fibre. Light emitted from the surface of the microresonator can be collected through an optical fibre.

References

1. Yu, J. H. *et al.* High-resolution three-photon biomedical imaging using doped ZnS nanocrystals. *Nat. Mater.* **12,** 359–366 (2013).
2. Astruc, M.-C. & Astruc, D. Gold nanoparticles: assembly, supramolecular chemistry, quantum-size-related properties, and applications toward biology, catalysis, and nanotechnology. *Chem. Rev.* **104,** 293–346 (2004).
3. Haase, M. & Schafer, H. Upconverting nanoparticles. *Angew. Chem. Int. Ed.* **50,** 5808–5829 (2011).
4. Hola, K. *et al.* Carbon dots–emerging light emitters for bioimaging, cancer therapy and optoelectronics. *Nano Today* **9,** 590–603 (2014).
5. Chen, X., Liu, L., Yu, P. Y. & Mao, S. S. Increasing solar absorption for photocatalysis with black hydrogenated titanium dioxide nanocrystals. *Science* **331,** 746–750 (2011).
6. Carter, K. P., Young, A. M. & Palmer, A. E. Fluorescent sensors for measuring metal ions in living systems. *Chem. Rev.* **114,** 4564–4601 (2014).
7. Idris, N. M., Jayakumar, M. K., Bansal, A. & Zhang, Y. Upconversion nanoparticles as versatile light nanotransducers for photoactivation applications. *Chem. Soc. Rev.* **44,** 1449–1478 (2015).
8. Su, L. T. *et al.* Photon upconversion in hetero-nanostructured photoanodes for enhanced near-infrared light harvesting. *Adv. Mater.* **25,** 1603–1607 (2013).
9. Yang, D. *et al.* Current advances in lanthanide ion (Ln^{3+})-based upconversion nanomaterials for drug delivery. *Chem. Soc. Rev.* **44,** 1416–1448 (2015).
10. Liu, J. *et al.* Metal-free efficient photocatalyst for stable visible water splitting via a two-electron pathway. *Science* **347,** 970–974 (2015).
11. Sang, Y. *et al.* From UV to near-infrared, WS_2 nanosheet: a novel photocatalyst for full solar light spectrum photodegradation. *Adv. Mater.* **27,** 363–369 (2015).
12. Alivisatos, A. Semiconductor clusters, nanocrystals, and quantum dots. *Science* **271,** 933–937 (1996).
13. Novoselov, K. S. *et al.* Novoselov-electric field effect in atomically thin carbon films. *Science* **306,** 666–669 (2004).
14. Deng, R. *et al.* Temporal full-colour tuning through non-steady-state upconversion. *Nat. Nanotechnol* **10,** 237–242 (2015).
15. Sun, Y., Zhu, X., Peng, J. & Li, F. Core–shell lanthanide upconversion nanophosphors as four-modal probes for tumor angiogenesis imaging. *ACS Nano* **7,** 11290–11300 (2013).
16. Wen, H. *et al.* Upconverting near-infrared light through energy management in core-shell-shell nanoparticles. *Angew. Chem. Int. Ed.* **52,** 13419–13423 (2013).
17. Sun, L. D., Wang, Y. F. & Yan, C. H. Paradigms and challenges for bioapplication of rare earth upconversion luminescent nanoparticles: small size and tunable emission/excitation spectra. *Acc. Chem. Res.* **47,** 1001–1009 (2014).
18. Li, X., Zhang, F. & Zhao, D. Lab on upconversion nanoparticles: optical properties and applications engineering via designed nanostructure. *Chem. Soc. Rev.* **44,** 1346–1378 (2015).
19. Tu, L., Liu, X., Wu, F. & Zhang, H. Excitation energy migration dynamics in upconversion nanomaterials. *Chem. Soc. Rev.* **44,** 1331–1345 (2015).
20. Rossetti, R., Nakahara, S. & Brus, L. E. Quantum size effects in the redox potentials, resonance Raman spectra, and electronic spectra of CdS crystallites in aqueous solution. *J. Chem. Phys.* **79,** 1086 (1983).
21. Bruchez, M., Moronne, M., Gin, P., Weiss, S. & Alivisatos, A. P. Semiconductor nanocrystals as fluorescent biological labels. *Science* **281,** 2013–2016 (1998).
22. Wu, S. *et al.* Non-blinking and photostable upconverted luminescence from single lanthanide-doped nanocrystals. *Proc. Natl Acad. Sci. USA* **106,** 10917–10921 (2009).
23. Gargas, D. J. *et al.* Engineering bright sub-10-nm upconverting nanocrystals for single-molecule imaging. *Nat. Nanotechnol* **9,** 300–305 (2014).
24. Wang, J. *et al.* Enhancing multiphoton upconversion through energy clustering at sublattice level. *Nat. Mater.* **13,** 157–162 (2013).
25. Punjabi, A. *et al.* Amplifying the red-emission of upconverting nanoparticles for biocompatible clinically used prodrug-induced photodynamic therapy. *ACS Nano* **8,** 10621–10630 (2014).
26. Shen, J. *et al.* Tunable near infrared to ultraviolet upconversion luminescence enhancement in (α-$NaYF_4$:Yb,Tm)/CaF_2 core/shell nanoparticles for in situ real-time recorded biocompatible photoactivation. *Small* **9,** 3213–3217 (2013).
27. Mahalingam, V., Vetrone, F., Naccache, R., Speghini, A. & Capobianco, J. A. Colloidal Tm^{3+}/Yb^{3+}-doped $LiYF_4$ nanocrystals: multiple luminescence spanning the UV to NIR regions via low-energy excitation. *Adv. Mater.* **21,** 4025–4028 (2009).
28. Krämer, K. W. *et al.* Hexagonal sodium yttrium fluoride based green and blue emitting upconversion phosphors. *Chem. Mater.* **16,** 1244–1251 (2004).
29. Boardman, E. A. *et al.* Deep ultraviolet (UVC) laser for sterilisation and fluorescence applications. *Sharp Tech. Rep.* **104,** 31–35 (2012).
30. Ostrowski, A. D. *et al.* Controlled synthesis and single-Particle imaging of bright, sub-10 nm lanthanide-doped upconverting nanocrystals. *ACS Nano* **6,** 2686–2692 (2012).
31. Wang, F. *et al.* Simultaneous phase and size control of upconversion nanocrystals through lanthanide doping. *Nature* **463,** 1061–1065 (2010).
32. Chen, G., Ohulchanskyy, T. Y., Kumar, R., Ågren, H. & Prasad, P. N. Ultrasmall monodisperse $NaYF_4$:Yb^{3+}/Tm^{3+} nanocrystals with enhanced near-infrared to near-infrared upconversion photoluminescence. *ACS Nano* **4,** 3163–3168 (2010).
33. Witzigmann, B. L. *et al.* Analysis of temperature-dependent optical gain in GaN-InGaN quantum-well structures. *IEEE Photon. Technol. Lett* **18,** 1600–1602 (2006).
34. Pecora, E. F., Sun, H., Negro, L. D. & Moustakas, T. D. Deep-UV optical gain in AlGaN-based graded-index separate confinement heterostructure. *Opt. Mater. Express* **5,** 809–817 (2015).
35. Haro-González, P., Martín, I. R., Lahoz, F. & Capuj, N. E. Optical gain by upconversion in Tm–Yb oxyfluoride glass ceramic. *Appl. Phys. B* **104,** 237–240 (2010).
36. Zhu, H. *et al.* Amplified spontaneous emission and lasing from lanthanide-doped up-conversion nanocrystals. *ACS Nano* **7,** 11420–11426 (2013).
37. Vahala, K. J. Optical microcavities. *Nature* **424,** 839–846 (2003).
38. Pöllinger, M., O'Shea, D., Warken, F. & Rauschenbeutel, A. Ultrahigh-Q tunable whispering-gallery-mode microresonator. *Phys. Rev. Lett.* **103,** 053901 (2009).
39. Yoshida, H., Yamashita, Y., Kuwabara, M. & Kan, H. A 342-nm ultraviolet AlGaN multiple-quantum-well laser diode. *Nat. Photon* **2,** 551–554 (2008).
40. Valenta, J., Pelant, I. & Linnros, J. Waveguiding effects in the measurement of optical gain in a layer of Si nanocrystals. *Appl. Phys. Lett.* **81,** 1396 (2002).

Acknowledgements

This work was supported by the Research Grants Council of Hong Kong (CityU 109413, 21300014 and PolyU 153036/14P) and the National Natural Science Foundation of China (Nos. 21303149, 51332008 and 21403182).

Author contributions

X.C., L.J., S.F.Y. and F.W. conceived the project and wrote the paper. X.C., W.K., T.S. and F.W synthesized and characterized the nanoparticles. L.J., W.Z. and S.F.Y. fabricated and tested the lasers. X.L. and J.F. solved the probability of finding the excitation energy. All authors contributed to the analysis of this manuscript.

Additional information

Competing financial interests: The authors declare no competing financial interests.

12

Giant photostriction in organic-inorganic lead halide perovskites

Yang Zhou[1,*], Lu You[1,*], Shiwei Wang[1], Zhiliang Ku[2], Hongjin Fan[2], Daniel Schmidt[3], Andrivo Rusydi[3], Lei Chang[1], Le Wang[1], Peng Ren[1], Liufang Chen[4], Guoliang Yuan[4], Lang Chen[5] & Junling Wang[1]

Among the many materials investigated for next-generation photovoltaic cells, organic-inorganic lead halide perovskites have demonstrated great potential thanks to their high power conversion efficiency and solution processability. Within a short period of about 5 years, the efficiency of solar cells based on these materials has increased dramatically from 3.8 to over 20%. Despite the tremendous progress in device performance, much less is known about the underlying photophysics involving charge-orbital-lattice interactions and the role of the organic molecules in this hybrid material remains poorly understood. Here, we report a giant photostrictive response, that is, light-induced lattice change, of >1,200 p.p.m. in methylammonium lead iodide, which could be the key to understand its superior optical properties. The strong photon-lattice coupling also opens up the possibility of employing these materials in wireless opto-mechanical devices.

[1] School of Materials Science and Engineering, Nanyang Technological University, Block N4.1-02-24, 50 Nanyang Avenue, Singapore 639798, Singapore. [2] School of Physical and Mathematical Sciences, Nanyang Technological University, Singapore 639798, Singapore. [3] Singapore Synchrotron Light Source, National University of Singapore, 5 Research Link, Singapore 117603, Singapore. [4] Department of Materials Science and Engineering, Nanjing University of Science and Technology, Nanjing 210094, China. [5] Department of Physics, South University of Science and Technology of China, Shenzhen 518055, China. * These authors contributed equally to this work. Correspondence and requests for materials should be addressed to L.Y. (email: mailyoulu@gmail.com) or to J.W. (email: jlwang@ntu.edu.sg).

The past few years witnessed the explosion of research on photovoltaic cells based on the hybrid organic–inorganic perovskites[1–5], in particular methylammonium lead iodide ($MAPbI_3$). Concomitantly, these materials have also been explored for lasers[6], light emitting diodes[7] and photodetectors[8]. Besides improving the photovoltaic cell efficiency, much work has been devoted to the mechanism behind their extraordinary performances. Anomalously long lifetime and diffusion length of photo-carriers have been observed and related to the high efficiency[9–12]. In solution processed thin films of mixed halide perovskites, carrier lifetime of longer than 1 μs[13] and diffusion length exceeding 1 μm[11] were reported. These values can be further increased in high-quality $MAPbI_3$ single crystals with greatly suppressed trap-state densities[14,15], suggesting an even higher attainable efficiency approaching the Shockley–Queisser limit[16]. However, to become commercially viable, the long-term stability of these materials has to be significantly improved. Despite the intensive research efforts, the origin of the long-carrier lifetime and diffusion length remains elusive. The bi-molecular recombination rate deviates from Langevin theory by orders of magnitude, which underlines the unique attribute of hybrid perovskites compared with conventional low-mobility semiconductors, but is reminiscent of disordered material systems such as amorphous Si and organic solar cells[17,18]. These clues lead us to ponder on the role of the organic molecules on the material's unusual photophysical properties. These molecules are dynamically tumbling inside the inorganic scaffold due to the small rotational energy barrier[19,20]. Even though the organic group is not directly involved in the electronic structure around the band edges, it may interact with the inorganic PbI_6 octahedron through its rotational degree of freedom, as revealed in recent density functional theory (DFT) calculations[21,22]. Experimentally, hydrogen bonding between the halides and the amine group at room temperature was confirmed by infrared spectroscopy[23], supporting the theoretical results.

Here we provide yet another evidence for the interaction between the organic and inorganic moieties in the hybrid perovskites. A giant photostrictive response, namely, light-induced lattice change, of $>$1,200 p.p.m. was observed in $MAPbI_3$. Careful analysis suggests that the strong photon-lattice coupling may arise from the weakening of the hydrogen bonding between N and I by the photo-generated carriers. Not only could it shed light on the anomalous photophysics, this discovery also opens up new possibilities for these fascinating materials, enabling novel device paradigms such as photo-driven microsensing and microactuation[24,25].

Results

Photostriction in $MAPbI_3$ single crystals. $MAPbI_3$ single crystals as large as $10 \times 8 \times 8$ mm are prepared using solution growth method. Photograph of a typical sample with well-defined facets used in this study is shown in Fig. 1a. Only those peaks corresponding to the tetragonal (*l*00) planes (Fig. 1b) are present in the X-ray diffraction scan (Fig. 1e), confirming the high quality of the single crystal. Based on the absorption coefficient deduced from the spectroscopic ellipsometry (Supplementary Fig. 1), an abrupt excitonic absorption at 1.575 eV has been determined (Fig. 1f), which is consistent with previous reports[14,15]. The topography and photo-induced dimension change are investigated using an atomic force microscope (AFM) over the atomically flat surface of the sample as shown in Fig. 1c,d.

Interestingly, when light is shining on the crystal, a sudden change in the dimension is observed. Figure 2a schematically depicts the experimental set-up for the measurements. Single crystals with various facets facing up are glued on the glass substrates using silver paint. The crystals are then illuminated from top surface using a halogen lamp with a continuous spectrum ranging from 400 to 750 nm (Supplementary Fig. 2). The light goes through the AFM built-in optical system onto the surface of the sample. By placing the AFM tip at the surface, the sample height is recorded as a function of time and illumination conditions (see Methods). As shown in Fig. 2b, under 100 mW cm^{-2} white-light illumination, a reproducible change in the sample height is clearly observed. Both $(100)_T$- and $(010)_T$-oriented (subscript T denotes tetragonal index) single crystals produce a similar elongation of approximately 50 nm along the vertical direction. Considering the crystal thickness of approximately 1 mm, this translates into a photostriction (defined as the height change divided by the sample thickness $\Delta H/H$) of 5×10^{-5} (or 50 p.p.m.). To confirm that the observed height change is an intrinsic material property instead of a light-induced measurement artifact, we also tested Si and $SrTiO_3$ (STO) single crystals (Fig. 2b and Supplementary Fig. 3a), both of which should show negligible photostrictive effects. Instead of a sudden change of height, both samples exhibit a much slower response, whose magnitude scales with the illumination time. This suggests a possible thermal effect. However, the very different optical absorption and thermal expansion coefficients of Si and STO are at odds with the similar responses. Furthermore, this slow response can also be seen for $MAPbI_3$ as indicated in Fig. 2b. Thus, we infer that this material-independent response is a measurement artifact that results from the heating-induced bending of the AFM tip rather than the samples under test. To exclude the possible contribution from the thermal expansion of the sample, the temperature of the sample surface was monitored under the same illumination conditions (Supplementary Fig. 3b). Clearly, the temperature profile does not match the sudden height change. Besides, a brief estimation based on the material parameters of $MAPbI_3$ gives a thermal expansion on the order of 0.1 nm (Supplementary Note 1), much smaller than the response observed.

Further investigation shows that the photostrictive response is proportional to the light intensity (Fig. 2c), and no saturation is observed up to 100 mW cm^{-2}, which suggests a possible correlation between the photostriction and photo-generated carriers. To check the photoconductivity of the sample, we have measured the current under the same illumination condition by applying 1 V bias to the Pt electrodes coated on the two opposite facets of the crystal. As shown in Fig. 2d, the sample height change follows the same profile as the current change on illumination. Since the conductivity is proportional to the amount of free carriers (provided there is no significant change in the carrier mobilities), it implies that the photostriction is directly related to the photo-generated carriers, that is, of an electronic origin. The long tail of the height signal, after the light is turned off, is attributed to the slow thermal relaxation of the AFM tip. This is supported by the fact that it scales with the illumination time (Fig. 2b).

Photon energy dependence. If the photo-generated free carriers are indeed responsible for the dimension change, it must depend on the incident photon energy. As revealed by the energy-dependent absorption coefficient (Fig. 1f), the excitonic absorption of $MAPbI_3$ single crystal is 1.575 eV with the true band gap being a few tens of meV above that. When the lasers with photon energy higher than or close to the band gap of $MAPbI_3$ are used (460, 650 and 808 nm), efficient generation of electron–hole pairs can be expected and thus large photo-conductivity is obtained (Fig. 3a, the bias applied is 1 V). When 980 nm laser is used, no photocurrent is observed since its photon

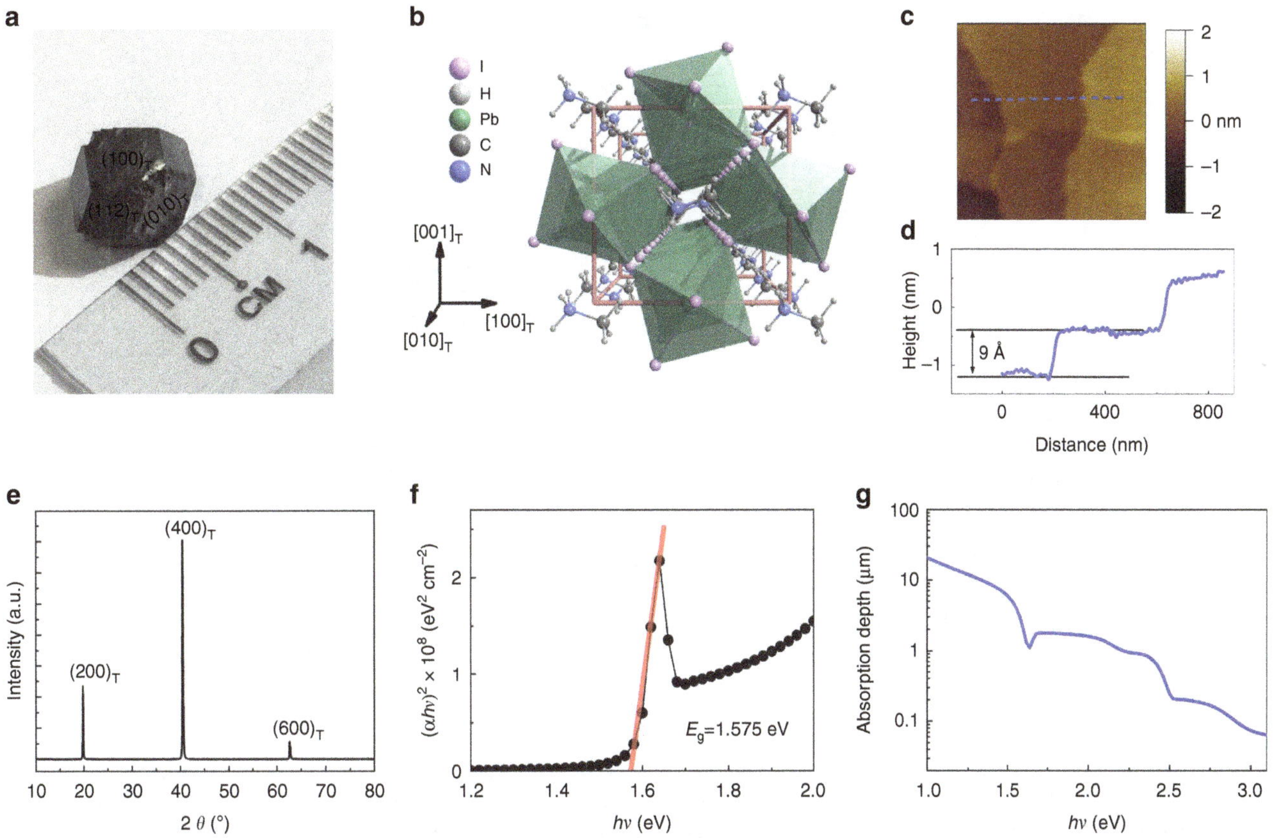

Figure 1 | Basic properties of the $MAPbI_3$ single crystals. (**a**) Photograph of a $MAPbI_3$ single crystal used in this study. (**b**) Perspective view of the unit cell of tetragonal $MAPbI_3$. (**c**) Typical topography of the single crystal obtained by AFM, showing smooth surface with unit-cell steps. The scan size is 1 × 1 μm. (**d**) Height profile along the blue dash line denoted in **c**. (**e**) X-ray diffraction pattern of the crystal, which only shows tetragonal (*l*00) type peaks, indicating the high quality of the single crystal. (**f**) Direct-transition tauc plot according to the absorption coefficient of the crystal, from which a direct band gap of 1.575 eV is extracted. The red solid line is the linear fit. (**g**) The penetration depth deduced from the absorption coefficient.

energy is below the band gap of $MAPbI_3$. Similarly, the photostrictive response shows clear photon energy dependence as well (Fig. 3b). When we plot the photocurrent and photostriction together as functions of the incident photon energy (Fig. 3c), the correlation is even clearer. This observation again confirms that the photostrictive response in $MAPbI_3$ single crystal is caused by photo-generated carriers.

Giant photostriction in $MAPbI_3$ film. According to the Beer–Lambert law, the intensity of incident light decays exponentially into the material's bulk. The strong absorption of $MAPbI_3$ in visible spectrum results in a penetration depth less than a few micrometres (Fig. 1g), leaving most of the crystal not illuminated and the effective photostriction ($\Delta H/H$) likely underestimated. To address this issue, we have carried out thickness-dependent photostriction measurement on small flakes cleaved from the same single crystal (Supplementary Fig. 4). It is found that for a 700-μm-thick flake, the height change of 50 nm is comparable to those of crystals with thicknesses greater than or equal to 1 mm, suggesting a saturated value. However, for a 150-μm-thick flake, the photostrictive response dramatically reduces to about 20 nm. Note that this thickness is already smaller than the carrier diffusion length estimated for single crystals[15]. Hence, we argue that although the light absorption depth is only a few micrometres, the photo-excited carriers can diffuse deep into the bulk, leading to a much thicker responsive layer. This conclusion helps us to further separate the photostriction signal into a fast near-surface response (absorption limited), a relatively slower contribution from the bulk limited by diffusion of the photo-carriers, and finally the much slower plateau due to the thermal bending of the AFM tip (Supplementary Note 2). The calculated photostriction of the 150-μm-thick flake is greater than 100 p.p.m. Last, we prepare $MAPbI_3$ thin films of 4 μm thick on fluorine-doped tin oxide (FTO)-coated glass and measure its photostrictive response. The quality of the film is confirmed by the performance of a testing cell ($FTO/TiO_2/MAPbI_3/$Spiro-OMeTAD/Ag), in which an efficiency of 12.5% is obtained (Fig. 4a). Since the slow height response is likely due to the thermal effect of the AFM tip, we subtract it from the height change profile and obtain a thickness change of about 5 nm under 100 mW cm^{-2} white light, which translates into a photostrictive response of approximately 1,250 p.p.m. in the $MAPbI_3$ film (Fig. 4b). This is much larger than any of the reported intrinsic photostrictive effect in ferroelectric materials, polar and non-polar semiconductors[24].

Discussion

In the literature, photostriction has been reported for several non-polar semiconductors, polar materials (including ferroelectrics), chalcogenide glasses and organic materials. Following the work of Kundys[24], we analyse the experimentally observed photostrictive coefficients (normalized to the light intensity) of known materials as shown in Table 1 (refs 24,26–40). In non-polar semiconductors, it is related to the pressure susceptibility of the energy gap based on the deformation potential theory[41]. This is justified by the opposite signs of dE_g/dP in Ge and Si and the corresponding photostrictive responses[26,27]. Recently, it is reported that the band gap of $CH_3NH_3PbBr_3$ ($MAPbBr_3$) reduces with increasing

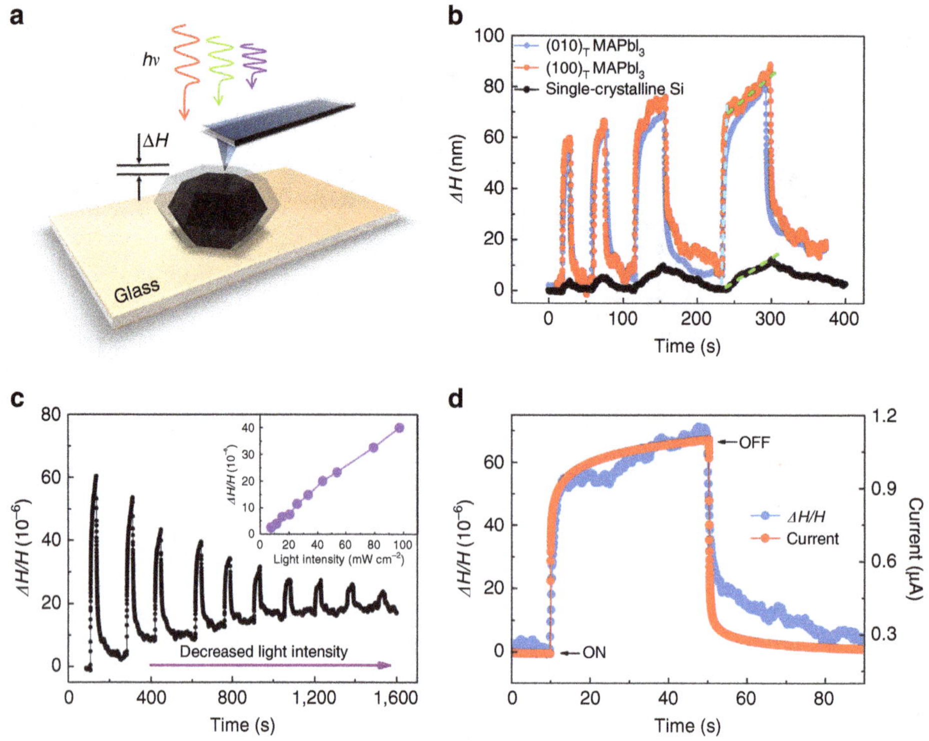

Figure 2 | Photostrictive effect of $MAPbI_3$ single crystals. (**a**) Schematic drawing of the experimental set-up for the photostrictive measurements. The AFM tip is fixed at one point on the sample to record the height as a function of time. (**b**) Photostriction in the $MAPbI_3$ single crystal. Both $(100)_T$- and $(010)_T$-oriented single crystals produce a similar elongation of ~50 p.p.m. under 100 mW cm^{-2} white-light illumination. The height change of a Si single crystal is also measured as a comparison. Cyan and green dash lines delineate the fast and slow components of the height change, respectively. (**c**) Light intensity dependence of the photostrictive effect. The inset shows the proportional relationship between the photostriction and light intensity. (**d**) A comparison between the height and current changes on light irradiation.

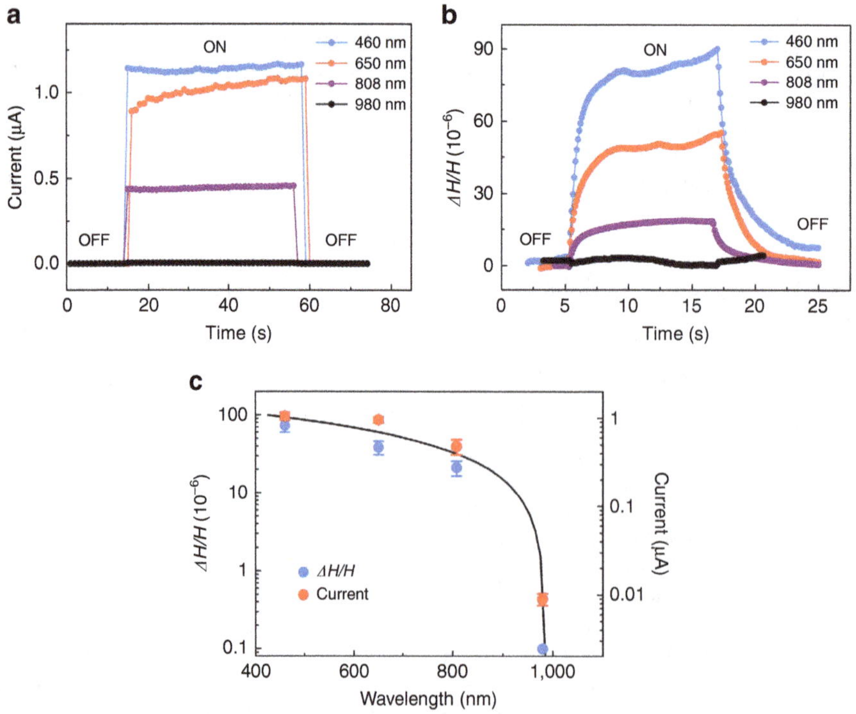

Figure 3 | Photon energy dependence of the photostriction. (**a**) When the photon energy is close to or above the band gap (that is, 460, 650 and 808 nm lasers), large current on illumination (under 1 V voltage) is observed, whereas negligible photocurrent is obtained when photon energy is below the band gap (that is, 980 nm laser). (**b**) The photostriction shows similar behaviour to that of photocurrent. (**c**) Photocurrents and height changes of the single crystal as functions of the incident photon energy, showing clear correlation between these two properties. The light intensity was kept at ~100 mW cm^{-2} for all lasers. The black solid line serves as a guide to the eye.

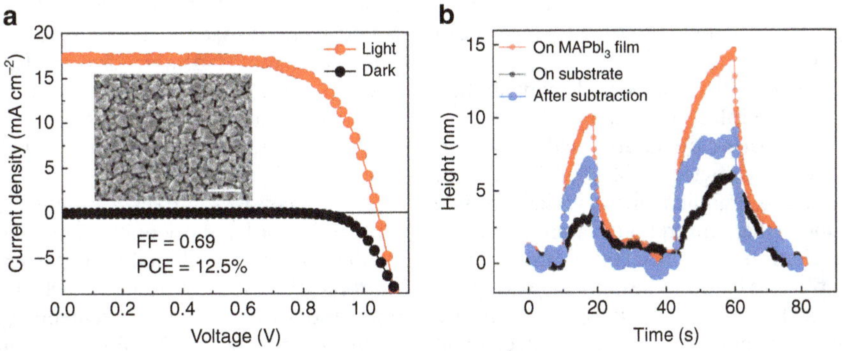

Figure 4 | Photovoltaic and photostrictive properties of $MAPbI_3$ thin films. (**a**) Typical current density-voltage characteristic of a $MAPbI_3$ thin-film photovoltaic cell (FTO/TiO_2/$MAPbI_3$/Spiro-OMeTAD/Ag) under simulated AM1.5 100 mW cm^{-2} illumination (red line) and under dark (black line). The power conversion efficiency can reach 12.5%. The inset shows the SEM image of the $MAPbI_3$ thin film. The scale bar, 1 µm. (**b**) Height change of the $MAPbI_3$ thin film (4 µm) on FTO-coated glass substrate under illumination. The net response from the film can be obtained by substracting the extrinsic contribution from the substrate. Under 100 mW cm^{-2} white light, about 5 nm height change can be observed, corresponding to a photostriction of 1,250 p.p.m.

Table 1 | Photostrictive coefficients of different materials.

		Photostriction, $\Delta L/L$	Light intensity, I (W m^{-2})	$(\Delta L/L)/I$ (m^2 W^{-1})	Refs
Non-polar semiconductors	Si crystal	-6.4×10^{-6}	$8.47\times10^{10\dagger}$	-7.56×10^{-17}	26
	Ge crystal	7.84×10^{-10}	1,000*	7.84×10^{-13}	27
Polar semiconductors	CdS crystal	7.5×10^{-5}	1,000*	7.5×10^{-8}	24,28
	GaAs crystal	4×10^{-7}	1,000*	4×10^{-10}	29
Ferroelectric materials	SbSI crystal	4×10^{-5}	1,000*	4×10^{-8}	30
	La doped $Pb(Zr_xTi_{1-x})O_3$ ceramics	10^{-4}	150	6.67×10^{-7}	31
	$BiFeO_3$ crystal	3×10^{-5}	326	9.2×10^{-8}	32
	$BiFeO_3$ film (35 nm)	4.6×10^{-3}	$\sim4\times10^{14\dagger}$	1.15×10^{-17}	33
	$PbTiO_3$ film (20 nm)	2.5×10^{-3}	$\sim10^{15\dagger}$	2.5×10^{-18}	34
Chalcogenide glasses	$As_{40}Se_{25}S_{25}Ge_{10}$ film	4.5×10^{-4}	1,000*	4.5×10^{-7}	35
	As_2Se_3 film	6.4×10^{-2}	400	1.6×10^{-4}	36
	As_2S_3 film	5.4×10^{-2}	400	1.35×10^{-4}	36
	$GeSe_2$ film	-5.6×10^{-2}	400	-1.4×10^{-4}	36
	GeS_2 film	-1.1×10^{-1}	400	-2.75×10^{-4}	36
Organic materials	Poly-(4,4′-diaminoazoben-zenepyromelliti-mide) films	-1.2×10^{-2}	1,000*	-1.2×10^{-5}	37
	Nematic elastomers	2×10^{-1}	1,000*	2×10^{-4}	38
	Poly(ethylacrylate) networks with azo-aromatic crosslinks	2.5×10^{-3}	1,000*	2.5×10^{-6}	39
	Diarylethenes	-7×10^{-2}	5,200	-1.35×10^{-5}	40
Hybrid perovskites	$MAPbI_3$ crystal	5×10^{-5}	1,000	5×10^{-8}	Current work
	$MAPbI_3$ film (4 µm)	1.25×10^{-3}	1,000	1.25×10^{-6}	

*The light intensity was not reported in the references and we use 1,000 W m^{-2} (1 Sun) to calculate the photostrictive coefficients.
†The light sources were high-energy laser pulses.

hydrostatic pressure[42], which should lead to the contraction of the lattice under illumination, contrary to our results. Besides, this effect is usually small, inconsistent with the giant photostriction we observed. In polar materials, the photostriction is closely linked to the bulk photovoltaic effect and commonly interpreted as a consequence of the converse piezoelectric effect[24]. Owing to the non-centrosymmetric structure, photo-generated carriers are spontaneously separated to produce an effective electric field that deforms the lattice through the piezoelectric tensor. However, despite many theoretical and experimental studies advocating for the ferroelectricity in $MAPbI_3$ (refs 43–46), a strong proof remains lacking[47]. While the structure of the low-temperature orthorhombic phase was refined to be centrosymmetric[48,49], the high-temperature structure is difficult to be determined due to the dynamic disorder of the $CH_3NH_3^+$ group. Considering the dynamic motion of the molecular dipole at room temperature, a long-range ferroelectric order can hardly exist, as confirmed by our polarization measurements, as well as piezoelectric force microscopy (Supplementary Figs 5 and 6). Even if polar nanoregions exist in $MAPbI_3$ as ferroelectric relaxors[50], a completely random distribution of the electric dipoles should cancel out the lattice change along all directions because of the volume conservation during piezoelectric deformation. For chalcogenide glasses and organic materials, both of them exhibit huge photostrictive response, which can be attributed to photo-induced bond modifications or photoisomerizations[51–53]. Strictly speaking, they should be classified as photochemical effects, most of which are irreversible after the irradiation. Besides, their responses are usually very slow, on the order of minutes or even hours, though some exceptions are recently reported[40,54]. It thus appears that none of these mechanisms can account for our observation.

Furthermore, it should be noted that the possible contribution from thermal expansion has been ruled out, and only the intrinsic effect is discussed here. By comparing the values listed in Table 1, it is clear that the photostrictive coefficient of $MAPbI_3$ happens to lie in between the inorganic and organic materials (chalcogenide glasses are an exception due to their amorphous nature), which is consistent with its hybrid character and mechanical properties[55]. The magnitude of the photostriction relies on how large a material's lattice can distort. In inorganic crystals, the atoms are closely packed with strong covalent or ionic bonds. Hence, it will be energetically costly for their lattices to deform. Organic materials, on the contrary, are at the other extreme. The small molecules or polymer chains are glued together by hydrogen bonds or van de Waals interactions, and their lattices can deform significantly due to the large intermolecular spacing. It suggests that the organic group may play a vital role in the giant photostriction effect observed in hybrid perovskites. In fact, hydrogen bonding between organic and inorganic moieties has long been studied in hybrid materials, which may lead to emergent properties[56–58]. Specifically in hybrid perovskites, hydrogen bonding between the amine group and the halide ions has been verified both theoretically and experimentally[23,59,60]. In this regard, we propose that it is the weakening of the hydrogen bonding by photo-generated carriers that results in the lattice dilation. As schematically shown in Fig. 5a, the hydrogen bonding between N and I is geometrically coupled to the buckling of the Pb–I–Pb bond and the tilting of the iodine octahedron. Under above-band-gap illumination, the first direct transition corresponds to the charge transfer from hybridized Pb *6s*–I *5p* orbital to the Pb *6p* orbital, forming weakly bound excitons that are easily dissociated by thermal energy[61–63]. The electronic transition directly leads to the reduction of electron density on the I site, and thereby reduces its Coulomb interaction with the amine group. This in turn straightens the Pb–I–Pb bond and results in a larger interatomic spacing (Fig. 5b).

If our analysis is correct, a couple of predictions can be made. First, the disordered orientations of the organic molecules, as well as the elastic multidomain of the crystal should lead to an isotropic photostrictive response, which is indeed observed in our orientation-dependent measurements shown in Fig. 5c. The result of the polycrystalline $MAPbI_3$ film provides another piece of evidence. Second, as the hydrogen bonding is intimately coupled to the octahedral tilt, divergence in the photostrictive response is expected around the tetragonal-cubic phase transition. Once again, our temperature-dependent measurements confirm this prediction (Fig. 5d,e). The photostriction measured at 60 °C is almost twice of that measured at room temperature, which can be interpreted by the enhanced lattice susceptibility at the structure transition boundary. The photostriction is reduced in the high-temperature cubic phase, but still with appreciable magnitudes. As pointed out by Quarti *et al.*[64], although the high-temperature phase appears cubic at a large scale, the local structure may strongly deviate from the nominal cubic one at any given time owing to the fluctuation of the organic molecule inside the inorganic framework. Thus, the interaction between them, though reduced, does not completely vanish.

The giant photostriction suggests strong photon-lattice coupling in $MAPbI_3$. It has important implications for understanding the exceptional photovoltaic performance of hybrid perovskites. For instance, the lattice expansion and reduced

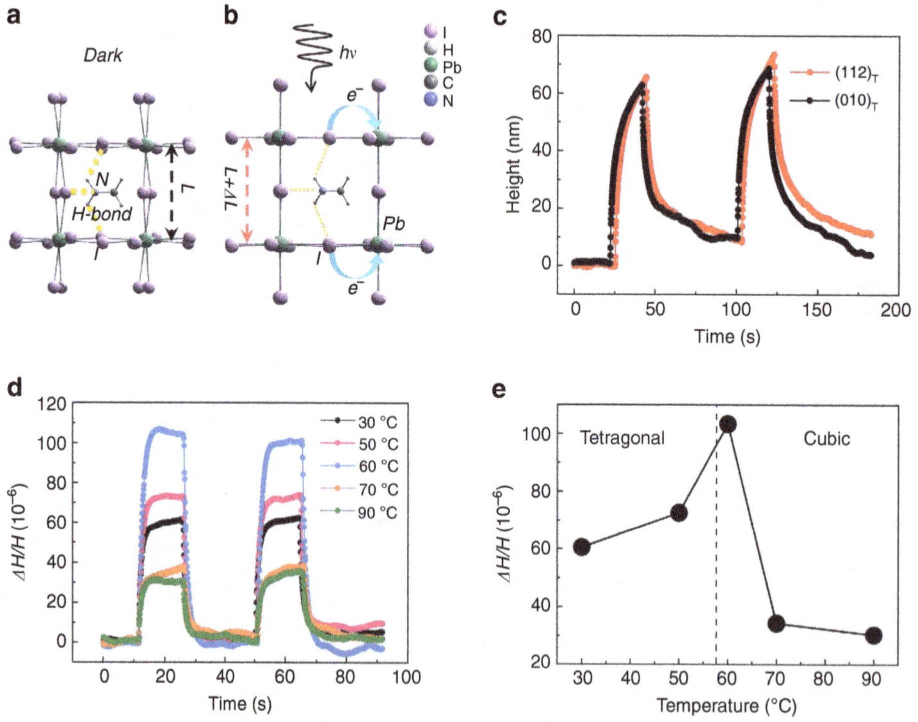

Figure 5 | Proposed mechansim for the giant photostriction in $MAPbI_3$. (**a**,**b**) Schematic illustrations (not to scale) showing that the weakening of the hydrogen bonding between the amine group and the iodine ion by photo-generated carriers leads to the lattice expansion. (**c**) Orientation dependence of photostriction of a $MAPbI_3$ single crystal shows similar magntidue due to the isotropic expansion of the lattice. (**d**,**e**) Temperature-dependent photostriction of a $MAPbI_3$ single crystal across the tetragonal-cubic phase transition under 100 mW cm^{-2} white light. The phase transition point is indicated by the black dash line. Enhanced lattice susceptibility around the phase transition boundary is likely responsible for the larger photostriction observed.

Coulomb interaction between the organic group and the inorganic framework will further reduce the rotational barrier for the molecule dipole. The enhanced tumbling of the electric dipole may lead to a dynamic change of the local band structure that suppresses the electron–hole recombination[22]. This scenario is in line with the decrease of the bi-molecular recombination rate at high temperatures[65]. Besides being important to explain the long-carrier lifetime and diffusion length in the hybrid lead halide perovskites, the giant photostriction also opens up new pathways for applications in opto-mechanical devices[24,25].

Methods

Growth of $MAPbI_3$ single crystal and structure characterization. CH_3NH_3I was first prepared according to previous report[66]. Typically, 27.8 ml methylamine (40% in methanol, Aldrich) was added into 30 ml hydroiodic acid (57 wt.% in water, Aldrich) and stirred for 2 h at 0 °C. After removing the solvent by rotary evaporating at 50 °C, the product was washed with diethyl ether and then recrystallized with ethanol. White crystals were obtained and dried in a vacuum box. PbI_2 (1.157 g, 99%, Aldrich) and the as prepared CH_3NH_3I (0.395 g) were mixed in 2 ml γ-butyrolactone at 60 °C with stirring. Then, 1 ml acetonitrile (Aldrich) was added into the solution and a clear pale yellow solution was obtained. The solution was then placed in a vial and kept in an oven at 70 °C for 20 min. Small $MAPbI_3$ single crystals then appeared at the bottom of the vial. One of these small crystals was picked out and put into another vial with the same solution for continuous growth. The crystals used for measurement were grown for 3 h. X-ray diffraction data were collected on a high-resolution diffractometer (Bruker D8 Discover) using Cu K_α radiation.

Thin-film solar cell testing. Photovoltaic cell with the structure of $FTO/TiO_2/MAPbI_3$/Spiro-OMeTAD/Ag was prepared to measure the power conversion efficiency. Under illumination of solar-simulated AM1.5 sunlight at 100 mW cm^{-2}, the current–voltage curve was obtained using a pA meter/direct current (DC) voltage source (Hewlett Package 4140B).

Spectroscopic ellipsometry. The spectroscopic ellipsometry measurements have been performed using a commercially available rotating analyser instrument with compensator (V-VASE; J.A. Woollam Co., Inc.) within the spectral range from 0.6 to 6 eV. Data has been collected at two incidence angles (50° and 70°), while the sample was continuously purged with nitrogen gas to avoid degradation. The absorption coefficient was then calculated from the pseudodielectric function (Supplementary Fig. 1).

Atomic force microscopy and piezoelectric force microscopy. Two commercial AFMs were used to measure the photostrictive responses: Asylum Research (AR) MFP 3D and Park XE 150 under ambient condition (20–30% relative humidity, 25 °C). Although the surface of the crystal degrades slowly with time in ambient condition possibly due to surface hydration, it does not affect the photostrictive response significantly. We believe this is because the hydration product is an optically transparent layer, which does not affect the absorption and thus the photostriction of the bulk. During the photostriction test, the light (from either a halogen lamp or laser diodes) was guided through an optical fibre into the built-in optical microscope of the AFM. For the AR AFM system, the light has to go through its internal optics, which include a cold mirror that reflects only visible light. As such, only the white-light tests were carried out using AR AFM. The wavelength-dependent tests were performed on a Park AFM system, whose optics system includes only an optical microscope that allows all the studied wavelengths (460, 650, 808 and 980 nm) to pass through. In both systems, the light was shined from above the AFM tip. However, because the light had gone through an optical fibre, then been focused by the optical lens, the light was no more collimated. Therefore, the region right beneath the AFM tip was not shadowed (Supplementary Note 3). Furthermore, the beam size was about 0.1 cm^2 for all light sources. The height profile as a function of time, with the light periodically irradiated on the surface, was acquired. Both contact mode and tapping mode measurements gave similar results (Supplementary Fig. 7). The light intensity at the sample location was carefully calibrated using a commercial energy meter (Newport, 91,150 V). Piezoelectric force microscopy was carried out on AR MFP 3D mode under dual AC resonance tracking (DART) mode using a Pt/Ir coated tip with a spring constant of 2 N m^{-1}. The imaging voltage $V_{ac} = 1$ V.

Photocurrent and ferroelectric polarization measurements. We fabricated a simple device with the two opposite facets of the single crystal coated by semi-transparent Pt electrodes and measured the current under illumination through the top electrode. A low-noise probe station and a pA meter/DC voltage source were used and the voltage applied was 1 V. The ferroelectric polarization measurement was conducted using a commercial ferroelectric tester (Radiant Technologies, Precision LC) at different temperatures on a low-temperature probe station.

References

1. Gratzel, M. The light and shade of perovskite solar cells. *Nat. Mater.* **13,** 838–842 (2014).
2. Green, M. A., Ho-Baillie, A. & Snaith, H. J. The emergence of perovskite solar cells. *Nat. Photon.* **8,** 506–514 (2014).
3. Kojima, A., Teshima, K., Shirai, Y. & Miyasaka, T. Organometal halide perovskites as visible-light sensitizers for photovoltaic cells. *J. Am. Chem. Soc.* **131,** 6050–6051 (2009).
4. Stranks, S. D. & Snaith, H. J. Metal-halide perovskites for photovoltaic and light-emitting devices. *Nat. Nanotechnol.* **10,** 391–402 (2015).
5. National Renewable Energy Labs (NREL). Efficiency chart http://www.nrel.gov/ncpv/images/efficiency_chart.jpg (2015).
6. Xing, G. C. *et al.* Low-temperature solution-processed wavelength-tunable perovskites for lasing. *Nat. Mater.* **13,** 476–480 (2014).
7. Tan, Z. K. *et al.* Bright light-emitting diodes based on organometal halide perovskite. *Nat. Nanotechnol.* **9,** 687–692 (2014).
8. Dou, L. T. *et al.* Solution-processed hybrid perovskite photodetectors with high detectivity. *Nat. Commun.* **5,** 5404 (2014).
9. Xing, G. C. *et al.* Long-range balanced electron- and hole-transport lengths in organic–inorganic $CH_3NH_3PbI_3$. *Science* **342,** 344–347 (2013).
10. Wehrenfennig, C., Eperon, G. E., Johnston, M. B., Snaith, H. J. & Herz, L. M. High charge carrier mobilities and lifetimes in organolead trihalide perovskites. *Adv. Mater.* **26,** 1584–1589 (2014).
11. Stranks, S. D. *et al.* Electron-hole diffusion lengths exceeding 1 micrometer in an organometal trihalide perovskite absorber. *Science* **342,** 341–344 (2013).
12. Ponseca, C. S. *et al.* Organometal halide perovskite solar cell materials rationalized: ultrafast charge generation, high and microsecond-long balanced mobilities, and slow recombination. *J. Am. Chem. Soc.* **136,** 5189–5192 (2014).
13. deQuilettes, D. W. *et al.* Impact of microstructure on local carrier lifetime in perovskite solar cells. *Science* **348,** 683–686 (2015).
14. Shi, D. *et al.* Low trap-state density and long carrier diffusion in organolead trihalide perovskite single crystals. *Science* **347,** 519–522 (2015).
15. Dong, Q. F. *et al.* Electron-hole diffusion lengths > 175 μm in solution-grown $CH_3NH_3PbI_3$ single crystals. *Science* **347,** 967–970 (2015).
16. Shockley, W. & Queisser, H. J. Detailed balance limit of efficiency of p-n junction solar cells. *J. Appl. Phys.* **32,** 510–519 (1961).
17. Adriaenssens, G. J. & Arkhipov, V. I. Non-Langevin recombination in disordered materials with random potential distributions. *Solid State Commun.* **103,** 541–543 (1997).
18. Lakhwani, G., Rao, A. & Friend, R. H. Bimolecular recombination in organic photovoltaics. *Annu. Rev. Phys. Chem.* **65,** 557–581 (2014).
19. Bakulin, A. A. *et al.* Real-time observation of organic cation reorientation in methylammonium lead iodide perovskites. *J. Phys. Chem. Lett.* **6,** 3663–3669 (2015).
20. Leguy, A. M. A. *et al.* The dynamics of methylammonium ions in hybrid organic–inorganic perovskite solar cells. *Nat. Commun.* **6,** 7124 (2015).
21. Park, J.-S. *et al.* Electronic structure and optical properties of α-$CH_3NH_3PbBr_3$ perovskite single crystal. *J. Phys. Chem. Lett.* **6,** 4304–4308 (2015).
22. Motta, C. *et al.* Revealing the role of organic cations in hybrid halide perovskite $CH_3NH_3PbI_3$. *Nat. Commun.* **6,** 7026 (2015).
23. Glaser, T. *et al.* Infrared spectroscopic study of vibrational modes in methylammonium lead halide perovskites. *J. Phys. Chem. Lett.* **6,** 2913–2918 (2015).
24. Kundys, B. Photostrictive materials. *Appl. Phys. Rev.* **2,** 011301 (2015).
25. Kreisel, J., Alexe, M. & Thomas, P. A. A photoferroelectric material is more than the sum of its parts. *Nat. Mater.* **11,** 260–260 (2012).
26. Buschert, J. R. & Colella, R. Photostriction effect in silicon observed by time-resolved X-ray-diffraction. *Solid State Commun.* **80,** 419–422 (1991).
27. Figielski, T. Photostriction effect in germanium. *Phys. Status Solidi* **1,** 306–316 (1961).
28. Lagowski, J. & Gatos, H. C. Photomechanical effect in noncentrosymmetric semiconductors CdS. *Appl. Phys. Lett.* **20,** 14–16 (1972).
29. Lagowski, J. & Gatos, H. C. Photomechanical vibration of thin crystals of polar semiconductors. *Surf. Sci.* **45,** 353–370 (1974).
30. Tatsuzak, I., Itoh, K., Ueda, S. & Shindo, Y. Strain along c axis of SbSI caused by illumination in dc electric field. *Phys. Rev. Lett.* **17,** 198–200 (1966).
31. Takagi, K. *et al.* Ferroelectric and photostrictive properties of fine-grained PLZT ceramics derived from mechanical alloying. *J. Am. Ceram. Soc.* **87,** 1477–1482 (2004).
32. Kundys, B. *et al.* Wavelength dependence of photoinduced deformation in $BiFeO_3$. *Phys. Rev. B* **85,** 092301 (2012).
33. Schick, D. *et al.* Localized excited charge carriers generate ultrafast inhomogeneous strain in the multiferroic $BiFeO_3$. *Phys. Rev. Lett.* **112,** 097602 (2014).
34. Daranciang, D. *et al.* Ultrafast photovoltaic response in ferroelectric nanolayers. *Phys. Rev. Lett.* **108,** 087601 (2012).
35. Igo, T., Noguchi, Y. & Nagai, H. Photoexpansion and thermal contraction of amorphous-chalcogenide glasses. *Appl. Phys. Lett.* **25,** 193–194 (1974).

36. Kuzukawa, Y., Ganjoo, A. & Shimakawa, K. Photoinduced structural changes in obliquely deposited As- and Ge-based amorphous chalcogenides: correlation between changes in thickness and band gap. *J. Non-Cryst. Solids* **227,** 715–718 (1998).
37. Vanderve, G. & Prins, W. Photomechanical energy conversion in a polymer membrane. *Nature* **230,** 70–72 (1971).
38. Finkelmann, H., Nishikawa, E., Pereira, G. G. & Warner, M. A new opto-mechanical effect in solids. *Phys. Rev. Lett.* **87,** 015501 (2001).
39. Eisenbach, C. D. Isomerization of aromatic azo chromophores in poly(ethyl acrylate) networks and photomechanical effect. *Polymer* **21,** 1175–1179 (1980).
40. Kobatake, S., Takami, S., Muto, H., Ishikawa, T. & Irie, M. Rapid and reversible shape changes of molecular crystals on photoirradiation. *Nature* **446,** 778–781 (2007).
41. Thomsen, C., Grahn, H. T., Maris, H. J. & Tauc, J. Surface generation and detection of phonons by picosecond light pulses. *Phys. Rev. B* **34,** 4129–4138 (1986).
42. Wang, Y. *et al.* Pressure-induced phase transformation, reversible amorphization, and anomalous visible light response in organolead bromide perovskite. *J. Am. Chem. Soc.* **137,** 11144–11149 (2015).
43. Frost, J. M. *et al.* Atomistic origins of high-performance in hybrid halide perovskite solar cells. *Nano Lett.* **14,** 2584–2590 (2014).
44. Kutes, Y. *et al.* Direct observation of ferroelectric domains in solution-processed $CH_3NH_3PbI_3$ perovskite thin films. *J. Phys. Chem. Lett.* **5,** 3335–3339 (2014).
45. Stroppa, A. *et al.* Tunable ferroelectric polarization and its interplay with spin–orbit coupling in tin iodide perovskites. *Nat. Commun.* **5,** 5900 (2014).
46. Fan, Z. *et al.* Ferroelectricity of $CH_3NH_3PbI_3$ perovskite. *J. Phys. Chem. Lett.* **6,** 1155–1161 (2015).
47. Xiao, Z. *et al.* Giant switchable photovoltaic effect in organometal trihalide perovskite devices. *Nat. Mater.* **14,** 193–198 (2015).
48. Swainson, I. P., Hammond, R. P., Soullière, C., Knop, O. & Massa, W. Phase transitions in the perovskite methylammonium lead bromide, $CH_3ND_3PbBr_3$. *J. Solid State Chem.* **176,** 97–104 (2003).
49. Baikie, T. *et al.* Synthesis and crystal chemistry of the hybrid perovskite $(CH_3NH_3)PbI_3$ for solid-state sensitised solar cell applications. *J. Mater. Chem. A* **1,** 5628–5641 (2013).
50. Xu, G. Y., Zhong, Z., Bing, Y., Ye, Z. G. & Shirane, G. Electric-field-induced redistribution of polar nano-regions in a relaxor ferroelectric. *Nat. Mater.* **5,** 134–140 (2006).
51. Kugler, S., Hegedüs, J. & Kohary, K. in *Optical Properties of Condensed Matter and Applications* 143–158 (John Wiley & Sons, Ltd, 2006).
52. Iqbal, D. & Samiullah, M. Photo-responsive shape-memory and shape-changing liquid-crystal polymer networks. *Materials* **6,** 116–142 (2013).
53. Yu, H. Recent advances in photoresponsive liquid-crystalline polymers containing azobenzene chromophores. *J. Mater. Chem. C* **2,** 3047–3054 (2014).
54. Camacho-Lopez, M., Finkelmann, H., Palffy-Muhoray, P. & Shelley, M. Fast liquid-crystal elastomer swims into the dark. *Nat. Mater.* **3,** 307–310 (2004).
55. Rakita, Y., Cohen, S. R., Kedem, N. K., Hodes, G. & Cahen, D. Mechanical properties of $APbX_3$ ($A = Cs$ or CH_3NH_3; $X = I$ or Br) perovskite single crystals. *MRS Commun.* **5,** 623–629 (2015).
56. Mitzi, D. B. Templating and structural engineering in organic–inorganic perovskites. *J. Chem. Soc. Dalton Trans.* 1–12 (2001).
57. Jain, P. *et al.* Multiferroic behaviour associated with an order – disorder hydrogen bonding transition in metal – organic frameworks (MOFs) with the perovskite ABX_3 architecture. *J. Am. Chem. Soc.* **131,** 13625–13627 (2009).
58. Zhang, W. & Xiong, R.-G. Ferroelectric metal–organic frameworks. *Chem. Rev.* **112,** 1163–1195 (2012).
59. Swainson, I. *et al.* Orientational ordering, tilting and lone-pair activity in the perovskite methylammonium tin bromide, $CH_3NH_3SnBr_3$. *Acta Crystallogr. Sect. B* **66,** 422–429 (2010).
60. Lee, J.-H., Bristowe, N. C., Bristowe, P. D. & Cheetham, A. K. Role of hydrogen-bonding and its interplay with octahedral tilting in $CH_3NH_3PbI_3$. *Chem. Commun.* **51,** 6434–6437 (2015).
61. D'Innocenzo, V. *et al.* Excitons versus free charges in organo-lead tri-halide perovskites. *Nat. Commun.* **5,** 3586 (2014).
62. Saba, M. *et al.* Correlated electron–hole plasma in organometal perovskites. *Nat. Commun.* **5,** 5049 (2014).
63. Miyata, A. *et al.* Direct measurement of the exciton binding energy and effective masses for charge carriers in organic–inorganic tri-halide perovskites. *Nat. Phys.* **11,** 582–587 (2015).
64. Quarti, C. *et al.* Structural and optical properties of methylammonium lead iodide across the tetragonal to cubic phase transition: implications for perovskite solar cells. *Energy Environ. Sci.* **9,** 155–163 (2016).
65. Milot, R. L., Eperon, G. E., Snaith, H. J., Johnston, M. B. & Herz, L. M. Temperature-dependent charge-carrier dynamics in $CH_3NH_3PbI_3$ perovskite thin films. *Adv. Funct. Mater.* **25,** 6218–6227 (2015).
66. Kim, H. S. *et al.* Lead iodide perovskite sensitized all-solid-state submicron thin film mesoscopic solar cell with efficiency exceeding 9%. *Sci. Rep.* **2,** 591 (2012).

Acknowledgements

This work is supported by the Ministry of Education, Singapore under project No. MOE2013-T2-1-052 and AcRF Tier 1 RG126/14. We thank Dr Pio John S. Buenconsejo and Dr Fucai Liu for the help on XRD and photocurrent measurements, respectively.

Author contributions

Y.Z., L.Y. and J.W. conceived and designed the work. Z.K. and H.F. synthesized the single crystals. S.W. fabricated the films and corresponding photovoltaic cells. Y.Z. and L.Y. conducted the photostriction and photocurrent measurements with help from Lei Chang, L.W. and P.R. D.S. and A.R. performed the spectroscopic ellipsometry measurements. Liufan Chen and G.Y. independently confirmed the photostrictive response. Y.Z., L.Y. and J.W. co-wrote the manuscript. J.W. supervised the project. All authors discussed the results and commented on the manuscript.

Additional information

Competing financial interests: The authors declare no competing financial interests.

13

Demonstration of a near-IR line-referenced electro-optical laser frequency comb for precision radial velocity measurements in astronomy

X. Yi[1], K. Vahala[1], J. Li[1], S. Diddams[2,3], G. Ycas[2,3], P. Plavchan[4], S. Leifer[5], J. Sandhu[5], G. Vasisht[5], P. Chen[5], P. Gao[6], J. Gagne[7], E. Furlan[8], M. Bottom[9], E.C. Martin[10], M.P. Fitzgerald[10], G. Doppmann[11] & C. Beichman[8]

An important technique for discovering and characterizing planets beyond our solar system relies upon measurement of weak Doppler shifts in the spectra of host stars induced by the influence of orbiting planets. A recent advance has been the introduction of optical frequency combs as frequency references. Frequency combs produce a series of equally spaced reference frequencies and they offer extreme accuracy and spectral grasp that can potentially revolutionize exoplanet detection. Here we demonstrate a laser frequency comb using an alternate comb generation method based on electro-optical modulation, with the comb centre wavelength stabilized to a molecular or atomic reference. In contrast to mode-locked combs, the line spacing is readily resolvable using typical astronomical grating spectrographs. Built using commercial off-the-shelf components, the instrument is relatively simple and reliable. Proof of concept experiments operated at near-infrared wavelengths were carried out at the NASA Infrared Telescope Facility and the Keck-II telescope.

[1] Department of Applied Physics and Materials Science, Pasadena, California 91125, USA. [2] National Institute of Standards and Technology, 325 Broadway, Boulder, Colorado 80305, USA. [3] Department of Physics, University of Colorado, 2000 Colorado Avenue, Boulder, Colorado 80309, USA. [4] Department of Physics, Missouri State University, 901 S National Avenue, Springfield, Missouri 65897, USA. [5] Jet Propulsion Laboratory, California Institute of Technology, 4800 Oak Grove Drive, Pasadena, California 91109, USA. [6] Division of Geological and Planetary Sciences, California Institute of Technology, Pasadena, California 91125, USA. [7] Department of Terrestrial Magnetism, Carnegie Institution of Washington, 5241 Broad Branch Road, Washington, District of Columbia 20015, USA. [8] NASA Exoplanet Science Institute, California Institute of Technology, Pasadena, California 91125, USA. [9] Department of Astronomy, California Institute of Technology, Pasadena, California 91125, USA. [10] Department of Physics and Astronomy, University of California Los Angeles, Los Angeles, California 90095, USA. [11] W.M. Keck Observatory, Kamuela, Hawaii 96743, USA. Correspondence and requests for materials should be addressed to K.V. (email: vahala@caltech.edu) or to C.B. (email: chas@ipac.caltech.edu).

The earliest technique for the discovery and characterization of planets orbiting other stars (exoplanets) is the Doppler or radial velocity (RV) method whereby small periodic changes in the motion of a star orbited by a planet are detected via careful spectroscopic measurements[1]. The RV technique has identified hundreds of planets ranging in mass from a few times the mass of Jupiter to less than an Earth mass, and in orbital periods from less than a day to over 10 years (ref. 2). However, the detection of Earth-analogues at orbital separations suitable for the presence of liquid water at the planet's surface, that is, in the 'habitable zone'[3], remains challenging for stars like the Sun with RV signatures $<0.1\,\mathrm{m\,s^{-1}}$ ($\Delta V/c < 3\times10^{-10}$) and periods of a year ($\sim10^8$ sec to measure three complete periods). For cooler, lower luminosity stars (spectral class M), however, the habitable zone moves closer to the star which, by application of Kepler's laws, implies that a planet's RV signature increases, $\sim0.5\,\mathrm{m\,s^{-1}}$ ($\Delta V/c < 1.5\times10^{-9}$), and its orbital period decreases, ~30 days ($\sim10^7$ s to measure three periods). Both of these effects make the detection easier. But for M stars, the bulk of the radiation shifts from the visible wavelengths, where most RV measurements have been made to date, into the near-infrared. Thus, there is considerable interest among astronomers in developing precise RV capabilities at longer wavelengths.

Critical to precision RV measurements is a highly stable wavelength reference[4]. Recently a number of groups have undertaken to provide a broadband calibration standard that consists of a 'comb' of evenly spaced laser lines accurately anchored to a stable frequency standard and injected directly into the spectrometer along with the stellar spectrum[5–9]. While this effort has mostly been focused on visible wavelengths, there have been successful efforts at near-IR wavelengths as well[10–12]. In all of these earlier studies, the comb has been based on a femtosecond mode-locked laser that is self-referenced[13–15], such that the spectral line spacing and common offset frequency of all lines are both locked to a radio frequency standard. Thus, laser combs potentially represent an ideal tool for spectroscopic and RV measurements.

However, in the case of mode-locked laser combs, the line spacing is typically in the range of 0.1–1 GHz, which is too small to be resolved by most astronomical spectrographs. As a result, the output spectrum of the comb must be spectrally filtered to create a calibration grid spaced by >10 GHz, which is more commensurate with the resolving power of a high-resolution astronomical spectrograph[8]. While this approach has led to spectrograph characterization at the $\mathrm{cm\,s^{-1}}$ level[16], it nonetheless increases the complexity and cost of the system.

In light of this, there is interest in developing photonic tools that possess many of the benefits of mode-locked laser combs, but that might be simpler, less expensive and more amenable to 'hands-off' operation at remote telescope sites. Indeed, in many RV measurements, other system-induced errors and uncertainties can limit the achievable precision, such that a frequency comb of lesser precision could still be equally valuable. For example, one alternative technique recently reported is to use a series of spectroscopic peaks induced in a broad continuum spectrum using a compact Fabry–Perot interferometer[17–19]. While the technique must account for temperature-induced tuning of the interferometer, it has the advantage of simplicity and low cost. Another interesting alternative is the so-called Kerr comb or microcomb, which has the distinct advantage of directly providing a comb with spacing in the range of 10–100 GHz, without the need for filtering[20]. While this new type of laser comb is still under development, there have been promising demonstrations of full microcomb frequency control[21,22] and in the future it could be possible to fully integrate such a microcomb on only a few square centimetres of silicon, making a very robust and inexpensive calibrator. Another approach that has been proposed is to create a comb through electro-optical modulation of a frequency-stabilized laser[23,24].

In the following, we describe a successful effort to implement this approach. We produce a line-referenced, electro-optical modulation frequency comb (LR-EOFC) ~1559.9 nm in the astronomical H band (1,500–1,800 nm). We discuss the experimental set-up, laboratory results and proof of concept demonstrations at the NASA Infrared Telescope Facility (IRTF) and the W. M. Keck observatory (Keck) 10 m telescope.

Results

Comb generation. A LR-EOFC is a spectrum of lines generated by electro-optical modulation of a continuous-wave laser source[25–29] which has been stabilized to a molecular or atomic reference (for example, $f_0 = f_{\mathrm{atom}}$). The position of the comb teeth ($f_N = f_0 \pm N f_m$, N is an integer) has uncertainty determined by the stabilization of f_0 and the microwave source that provides the modulation frequency f_m. However, the typical uncertainty of a microwave source can be sub-Hertz when synchronized with a compact Rb clock and moreover can be global positioning system (GPS)-disciplined to provide long-term stability[12]. Thus, the dominant uncertainty in comb tooth frequency in the LR-EOFC is that of f_0.

The schematic layout for LR-EOFC generation is illustrated in Fig. 1 and a detailed layout is shown in Fig. 2. All components are commercially available off-the-shelf telecommunications components. Pictures of the key components are shown in the left column of Fig. 1. The frequency-stabilized laser is first pre-amplified to 200 mW with an Erbium-Doped Fibre Amplifier (EDFA, model: Amonics, AEDFA-PM-23-B-FA) and coupled into two tandem lithium niobate ($LiNbO_3$) phase modulators ($V_\pi = 3.9$ V at 12 GHz, RF input limit: 33 dBm). The phase modulators are driven by an amplified 12 GHz frequency signal at 32.5 and 30.7 dBm, and synchronized by using microwave phase shifters. This initial phase modulation process produces a comb having ~40 comb lines ($\approx 2\pi \times V_{\mathrm{drive}}/V_\pi$), or equivalently 4 nm bandwidth. This comb is then coupled into a $LiNbO_3$ amplitude modulator with 18–20 dB distinction ratio, driven at the same microwave frequency by the microwave power recycled from the phase modulator external termination port. The modulation index of $\pi/2$ is set by an attenuator and the phase offset of the two amplitude modulator arms is set and locked to $\pi/2$. Microwave phase shifters are used to align the drive phase so that the amplitude modulator gates-out only those portions of the phase modulation that are approximately linearly chirped with one sign (that is, parabolic phase variation in time). A nearly transform-limited pulse is then formed when this parabolic phase variation is nullified by a dispersion compensation unit using a chirped fibre Bragg grating with $8\,\mathrm{ps\,nm^{-1}}$ dispersion. A 2 ps full-width at half-maximum pulse is measured after the fibre grating using an autocorrelator. Owing to this pulse formation, the duty cycle of the pulse train reaches below 2.5%, boosting the peak intensity of the pulses. These pulses are then amplified in a second EDFA (IPG Photonics, EAR-5 K-C-LP). For an average power of 1 W, peak power (pulse energy) is 40 W (83 pJ). The amplified pulses are then coupled into a 20 m length of highly nonlinear fibre with $0.25\pm0.15\,\mathrm{ps\,nm^{-1}\,km^{-1}}$ dispersion and dispersion slope of $0.006\pm0.004\,\mathrm{ps\,nm^{-2}\,km^{-1}}$. Propagation in the highly nonlinear fibre causes self-phase modulation and strong spectral broadening of the comb[30]. Comb spectra span and envelope can be controlled by the pump power launched into the highly nonlinear fibre. A typical comb spectrum with >600 mW pump power from the 1,559.9 nm laser is shown in Fig. 3a, with >100 nm spectral span. Moreover, by using various nonlinear fibre and spectral flattening methods, broad combs with level power are possible[31].

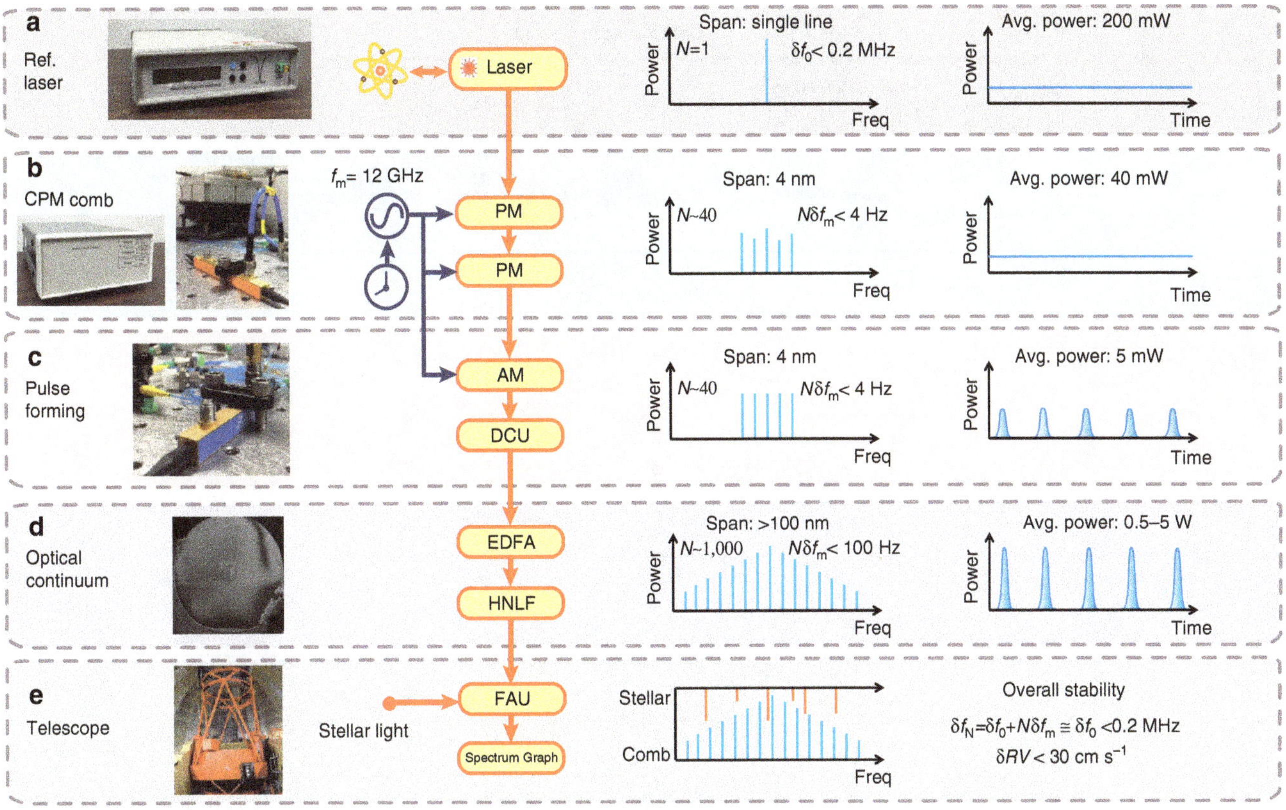

Figure 1 | Conceptual schematics of the line-referenced electro-optical frequency comb for astronomy. Vertically, the first column contains images of key instruments. (**a–e**) The images are reference laser, Rb clock (left) and phase modulator (right), amplitude modulator, highly nonlinear fibre and telescope. A simplified schematic set-up is in the second column. Third and fourth columns present the comb state in the frequency and temporal domains. The frequency of N-th comb tooth is expressed as $f_N = f_0 + N \times f_m$, where f_0 and f_m are the reference laser frequency and modulation frequency, respectively. N is the number of comb lines relative to the reference laser (taken as comb line $N=0$), RV is radial velocity and δf_N, δf_0 and δf_m are the variance of f_N, f_0 and f_m. (**a**) The reference laser is locked to a molecular transition, acquiring stability of 0.2 MHz, corresponding to 30 cm s^{-1} RV. (**b**) Cascaded phase modulation (CPM) comb: the phase of the reference laser is modulated by two phase modulators (PM), creating several tens of sidebands with spacing equal to the modulation frequency. The RF frequency generator is referenced to a Rb clock, providing stability at the sub-Hz level ($\delta f_m < 0.03$ Hz at 100 s). (**c**) Pulse forming is then performed by an amplitude modulator (AM) and dispersion compensation unit (DCU), which could be a long single mode fibre (SMF) or chirped fibre Bragg grating (FBG). (**d**) After amplification by an erbium-doped fibre amplifier (EDFA), the pulse undergoes optical continuum broadening in a highly nonlinear fibre (HNLF), extending its bandwidth >100 nm. (**e**) Finally the comb light is combined with stellar light using a fibre acquisition unit (FAU) and is sent into the telescope spectrograph. The overall comb stability is primarily determined by the pump laser.

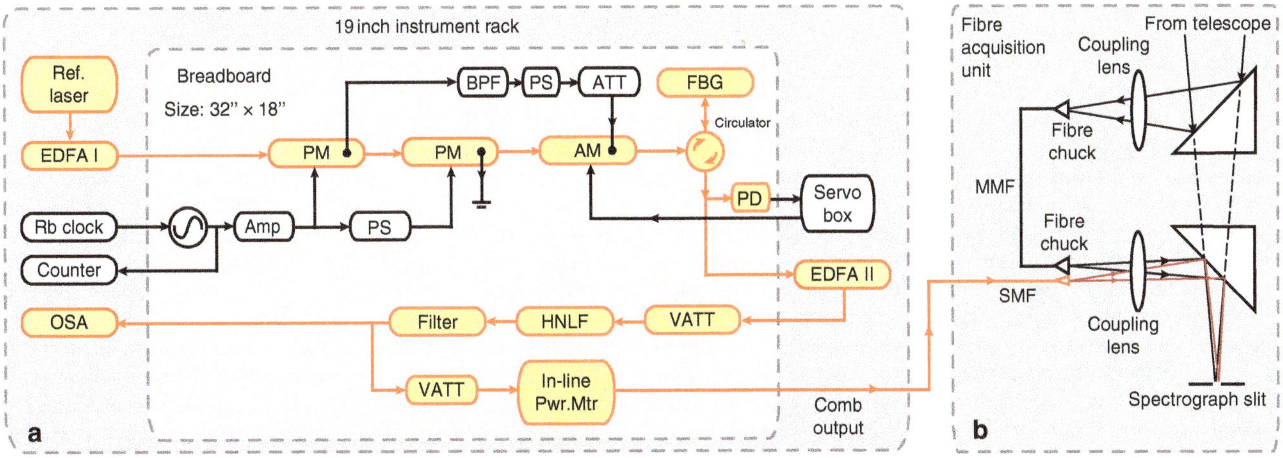

Figure 2 | Detailed set-up of line-referenced electro-optical frequency comb. (**a**) The entire LR-EOFC system sits in a 19 inch instrument rack. Optics and microwave components in the rack are denoted in orange and black, respectively. Small components were assembled onto a breadboard. These included the phase modulators (PM), amplitude modulator (AM), fibre Bragg grating (FBG), photodetector (PD), variable attenuator (VATT), attenuator (ATT), highly nonlinear fibre (HNLF), microwave source, microwave amplifier (Amp), phase shifter (PS) and band-pass filter (BPS). The reference laser, erbium-doped fiber amplifier (EDFA), rubidium (Rb) clock, counter, optical spectrum analyser (OSA) and servo lock box are separately located in the instrument rack. (**b**) A simplified schematic of the fibre acquisition unit (FAU) is also shown. Stellar light is focused and coupled into a multimode fibre (MMF). The comb light from a single mode fibre (SMF), together with the stellar light in the MMF, are focused on the spectrograph slit and sent into the spectrograph.

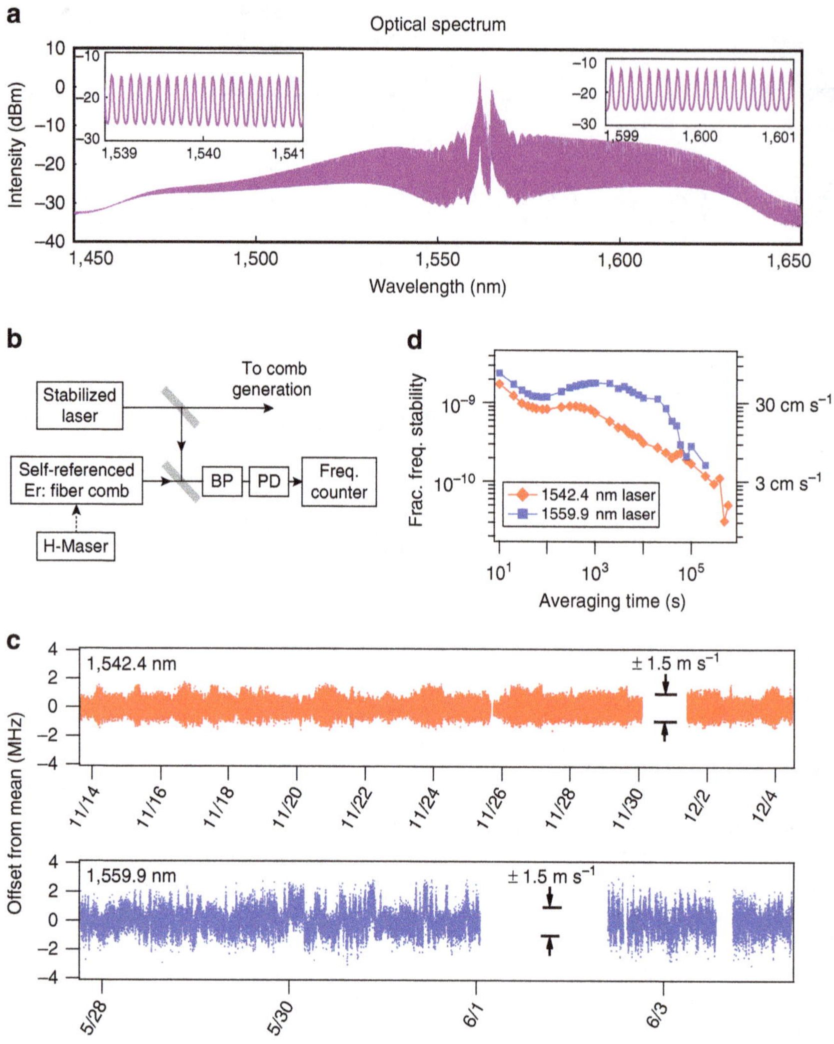

Figure 3 | Comb spectra and stability of the C_2H_2 and HCN reference lasers. (**a**) A typical comb spectrum from the 1,559.9 nm laser with >100 nm span generated with 600 mW pump power. The insets show the resolved line spacing of 12 GHz or ~0.1 nm. (**b**) Experimental set-up: BP, optical band-pass filter; PD, photodiode. All beam paths and beam combiners are in single mode fibre. (**c**) Time series of measured beat frequencies for the two frequency-stabilized lasers with 10 s averaging per measurement. The *x* axes are the dates in November of 2013 and May/June of 2014, respectively. (**d**) Allan deviation, which is a measure of the fractional frequency stability, computed from the time series data of **c**. Right-side scale gives the radial velocity precision.

The LR-EOFC system is mounted on an aluminum breadboard (18" × 32", or equivalently 45.7 × 81.3 cm) in a standard 19-inch instrument rack (see Fig. 2) for transport and implementation with the spectrograph at the NASA IRTF and at Keck II on Mauna Kea in Hawaii. The system is designed to provide operational robustness matching the requirements of astronomical observation. All optical components before the highly nonlinear fibre are polarization maintaining fibre-based, so as to eliminate the effect of polarization drift on spectral broadening in the highly nonlinear fibre. Moreover, no temperature control is required at the two telescope facilities. As a result, the comb is able to maintain its frequency, bandwidth and intensity without the need to adjust any parameters. During a 5 day run at IRTF, the comb had zero failures and the intensity of individual comb teeth was measured to deviate less than 2 dB, including multiple power-off and on cycling of the optical continuum generation system (see Fig. 4b).

Comb stability. As noted above, the frequency stability of the LR-EOFC is dominated by the stability of the reference laser frequency f_0. We explored the use of two different commercially available lasers (Wavelength References) that were stabilized, respectively, to Doppler- and pressure-broadened transitions in acetylene (C_2H_2) at 1,542.4 nm, and in hydrogen cyanide ($H^{13}C^{15}N$) at 1,559.9 nm. We note that the spectroscopy related to the locking of the reference laser to the molecular resonances is done internally to the laser system, so that our experiments only assess the stability of these commercial off-the-shelf lasers. To assess the stability, the stabilized laser frequencies were measured relative to an Er:fibre-based self-referenced optical frequency comb[11,32]. Fibre-coupled light from a reference laser was combined into a common optical fibre with light from the Er:fibre comb. Then the heterodyne beat between a single-comb line and the line-stabilized reference was filtered, amplified and counted with a 10 s gate time using a frequency counter that was referenced to a hydrogen maser (see Fig. 3b). The Er:fibre comb was stabilized relative to the same hydrogen maser, such that the fractional frequency stability of the measurement was $<2\times10^{-13}$ at all averaging times. The drift of the hydrogen maser frequency is $<1\times10^{-15}$ per day, thereby providing a stable reference at levels corresponding to a RV uncertainty $\ll 1\,\mathrm{cm\,s^{-1}}$. Thus, the frequency of the counted heterodyne beat accurately represents the fluctuations in the reference laser.

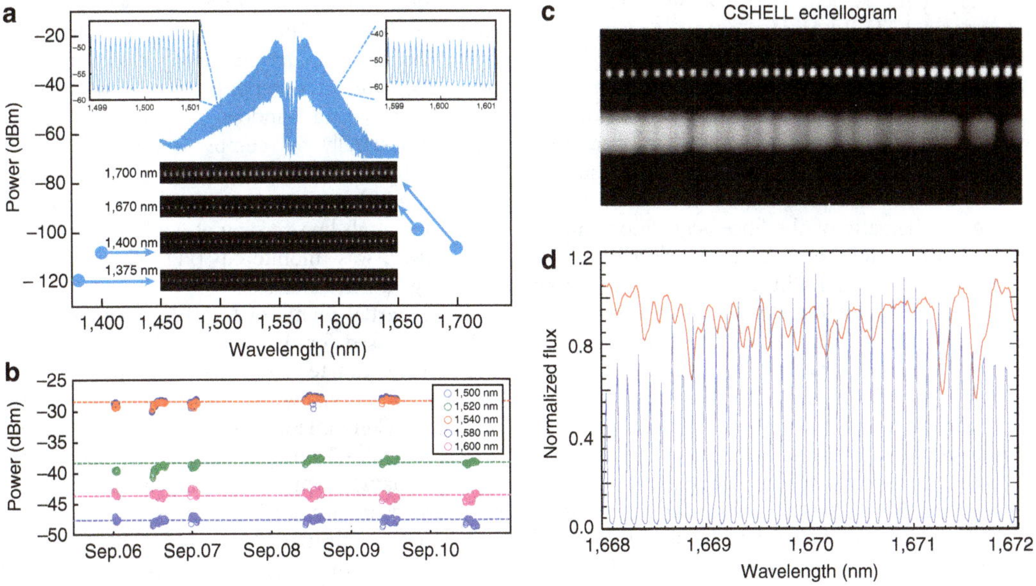

Figure 4 | Experimental results at IRTF. (**a**) Comb spectrum produced using 1,559.9 nm reference laser. The insets on top left and right show the resolved comb lines on the optical spectrum analyser. Comb spectra taken by the CSHELL spectrograph at 1,375, 1,400, 1,670 and 1,700 nm are presented as insets in the lower half of the figure. The blue circles mark the estimated comb line power and centre wavelength for these spectra. Comb lines are detectable on CSHELL at fW power levels. (**b**) Comb spectral line power versus time is shown at five different wavelengths. During the 5 day test at IRTF, no parameter adjustment was made, and comb intensity was very stable even with multiple power-on and -off cycling of the optical continuum generation system. (**c**) An image of the echelle spectrum from CSHELL on IRTF showing a 4 nm portion of spectrum ~1,670 nm. The top row of dots are the laser comb lines, while the broad spectrum at the bottom is from the bright M2 II-III giant star β Peg seen through dense cloud cover. (**d**) Spectra extracted from **c**. The solid red curve denotes the average of 11 individual spectra of β Peg (without the gas cell) obtained with CSHELL on the IRTF. The regular sine-wave like blue lines show the spectrum from the laser comb obtained simultaneously with the stellar spectrum. The vertical axis is normalized flux units.

The series of 10 s measurements of the heterodyne beat was recorded over 20 days in 2013 for the case of the 1,542.4 nm laser and more than 7 days in 2014 for the case of the 1,559.9 nm laser, as shown in Fig. 3c. Gaps in the measurements near 11/31 and 6/4 are due to unlocking of the Er:fibre comb from the hydrogen maser reference. From these time series, we calculate the Allan deviation, which is a measure of the fractional frequency fluctuations (instability) of the reference laser as a function of averaging time. As seen in Fig. 3d, the instability of the 1,542.2 nm laser is $<10^{-9}$ (30 cm s^{-1} RV, or corresponding to 200 kHz in frequency) at all averaging times greater than ~30 s. The 1,559.9 nm laser is less stable, but provides a corresponding RV precision of <60 cm s^{-1} for averaging times greater than 20 s. This different instability was to be expected because of the difference in relative absorption line strength between the acetylene and HCN-stabilized lasers. In both cases, the stability improves with averaging time, although at a rate slower than predicted for white frequency noise. As an aside, we note that despite the lower stability of the 1,559.9 nm laser, this wavelength ultimately produced wider and flatter comb spectra owing to the better gain performance of the fibre amplifier used in this work. We did not explore the noise mechanisms that lead to the observed Allan deviation, as they arise from details of the spectroscopy internal to the commercial off-the-shelf laser, to which we did not have access.

Additional analysis included an estimate of the drift of the frequencies of the two reference lasers obtained by fitting a line to the full multi-day counter time series. From these linear fits, an upper limit of the drift over the given measurement period was determined to be $<9\times10^{-12}$ per day for the acetylene-referenced laser and $<4\times10^{-11}$ for the hydrogen cyanide-referenced laser. This corresponds to equivalent RV drifts of <0.27 and <1.2 cm s^{-1} per day for the two references. Finally, we attempted to place a bound on the repeatability of the 1,542.4 nm reference laser during re-locking and power cycling. Although only evaluated for a limited number of power cycles and re-locks, in all cases, we found that the laser frequency returned to its predetermined value within <100 kHz, or equivalently, with a RV precision of <15 cm s^{-1}.

While these calibrations are sufficient for the few-day observations reported below, confidence in the longer term stability of the molecularly referenced continuous-wave lasers would be required for observations that could extend over many years. Likewise, frequency uncertainty of the molecular references should be examined. Properly addressing the potential frequency drifts on such a multi-year time scale would require a more thorough investigation of systematic frequency effects due to a variety of physical and operational parameters (for example, laser power, pressure, temperature and electronic offsets). Alternatively, narrower absorption features, as available in nonlinear Doppler-free saturation spectroscopy, could provide improved performance. For example, laboratory experiments have shown fractional frequency instability at the level of 10^{-12} and reproducibility of 1.5×10^{-11} for lasers locked to a Doppler-free transition in acetylene[33]. Most promising of all, self-referencing of an EOFC comb has been demonstrated recently[34], enabling full stabilization of the frequency comb to a GPS-disciplined standard. This would eliminate the need for the reference laser to define f_0, and thereby provide comb stability at the level of the GPS reference (for example, $<10^{-11}$ or equivalently <0.3 cm s^{-1}) on both long and short timescales.

IRTF telescope demonstration. To demonstrate that the laser comb is portable, robust and easy-to-use as a wavelength calibration standard, we shipped the laser comb to the NASA IRTF. IRTF is a 3 m diameter infrared-optimized telescope located at the summit of Mauna Kea, Hawaii. The telescope is equipped with a cryogenic echelle spectrograph (CSHELL) operating from 1–5.4 µm. CSHELL is a cryogenic, near-infrared traditional slit-fed spectrograph, with a resolution[35,36] of $R \sim \lambda/\Delta\lambda = 46{,}000$ and it images an adjustable single ~5-nm-wide order spectrum on a 256 × 256 InSb detector. We have modified the CSHELL spectrograph to permit the addition of a fibre acquisition unit for the injection of starlight and laser frequency comb light into a fibre array and focusing on the spectrograph entrance slit. A simple schematic of the fibre acquisition unit is shown in Fig. 2 and the details are described elsewhere[37,38]. Before the starlight reaches the CSHELL entrance slit, it can be switched to pass through an isotopic methane absorption gas cell to introduce a common optical path wavelength reference[38]. A pickoff mirror is next inserted into the beam to re-direct the near-infrared starlight to a fibre via a fibre-coupling lens. A dichroic window re-directs the visible light to a guide camera to maintain the position of the star on the entrance of the fibre tip. For the starlight, we made use of a specialized non-circular core multi-mode fibre, with a 50 × 100 µm rectangular core. These fibres 'scramble' the near-field spatial modes of the fibre, so that the spectrograph is evenly illuminated by the output from the fibre, regardless of the alignment, focus or weather conditions of the starlight impinging upon the input to the fibre. We additionally made use of a dual-frequency agitator motor to vibrate the 10 m length of the fibre to provide additional mode mixing, distributing the starlight evenly between all modes. Finally, a lens and a second pickoff mirror are used to relay the output of the starlight from the fibre output back to the spectrograph entrance slit. A single-mode fibre carrying the laser comb is added next to the non-circular core fibre carrying the starlight. This was accomplished by replacing the output single-fibre SMA-fibre chuck with a custom three-dimensional printed V-groove array ferrule. This allowed us to send the light from both the star and frequency comb to the entrance slit of the CSHELL spectrograph when rotated in the same orientation as the slit.

Finally, the laser comb and associated electronics rack were set-up in the room temperature (~ ±5 °C) control room. A 50 m length of single mode fibre was run from the control room to the telescope dome floor, and along the telescope mount to the CSHELL spectrograph to connect to the V-groove array and the fibre acquisition unit. The unpacking, set-up and integration of the comb fibre with CSHELL were straightforward, and required only 2 days working at an oxygen-deprived elevation of 14,000 feet in preparation for the observing run. Because the CSHELL spectrometer has a spectral window <5 nm, there was no effort made to generate spectrally flat combs. Comb lines are well resolved on CSHELL from 1,375 to 1,700 nm (Fig. 4a), with power adjusted by tunable optical attenuators to match the power of starlight and 6.7 pixels per comb line spacing at 1,670 nm wavelength. Also, comb line power was monitored (Fig. 4b) periodically during the observing run and was stable.

Three partial nights of CSHELL telescope time in September 2014 were used for this first on-sky demonstration of the laser comb. Unfortunately, the observing run was plagued by poor weather conditions, with 5–10 magnitudes of extinction because of clouds. Consequently, we observed the bright M2 II–III star β Peg (H = −2.1 mag), which is a pulsating variable star ($P = 43.3$ days). Typical exposure times were 150 s, and multiple exposures were obtained in sequence.

The star was primarily observed at 1,670 nm, with and without the isotopic methane gas cell to provide a wavelength calibration comparison for the laser comb. Other wavelengths were also observed to demonstrate that the spectral grasp of the comb is much larger than the spectral grasp of the spectrograph itself. Given the low SNR (signal-to-noise ratio) on β Peg from the high extinction because of clouds and CSHELL's limited spectral grasp, the SNR of these data is inadequate to demonstrate that the comb is more stable than the gas cell, as shown above.

One critical aspect of demonstrating the usability of the comb for astrophysical spectrographs is the comb line spacing. As seen in Fig. 4a,c,d, the spectra clearly demonstrate that the individual comb lines are resolved with the CSHELL spectrograph without the need for additional line filtering[39]. Thus this comb operates at a frequency that is natively well-suited for astronomical applications with significantly less hardware complexity compared with 'traditional' laser frequency combs.

Keck telescope demonstration. We were able to use daytime access to the near-infrared cryogenic echelle spectrograph

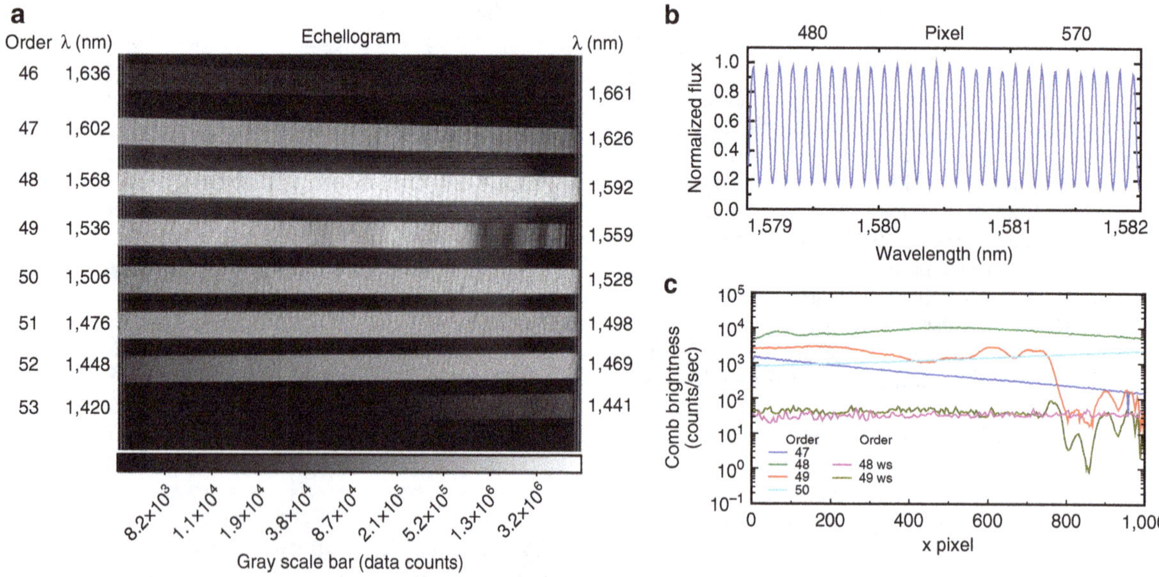

Figure 5 | Data from testing at Keck II. (**a**) Reduced NIRSPEC image from echelle order 46–53, displaying the stabilized laser comb using the 1,559.9 nm reference laser. Line brightness represents data counts. (**b**) A portion of the extracted comb spectrum from order 48 is plotted versus wavelength. (**c**) Comb brightness envelope of orders 47–50 and orders 48 and 49 when flattened by a waveshaper (ws).

(NIRSPEC) on the Keck-II telescope[40] to demonstrate our laser comb. NIRSPEC is a cross-dispersed echelle capable of covering a large fraction of the entire H-band in a single setting with a spectral resolution of R ~ 25,000. Observations were taken on 18 and 19 May 2015, with the comb set-up in the Keck-II control room in the same configuration as at the IRTF. The apparatus was reassembled after almost 8 months of storage from the time of the IRTF experiment and was fully operational within a few hours. The fibre output from the comb was routed through a cable wrap up to the Nasmyth platform where NIRSPEC is located. We injected the comb signal using a fibre feed into the integrating sphere at the input to the NIRSPEC calibration subsystem. While this arrangement did not allow for simultaneous stellar and comb observations, we were able to measure the comb lines simultaneously with the arc lamps normally used for wavelength calibration and to make hour-long tests of the stability of the NIRSPEC instrument at the sub-pixel level.

Figure 5a shows the laser comb illuminating more than six orders of the high-resolution echellogram. The echelle data were reduced in standard fashion, correcting for dark current and flat-field variations. Under this comb setting, a spectral grasp of ~ 200 nm is covered, from 1,430 to 1,640 nm. A zoomed-in spectral extraction (Fig. 5b) shows that individual comb lines are well resolved at NIRSPEC's resolution and spaced approximately 4 pixels apart (0.1 nm), consistent with the higher resolution IRTF observations described above. The spectral intensity of the comb lines can be made more uniform with a flattening filter to allow constant illumination over the entire span. In this demonstration, we were also able to implement a programmable optical filter (Waveshaper 1000s) from 1,530 to 1,600 nm, greatly reducing comb intensity variation (Plots 48ws and 49ws in Fig. 5c). If desired, a customized filter could increase the bandwidth of the flattened regime to cover the entire comb span.

We used a series of 600 spectra taken over a ~ 2 h time period to test the instrumental stability of NIRSPEC. Order 48, which had the highest SNR comb lines, was reduced following a standard procedure to correct for dark current and flat-field variations. Due to the quasi-Littrow configuration of the instrument, the slits appear tilted on the detector and the spectra have some curvature. We performed a spatial rectification using a flat-field image taken with a pinhole slit to mimic a bright compact object on the spectrum in order to account for this curvature. Wavelength calibration and spectral rectification to account for slit tilting were applied using the Ne, Kr, Ar and Xe arc lamps and the rectification procedure in the REDSPEC software written for NIRSPEC.

Instrumental stability was tested by performing a cross-correlation between the first comb spectrum in the 600 image series and each successive comb spectrum. The peak of the cross-correlation function corresponded to the drift, measured in pixels, between the images. Figure 6a demonstrates the power of the laser comb to provide a wavelength standard for the spectrometer. Over a period of roughly an hour the centroid of each comb line in Order 48 moved by about 0.05 pixel, equivalent to 0.0114 Å. By examining various internal NIRSPEC temperatures it is possible to show that this drift correlates to changes inside the instrument. Figure 6b shows changes in the temperatures measured at five different points within the instrument: the grating mechanism motor, an optical mounting plate, the top of the grating rotator mechanism, the base of the (unused) LN_2 container and the three mirror anastigmat assembly[40]. At these locations the temperatures range from 50 to 75 K and have been standardized to fit onto a single plot: $\Theta_i(t) = (T_i(t) - <T>)/\sigma(T)$. Average values of each temperature are given in Table 1 and show drifts of order 15–35 mK over this 1 h period. In its present configuration NIRSPEC is cooled using a closed cycle refrigerator without active temperature control—only the detector temperature is maintained under closed cycle control to ~ 1 mK.

Examination of the wavelength and temperature drifts in the two figures reveals an obvious correlation. A simple linear fit of the wavelength drift to the five standardized temperatures reduces the temperature-induced wavelength drifts from 0.05 pixel per hour to a near-constant value with a s.d. of $\sigma = 0.0017$ pixel for a single-comb line (bottom curve in Fig. 6a). While other

Table 1 | Internal NIRSPEC temperatures (K).

Rotator motor	54.944 ± 0.015	Optics plate	52.887 ± 0.023
Top of rotator	74.778 ± 0.035	LN_2 Can	53.663 ± 0.021
TMA	53.866 ± 0.022		

NIRSPEC, near-infrared cryogenic echelle spectrograph; TMA, three mirror anastigmat.

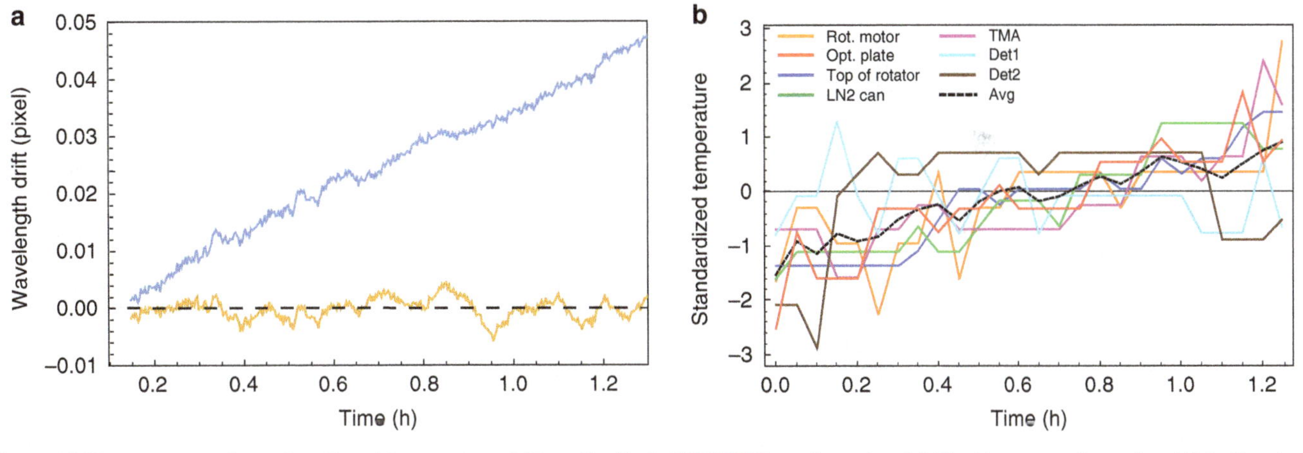

Figure 6 | Measurement of wavelength and temperature drift on the Keck II NIRSPEC spectrometer. (**a**) The blue curve shows the drift in the pixel location of individual comb lines in order 48 as measured with the cross-correlation techniques described in the text. The yellow curve shows the residual shifts after de-correlating the effects of the internal NIRSPEC temperatures. (**b**) Five internal NIRSPEC temperatures are shown as a function of time. For ease of plotting, the individual temperatures have been standardized with respect to the means and s.d. of each sensor (Table 1). The black dashed curve shows the average of these standardized temperatures. The effect of the quantization of the temperature data at the 10 mK level (as recorded in the available telemetry) is evident in the individual temperature curves.

mechanical effects may manifest themselves in other or longer time series, this small data set indicates the power of the laser comb to stabilize the wavelength scale of the spectrometer. At the present spectral resolution of NIRSPEC, $R \sim 25{,}000$, and with over 240 comb lines in just this one order, we can set a limit on the velocity drift due to drifts within NIRSPEC of $\sim c/R \times \sigma / \sqrt{\# lines} < 1.5\,\mathrm{m\,s^{-1}}$ where c is the speed of light.

Thus, operation with a laser comb covering over 200 nm with more than 2,000 lines in the H-band would allow much higher RV precision than is presently possible using, for example, atmospheric OH lines, as a wavelength standard. NIRSPEC's ultimate RV precision will depend on many factors, including the brightness of the star, NIRSPEC's spectral resolution (presently 25,000 but increasing to 37,500 after a planned upgrade) and the ability to stabilize the input stellar light against pointing drifts and line profile variations. We anticipate that in an exposure of 900 s NIRSPEC should be able to achieve an RV precision $\sim 1\,\mathrm{m\,s^{-1}}$ for stars brighter than $H = 7$ mag and $< 3\,\mathrm{m\,s^{-1}}$ for a stars brighter than $H < 9$ mag. A detailed discussion of the NIRSPEC error budget is beyond the scope of this paper, but a stable wavelength reference, observed simultaneously with the stellar spectrum, is critical to achieving this precision.

Discussion

Many challenges remain to achieving the high precision RV capability needed for the study of exoplanets orbiting late M dwarfs, jitter-prone hotter G and K spectral types, or young stars exhibiting high levels of RV noise in the visible. Achieving adequate signal-to-noise on relatively faint stars requires a large spectral grasp on a high-resolution spectrometer on a large aperture telescope. Injecting both the laser comb and starlight into the spectrograph with a highly stable line spread function demands carefully designed interfaces between the comb light and starlight at the entrance to the spectrograph. Extracting the data from the spectrometer requires careful attention to flat-fielding and other detector features. Finally, reducing the extracted spectra to produce RV measurements at the required level of precision requires sophisticated modelling of complex stellar atmospheres and telluric atmospheric absorption. The research described here addresses only one of these steps, namely the generation of a highly stable wavelength standard in the near IR suitable for sub $\mathrm{m\ s^{-1}}$ RV measurements.

References

1. Perryman, M. *The Exoplanet Handbook* (Cambridge University Press, 2011).
2. Marcy, G. & Howard, A. The astrophysics of planetary systems: formation, structure, and dynamical evolution. in *Proceedings IAU Symposium*. vol. 276, 3–12 (2011).
3. Kasting, J. F., Whitmire, D. P. & Reynolds, R. T. Habitable zones around main sequence stars. *Icarus* **101**, 108–128 (1993).
4. Pepe, F., Ehrenreich, D. & Meyer, M. R. Instrumentation for the detection and characterization of exoplanets. *Nature* **513**, 358–366 (2014).
5. Murphy, M. *et al.* High-precision wavelength calibration of astronomical spectrographs with laser frequency combs. *Mon. Not. R. Astron. Soc.* **380**, 839–847 (2007).
6. Osterman, S. *et al. Optical Engineering + Applications* 66931G–66931G (International Society for Optics and Photonics, 2007).
7. Li, C.-H. *et al.* A laser frequency comb that enables radial velocity measurements with a precision of 1 cm/s. *Nature* **452**, 610–612 (2008).
8. Braje, D., Kirchner, M., Osterman, S., Fortier, T. & Diddams, S. Astronomical spectrograph calibration with broad-spectrum frequency combs. *Eur. Phys. J. D* **48**, 57–66 (2008).
9. Glenday, A. G. *et al.* Operation of a broadband visible-wavelength astro-comb with a high-resolution astrophysi-cal spectrograph. *Optica* **2**, 250–254 (2015).
10. Steinmetz, T. *et al.* Laser frequency combs for astronomical observations. *Science* **321**, 1335–1337 (2008).
11. Yeas, G. G. *et al.* Demonstration of on-sky calibration of astronomical spectra using a 25 Ghz near-IR laser frequency comb. *Opt. Express* **20**, 6631–6643 (2012).
12. Quinlan, F., Yeas, G., Osterman, S. & Diddams, S. A 12.5 Ghz-spaced optical frequency comb spanning> 400 nm for near-infrared astronomical spectrograph calibration. *Rev. Sci. Instrum.* **81**, 063105 (2010).
13. Jones, D. J. *et al.* Carrier-envelope phase control of femtosecond mode-locked lasers and direct optical frequency synthesis. *Science* **288**, 635–639 (2000).
14. Cundiff, S. T. & Ye, J. Colloquium: femtosecond optical frequency combs. *Rev. Mod. Phys.* **75**, 325 (2003).
15. Diddams, S. A. The evolving optical frequency comb [invited]. *JOSA B* **27**, B51–B62 (2010).
16. Wilken, T. *et al.* A spectrograph for exoplanet observations calibrated at the centimetre-per-second level. *Nature* **485**, 611–614 (2012).
17. Wildi, F., Pepe, F., Chazelas, B., Curto, G. L. & Lovis, C. *SPIE Astronomical Telescopes I Instrumentation* 77354X–77354X (International Society for Optics and Photonics, 2010).
18. Halverson, S. *et al.* Development of fiber fabry-perot interferometers as stable near-infrared calibration sources for high resolution spectrographs. *Publ. Astron. Soc. Pac.* **126**, 445–458 (2014).
19. Bauer, F. F., Zechmeister, M. & Reiners, A. Calibrating echelle spectrographs with fabry-perot etalons, Astronomy & Astrophysics. **581**, A117 (2015).
20. Kippenberg, T. J., Holzwarth, R. & Diddams, S. Microresonator-based optical frequency combs. *Science* **332**, 555–559 (2011).
21. Del'Haye, P., Arcizet, O., Schliesser, A., Holzwarth, R. & Kippenberg, T. J. Full stabilization of a microresonator-based optical frequency comb. *Phys. Rev. Lett.* **101**, 053903 (2008).
22. Papp, S. B. *et al.* Microresonator frequency comb optical clock. *Optica* **1**, 10–14 (2014).
23. Suzuki, S. *et al. Nonlinear Optics* NM3A–NM33 (Optical Society of America, 2013).
24. Kotani, T. *et al. SPIE Astronomical Telescopes + Instrumentation* 914714–914714 (International Society for Optics and Photonics, 2014).
25. Imai, K., Kourogi, M. & Ohtsu, M. 30-thz span optical frequency comb generation by self-phase modulation in an optical fiber. *IEEE J. Quantum Electron.* **34**, 54–60 (1998).
26. Fujiwara, M., Kani, J., Suzuki, I. I., Araya, K. & Teshima, M. Flattened optical multicarrier generation of 12.5 Ghz spaced 256 channels based on sinusoidal amplitude and phase hybrid modulation. *Electron. Lett.* **37**, 967–968 (2001).
27. Huang, C.-B., Park, S.-G., Leaird, D. E. & Weiner, A. M. Nonlinearly broadened phase-modulated continuous-wave laser frequency combs characterized using dpsk decoding. *Opt. Express* **16**, 2520–2527 (2008).
28. Morohashi, I. *et al.* Widely repetition-tunable 200 fs pulse source using a mach-zehnder-modulator-based fiat comb generator and dispersion-flattened dispersion-decreasing fiber. *Opt. Lett.* **33**, 1192–1194 (2008).
29. Ishizawa, A. *et al.* Phase-noise characteristics of a 25-ghz-spaced optical frequency comb based on a phase-and intensity-modulated laser. *Opt. Express* **21**, 29186–29194 (2013).
30. Dudley, J. M., Genty, G. & Coen, S. Supercontinuum generation in photonic crystal fiber. *Rev. Mod. Phys.* **78**, 1135 (2006).
31. Mori, K. Supercontinuum lightwave source employing fabry-perot filter for generating optical carriers with high signal-to-noise ratio. *Electron. Lett.* **41**, 975–976 (2005).
32. Ycas, G., Osterman, S. & Diddams, S. Generation of a 660–2100 nm laser frequency comb based on an erbium fiber laser. *Opt. Lett.* **37**, 2199–2201 (2012).
33. Edwards, C. S. *et al.* Absolute frequency measurement of a 1.5-μm acetylene standard by use of a combined frequency chain and femtosecond comb. *Opt. Lett.* **29**, 566–568 (2004).
34. Beha, K. *et al.* Self-referencing a continuous-wave laser with electro-optic modulation. Preprint at arXiv:1507.06344 (2015).
35. Greene, T. P., Tokunaga, A. T., Toomey, D. W. & Carr, J. B. in *Optical Engineering and Photonics in Aerospace Sensing* 313–324 (International Society for Optics and Photonics, 1993).
36. Tokunaga, A. T., Toomey, D. W., Carr, J. B., Hall, D. N. & Epps, H. W. *Astronomy'90, Tucson AZ, 11-16 Feb 90,* 131–143 (International Society for Optics and Photonics, 1990).
37. Plavchan, P. P. *et al. SPIE Optical Engineering + Applications* 88641J–88641J (International Society for Optics and Photonics, 2013).
38. Plavchan, P. P. *et al. SPIE Optical Engineering + Applications* 88640G–88640G (International Society for Optics and Photonics, 2013).
39. Osterman, S. *et al. EPJ Web of Conferences* vol. 16, 02002 (EDP Sciences, 2011).
40. McLean, I. S. *et al. Astronomical Telescopes and Instrumentation* 566–578 (International Society for Optics and Photonics, 1998).

Acknowledgements

Three IRTF nights were donated in September 2014 to integrate and test the laser comb with CSHELL. One of these nights came from IRTF engineering time and the other two came from Peter Plavchan's CSHELL program to observe nearby M dwarfs with the absorption gas cell to obtain precise radial velocities. We are grateful to the leadership of the IRTF, Director Alan Tokunaga and Deputy Director John Rayner, as well as to the daytime and night time staff at the summit for their support. We further thank Jeremy Colson at Wavelength References for his assistance with the molecular-stabilized lasers. On-sky observations were obtained at the Infrared Telescope Facility, which is operated by the University of Hawaii under Cooperative Agreement no. NNX-08AE38A with the National Aeronautics and Space Administration, Science Mission Directorate, Planetary Astronomy Program. Daytime operations at the Keck-II telescope were carried out with the assistance of Sean Adkins and Steve Milner. We greatfully acknowledge the support of the entire Keck summit team in making these tests possible. We recognize and acknowledge the very significant cultural role and reverence that the summit of Mauna Kea has always had within the indigenous Hawaiian community. We are most fortunate to have the opportunity to conduct observations from this mountain. The data presented herein were obtained at the W.M. Keck Observatory, which is operated as a scientific partnership among the California Institute of Technology, the University of California and the National Aeronautics and Space Administration. The Observatory was made possible by the generous financial support of the W.M. Keck Foundation. We also acknowledge support from NIST and the NSF grant AST-1310875. This research was carried out at the Jet Propulsion Laboratory and the California Institute of Technology under a contract with the National Aeronautics and Space Administration and funded through the President's and Director's Fund Program.

Author contributions

X.Y., K.V., J.L., S.D., P.P., S.L., G.V., P.C. and C.B. conceived the experiments. All co-authors designed and performed experiments. X.Y. and K.V. prepared the manuscript with input from all co-authors.

Additional information

Competing financial interests: The authors declare no competing financial interests.

14

Widely tunable two-colour seeded free-electron laser source for resonant-pump resonant-probe magnetic scattering

Eugenio Ferrari[1,2,*], Carlo Spezzani[1,3,*], Franck Fortuna[4], Renaud Delaunay[5], Franck Vidal[6], Ivaylo Nikolov[1], Paolo Cinquegrana[1], Bruno Diviacco[1], David Gauthier[1], Giuseppe Penco[1], Primož Rebernik Ribič[1], Eleonore Roussel[1], Marco Trovò[1], Jean-Baptiste Moussy[7], Tommaso Pincelli[8], Lounès Lounis[6,9], Michele Manfredda[1], Emanuele Pedersoli[1], Flavio Capotondi[1], Cristian Svetina[1,10], Nicola Mahne[1], Marco Zangrando[1,11], Lorenzo Raimondi[1], Alexander Demidovich[1], Luca Giannessi[1,12], Giovanni De Ninno[1,13], Miltcho Boyanov Danailov[1], Enrico Allaria[1] & Maurizio Sacchi[6,14]

The advent of free-electron laser (FEL) sources delivering two synchronized pulses of different wavelengths (or colours) has made available a whole range of novel pump-probe experiments. This communication describes a major step forward using a new configuration of the FERMI FEL-seeded source to deliver two pulses with different wavelengths, each tunable independently over a broad spectral range with adjustable time delay. The FEL scheme makes use of two seed laser beams of different wavelengths and of a split radiator section to generate two extreme ultraviolet pulses from distinct portions of the same electron bunch. The tunability range of this new two-colour source meets the requirements of double-resonant FEL pump/FEL probe time-resolved studies. We demonstrate its performance in a proof-of-principle magnetic scattering experiment in Fe-Ni compounds, by tuning the FEL wavelengths to the Fe and Ni 3*p* resonances.

[1] ELETTRA—Sincrotrone Trieste, Area Science Park, 34149 Trieste, Italy. [2] Dipartimento di Fisica, Università degli Studi di Trieste, 34127 Trieste, Italy. [3] Laboratoire de Physique des Solides, Université Paris-Sud, CNRS-UMR 8502, Bât. 510, 91405 Orsay, France. [4] Centre de Sciences Nucléaires et de Sciences de la Matière, Université Paris-Sud, CNRS UMR 8609, Bât. 104-108, 91405 Orsay, France. [5] Laboratoire de Chimie Physique Matière et Rayonnement, Sorbonne Universités, UPMC Univ Paris 06, CNRS UMR 7614, 75005 Paris, France. [6] Institut des NanoSciences de Paris, Sorbonne Universités, UPMC Univ Paris 06, CNRS UMR 7588, 75005 Paris, France. [7] Service de Physique de l'Etat Condensé, DSM/IRAMIS/SPEC, CNRS UMR 3680, CEA Saclay, 91191 Gif-sur-Yvette, France. [8] Dipartimento di Fisica, Università degli Studi di Milano, 20133 Milano, Italy. [9] Ecole Normale Supérieure, PSL Research University, 75231 Paris, France. [10] Graduate School of Nanotechnology, Università degli Studi di Trieste, 34127 Trieste, Italy. [11] Istituto Officina dei Materiali, Consiglio Nazionale delle Ricerche, 34149 Trieste, Italy. [12] ENEA, Centro Ricerche Frascati, Via E. Fermi 45, 00044 Frascati, Italy. [13] Laboratory of Quantum Optics, University of Nova Gorica, 5001 Nova Gorica, Slovenia. [14] Synchrotron SOLEIL, L'Orme des Merisiers, Saint-Aubin, B.P. 48, 91192 Gif-sur-Yvette, France. * These authors contributed equally to this work. Correspondence and requests for materials should be addressed to E.A. (email: enrico.allaria@elettra.eu) or to M.S. (email: maurizio.sacchi@insp.jussieu.fr).

Free-electron laser (FEL) sources covering the wide spectral range from extreme ultraviolet to hard X-rays represent a breakthrough in photon science, with applications in physics, chemistry and biology. Many aspects of the spectral and temporal characteristics of the FEL pulses can be tailored to specific experimental needs by an accurate control of the lasing process, in the so-called beam by design approach[1]. The ability to run the FEL source in two-colour configuration, that is, to create two synchronized FEL pulses of differing wavelengths, has enormous potential for femtosecond time-resolved studies[2,3] as it opens up unique opportunities for studying the dynamic response in atomic, molecular and solid state systems by selectively tuning electron resonances in atoms. As a consequence it has engendered major research[4–8] and development[9–15] efforts at all FEL facilities worldwide, with the ambition of attaining wide-ranging colour tunability and timing control.

Various two-colour schemes have been proposed, both for seeded[2,9,10,14] and for self-amplified spontaneous emission (SASE)[11–13] FEL sources. Initial configurations delivered two short FEL pulses with a controlled temporal separation in the range of a few hundred femtoseconds and a small photon wavelength separation ($\sim 1\%$). Such configurations, where a single electron bunch generates the two FEL pulses, have served users for experiments both at seeded[2] and at SASE[11–13] facilities. In the case of SASE, differing photon wavelengths are obtained by dividing the radiator in two slightly detuned sections[11]. In the case of external seeding, the FEL wavelength separation is controlled by acting on the seed laser wavelength and by taking advantage of a residual controllable energy chirp on the electron beam[2,10]. For self-seeding schemes, it has been demonstrated[14] that two seeded FEL pulses can be generated using two distinct Bragg diffraction lines in the self-seeding crystal recombined within the taper-tuned undulators. The possibility of producing two colours with a wider spectral separation (up to 30%) has been demonstrated recently at the SACLA hard X-ray SASE source by using the capabilities of a variable gap undulator[13].

Until now, no configuration that generates two pulses with independently tunable wavelengths over a wide spectral range had been designed for externally seeded FELs. A whole new class of pump–probe experiments that require both pump and probe to be element selective is created by combining the full coherence of seeded FELs with a broad and independent tunability of the two colours.

Over the last decade, time-resolved studies made frequent use of short X-ray pulses as a probe that is coupled to an optical laser pump. Femto-slicing at synchrotrons[16–22], high harmonic generation in gases[23–26] and FEL sources[27–32] deliver extreme ultraviolet and X-ray pulses with sub-100-femtoseconds duration that have been used for studying the ultrafast dynamics of magnetic[16–27,29,31] and structural[20,28–30] order in optical-laser pump/X-ray probe experiments. Tuning the wavelength to an atomic resonance provides the probe with element selectivity, which is of considerable interest especially for magnetic studies.

Developing FEL sources that can produce two pulses with independently selectable wavelengths for the pump and the probe and with a well-defined time separation obviously widens the potential of FEL radiation for studying the dynamics of a process and makes it possible to associate the pump energy to a specific electronic excitation of a given element. One field that will surely profit from this new tool is magnetization dynamics in $3d$-transition-metal and rare-earth based oxides and compounds[18,19,22,33–37]: the presence of highly localized $3d$ and $4f$ orbitals and of mediated exchange interactions suggests that associating the pump energy to a specific electronic excitation will influence the magnetization dynamics profoundly, compared with using a non-resonant pump.

In the proof-of-principle time-resolved scattering experiment on Fe–Ni compounds described here, we use the new two-colour configuration of the externally seeded FERMI FEL source to generate, from the same electron bunch, two synchronized pulses with up to 30% spectral separation. The pump FEL pulse excites the Fe $3p \rightarrow 3d$ transition resonantly, while the second FEL pulse, tuned to the Ni $3p \rightarrow 3d$ resonance, probes the ultrafast Ni magnetization dynamics. The experiment successfully reveals the potential of this new source for investigating structural, electronic and magnetization dynamics in the fields of condensed matter as well as atomic and molecular physics.

Results

Two-colour seeded FEL with wide wavelength tunability. The experiment was performed at the FERMI facility[38,39], which is a seeded FEL operated in the high-gain harmonic generation (HGHG) mode[40,41]. The chosen configuration (see Methods) provided a relatively long (~ 1 ps) electron bunch interacting with a short (~ 100 fs) ultraviolet laser pulse (seed laser) in the first undulator section called the modulator (Mod in Fig. 1a). As a consequence of this interaction, the electron beam energy is modulated with a periodicity imposed by the seed laser wavelength λ_{seed}. Following a magnetic chicane that works as a dispersive section (DS in Fig. 1a), the energy modulation is converted into a density modulation (bunching), which has strong harmonic components. Finally, in a second long undulator section called the radiator (Rad in Fig. 1a), the bunched electrons generate coherent FEL emission at one of the harmonics of the seed laser which is selected by setting the undulator gap. The advantages of HGHG with respect to SASE FEL stem from the fine control of the initial bunching, making it possible to generate FEL pulses with a high degree of longitudinal coherence[42]. Moreover, since only electrons interacting with the seed laser are bunched, this scheme provides a good control of the FEL temporal properties[43].

Two FEL pulses with a controlled delay can be produced by seeding the same electron bunch with two seed pulses[2]. Since in the HGHG seeding process the final FEL wavelength is determined mainly by λ_{seed} and it must be close to one of its harmonics, a way for delivering two-colour FEL pulses with very different wavelengths ($>10\%$ separation) relies on seeding the electron beam with two laser pulses and on sustaining the amplification process at both wavelengths independently (Fig. 1a).

To achieve this, some constraints have to be dealt with. Both seed wavelengths λ_{seed_1} (for the probe) and λ_{seed_2} (for the pump) have to modulate the electron energy in the interaction region efficiently so their separation must be within the modulator working bandwidth. The two seed pulses modulate the energy in distinct regions of the electron beam. For each region, the dispersive section converts the energy modulation into an electron density modulation that carries all the harmonic components of the corresponding seed wavelength, either λ_{seed_1} or λ_{seed_2}. The electron beam is now ready for the amplification of one of these harmonics, selected by the resonance condition of the radiator (undulator gap). A large separation between the two colours can be obtained by dividing the radiator into two subsections (Rad_1 and Rad_2 in Fig. 1a), one resonant at $\lambda_{FEL\_1} = \lambda_{seed\_1}/m$ and the other at $\lambda_{FEL\_2} = \lambda_{seed\_2}/n$, with m and n integers. Since the radiator bandwidths are markedly narrower than the modulator one, we can emit efficiently the pump (or the probe) beam from one radiator subsection only, while suppressing its amplification in the other, selectively

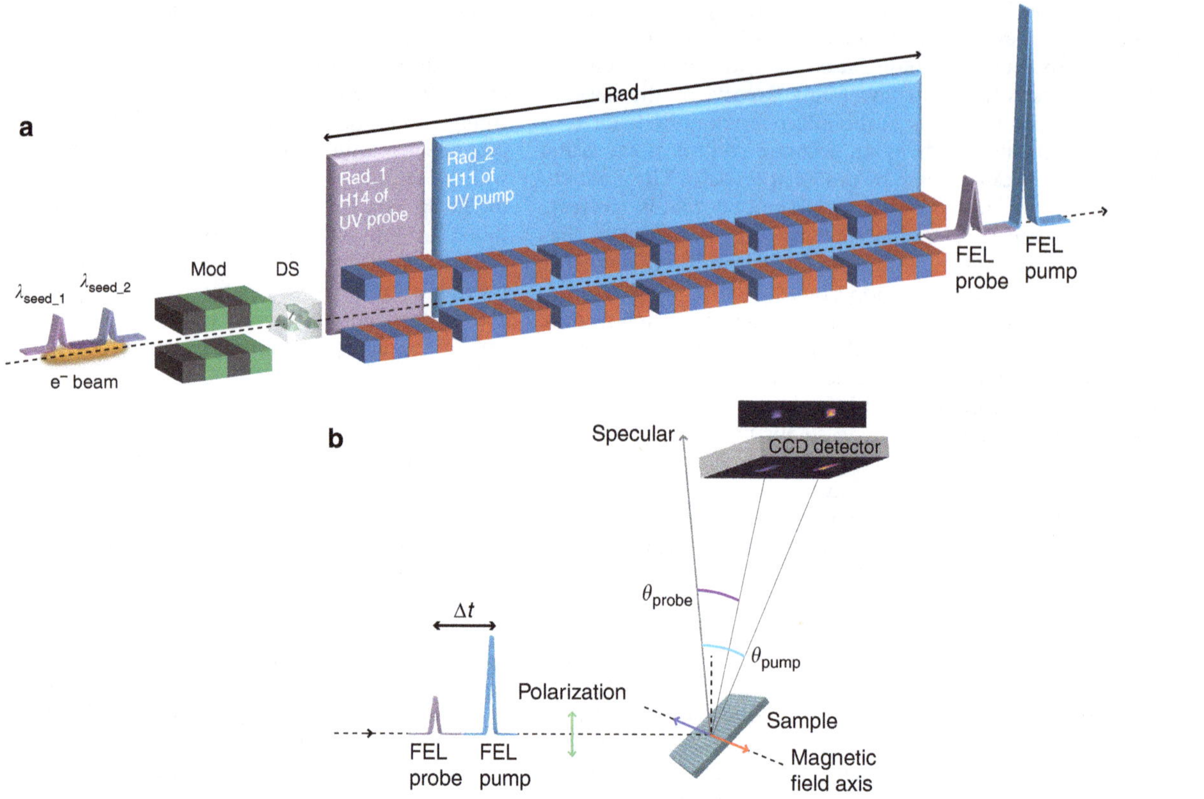

Figure 1 | Schematic setup for a two-colour double resonance FEL experiment. (**a**) Two-colour seeded FEL source configuration: the modulator (Mod), dispersive (DS) and radiator (Rad) sections of the FEL source are outlined. In the modulator section, two ultraviolet (UV) laser pulses of wavelength λ_{seed_1} and λ_{seed_2} delayed by Δt interact with the same electron bunch, imposing an energy modulation that is converted into density modulation in DS. The first radiator subsection Rad_1 is tuned to the 14th harmonic of λ_{seed_1} and the second subsection Rad_2 is tuned to the 11th harmonic of λ_{seed_2}, generating the FEL probe and pump pulses, respectively (see Methods). (**b**) Magnetic scattering experiment: the two linear p-polarized FEL pulses reach the magnetic grating sample and diffract at different angles according to their wavelengths. The diffracted intensities are recorded by a two-dimensional detector (CCD camera). The wavelength separation between pump and probe is detected as a spatial separation at the CCD, while their time separation Δt is defined by the delay between the two seed pulses.

(see Methods). Finally, constraints on the temporal separation Δt between the two FEL pulses are set by the need to avoid interference between the laser seeds (lower limit) and by the electron bunch duration (upper limit). In the example reported below, we spanned delays ranging from 300 to 800 fs.

The Fe-3*p* resonant-pump and Ni-3*p* resonant-probe test experiment (Fig. 1b) used two FEL pulses tuned to $\lambda_{FEL\_2} = 23.2$ nm and $\lambda_{FEL\_1} = 18.7$ nm, corresponding to the 11th harmonic of $\lambda_{seed\_2} = 255$ nm and to the 14th harmonic of $\lambda_{seed\_1} = 261.5$ nm, respectively. To this purpose, a special configuration of the FERMI seed laser was implemented based on the combined use of two ultraviolet pulses originating from a common infra-red source through two separated generation channels. One made use of an optical parametric amplifier (OPA) for producing the 255 nm seed, the other of a third harmonic generation (THG) setup for the 261.5 nm seed. This approach made the twin seeding possible at two different ultraviolet wavelengths, one of them tunable via the OPA (see Methods). We tested different distributions of the six undulator modules over the Rad_1 and Rad_2 radiator subsections (see Fig. 1a). We obtained different power distributions between pump and probe by going from one module in Rad_1 and five in Rad_2 to three modules in each subsection. All the configurations provided satisfactory stable conditions for producing two-colour FEL pulses. Since in our test experiment the pump is required to be more energetic than the probe, five of the six available radiator modules were tuned to produce 23.2 nm pulses (Rad_2 in Fig. 1a), while the remaining module (Rad_1) was tuned to the probe wavelength. It was important, in this configuration, that Rad_1 was the first of the undulator modules, to prevent the smearing of the electron density modulation along the radiator section to degrade its performance. We verified also that one can switch readily the FEL pump and probe wavelengths, by reversing the time delay between OPA and THG generated seed pulses and inverting the gap settings of the Rad_1 and Rad_2 radiator subsections.

Figure 2a shows the spectral distribution of the two ultraviolet seed laser pulses and Fig. 2b shows the FEL pulse energy as a function of the modulator gap, when using only the ultraviolet -probe or only the ultraviolet -pump seeds. The two curves of Fig. 2b, which are normalized to the same amplitude, illustrate at each wavelength the extreme sensitivity of the FEL intensity to the modulator setting. A modulator gap of 19.94 mm optimizes the FEL pump emission when seeding at λ_{seed_2}, while a gap of 19.60 mm is best when seeding at λ_{seed_1} to produce the FEL probe pulse. The gap can be used as an adjustable parameter for the fine control of the relative efficiency in the generation of the pump and probe FEL pulses, thanks to the ~3% resonance bandwidth of the modulator (Supplementary Fig. 1). In our case, a good compromise was found at a gap of 19.75 mm, which made it possible to generate both $\lambda_{FEL\_1} = 18.7$ nm and $\lambda_{FEL\_2} = 23.2$ nm pulses, albeit with a reduced intensity. For the Fe–Ni experiment, the FERMI FEL source was characterized by pulse energies of up to ~10 μJ at the pump wavelength and ~1 μJ at the probe wavelength using these parameters. Once converted into a fluence *F* at the sample surface (see Methods), these values were sufficient to reach, in our experiment, the damage threshold

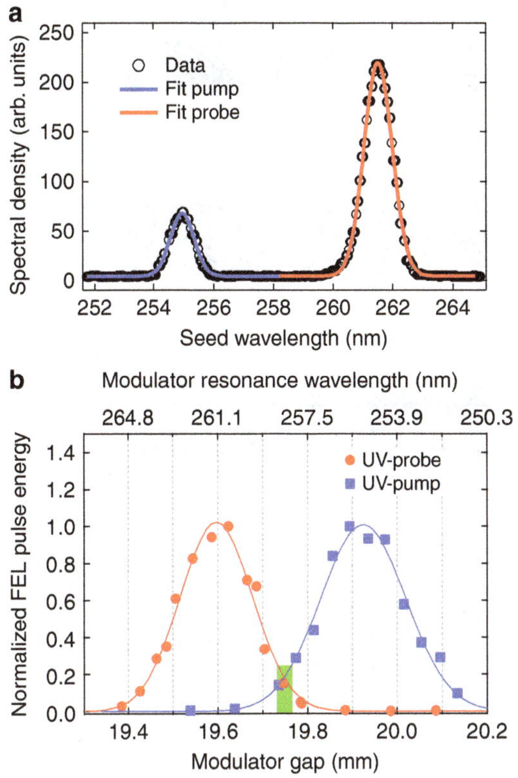

Figure 2 | Seed pulses and modulator setting. (**a**) Spectral properties of the ultraviolet (UV) laser twin-seed source. Lines are Gaussian fits to the $\lambda_{seed\_1} = 261.5$ nm (red line) and $\lambda_{seed\_2} = 255$ nm (blue line) probe and pump contributions, respectively. (**b**) Modulator gap dependence of the FEL output for the two seed wavelengths. Circles and squares refer to seeding at 261.5 and 255 nm, respectively. Each point is the average of 100 consecutive FEL shots. Lines represent Gaussian fits to the intensity distributions. The curves are normalized to the same average maximum, showing that tuning the modulator gap to 19.75 mm (vertical green bar) makes it possible to seed with two colours simultaneously, preserving a fraction of the maximum pulse energy.

and single shot detection conditions for the pump and the probe pulses, respectively.

Resonant-pump/resonant-probe magnetic scattering experiment. We tested the two-colour twin-seeded FEL source by studying the resonant-pump/resonant-probe magnetization dynamics in Fe-Ni samples, using the IRMA reflectometer[44] installed at the DiProI beamline[45,46]. The samples were a 20-nm-thick permalloy ($Ni_{0.81}Fe_{0.19}$ alloy) film deposited on a Si grating and a 12.5-nm-thick $NiFe_2O_4$ layer epitaxially grown on $MgAl_2O_4(001)$. Both samples were structured as line gratings with a period of ~600 nm (see Methods). They worked as dispersive elements, separating different wavelengths at the level of the two-dimensional in-vacuum charge-coupled device (CCD) detector[23,24]. All Bragg peaks generated by the grating samples at different wavelengths fell within the angular acceptance of the detector and could be collected simultaneously (see Fig. 3).

The FEL polarization was set to linear vertical to optimize the sensitivity to the sample magnetization in transverse geometry[23,24,47–49], that is, with the external magnetic field applied normal to the scattering plane and parallel to the lines of the grating sample (see Fig. 1b). After an initial 80 mT magnetic pulse, the scattered intensity was collected in an applied field of 20 mT, guaranteeing the sample magnetic saturation (see Methods). In the following, the magnetic signal is defined as an asymmetry ratio, that is, as the difference between scattered intensities measured for opposite signs of the applied field divided by their sum, as shown in Fig. 4. At each given delay Δt, the Ni magnetic signal was measured as a function of the pump fluence F (see Methods for the relationship between FEL pulse energy and fluence at the sample). The pump wavelength was tuned either to the Fe-3p resonance ($\lambda_{FEL\_2} = 23.2$ nm) or off-resonance ($\lambda_{FEL\_2} = 25.5$ nm), the latter being obtained simply by tuning the radiator subsection Rad_2 to the 10th harmonic of the λ_{seed_2} seed laser wavelength, instead of the 11th. It is worth underlining that, according to calculations based on tabulated optical constants[50] (see also http://henke.lbl.gov/optical_constants/), the fraction of pump energy absorbed by the sample at 23.2 nm and at 25.5 nm differs by less than 2% for both permalloy and ferrite films.

First, we explored the ultrafast Ni demagnetization while varying the delay Δt between the FEL probe and pump by adjusting the delay between the corresponding seed laser pulses. An example of delay dependence spanning the 300–800 fs range is shown in Fig. 5 where the Ni magnetic signal is reported after a Fe-3p resonant pump pulse with fluence $F = 10$ mJ cm^{-2} (dots and squares refer to Ni-ferrite and permalloy samples, respectively). The asymmetry ratio in the Bragg peak intensity is calculated over a limited detector area of ~100 × 100 μm^2 to ensure homogeneous pump fluence and the Ni magnetic signal is normalized to its static value measured with no pump.

The main advantage of this novel two-colour scheme over those developed previously at the FERMI seeded source[2] is its ability to tune both λ_{FEL_1} and λ_{FEL_2} to selected values over a broad range. It is also important to stress that this scheme makes the switching between on- and off-resonance pumping fast and easy. As mentioned before, this can be achieved simply by changing the gap of the Rad_2 radiator subsection for selecting a different harmonic of the λ_{seed_2} wavelength. An example of on/off-resonance pumping is given in Fig. 6. It shows the Ni magnetic signal (normalized to its static value) measured at a fixed time delay of ~400 fs for a FEL pump wavelength tuned to the Fe-3p resonance ($\lambda_{FEL\_2} = 23.2$ nm, red circles) or off-resonance ($\lambda_{FEL\_2} = 25.5$ nm, blue squares) as a function of the pump fluence F. The permalloy results (Fig. 6a) do not reveal a measurable effect of the pump wavelength: both curves show the same F-dependence of the Ni magnetic signal, which attains a ~50% reduction at F~10 mJ cm^{-2}. On the contrary, pumping at the two on/off-resonance wavelengths results in an apparent difference in Ni demagnetization behaviour when F exceeds ~5 mJ cm^{-2} in the case of Ni-ferrite (Fig. 6b).

Although a detailed discussion of the results reported in Figs 5 and 6 is not within the scope of this communication, the observed differences between ferrite and permalloy behaviour can be ascribed to the direct hybridization of delocalized Fe and Ni 3d orbitals in ferromagnetic permalloy versus indirect exchange (via oxygen) of more localized 3d orbitals in ferrimagnetic $NiFe_2O_4$. These early results are intriguing and more studies are under consideration to shed light on the observed pump wavelength dependence.

Discussion

We have developed and tested a new FEL setup capable of delivering two-colour time-delayed pulses with independent wavelength tunability over a wide spectral range (18.7–25.5 nm). Combined with the seeded nature of the FERMI source[39], this provides improved conditions for two-colour FEL experiments that require tuning both the pump and the probe to selected atomic resonances. The potential of this two-colour

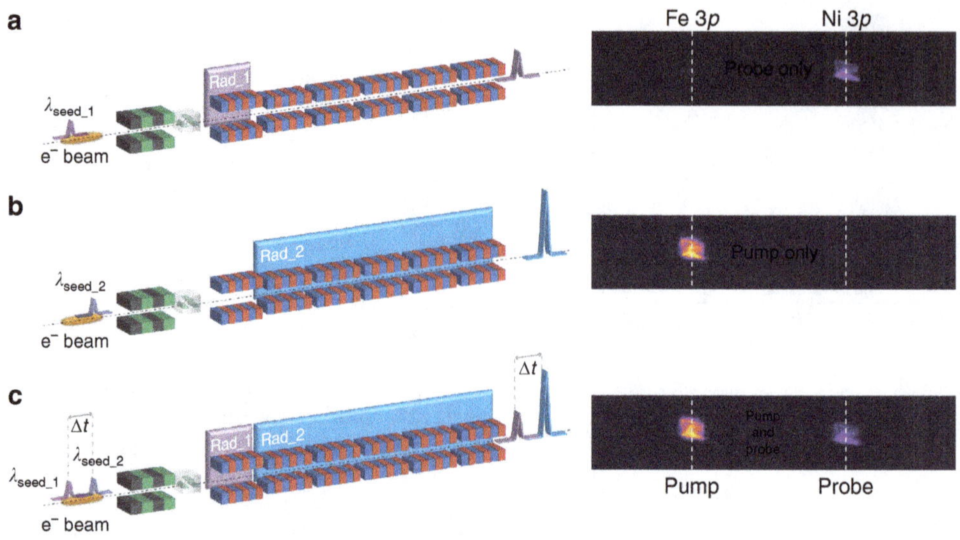

Figure 3 | FEL source configuration and scattering data recording. Diffracted intensity from the 20-nm-thick permalloy grating sample at 46.3° incidence. Data are collected under different seeding conditions (schematics on the left) using a position-sensitive CCD detector (images on the right); the 1,025 × 202 pixel images correspond to $13.84 \times 2.73\,mm^2$ and cover ~1.48° in scattering angle. (**a**) The $\lambda_{seed\_1} = 261.5$ nm laser pulse is sent through the modulator, turning on the Ni-3*p* resonant FEL emission at $\lambda_{FEL\_1} = 18.7$ nm in Rad_1 (14th harmonic) and no emission from Rad_2. (**b**) The $\lambda_{seed\_2} = 255$ nm laser pulse generates the Fe-3*p* resonant FEL emission at $\lambda_{FEL\_2} = 23.2$ nm in the radiator section Rad_2 (11th harmonic) and no emission from Rad_1. (**c**) Both seed laser pulses, delayed by Δt, interact with the electron bunch, generating Fe-3*p* resonant pump and Ni-3*p* resonant probe FEL pulses, also delayed by Δt.

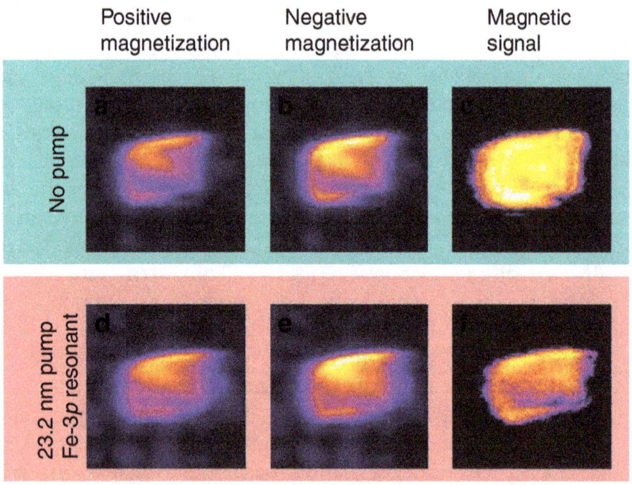

Figure 4 | Magnetic signal in the scattering data. Diffracted intensity at the Ni-3*p* resonant probe wavelength with no pump (**a–c**) and following a Fe-3*p* resonant pump pulse (**d–f**). The pump fluence is $F = 8\,mJ\,cm^{-2}$, the delay Δt is 450 fs. (**a**,**d**) and (**b**,**e**) Diagrams refer to a positive and negative saturating magnetic field, respectively. The magnetic signal, expressed as the asymmetry ratio, is shown in **c** and **f**. Each picture is 128 × 128 pixels, corresponding to $1.73 \times 1.73\,mm^2$.

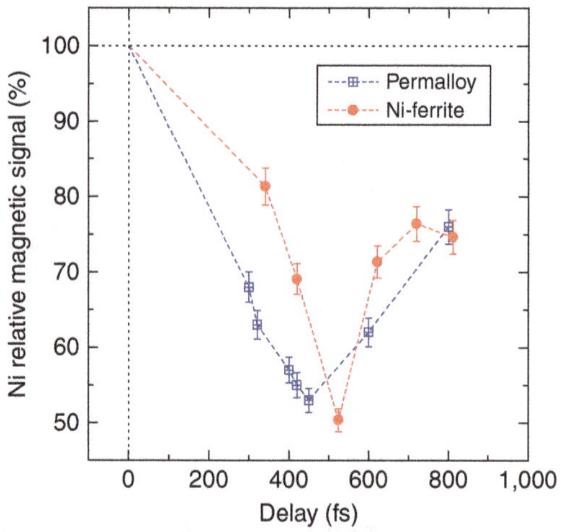

Figure 5 | Time-dependent magnetic signal. Ni demagnetization in the permalloy (blue squares) and the Ni-ferrite (red circles) samples at several delays Δt between probe and pump pulses ($F = 10\,mJ\,cm^{-2}$). Vertical error bars represent s.d. (see Methods). The maximum fluctuation in the pump-probe delay over the measurement duration (±5 fs, see Methods) is smaller than the point width. Lines are a guide to the eye.

scheme has been demonstrated by a scattering experiment that probes the magnetization dynamics in systems containing two magnetic elements, Fe and Ni. Undoubtedly, it can find original applications in many other fields of condensed matter, atomic and molecular physics.

From a technical point of view, the solution that we propose is based on seeding the same electron bunch with two independent laser pulses and on splitting the FEL radiator into two subsections. On one hand, this solution offers the possibility of selectively tuning the two FEL colours over a very wide range. It may go well beyond the 30% bandwidth demonstrated here, by amplifying different harmonics of the seed wavelengths in each radiator subsection. On the other hand, using two laser seeds that modulate the same electron bunch, and two radiators impose some constraints on the relationship between the λ_{FEL_1} and λ_{FEL_2} wavelengths, both in terms of FEL intensity and of possible gaps in the range of wavelengths that can be spanned.

Figure 7 summarizes the calculated source performance when λ_{FEL_1} and λ_{FEL_2} span the 16–28 nm range, showing that marked intensity variations are present. The colour code represents the relative modulator efficiency for each couple of wavelengths, calculated assuming that the modulator resonance is set to the average value of λ_{FEL_1} and λ_{FEL_2}. Both seed wavelengths are

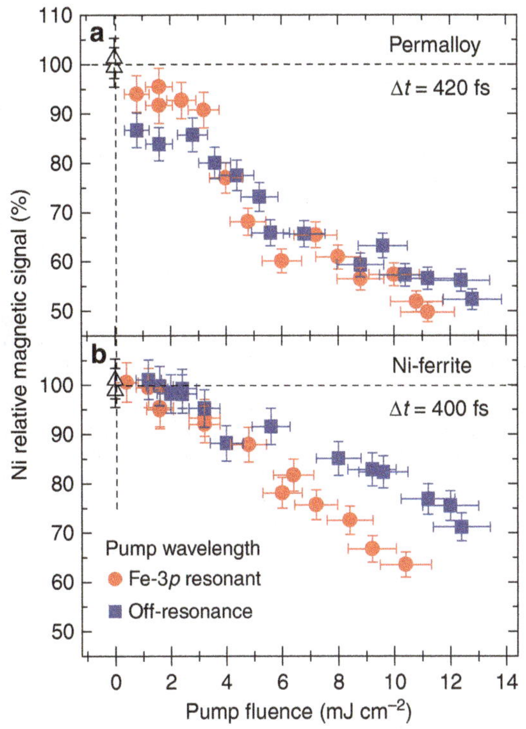

Figure 6 | Resonant versus non-resonant pumping. Pump fluence dependence of the Ni demagnetization in permalloy (**a**) and Ni-ferrite (**b**) at ~400 fs delay, comparing the results for Fe-3p resonant ($\lambda_{FEL\_2}=23.2$ nm, red circles) and non-resonant ($\lambda_{FEL\_2}=25.5$ nm, blue squares) FEL pump pulses. The Ni magnetic signal is reported as the asymmetry ratio in the Bragg peak intensity, normalized to the value measured with no pump (open triangles). Error bars represent s.d. (see Methods).

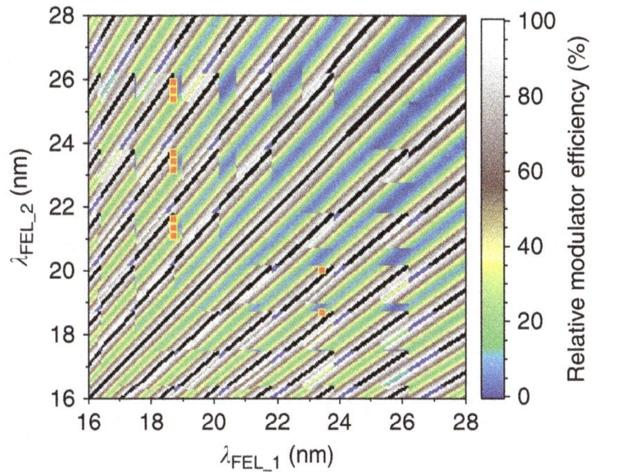

Figure 7 | Calculated seeding efficiency over the 16–28 nm range. The colour code represents the relative modulator efficiency at λ_{FEL_1} and λ_{FEL_2} when the modulator gap is set to resonate with their average value. The calculation uses radiator harmonics from 9 to 16 and λ_{seed} values between 228 and 262 nm. Black dots correspond to (λ_{FEL_1}, λ_{FEL_2}) couples whose λ_{seed} values are within the radiator bandwidth and cannot be produced using the proposed source scheme. Red squares identify the couples of wavelengths explored during the test experiment (Supplementary Figs 2 and 3).

allowed to span the 228–262 nm range covered by the OPA, and radiator harmonics from 9 to 16 are considered. The finite modulator bandwidth defines the maximum intensity that can be obtained for each (λ_{FEL_1}, λ_{FEL_2}) combination, hence the efficiency of the two-colour process. The radiator bandwidth imposes limitations on the independent tunability of λ_{FEL_1} and λ_{FEL_2}: black dots forming diagonal lines in Fig. 7 mark couples of wavelengths whose corresponding λ_{seed} values are close enough to be amplified in both radiator subsections. In this case, four FEL pulses, and not two, would be generated and the proposed two-colour scheme does not work properly.

Figure 7 shows the tuning capabilities and limitations of the adopted two-colour FEL scheme over the 16–28 nm range that broadly covers the wavelengths used in our test experiment. The red squares in Fig. 7 indicate the pairs of FEL wavelengths that were actually explored for the Fe-Ni double-resonant pump–probe measurements (Supplementary Fig. 2 and Supplementary Fig. 3). In principle, a much wider range of λ_{FEL} values extending up to 90 nm can be covered by using the full set of harmonics available at FERMI[39,51]. At wavelengths longer than ~45 nm, though, the limited range of the OPA and the low radiator harmonic numbers introduce gaps in the (λ_{FEL_1}, λ_{FEL_2}) values that can be covered by this two-colour FEL scheme (Supplementary Fig. 4).

The accessible delay range between the pump and the probe is limited by the generation of the two FEL pulses from the same electron bunch of finite temporal length. In our experiment, we spanned the 300–800 fs range and an extension to 200–1,000 fs can be envisaged. This remains a strong constraint on the class of dynamic phenomena that can be addressed. Concerning ultra-fast demagnetization, in particular, many systems of interest feature response times of the order of 200 fs (refs 21,26,52), at the limit of the accessible range.

Further developments can be envisaged for improving the source characteristics, such as the twin-bunch mode recently demonstrated in SASE configuration[15]. The implementation of a similar scheme at FERMI would provide a more efficient bunching at the two wavelengths, a more efficient coupling in the radiator sections and, in fine, a significant increase in the energy per pulse, which could attain tens of microjoules for both the pump and the probe. Moreover, using two independent bunches would provide additional flexibility for tuning the two λ_{seed} wavelengths and would soften the constraints on the temporal separation between pump and probe pulses. Another significant improvement, already planned at FERMI, implies a second OPA for tuning both λ_{seed} wavelengths independently, as used for computing the tuning range reported in Fig. 7 and Supplementary Fig. 4. The desired resonant condition for both the pump and the probe FEL pulses could be finely matched.

Finally, it is worth remembering that the FERMI radiator section is composed of Apple-II type undulators[53] delivering radiation of selectable polarization, either circular (right/left) or linear (vertical/horizontal). Therefore our two-colour source offers the possibility of choosing the polarization state of each pulse independently, which may be especially important in atomic and molecular physics studies.

The two-colour extreme ultraviolet source that we have developed at FERMI already has potential for many interesting and original studies in magnetization dynamics and beyond. For instance, it can cover the 3*p* resonances of any couple of elements among Mn, Fe, Co and Ni, making a wide class of relevant magnetic materials accessible to resonant FEL pump/resonant FEL probe experiments. More generally, it enables the excitation of a particular energy and polarization-selected resonance on a well-defined atomic site in a complex system and makes it possible to study its dynamics with the second FEL pulse, by choosing for the probe another electronic subshell or another atomic site. This new source will provide unprecedented opportunities for probing in a highly selective way the dynamics of complex relaxation processes, such as Auger cascades or sequential multiple ionization, and of charge transfer processes in large molecules and clusters.

Methods

Accelerator. The FERMI linac[54] was operated at 1.3 GeV electron beam energy and 700 pC nominal charge. A moderate compression produced almost flat 500 A current electron bunches. The bunch length provided the conditions for an effective twin seeding with temporal separation of up to ~900 fs. The longitudinal phase space of the electron beam (energy versus time) was characterized by a chirp with both linear and quadratic components[55] that can be exploited to further enhance the difference between the resonant wavelengths in the two parts of the beam.

Seed lasers. The special twin-seed laser configuration for wide tunability two-colour FEL was based on the standard FERMI seed laser system described earlier[56–58]. The output of a Ti:Sapphire amplifier (5–7 mJ per pulse, 100 fs pulse duration, 784 nm central wavelength) was shared between an infrared OPA and a THG setup. Inside the OPA box, the signal pulses delivered by a two-stage white-light-seeded OPA process were frequency mixed with a residual pump pulse and the generated visible light was further up-converted by second harmonic generation to obtain ultraviolet pulses in the 228–262 nm range. The second ultraviolet pulse, generated in a time-plate-type BBO crystal-based THG setup, was adjusted to generate pulses with a central wavelength of 261.5 nm. The intensities of the two pulses could be varied independently through remotely controlled waveplates. The time delay between the two seed pulses was measured using an optical cross-correlator, where each ultraviolet pulse was cross-correlated with an IR pulse derived from the ultrafast oscillator that seeds the Ti:Sapphire amplifier. A remotely controlled delay stage on the THG path was used to set the time delay between the two seed pulses, before recombining them through a 50% beam splitter. Both seed pulses originate from the same source (laser oscillator and regenerative amplifier) and their relative time delay is very stable[56,57]. It has been verified that once set, the relative time delay between the two ultraviolet pulses (hence between the two FEL pulses) remains stable within less than ±5 fs over a time span of 2 h. This includes both short-term timing jitter and slow timing drifts. The adjustment and long-term stabilization of the spatial coincidence and collinearity of the two seed beams inside the FEL undulator, which are essential for obtaining the coincidence of the two FEL pulses on the sample, were obtained by using a dedicated feedback loop based on independent steering optics for each beam.

Undulators. The modulator is a 100 mm period 3 m long planar undulator with ~3% nominal resonance bandwidth. The radiator comprises six independent 55 mm period 2.42 m long undulators based on the APPLE–II design[53] that provide adjustable polarization[59,60]. The radiator was divided into two subsections, Rad_1 and Rad_2, set to resonate with harmonic 14 of λ_{seed_1} and harmonic 11 of λ_{seed_2}, respectively (Fig. 1a). The $\lambda_{seed\_2} = 255$ nm OPA seed produces a localized bunching at all the harmonics including the 11th that matches the resonance in Rad_2, generating 23.2 nm coherent emission which is amplified along the radiator. However, the beam has also bunching at the 18.2 nm 14th harmonic close to the resonant wavelength of Rad_1 (18.7 nm), which may produce unwanted emission. Similarly, the $\lambda_{seed\_1} = 261.5$ nm THG seed induces a bunching at 18.7 nm (14th harmonic), which generates the FEL probe pulse in Rad_1, but also at 23.8 nm (11th harmonic) which may excite emission from Rad_2 tuned at 23.2 nm. In both cases, though, the separation between the undesired bunching wavelength and the radiator resonant wavelength is >2%, that is, larger than the ~0.7% gain bandwidth measured for the radiators (Allaria *et al.* FEL-1 current status and recent achievements, FERMI Machine Advisory Committee, Sincrotrone Trieste, April 2014, unpublished). It is the narrow bandwidth of the radiators compared with the modulator that makes it possible to produce time-delayed single-frequency pump and probe FEL pulses from the same electron bunch.

Samples. The 20 nm permalloy film was sputter-deposited from a $Fe_{19}Ni_{81}$ target onto a commercial Si grating (605 nm period, 190 nm groove depth), with 3 nm Al buffer and capping layers. Room temperature magneto-optical Kerr effect measurements showed 100% remanence and ~8 mT coercive field along the grating lines. The 12.5-nm-thick $NiFe_2O_4$ layer was grown on $MgAl_2O_4$(001) by molecular beam epitaxy in atomic oxygen plasma. A 100 × 400 μm² area of the Ni-ferrite layer was ruled by focused ion beam etching with a set of ~350 nm wide stripes with a ~600 nm period. The magnetic signal at the Fe-3*p* resonance measured on the patterned area at the FEL source showed an ~50 mT coercive field with 100% remanence along the stripes.

Scattering setup. The experiment was performed using the IRMA vertical-scattering-plane reflectometer[44]. A horseshoe electromagnet applied variable fields (±150 mT) parallel to the sample surface and normal to the scattering plane (Fig. 1b). The FEL beam was refocused at the sample position by two bendable mirrors in Kirkpatrick–Baez configuration, using an extreme ultraviolet imager at the sample position. The final spot size (~80 μm) was estimated by scanning a movable pin-hole while measuring the transmitted intensity. The reflectometer allowed for a precise alignment of the sample with respect to the FEL beam using a slitted photodiode mounted on the detector arm. The vertically scattered intensity was detected by an in-vacuum CCD camera (2,048 × 2,048 pixels, pixel size 13.5 × 13.5 μm²) shielded from visible light by a 100-nm-thick Al filter. The CCD was mounted at 90° from the incoming FEL beam and at 535 mm from the sample. The pump fluence *F* at the sample was evaluated by correcting the pump energy measured at the source for the transport-line transmission (six reflections and a 200-nm-thick Al filter), focal spot size (~80 × 80 μm²) and angle of incidence (46.5°). Error bars on fluence (Fig. 6) account for both the pump energy measurement accuracy and for the source intensity fluctuations. The maximum fluence at the sample was ~40 and ~3.5 mJ cm^{-2} for the pump and the probe, respectively. *F* values could be adjusted rapidly and continuously by attenuating the pump seed laser. The scattering of the p-polarized FEL radiation was measured near the Brewster extinction condition, reducing non-magnetic contributions and maximizing the magnetic contrast[23,24,47–49]. All the data reported here were collected at 46.5° incidence of the FEL radiation. Magnetization-dependent data were collected following the same protocol for both samples and for all the measurements: (a) application of +80 mT pulse of ~10 ms duration, exceeding the saturation field; (b) +20 mT applied while collecting the scattered intensity at the CCD detector during a given acquisition time (1–10 s per frame); (c) repeat (a,b) for negative field values; (d) repeat the whole (a–c) sequence 50 times. The magnetic signal was then defined as an asymmetry ratio, that is, as the difference divided by the sum of two images collected for opposite signs of the applied field. Data reported in Figs 5 and 6 represent average values taken over a 7 × 7 pixels area.

References

1. Hemsing, E., Stupakov, G. & Xiang, D. Beam by design: laser manipulation of electrons in modern accelerators. *Rev. Mod. Phys.* **86,** 897–941 (2014).
2. Allaria, E. *et al.* Two-colour pump–probe experiments with a twin-pulse-seed extreme ultraviolet free-electron laser. *Nat. Commun.* **4,** 2476 (2013).
3. Bencivenga, F. *et al.* Multi-colour pulses from seeded free-electron-lasers: towards the development of non-linear core-level coherent spectroscopies. *Faraday Discuss.* **171,** 487–503 (2014).
4. Ciocci, F. *et al.* Two color free-electron laser and frequency beating. *Phys. Rev. Lett.* **111,** 264801 (2013).
5. Marcus, G., Penn, G. & Zholents, A. A. Free electron laser design for four-wave mixing experiments with soft X-ray pulses. *Phys. Rev. Lett.* **113,** 024801 (2014).
6. Campbell, L. T., McNeil, B. W. J. & Reiche, S. Two-colour free electron laser with wide frequency separation using a single monoenergetic electron beam. *New J. Phys.* **16,** 103019 (2014).
7. Chiadroni, E. *et al.* Two color FEL driven by a comb-like electron beam distribution. *Phys. Procedia* **52,** 27–35 (2014).
8. Dattoli, G., Mirian, N. S., DiPalma, E. & Petrillo, V. Two-color free-electron laser with two orthogonal undulators. *Phys. Rev. Spec. Top. Accel. Beams* **17,** 050702 (2014).
9. De Ninno, G. *et al.* Chirped seeded free-electron lasers: self-standing light sources for two-color pump-probe experiments. *Phys. Rev. Lett.* **110,** 064801 (2013).
10. Mahieu, B. *et al.* Two-colour generation in a chirped seeded free-electron laser: a close look. *Opt. Express* **21,** 22728–22741 (2013).
11. Lutman, A. A. *et al.* Experimental demonstration of femtosecond two-color X-ray free-electron lasers. *Phys. Rev. Lett.* **110,** 134801 (2013).
12. Marinelli, A. *et al.* Multicolor operation and spectral control in a gain-modulated X-ray free-electron laser. *Phys. Rev. Lett.* **111,** 134801 (2013).
13. Hara, T. *et al.* Two-colour hard X-ray free-electron laser with wide tunability. *Nat. Commun.* **4,** 2919 (2013).
14. Lutman, A. A. *et al.* Demonstration of single-crystal self-seeded two-color X-ray free-electron lasers. *Phys. Rev. Lett.* **113,** 254801 (2014).
15. Marinelli, A. *et al.* High-intensity double-pulse X-ray free-electron laser. *Nat. Commun.* **6,** 6369 (2015).
16. Stamm, C. *et al.* Femtosecond modification of electron localization and transfer of angular momentum in nickel. *Nat. Mater.* **6,** 740–743 (2007).
17. Boeglin, C. *et al.* Distinguishing the ultrafast dynamics of spin and orbital moments in solids. *Nature* **465,** 458–461 (2010).
18. Radu, I. *et al.* Transient ferromagnetic-like state mediating ultrafast reversal of antiferromagnetically coupled spins. *Nature* **472,** 205–208 (2011).
19. Wietstruk, M. *et al.* Hot-electron-driven enhancement of spin-lattice coupling in Gd and Tb 4f ferromagnets observed by femtosecond X-ray magnetic circular dichroism. *Phys. Rev. Lett.* **106,** 127401 (2011).
20. Mariager, S. O. *et al.* Structural and magnetic dynamics of a laser induced phase transition in FeRh. *Phys. Rev. Lett.* **108,** 087201 (2012).
21. Eschenlohr, A. *et al.* Ultrafast spin transport as key to femtosecond demagnetization. *Nat. Mater.* **12,** 332–336 (2013).
22. Bergeard, N. *et al.* Ultrafast angular momentum transfer in multisublattice ferrimagnets. *Nat. Commun.* **5,** 3466 (2014).
23. La-O-Vorakiat, C. *et al.* Ultrafast demagnetization dynamics at the *M* edges of magnetic elements observed using a tabletop high-harmonic soft X-ray source. *Phys. Rev. Lett.* **103,** 257402 (2009).

24. La-O-Vorakiat, C. *et al.* Ultrafast demagnetization measurements using extreme ultraviolet light: comparison of electronic and magnetic contributions. *Phys. Rev. X* **2,** 011005 (2012).
25. Mathias, S. *et al.* Probing the timescale of the exchange interaction in a ferromagnetic alloy. *Proc. Natl Acad. Sci. USA* **109,** 4792–4797 (2012).
26. Günther, S. *et al.* Testing spin-flip scattering as a possible mechanism of ultrafast demagnetization in ordered magnetic alloys. *Phys. Rev. B* **90,** 180407 R (2014).
27. Pfau, B. *et al.* Ultrafast optical demagnetization manipulates nanoscale spin structure in domain walls. *Nat. Commun.* **3,** 1100 (2012).
28. Zhang, W. *et al.* Tracking excited-state charge and spin dynamics in iron coordination complexes. *Nature* **509,** 345–348 (2014).
29. Beaud, P. *et al.* A time-dependent order parameter for ultrafast photoinduced phase transitions. *Nat. Mater.* **13,** 923–927 (2014).
30. Clark, J. N. *et al.* Imaging transient melting of a nanocrystal using an X-ray laser. *Proc. Natl Acad. Sci. USA* **112,** 7444–7448 (2015).
31. Först, M. *et al.* Spatially resolved ultrafast magnetic dynamics initiated at a complex oxide heterointerface. *Nat. Mater.* **14,** 883–888 (2015).
32. Wernet, P. *et al.* Orbital-specific mapping of the ligand exchange dynamics of $Fe(CO)_5$ in solution. *Nature* **520,** 78–81 (2015).
33. Ostler, T. A. *et al.* Ultrafast heating as a sufficient stimulus for magnetization reversal in a ferrimagnet. *Nat. Commun.* **3,** 666 (2012).
34. Graves, C. E. *et al.* Nanoscale spin reversal by non-local angular momentum transfer following ultrafast laser excitation in ferrimagnetic GdFeCo. *Nat. Mater.* **12,** 293–298 (2013).
35. Finazzi, M. *et al.* Laser-induced magnetic nanostructures with tunable topological properties. *Phys. Rev. Lett.* **110,** 177205 (2013).
36. Mangin, S. *et al.* Engineered materials for all-optical helicity-dependent magnetic switching. *Nat. Mater.* **13,** 286–292 (2014).
37. Le Guyader, L. *et al.* Nanoscale sub-100 picosecond all-optical magnetization switching in GdFeCo microstructures. *Nat. Commun.* **6,** 5839 (2015).
38. Bocchetta, C. J. *et al.* FERMI@Elettra FEL conceptual design report (Sincrotrone Trieste, 2007).
39. Allaria, E. *et al.* Highly coherent and stable pulses from the FERMI seeded free-electron laser in the extreme ultraviolet. *Nat. Photon.* **6,** 699–704 (2012).
40. Yu, L. H. Generation of intense UV radiation by sub-harmonically seeded single-pass free-electron lasers. *Phys. Rev. A* **8,** 5178–5193 (1991).
41. Yu, L. H. *et al.* High-gain harmonic-generation free-electron laser. *Science* **289,** 932–934 (2000).
42. De Ninno, G. *et al.* Single-shot spectro-temporal characterization of XUV pulses from a seeded free-electron laser. *Nat. Commun.* **6,** 8075 (2015).
43. Gauthier, D. *et al.* Spectrotemporal shaping of seeded free-electron laser pulses. *Phys. Rev. Lett.* **115,** 114801 (2015).
44. Sacchi, M. *et al.* Ultra-high vacuum soft X-ray reflectometer. *Rev. Sci. Instrum.* **74,** 2791–2795 (2003).
45. Pedersoli, E. *et al.* Multipurpose modular experimental station for the DiProI beamline of Fermi@Elettra free electron laser. *Rev. Sci. Instrum.* **82,** 043711 (2011).
46. Capotondi, F. *et al.* Coherent imaging using seeded free-electron laser pulses with variable polarization: first results and research opportunities. *Rev. Sci. Instrum.* **84,** 051301 (2013).
47. Spezzani, C. *et al.* Magnetization and microstructure dynamics in Fe/MnAs/GaAs(001): Fe magnetization reversal by a femtosecond laser pulse. *Phys. Rev. Lett.* **113,** 247202 (2014).
48. Sacchi, M., Panaccione, G., Vogel, J., Mirone, A. & van der Laan, G. Magnetic dichroism in reflectivity and photoemission using linearly polarized light: 3*p* core level of Ni(110). *Phys. Rev. B* **58,** 3750–3754 (1998).
49. Hecker, M., Oppeneer, P. M., Valencia, S., Mertins, H. C. & Schneider, C. M. Soft X-ray magnetic reflection spectroscopy at the 3*p* absorption edges of thin Fe films. *J. Electron Spectros. Relat. Phenomena* **144-147,** 881–884 (2005).
50. Henke, B. L., Gullikson, E. M. & Davis, J. C. X-ray interactions: photoabsorption, scattering, transmission, and reflection at E = 50-30,000 eV, Z = 1-92. *Atom. Data Nucl. Data Tables* **54,** 181–342 (1993).
51. Allaria, E. *et al.* The FERMI free-electron lasers. *J. Synchrotron Radiat.* **22,** 485–491 (2015).
52. Vodungbo, B. *et al.* Laser-induced ultrafast demagnetization in the presence of a nanoscale magnetic domain network. *Nat. Commun.* **3,** 999 (2012).
53. Sasaki, S. Analyses for a planar variably-polarizing undulator. *Nucl. Instrum. Methods A* **347,** 83–86 (1994).
54. Di Mitri, S. *et al.* Design and simulation challenges for FERMI@ELETTRA. *Nucl. Instrum. Methods A* **608,** 19–27 (2009).
55. Penco, G. *et al.* Experimental demonstration of electron longitudinal-phase-space linearization by shaping the photoinjector laser pulse. *Phys. Rev. Lett.* **112,** 044801 (2014).
56. Danailov, M. B. *et al.* Design and first experience with the FERMI seed laser. *Proceedings of the 33rd International Free Electron Laser Conference (FEL 2011),* (eds Zhao, Z. & Wang, D.) 183–186 (SINAP, Shanghai, China TUOC4, 2012).
57. Danailov, M. B. *et al.* Towards jitter-free pump-probe measurements at seeded free electron laser facilities. *Opt. Express* **22,** 12869–12879 (2014).
58. Cinquegrana, P. *et al.* Optical beam transport to a remote location for low jitter pump-probe experiments with a free electron laser. *Phys. Rev. Spec. Top. Accel. Beams* **17,** 040702 (2014).
59. Schmidt, T. & Zimoch, D. About APPLE II Operation. *AIP Conf. Proc.* **879,** 404–407 (2007).
60. Allaria, E. *et al.* Control of the polarization of a vacuum-ultraviolet, high-gain, free-electron laser. *Phys. Rev. X* **4,** 041040 (2014).

Acknowledgements

We are grateful to Maya Kiskinova (Sincrotrone Trieste), Jan Vogel (Institut Néel, Grenoble), Giancarlo Panaccione (CNR-IOM, Trieste), Fausto Sirotti (Synchrotron SOLEIL), Nicolas Moisan (LPS, Orsay), Michael Meyer (European XFEL, Hamburg) and Coryn F. Hague (LCPMR, Paris) for useful discussions and suggestions. This research received financial support from the European Community 7th Framework Programme under grant agreement n° 312284, and from CNRS (France) via the PEPS_SASLELX program. The FERMI project at Elettra—Sincrotrone Trieste is supported by MIUR under grants FIRB-RBAP045JF2 and FIRB-RBAP06AWK3.

Author contributions

E.F., C.Sp., G.D.N., M.B.D., E.A. and M.S. devised and coordinated the experiment. E.F., C.Sp., L.G., G.D.N., M.B.D. and E.A. designed and optimized the two-colour FEL source. I.N., P.C., A.D. and M.B.D. reconfigured and operated the twin-seed laser source. E.F., B.D., D.G., G.P., P.R.R., E.R., M.T., L.G., G.D.N. and E.A. operated the FEL source during the experiment. F.F., R.D., F.V., J.B.M. and L.L. fabricated and characterized the samples. M.M., E.P., F.C., C.Sv., N.M., M.Z. and L.R. contributed to the integration of the IRMA experimental chamber at the DIPROI beamline. C.Sp., F.F., R.D., T.P. and M.S. performed the scattering experiment. E.F., C.Sp., F.V., M.B.D., G.D.N., E.A. and M.S. analysed the data and wrote the manuscript, with contributions from all the authors.

Additional information

Competing financial interests: The authors declare no competing financial interests.

15

Nonlinear optomechanical measurement of mechanical motion

G.A. Brawley[1,*], M.R. Vanner[1,2,*], P.E. Larsen[3], S. Schmid[3], A. Boisen[3] & W.P. Bowen[1]

Precision measurement of nonlinear observables is an important goal in all facets of quantum optics. This allows measurement-based non-classical state preparation, which has been applied to great success in various physical systems, and provides a route for quantum information processing with otherwise linear interactions. In cavity optomechanics much progress has been made using linear interactions and measurement, but observation of nonlinear mechanical degrees-of-freedom remains outstanding. Here we report the observation of displacement-squared thermal motion of a micro-mechanical resonator by exploiting the intrinsic nonlinearity of the radiation-pressure interaction. Using this measurement we generate bimodal mechanical states of motion with separations and feature sizes well below 100 pm. Future improvements to this approach will allow the preparation of quantum superposition states, which can be used to experimentally explore collapse models of the wavefunction and the potential for mechanical-resonator-based quantum information and metrology applications.

[1] ARC Centre for Engineered Quantum Systems, School of Mathematics and Physics, The University of Queensland, Brisbane, Queensland 4072, Australia. [2] Clarendon Laboratory, Department of Physics, University of Oxford, Oxford OX1 3PU, UK. [3] Department of Micro- and Nanotechnology, Technical University of Denmark, DTU Nanotech, DK-2800 Kongens Lyngby, Denmark. * These authors contributed equally to this work. Correspondence and requests for materials should be addressed to M.R.V. (email: michael.vanner@physics.ox.ac.uk).

A key tool in quantum optics is the use of measurement to conditionally prepare quantum states. This technique, often simply referred to as 'conditioning', has been applied to generate non-Gaussian quantum states for confined microwave fields[1], travelling optical fields[2,3] and superconducting systems[4]. In addition, quantum measurements are of vital importance to many quantum computation protocols[5]. In cavity optomechanics, light circulating inside an optical resonator is used to manipulate and measure the motion of a mechanical element via the radiation-pressure interaction[6]. After an optomechanical interaction performing a measurement on the light can then be used to conditionally prepare mechanical states of motion. This form of state preparation can be understood as a combination of Bayesian inference, that is, updating our knowledge of the system, and back-action (see Supplementary Note 1 for an introduction). Subsequent measurements following the conditioning step can be used to characterize the state prepared. Such mechanical conditioning has been performed with measurements of the mechanical position[7–9], however, thus far, conditioning has not been performed with a measurement of a nonlinear mechanical degree of freedom. One exciting prospect is to dispersively couple the optical field to the mechanical position squared[10], which could enable the detection of phonon number jumps[10,11] and thus demonstrate mechanical energy quantization. Here we implement an alternative approach that instead utilises the optical nonlinearity of the radiation-pressure coupling, as was proposed in ref. 12. Such a position-squared measurement can ultimately be used to prepare a mechanical superposition state[13] as the measurement does not reveal the sign of the mechanical position. Studying the dynamics of superposition states can be used to test models of decoherence beyond standard quantum mechanics[12,14–19] and for the development of mechanical quantum sensors.

For the optomechanical system of interest in this work, the intracavity Hamiltonian in a frame rotating at the optical carrier frequency is $H/\hbar = \omega_M b^\dagger b + g_0 a^\dagger a(b + b^\dagger)$ where $a(b)$ is the optical (mechanical) annihilation operator, ω_M is the mechanical angular frequency and g_0 is the zero-point optomechanical coupling rate. Quite generally, optomechanics experiments to-date have focused on dynamics describable by a linearized model of the radiation-pressure interaction[6], where the photon number operator is approximated by $a^\dagger a \simeq N + \sqrt{N}(a^\dagger + a)$ and N is the mean intracavity photon number. In this approximation, mechanical displacements give rise to displacements of the optical phase quadrature leaving the optical amplitude quadrature unchanged. Fundamentally, however, the radiation-pressure interaction is nonlinear[12,20], and generates mechanical position-dependent rotations of the intracavity optical field. For small changes in the mechanical position, the optical phase quadrature changes linearly in proportion to the mechanical displacement, and the optical amplitude quadrature reduces in proportion to the mechanical displacement squared. By choosing which optical quadrature to observe with homodyne detection, one may selectively measure the mechanical displacement or displacement-squared[12]. Since a displacement-squared measurement does not distinguish between positive and negative displacement, mechanical superposition states may be prepared by measurement[13]. A necessary requirement for the optical interaction to effect a direct measurement of the displacement-squared is that the mechanical motion is negligible during the intracavity photon lifetime, that is, operation in the bad-cavity regime ($\kappa \gg \omega_M$, where κ is the optical amplitude decay rate of the cavity). This should be contrasted to other approaches operating in the resolved sideband regime ($\kappa \ll \omega_M$), where cavity-averaged displacement-squared interactions have been predicted to allow the observation of the mechanical phonon-number[10,11]. In this work we observe optomechanical dynamics arising from the nonlinearity of the radiation-pressure interaction and, utilizing this nonlinearity, perform non-Gaussian state generation by measurement of displacement-squared mechanical motion.

In the bad-cavity regime, the intracavity field can be approximated as $a \simeq \sqrt{N}/(1 + i\lambda X_M) + \xi$, where $\lambda = \sqrt{2}g_0/\kappa$ quantifies the optomechanical interaction strength, ξ is the intracavity noise term and $X_M = (b + b^\dagger)/\sqrt{2}$ is the mechanical displacement in units of the mechanical quantum noise (Supplementary Note 2). Taylor expanding the intracavity field, the time-dependent optical output quadratures are then

$$X_L^{out} = X_L^{in} - 2\sqrt{\kappa N}\left[1 - \lambda^2 X_M^2 + \ldots\right], \tag{1}$$

$$P_L^{out} = P_L^{in} - 2\sqrt{\kappa N}\left[-\lambda X_M + \lambda^3 X_M^3 - \ldots\right], \tag{2}$$

where $X_L = (a + a^\dagger)/\sqrt{2}$, and $P_L = i(a^\dagger - a)/\sqrt{2}$. Conventionally, experimental optomechanics have focused on the leading, linear term in the expansion of the phase quadrature. In this linearized picture, only a single spectral peak at the mechanical resonance frequency is expected. Higher-order terms in mechanical displacement, however, give rise to spectral peaks at the respective multiples of the mechanical resonance frequency which are only described by the full nonlinear optomechanical Hamiltonian. Precision measurement of these higher-order terms enables the conditional preparation of non-Gaussian states which, in a quantum regime, produces highly non-classical states[21].

Here we use the intrinsic nonlinearity of the radiation-pressure interaction to effect a displacement-squared measurement of the motion of a mechanical oscillator. We use this measurement to conditionally prepare classical bimodal states of mechanical motion from an initial room-temperature thermal state. Further we theoretically show that this continuous measurement approach, with the introduction of feed-back, may be extended to a quantum regime, allowing preparation of macroscopic quantum superposition states of mechanical motion.

Results

Optomechanical system. A schematic of our nonlinear optomechanics experiment is shown in Fig. 1. We use a near-field cavity optomechanical set-up[22], where a mechanical SiN nanostring oscillator[23] is placed in close proximity to a 60-μm diameter optical microsphere resonator and interacts with the optical cavity field via the optical evanescent field (Fig. 1c). The nanostring has dimensions $1{,}000 \times 10 \times 0.054$ μm (length × width × thickness) and a fundamental mechanical resonance frequency of $\omega_M/2\pi = 100.2$ kHz. From the known dimensions and density we estimate an effective mass of $m = 0.86$ ng. A continuous-wave fibre laser, operating at 1,559 nm, is locked on resonance with a whispering-gallery mode of the microsphere. We measure an optical amplitude decay rate of $\kappa/2\pi = 25.6$ MHz and a mechanical linewidth of $\gamma/2\pi = 0.7$ Hz. An evanescent optomechanical coupling of 7.6 MHz/nm was determined (Methods section), corresponding to a coupling rate of $g_0/2\pi = 75$ Hz. We use ~ 2 μW of optical drive power resulting in an intracavity photon number $N = 2.4 \times 10^4$. A fibre-based Mach–Zehnder interferometer is used to perform homodyne detection and thereby selectively measure a quadrature of the optical output field.

Figure 2a,b show the observed homodyne noise power spectra for both optical phase and amplitude quadratures at the mechanical frequency and the second harmonic, respectively. At ω_M (Fig. 2a) we observe a Lorentzian peak in the phase quadrature from the thermal motion of the oscillator, which corresponds to a root-mean-square (RMS) displacement of

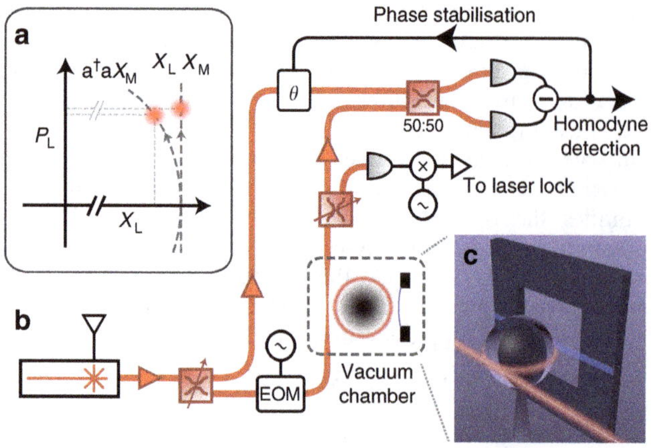

Figure 1 | Concept and experimental apparatus. (**a**) The intrinsic optical nonlinearity of an optomechanical interaction gives rise to rotations of the optical field in phase-space that can be observed in both the phase (P_L) and amplitude (X_L) quadratures. Conventionally this interaction is linearized in the weak coupling regime, leading to optical phase quadrature displacements only. Our optical set-up (**b**) can measure an arbitrary optical quadrature of light from the optomechanical system using homodyne interferometry and is capable of observing the higher-order terms in displacement, described by equations (1 and 2). The optomechanical system consists of a nanostring mechanical resonator evanescently coupled to an optical microsphere resonator (**c**) (not shown to scale), which is mounted in a high-vacuum chamber ($<10^{-6}$ mbar). The drive laser is stabilised to the cavity resonance using the Pound Drever Hall technique. Polarization control is not shown for clarity.

124 pm corresponding to a thermal occupation of $\bar{n} \simeq 10^8$. The thermal noise is resolved with 85 dB of signal relative to the homodyne noise power when the signal is blocked, which corresponds to an ideal displacement sensitivity of 1.3×10^{-15} m $\text{Hz}^{-1/2}$. In practice, the signal beam is not shot noise limited due to cavity, acoustic and laser noise which raise the measurement imprecision by roughly an order of magnitude. By setting the interferometer phase to measure the optical amplitude quadrature, the linear measurement of mechanical motion is suppressed by ~45 dB. At this quadrature, information about the displacement-squared mechanical motion is observed in a frequency band centred at $2\omega_M$ (Fig. 2b). We observe a Lorentzian peak with a linewidth of 1.5 Hz, which to within the measurement uncertainty, is equal to twice the linewidth at ω_M (Supplementary Note 3 and Supplementary Fig. 1). The signal-to-noise at this frequency is 65 dB relative to the homodyne noise, which corresponds to a calibrated ideal displacement-squared sensitivity of $3.3 \times 10^{-24}\,\text{m}^2\,\text{Hz}^{-1/2}$.

Figure 2c shows the band power in the first and second harmonics as a function of the interferometer phase. The powers in each band are expected to follow sine and cosine squared functions (fitted). The observed suppression of the linear measurement allows an upper bound to be placed on the phase instability of the cavity and interferometer locks of at most 5×10^{-3} rad. Figure 2d shows the observed relative noise powers up to the fourth harmonic of the mechanical frequency. The expected noise powers can be computed with the Isserlis–Wick theorem (Supplementary Notes 3 and 11), which show excellent agreement with experiment.

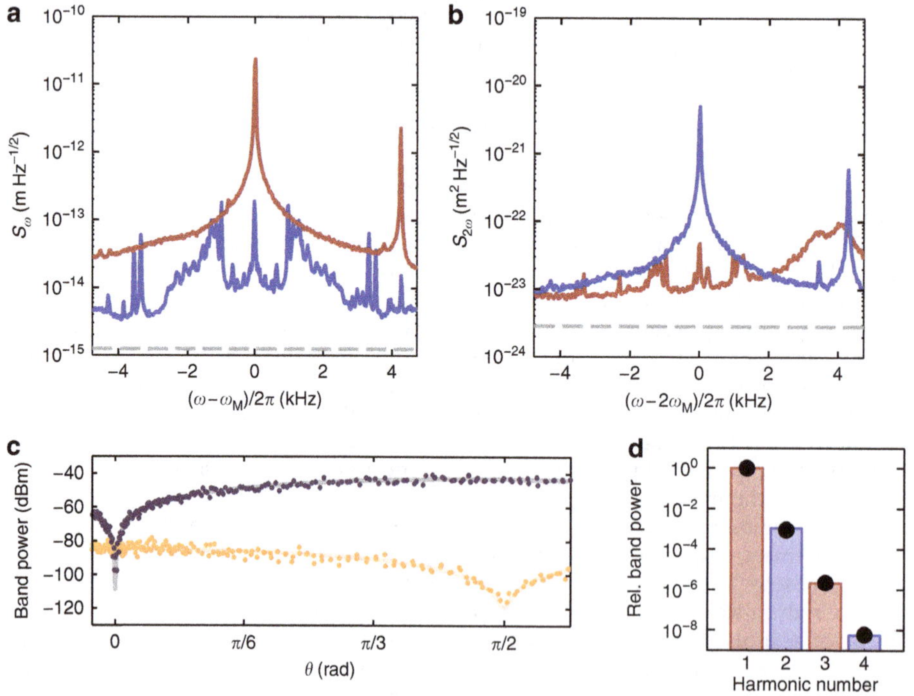

Figure 2 | Observation of linear and quadratic motion of a mechanical oscillator. (**a**) Measurement of the optical phase quadrature (red trace) at the mechanical frequency, ω_M, shows a Lorentzian mechanical displacement spectrum, which is strongly suppressed when measuring the optical amplitude quadrature (blue trace). Sidebands appear in the amplitude quadrature measurement due to mix-up of low-frequency noise. (**b**) Measurement of the optical amplitude quadrature at $2\omega_M$ (blue trace), shows the Lorentzian mechanical displacement-squared noise, which is suppressed when measuring the optical phase quadrature (red trace). Note the second flexural mode of the string, located within a few kHz of $2\omega_M$, is not transduced due to the positioning of the microsphere. In **a**,**b** the optical shot noise is shown as a grey dashed line, using a measurement bandwidth of 20 Hz. (**c**) The optical noise power measured over a 51-Hz bandwidth at ω_M (purple) and $2\omega_M$ (yellow) as a function of the optical homodyne angle. (**d**) The relative observed powers in each of the mechanical harmonics when measuring the optical phase (red) and amplitude quadratures (blue); bars, theory; dots, experimental data. Note the error bars in the power measurements are smaller than the dot size.

State preparation and read-out. Of primary interest in this work is the lowest order nonlinear measurement term in the optical amplitude quadrature, proportional to X_M^2. To describe this quantitatively we introduce the slowly varying quadratures of motion, X and Y, defined via $X_M(t) = X(t)\cos\omega_M t + Y(t)\sin\omega_M t$. The mechanical displacement-squared signal can then be written as:

$$X_M^2 = \tfrac{1}{2}(X^2+Y^2) + \tfrac{1}{2}(X^2-Y^2)\cos(2\omega_M t) + \tfrac{1}{2}(XY+YX)\sin(2\omega_M t) \quad (3)$$

where for later convenience, the displacement-squared quadratures of motion are defined $P=\frac{1}{2}(X^2-Y^2)$ and $Q=\frac{1}{2}(XY+YX)$. By inspection, it can be seen that the quadratic measurement has spectral components both at DC and $2\omega_M$. Higher-order terms in the expansion equations (1 and 2) can in principle contribute to the signal at $2\omega_M$, however since $\lambda^2\bar{n} \ll 1$, the quadratic term is the only term to contribute substantial power at $2\omega_M$. Consequently, linear and quadratic components of the measurement can be spectrally separated and therefore, at an appropriate homodyne angle, measured simultaneously.

To perform both state preparation and state reconstruction we set a homodyne angle of $\pi/4$, which allows simultaneous high-fidelity linear measurement (for state reconstruction) and quadratic measurements (for state preparation). The photocurrent generated at the homodyne output is digitized into 4 second blocks at a sample rate of $5\times10^6\,\mathrm{s}^{-1}$, which are then filtered numerically at ω_M and $2\omega_M$ to obtain the respective quadratures of motion in each frequency band as detailed in Supplementary Note 4. For a large signal-to-noise ratio, the squares of each quadrature of motion can be estimated from the measurements of P and Q via the nonlinear transformations $X^2 \simeq \tilde{X}_{2\omega}^2 = \sqrt{\tilde{P}^2+\tilde{Q}^2}+\tilde{P}$ and $Y^2 \simeq \tilde{Y}_{2\omega}^2 = \sqrt{\tilde{P}^2+\tilde{Q}^2}-\tilde{P}$, where the tildes denote the (noise inclusive) measurement outcomes of the respective quantity. These transformations allow the recovery of a classical estimate of X_M^2 without knowledge of the signal at DC. Figure 3a plots the $\tilde{X}_{2\omega}^2$ estimates thus obtained from the $2\omega_M$ signal against the cosine mechanical position quadrature, $\tilde{X}$, obtained from the measurement at ω_M. A clear quadratic relationship between the two measurements is observed, validating the displacement-squared nature of the $2\omega_M$ peak.

Conditioning based on the outcome of the quadratic measurement can be used to prepare non-Gaussian states. In the most basic approach, conditioning $\tilde{Q}=0$ and $\tilde{P}=C$, for some constant C, will produce a bimodal state with a separation of $2\sqrt{2C}$. However, to make more efficient use of the available data, we additionally perform a mechanical phase rotation for each sample in the measurement record. First, at each discrete sample we find a rotation by a phase angle 2ϕ such that the new rotated variable $\tilde{Q}^{2\phi} = \tilde{Q}\cos(2\phi) - \tilde{P}\sin(2\phi)$ is equal to zero. As a result, in the frame rotated by the half angle, ϕ, correlation between the two mechanical quadratures $\tilde{X}$ and $\tilde{Y}$ is conditionally eliminated. Second, we condition on a particular magnitude of $\tilde{P}^{2\phi} = \tilde{P}\cos(2\phi) + \tilde{Q}\sin(2\phi)$ (Methods section). This operation localizes the phase-space distribution of the reconstructed state to two small regions as shown in Fig. 3b, with a separation dependent on the conditioning value. These states, although classical, are evidently bimodal and non-Gaussian. Further details of this protocol are contained in Supplementary Note 5 and Supplementary Fig. 2. Extending this protocol to a regime where the quadratic measurement rate dominates all decoherence processes, a macroscopic quantum superposition state can be generated. Indeed, as detailed later, a simulated state prepared in this way is shown in Fig. 4b.

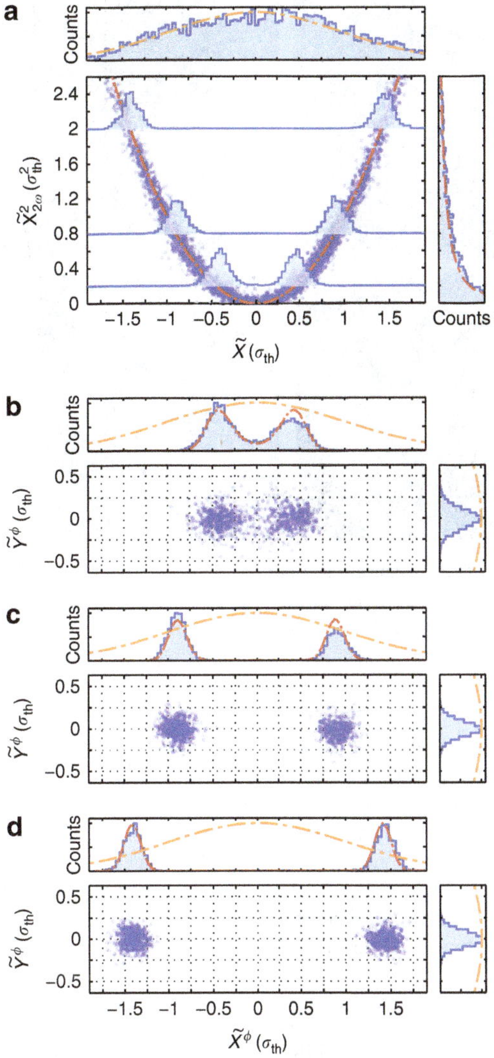

Figure 3 | Bimodal state preparation via nonlinear measurement. Using a homodyne angle of $\pi/4$, a high-fidelity measurement of both linear and quadratic motion of the mechanical oscillator at frequencies of ω_M and $2\omega_M$, respectively, is obtained. In **a** the quadratic measurement outcomes ($\tilde{X}_{2\omega}^2$) obtained from the $2\omega_M$ signal are plotted against the linear outcomes ($\tilde{X}$) obtained from the signal at ω_M. The histogram of $\tilde{X}$ measurements (above) is well described by a Gaussian thermal distribution with s.d. $\sigma_{th} = 124$ pm, while the histogram of the $\tilde{X}_{2\omega}^2$ measurements (right) forms a χ^2 distribution. Figures (**b-d**) show the phase-space distributions (and associated histograms) of states conditionally prepared using data at $2\omega_M$ and obtained using a read-out at ω_M. The conditionally rotated read-out data is decomposed into conjugate quadratures, labelled $\tilde{X}^\phi$ and $\tilde{Y}^\phi$. The chosen quadratic conditioning values are (**b**) ($2\tilde{P}^{2\phi} = 0.2$), (**c**) ($2\tilde{P}^{2\phi} = 0.8$) and (**d**) ($2\tilde{P}^{2\phi} = 2.0$). The same quadratic conditioning values are indicated as overlay histograms in **a**. The histograms in **b-d** are normalized to their peak value to more easily allow the width of the features to be compared with the initial thermal state (orange dash-dot curve). The red curves overlayed in the histograms in $\tilde{X}^\phi$ are determined via numerical simulation, with the signal-to-noise ratios for the linear and quadratic measurements as fitting parameters.

Comparison of effective coupling rates. At present, in opto- and electro-mechanics, techniques towards measurement of nonlinear observables of mechanical motion include coupling to two-level systems[24] and radiation-pressure interactions coupling to the displacement-squared, such as the 'membrane-in-the-middle'

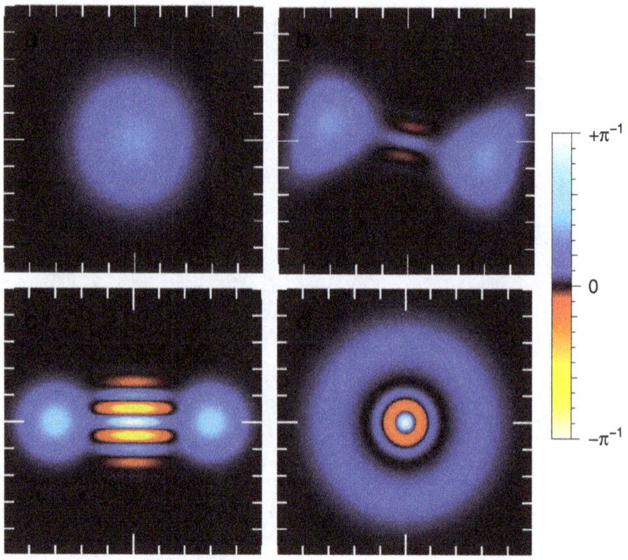

Figure 4 | Open quantum system simulations. Figures (**a**,**b**) show the conditional state evolution of the mechanical oscillator under continuous measurement of the AC component of the X^2 signal, as computed from a master equation simulation. The initial state with $\bar{n}$=1 is shown in **a**. As the state evolves, negativity appears in the Wigner function. This is illustrated in **b** at time $t = 6.4 \times 2\pi/\omega_M$. This state may be compared with the canonical cat-sate of $|\psi\rangle = (|3/\sqrt{2}\rangle + |-3/\sqrt{2}\rangle)/\sqrt{2}$, shown in **c**. See Supplementary Movie 1 for an animation of the Wigner function time evolution at intermediate times. Figures (**c**,**d**) show the effect of phonon number decoherence on an initial even cat state. Notably after tracing over a strong phonon number measurement (**d**) negativity is still preserved in the Wigner function. Further details of both the master equation simulation and analysis of the decoherence processes are contained in the Supplementary Information.

(MiM) approach[10]. In the latter approach, a mechanically vibrating element is appropriately placed within an optical standing wave in a cavity to give a displacement-squared dispersive coupling of the form $H_{int}/\hbar = \mu_0 a^\dagger a(b + b^\dagger)^2$, where μ_0 is the zero-point quadratic coupling rate. This interaction, when operating in the resolved sideband regime, in principle allows for the observation of quantum jumps in the mechanical phonon number. Experiments exhibiting this type of coupling include dielectric membrane systems[10,25,26], trapped cold atoms[27], trapped microspheres[28] or double-disk structures[29]. However it should be noted that in these systems, quadratic coupling rates μ_0 are typically orders of magnitude smaller than attainable linear coupling rates g_0.

In contrast to a displacement-squared Hamiltonian coupling, the scheme employed here gives an effective quadratic coupling rate of g_0^2/κ (ref. 12), which should be compared with μ_0 defined above. For the modest linear coupling achieved in the present work, we have a $g_0^2/2\pi\kappa = 2.2 \times 10^{-4}$ Hz. Crucially, since the coupling rate in our scheme scales as g_0^2, substantial gains are possible by improving g_0. For instance, the coupling rate for a state-of-the-art evanescently coupled nanostring-microcavity system as described in ref. 30 is $g_0^2/2\pi\kappa \simeq 5\times10^{-2}$ Hz and for a state-of-the-art electro-mechanical system[31] is $g_0^2/2\pi\kappa \simeq 2\times10^{-1}$ Hz. Furthermore, other optical systems with exceptional linear coupling rates[32] should have quadratic coupling rates using our scheme as high as 160 Hz. In comparison, a state-of-the-art MiM system as described in ref. 26 has a quadratic coupling rate of $\mu_0/2\pi = 6.0 \times 10^{-3}$ Hz. Thus a significantly larger effective quadratic coupling is possible using our protocol. Noteably, the quadratic measurement rate resulting from this fundamental coupling may be boosted by a coherent optical drive, which makes entering the quantum regime more feasible.

Requirements for non-classical behaviour. In all measurement-based quantum state preparation schemes, the measurement rate must dominate the sum of all decoherence process rates due to coupling of the system to the environment. In MiM displacement-squared coupling protocols, even in a zero-temperature environment, this introduces the challenging requirement of single-photon strong coupling ($g_0/\kappa > 1$)[11]. By contrast, our scheme offers a route to relax this stringent criterion. In our scheme, when state conditioning is performed with only the quadratic motion component of the detected signal, decoherence from the linearized radiation-pressure noise on the mechanics precludes non-classical state generation outside of a single-photon strong coupling regime, similar to MiM (Supplementary Note 6). However, by including feed-back, this form of decoherence can, in the limit of perfect detection efficiency, be completely eliminated (Supplementary Note 7). This is because the amplitude quadrature measurement records not only the X^2 mechanical motion, but also the optical intracavity amplitude fluctuations near the mechanical resonance frequency. Since these fluctuations drive the linearized radiation pressure back-action on the mechanics, suitable feed-back to the motion of the mechanical oscillator can cancel this radiation-pressure noise. Additionally, with this feed-back, the mechanical dynamics reduce to a similar form as with the displacement-squared dispersive Hamiltonian coupling. In the realistic case of imperfect detection efficiency η, the decoherence can be suppressed up to a factor of $1 - \eta$. This results in the coupling strength requirement $g_0^2/\kappa^2 > (1 - \eta)/2\eta$ to reach a quantum regime (see Supplementary Notes 7,9,10 and Supplementary Figs 3–5 which quantify the effect of decoherence arising from linear interactions). For example, with a detection efficiency of $\eta = 0.98$ the single-photon coupling rate need only be one tenth of the strong coupling requirement. Additionally, the quadratic measurement rate must dominate rethermalisation, that is, $4\eta N g_0^4/\kappa^3 > \gamma(\bar{n} + 1/2)$. Provided the coupling strength criterion is satisfied, rethermalisation can be made insignificant with only modest intracavity photon numbers in cryogenic systems.

Based on these criteria, the quantum regime of our scheme can be achieved with current atom-optomechanical systems. For instance, quantum superposition states of motion could immediately be implemented with the approach in ref. 33, provided the detection efficiency exceeds 10%. Furthermore, solid-state optomechanical devices have seen rapid gains in performance over the past decade, with both optical and microwave systems now operating within three orders of magnitude of the single-photon strong coupling regime. For example, an effective coupling rate of $g_0/\kappa = 0.04$ has recently been achieved in a superconducting microwave optomechanical device[34], and with modest modifications, it is expected the system will approach the strong coupling regime. When $g_0/\kappa = 0.04$, the generation of quantum superposition states using our protocol requires detection efficiency on the order of 99.7%. However, with a one order of magnitude improvement in the coupling rate, the required detection efficiency drops to 76%, such that in combination with state of the art amplifiers[35], non-classical state generation using our protocol could be realised. A full list of parameters is provided in Supplementary Table 1.

Technical limitations may also play a role in the implementation of our protocol in a quantum regime. For example, fluctuations or offsets in the interferometer phase or cavity lock will result in linear coupling to the environment, and therefore an additional source of decoherence. Linear coupling can also be introduced undesirably due to the presence of other mechanical modes in the system, which mix via the optical nonlinearity with the mode of interest. Indeed the sum beat between the two mechanical modes in Fig. 2a is observed as the +4.3 kHz peak in Fig. 2b. These additional linear decoherence channels are expected to be negligible compared with the decoherence due to linear radiation pressure back-action as detailed Supplementary Note 8. The currently un-utilised DC component of the homodyne signal constitutes an additional technical decoherence channel. However, unlike the decoherence mechanisms discussed above, this channel is nonlinear, carrying information about phonon number rather than mechanical position. This can be seen from the expansion of the quadratic motion in terms of the creation and annihilation operators, $X_M^2 = b^\dagger b + \frac{1}{2} + \frac{1}{2}(bb + b^\dagger b^\dagger)$, and identifying the number operator $n = b^\dagger b$ plus a constant as the DC part. As a result, loss of the DC information generates phase diffusion on the mechanical state. Somewhat strikingly, the non-classicality of states generated by X^2 measurement can in fact be quite robust against this form of decoherence. In Supplementary Note 9, the effect of phonon-number decoherence of an initial superposition state is analysed, showing that Wigner function negativity is preserved even in the presence of a complete loss of phonon number information to the environment. This result is illustrated in Fig. 4c,d and in further detail in Supplementary Fig. 6.

Quantum trajectory simulation. Finally, to elucidate the precise effect of the combination of all identified decoherence processes on the state conditioned via continuous quadratic measurement, a master equation simulation of our system was performed. The results for a particular trajectory are briefly summarized in the Wigner functions presented in Fig. 4a,b. Shown in Fig. 4a is an initial thermal state of the mechanical oscillator. After a period of continuous measurement of the AC component of X_M^2, and in the presence of DC and thermal decoherence, this initially symmetric Gaussian state evolves into a non-Gaussian bimodal quantum state, exhibiting Wigner negativity near the origin, as shown in Fig. 4b. See Supplementary Movie 1 for an animation of the Wigner function time evolution. Notably, qualitatively similar states have previously been shown to form in a different system under continuous position-squared measurement and conditioning[13]. The states prepared by our protocol exhibit many of the properties of the canonical Schrödinger cat state of Fig. 4c and are highly non-classical. As a result, even in the presence of the identified decoherence mechanisms, we can conclude our protocol can give rise to interesting non-classical states. We would like to highlight that initialization of the mechanical oscillator near its ground state is not required to generate non-classical mechanical states. This insensitivity to initial thermal occupation is because the continuous position-squared measurement also serves to purify the state. Further details of our simulation are found in Supplementary Note 10 and Supplementary Figs 7 and 8.

Discussion

To summarize, by exploiting the nonlinearity inherent in the radiation-pressure interaction, we report nonlinear measurement of thermo-mechanical motion in an optomechanical system. Utilising the measurement of displacement-squared motion, we demonstrate the first measurement-based state preparation of mechanical non-Gaussian states. Furthermore, we propose a method using feed-back to extend this protocol to a quantum regime without requiring single-photon strong coupling. Favourable scaling of the coupling rate in our approach makes realistic the possibility of observing the displacement-squared fluctuations at the level of the mechanical ground state in the near future. With sufficiently high detection efficiency, this would allow for mechanical quantum superposition state preparation. As a result, this experiment paves the way for quantum non-Gaussian state preparation of mechanical motion via measurement with applicability to a number of other physical systems, such as cold atoms[33], atomic spin ensembles[36], optomechanical systems[32] and superconducting microwave circuits[31,34,37,38].

Methods

Linear calibration procedure. We determine the evanescent optomechanical coupling by displacing the nanostring by a known distance using a piezoelectric element and measuring the resulting frequency shift on the optical resonance. The frequency shift is calibrated via modulation of known frequency applied to the laser. We then establish the response of the homodyne by sweeping the laser detuning over the optical resonance and measuring the slope of the phase response. This parameter combined with the previously determined optomechanical coupling rate gives the total response of the combined cavity interferometer system ($\mathrm{V\,nm^{-1}}$), allowing direct calibration of the time domain data (nm). We calibrate the response of our spectrum analyser by applying a test tone of known amplitude, which using the time domain calibration gives a spectral peak of known displacement spectral density.

Quadratic calibration procedure. Frequency domain calibration of the quadratic measurement is performed by ensuring the calibrated RMS displacement, obtained from the linear measurement, which is consistent with the noise power of the $2\omega_M$ peak, in accordance with the Isserlis–Wick theorem. In the time domain, a simple regression is used between the square of the linear measurement $(\tilde{X})$ and the quadratic measurement $(\tilde{X}^2_{2\omega})$. We verify that these procedures are consistent, to within known uncertainties, with one another and with the value of $\lambda^2\bar{n}$ computed from the independently measured system parameters.

State conditioning. From the continuously acquired data, estimates of the quadratures at $2\omega_M$ (ω_M) are obtained with the use of causal (acausal) decaying exponential filters, to time separate the conditioning and read-out phases. From the filtered data at each discrete time step, we rotate the vector $\{\tilde{P},\tilde{Q}\}$ by an angle 2ϕ, such that a new vector $\{\tilde{P}^{2\phi},\tilde{Q}^{2\phi}\}=\left\{\left(\tilde{P}^2+\tilde{Q}^2\right)^{\frac{1}{2}},0\right\}$ is obtained. The simultaneously acquired linear data $\{\tilde{X},\tilde{Y}\}$ is then rotated through the half angle, ϕ, to obtain $\{\tilde{X}^\phi,\tilde{Y}^\phi\}=\{\tilde{X}\cos(\phi)+\tilde{Y}\sin(\phi),\tilde{Y}\cos(\phi)-\tilde{X}\sin(\phi)\}$. For state preparation, the rotated linear data is conditioned on the value of $\tilde{P}^{2\phi}$, which is proportional to $\frac{1}{2}\left(\tilde{X}^\phi\right)^2$. We choose a conditioning window four times smaller than the quadratic measurement uncertainty. When the conditioning criterion is satisfied, the state is read-out using the rotated linear data $\{\tilde{X}^\phi,\tilde{Y}^\phi\}$. All the data presented here have been generated from three 4 s blocks of sampled homodyne output.

References

1. Deléglise, S. *et al.* Reconstruction of non-classical cavity field states with snapshots of their decoherence. *Nature* **455,** 510 (2008).
2. Ourjoumtsev, A., Jeong, H., Tualle-Brouri, R. & Grangier, P. Generation of optical Schrodinger cats from photon number states. *Nature* **448,** 784 (2007).
3. Bimbard, E., Jain, N., MacRae, A. & Lvovsky, A. I. Quantum-optical state engineering up to the two-photon level. *Nat. Photon.* **4,** 243–247 (2010).
4. Risté, D. *et al.* Deterministic entanglement of superconducting qubits by parity measurement and feedback. *Nature* **502,** 350–354 (2013).
5. Knill, E., Laflamme, R. & Milburn, G. J. A scheme for efficient quantum computation with linear optics. *Nature* **409,** 46–52 (2001).
6. Aspelmeyer, M., Kippenberg, T. J. & Marquardt, F. Cavity optomechanics. *Rev. Mod. Phys.* **86,** 1391 (2014).
7. Vanner, M. R. *et al.* Pulsed quantum optomechanics. *Proc. Natl Acad. Sci. USA* **108,** 16182–16187 (2011).
8. Vanner, M. R., Hofer, J., Cole, G. D. & Aspelmeyer, M. Cooling-by-measurement and mechanical state tomography via pulsed optomechanics. *Nat. Commun.* **4,** 2295 (2013).
9. Szorkovszky, A. *et al.* Strong thermomechanical squeezing via weak measurement. *Phys. Rev. Lett.* **110,** 183401 (2013).
10. Thompson, J. D. *et al.* Strong dispersive coupling of a high-finesse cavity to a micromechanical membrane. *Nature* **452,** 72–75 (2008).
11. Miao, H., Danilishin, S., Corbitt, T. & Chen, Y. Standard quantum limit for probing mechanical energy quantization. *Phys. Rev. Lett.* **103,** 100402 (2009).
12. Vanner, M. R. Selective linear or quadratic optomechanical coupling via measurement. *Phys. Rev. X* **1,** 021011 (2011).
13. Jacobs, K., Tian, L. & Finn, J. Engineering superposition states and tailored probes for nanoresonators via open-loop control. *Phys. Rev. Lett.* **102,** 057208 (2009).

14. Ghirardi, G. C., Rimini, A. & Weber, T. Unified dynamics for microscopic and macroscopic systems. *Phys. Rev. D* **34,** 470–491 (1986).
15. Diósi, L. Models for universal reduction of macroscopic quantum fluctuations. *Phys. Rev. A* **40,** 1165–1174 (1989).
16. Penrose, R. On gravity's role in quantum state reduction. *Class. Quantum Gravity* **28,** 581–600 (1996).
17. Klecker, D. *et al.* Creating and verifying a quantum superposition in a micro-optomechanical system. *N. J. Phys.* **10,** 095020 (2008).
18. Romero-Isart, O. Quantum superposition of massive objects and collapse models. *Phys. Rev. A* **84,** 052121 (2011).
19. Blencowe, M. P. Effective field theory approach to gravitationally induced decoherence. *Phys. Rev. Lett.* **111,** 021302 (2013).
20. Børkje, K., Nunnenkamp, A., Teufel, J. D. & Girvin, S. M. Signatures of nonlinear cavity optomechanics in the weak coupling regime. *Phys. Rev. Lett.* **111,** 053603 (2013).
21. Hudson, R. L. When is the wigner quasi-probability density non-negative? *Rep. Math. Phys.* **6,** 249–252 (1974).
22. Anetsberger, G. *et al.* Near-field cavity optomechanics with nanomechanical oscillators. *Nat. Phys.* **5,** 909–914 (2009).
23. Schmid, S., Jensen, K. D., Nielsen, K. H. & Boisen, A. Damping mechanisms in high-Q micro and nanomechanical string resonators. *Phys. Rev. B* **84,** 165307 (2011).
24. O'Connell, A. D. *et al.* Quantum ground state and single-phonon control of a mechanical resonator. *Nature* **464,** 697–703 (2010).
25. Sankey, J. C., Yang, C., Zwickl, B. M., Jayich, A. M. & Harris, J. G. E. Strong and tunable nonlinear optomechanical coupling in a low-loss system. *Nat. Phys.* **6,** 707–712 (2010).
26. Flowers-Jacobs, N. E. *et al.* Fiber-cavity-based optomechanical device. *Appl. Phys. Lett.* **101,** 221109 (2012).
27. Purdy, T. P. *et al.* Tunable cavity optomechanics with ultracold atoms. *Phys. Rev. Lett.* **105,** 133602 (2010).
28. Li, T., Kheifets, S. & Raizen, M. G. Millikelvin cooling of an optically trapped microsphere in vacuum. *Nat. Phys.* **7,** 527 (2011).
29. Lin, Q., Rosenberg, J., Jiang, X., Vahala, K. J. & Painter, O. Mechanical oscillation and cooling actuated by the optical gradient force. *Phys. Rev. Lett.* **103,** 103601 (2009).
30. Anetsberger, G. *et al.* Cavity optomechanics and cooling nanomechanical oscillators using microresonator enhanced evanescent near-field coupling. *Comptes Rendus Phys.* **12,** 800–816 (2011).
31. Teufel, J. D. *et al.* Sideband cooling of micromechanical motion to the quantum ground state. *Nature* **475,** 359–363 (2011).
32. Safavi-Naeini, A. H. *et al.* Squeezed light from a silicon micromechanical resonator. *Nature* **500,** 185–189 (2013).
33. Brennecke, F., Ritter, S., Donner, T. & Esslinger, T. Cavity optomechanics with a Bose-Einstein condensate. *Science* **322,** 235–238 (2008).
34. Pirkkalainen, J.-M. *et al.* Cavity optomechanics mediated by a quantum two-level system. *Nat. Commun.* **6,** 6981 (2015).
35. Macklin, C. *et al.* A near quantum-limited Josephson traveling-wave parametric amplifier. *Science* **350,** 307–310 (2015).
36. Hammerer, K., Sørensen, A. S. & Polzik, E. S. Quantum interface between light and atomic ensembles. *Rev. Mod. Phys.* **82,** 1041 (2010).
37. Hatridge, M. *et al.* Quantum back-action of an individual variable-strength measurement. *Science* **339,** 178–181 (2013).
38. Murch, K. W., Weber, S. J., Macklin, C. & Siddiqi, I. Observing single quantum trajectories of a superconducting quantum bit. *Nature* **502,** 211–214 (2013).

Acknowledgements

We would like to thank K.E. Khosla, G.J. Milburn and T.M. Stace for useful discussion. This research was supported primarily by the ARC CoE for Engineered Quantum Systems (CE110001013). M.R.V. acknowledges support provided by an ARC Discovery Project (DP140101638). P.E.L., S.S. and A.B. acknowledge funding from the Villum Foundation VKR Centre of Excellence NAMEC (Contract No. 65286) and Young Investigator Programme (Project No. VKR023125).

Author contributions

G.A.B. and M.R.V. contributed equally to this work. This quadratic measurement research programme was conceived by M.R.V. with refinements from G.A.B. and W.P.B. The optomechanical evanescent coupling set-up was designed by G.A.B. and W.P.B. with later input from M.R.V. G.A.B. was the main driving force behind building the experiment and performing the data analysis with important input from M.R.V. and W.P.B. Micro-fabrication of the SiN nanostring mechanical resonators was performed by P.E.L., S.S. and A.B. This manuscript was written by M.R.V. and G.A.B. with important contributions from W.P.B. Overall laboratory leadership was provided by W.P.B. and substantial supervision for this project was performed by M.R.V.

Additional information

Competing financial interests: The authors declare no competing financial interests.

16

Subwavelength nonlinear phase control and anomalous phase matching in plasmonic metasurfaces

Euclides Almeida[1], Guy Shalem[1] & Yehiam Prior[1]

Metasurfaces, and in particular those containing plasmonic-based metallic elements, constitute an attractive set of materials with a potential for replacing standard bulky optical elements. In recent years, increasing attention has been focused on their nonlinear optical properties, particularly in the context of second and third harmonic generation and beam steering by phase gratings. Here, we harness the full phase control enabled by subwavelength plasmonic elements to demonstrate a unique metasurface phase matching that is required for efficient nonlinear processes. We discuss the difference between scattering by a grating and by subwavelength phase-gradient elements. We show that for such interfaces an anomalous phase-matching condition prevails, which is the nonlinear analogue of the generalized Snell's law. The subwavelength phase control of optical nonlinearities paves the way for the design of ultrathin, flat nonlinear optical elements. We demonstrate nonlinear metasurface lenses, which act both as generators and as manipulators of the frequency-converted signal.

[1] Department of Chemical Physics, Weizmann Institute of Science, Rehovot 76100, Israel. Correspondence and requests for materials should be addressed to E.A. (email: Euclides.almeida@weizmann.ac.il) or to Y.P. (email: Yehiam.prior@weizmann.ac.il).

Metamaterials are a class of artificial materials whose optical properties can be tailored to exhibit phenomena not commonly found in nature, such as negative[1,2] or anomalous refraction[3,4]. These unique optical properties are frequently engineered by single- or multi-layered nanometric objects, often metallic, fabricated on the surface of 'classical' standard materials. Metasurfaces constitute a particularly interesting and attractive subset of such materials leading to the possibility of designing and creating, by means of modern nanolithographic fabrication techniques, flat and ultrathin optical elements[4–8]. Nano-plasmonic-based metallic elements are the commonly utilized building blocks, and their (linear) optical properties are quite well understood[6,7]. The significance of phase changes across a metasurface has been recognized early on[9] and the laws of refraction across such surfaces have been recently reformulated by Yu *et al.*[4] in terms of a generalized Snell's law. For metasurfaces with linear phase gradient[5–7], it was shown that the transverse phase gradient $d\Phi/dx$ must be accounted for as light crosses the metasurface (ref. 4, equation (2)). Based on these principles, the same group[10] demonstrated achromatic lenses by proper design of the phase elements in metasurfaces.

Metal-based metasurfaces can enhance optical nonlinearities at plasmonic resonances by orders of magnitude[11]. Negative refraction[12] and zero index materials[13] were demonstrated, enhancement in clusters and Fano resonances[14] were shown, and the potential for super resolution in plasmonically enhanced four-wave mixing (FWM) was also discussed[15]. As for nonlinear harmonic generation, second harmonic generation (SHG) was studied[16–18] and recently Segal *et al.*[19] and Li *et al.*[20] discussed the SHG signal generated on metasurfaces and demonstrated beam bending and focusing of such light. Other works, Lee *et al.*[21] and Wolf *et al.*[22], reported the use of metasurfaces to gain control over a SHG signal generated within the nonlinear substrate on which the metasurface is located. The next, third order, nonlinearity was investigated even less, although, in principle, it exists for any structure and for any surface symmetry. As an example, FWM was demonstrated for cavities perforated on gold films[23]. However, to the best of our knowledge, the fundamental issue of phase matching across metasurfaces has not yet been thoroughly addressed. This is, in part, due to the somewhat limited phase and amplitude control over the nonlinearities of individual plasmonic element, which in many cases drives researchers to resort to periodic structures (gratings) that impose specific angles of diffraction[24].

Based on the linear results, one may anticipate that transverse phase gradients at the interface provide an additional momentum that must be included in any nonlinear phase-matching scheme. Here, we demonstrate full phase control over nonlinear optical interactions in plasmonic metasurfaces. This control is achieved by introducing a spatially varying phase response of a metallic metasurface consisting of subwavelength nonlinear nanoantennas designed specifically for the frequency of the nonlinear signal. For such metasurfaces, we derive a new, anomalous nonlinear phase-matching condition that differs from the conventional phase-matching schemes in nonlinear optics. The complete phase control over the nonlinear emission enables us to design flat nonlinear optical components, such as nonlinearly blazed elements and ultrathin frequency-converting lenses with tight focusing.

Results

Finite-differences time-domain calculations. A coherent wave-mixing process obeys the phase-matching condition $\Delta\mathbf{k}=0$, where $\Delta\mathbf{k}$ is the vectorial sum of the momenta of all photons, incoming and generated, participating in the nonlinear mixing. This condition determines the direction of the coherent emission. In FWM, for collinear input beams in a homogeneous and non-dispersive medium, the phase-matched generated beam propagates in the same direction, and this is also the case for the quasi-phase-matching scheme[25,26], where a nonlinear material is designed with periodic reversal of the sign of the nonlinearity in the propagation direction. For non-collinear input beams, a more elaborate phase-matching scheme is required, often based on birefringence or temperature tuning. For our metasurfaces, the nonlinear phase gradient imposed by the plasmonic antennas determines the phase-matching conditions.

Optical nanoantennas, like other driven oscillators, reradiate the incoming light at the same frequency but with a shifted phase. In this work, we use rectangular nanocavities in thin gold films as optical antennas, and to convey the new physical principles more clearly, the aspect ratio (AR) of the rectangles was maintained as our single tuning parameter.

In Fig. 1, we present nonlinear finite-differences time-domain (FDTD) calculations (details given in the Methods section) performed for individual cavities of varying ARs. These separate calculations are combined to depict the transmission and nonlinear interactions in phased arrays of such cavities. In Fig. 1a,b, we show the calculated linear transmittance spectrum (intensity and phase) for light polarized along the short axis of the rectangles for a set of rectangular nanocavities of different ARs within a free-standing 250-nm thick gold film. Two distinct cavity resonances are seen[27], and the correlation between the intensity and the phase shift acquired by the transmitted wave is clear.

Consider now a FWM configuration where two transform-limited laser pulses (with ω_j, $\mathbf{k}_j$ and $\mathbf{E}_j(\mathbf{r},t)=\boldsymbol{\varepsilon}_j(\mathbf{r},t)\,e^{i(\mathbf{k}_j\bullet\mathbf{r}-\omega_j t+\Phi_j)}$ where $j=1,2$, respectively) interact with a metallic nanoantenna to generate a FWM signal $\mathbf{E}_{\mathrm{FWM}}(\mathbf{r},t)=\boldsymbol{\varepsilon}_{\mathrm{FWM}}(\mathbf{r},t)\,e^{i(\mathbf{k}_{\mathrm{FWM}}\bullet\mathbf{r}-\omega_{\mathrm{FWM}}t+\Phi_{\mathrm{FWM}})}$ travelling at the $\mathbf{k}_{\mathrm{FWM}}$ direction and with frequency $\omega_{\mathrm{FWM}}=2\omega_1-\omega_2$. The third-order polarization induced at position $\mathbf{r}$ is given by[28]:

$$\mathbf{P}^{(3)}(\mathbf{r},t)=\frac{1}{(2\pi)^3}\int d\omega_1\int d\omega_1\int d\omega_2\chi^{(3)}(\omega_{\mathrm{FWM}},2\omega_1,-\omega_2)\,\mathbf{E}_1(\mathbf{r},\omega_1)\mathbf{E}_1(\mathbf{r},\omega_1)\mathbf{E}_2^*(\mathbf{r},\omega_2) \quad (1)$$

Where $\chi^{(3)}$ is the third-order susceptibility of the metal and the fields $\mathbf{E}_i(\mathbf{r},\omega_i)$ are the position-dependent electric fields. An antenna much smaller than the wavelength can be approximated by a point dipole (eliminating the position dependence within the antenna) and this leads to effective fields[29], $\mathbf{E}_i(\omega_i)=A_i(\omega_i)\boldsymbol{\varepsilon}_1(\omega_i)e^{i\Phi(\omega_i)}$ where $A_i(\omega_i)$ is the field enhancement (a real quantity) and $\Phi(\omega_i)$ is the phase response. Therefore, we can rewrite equation (1) as

$$\mathbf{P}^{(3)}(t)\propto\int d\omega_1\int d\omega_1\int d\omega_2 S^{(3)}(\omega_{\mathrm{FWM}},2\omega_1,-\omega_2)\,|A_1(\omega_1)|^2A_2(\omega_2)\boldsymbol{\varepsilon}_1(\omega_1)\boldsymbol{\varepsilon}_1(\omega_1)\boldsymbol{\varepsilon}_2^*(\omega_2)e^{i(2\Phi(\omega_1)-\Phi(\omega_2))} \quad (2)$$

Where $S^{(3)}$ is the effective third-order nonlinear susceptibility. The nonlinear FWM signal carries the frequency response at the fundamental frequencies through the phase factor $e^{i(2\Phi(\omega_1)-\Phi(\omega_2))}$, which changes sharply for excitation close to the nanoantenna resonance. The nonlinear phase response of the nanoantennas can be directly calculated using full-wave nonlinear finite-differences time-domain (NL-FDTD) calculations.

In Fig. 1c,d, we show the generated field at ω_{FWM} for a set of nanorectangles with varying AR. In the NL-FDTD calculation, two 60-fs long transformed-limited co-propagating pulses, with centre frequencies $\omega_1=800$ nm and $\omega_2=1{,}088$ nm, respectively, are temporally and spatially overlapped at a single nanocavity. Figure 1d depicts the linear phase accumulated by individually propagating waves at 800, 1,088 and 633 nm, and the nonlinear phase accumulated by the generated FWM beam at 633 nm. The

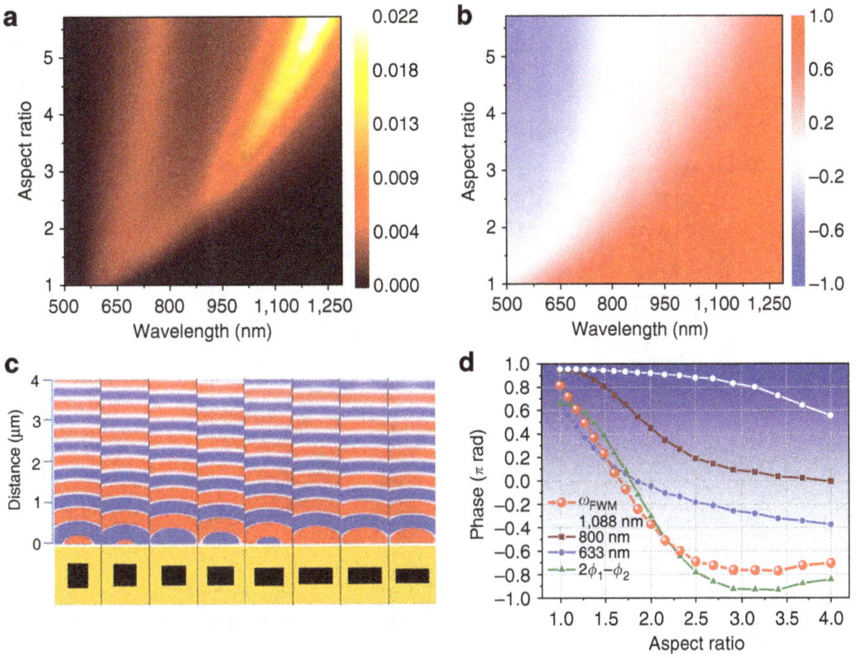

Figure 1 | Linear and FWM transmission, through rectangular gold nanocavities with varying aspect ratios. (**a**) Linear transmittance spectral intensity and (**b**) corresponding linear spectral phase (colour code is in units of π) (**c**) The calculated phase of the generated FWM field $E_y(\omega_{FWM})/|E_y(\omega_{FWM})|$ at the exit from the film as it propagates away from the surface. (**d**) The phases accumulated by the different fields (input and generated) on crossing the metasurface as a function of the aspect ratio: 1,088 nm—white circles, 800 nm—brown squares, 633 nm (independently propagating)—blue circles, generated FWM at 633 nm—red circles and the phase accumulated at $2\Phi_1 - \Phi_2$—green triangle (see text).

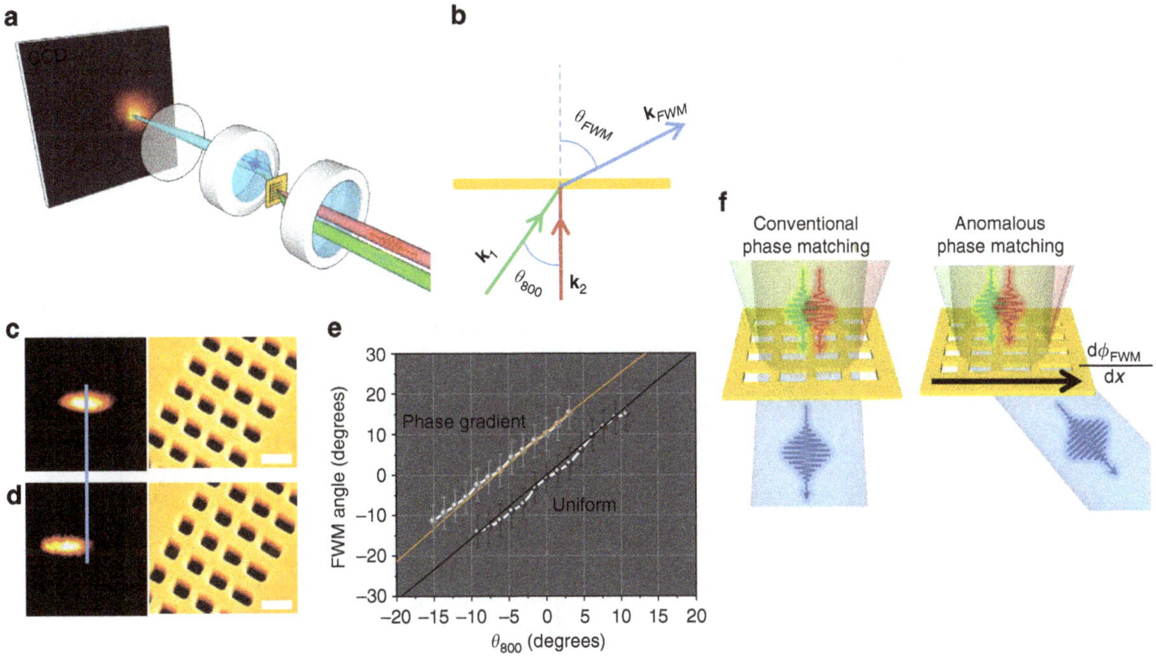

Figure 2 | k-Space analysis of the FWM. (**a**) Optical arrangement for measuring the FWM angle dependence. (**b**) The position of the 800 nm (green) beam on the focusing lens, as determined by a translation stage, controls the input angle. The θ_{FWM} was determined in relation to the beam generated by a uniform structure. (**c**) CCD image of a signal from a uniform unit cell, and (**d**) from a phase gradient unit cell (the scale bars in the SEM are 500 nm). (**e**) Input angle dependence of the phase-matching angle for the uniform and phase-gradient metasurfaces. The orange line is the line fit to the anomalous phase-matching condition (equation (4)), while the black line depicts the conventional phase-matching condition. (**f**) Illustration of the anomalous phase-matching condition for phase-gradient metasurfaces.

calculated Φ_{FWM} does not fit the phase of the 633-nm wave. The calculated phase difference, $2\Phi_1 - \Phi_2$, provides a much better estimate for the phase of the generated FWM signal, but the fit is still not perfect.

For a better description of the generated phase two additional factors need be included. The first, and less critical one, is integration over the laser bandwidth, which for these ultrashort pulses is not negligible. The more important factor is the fact that

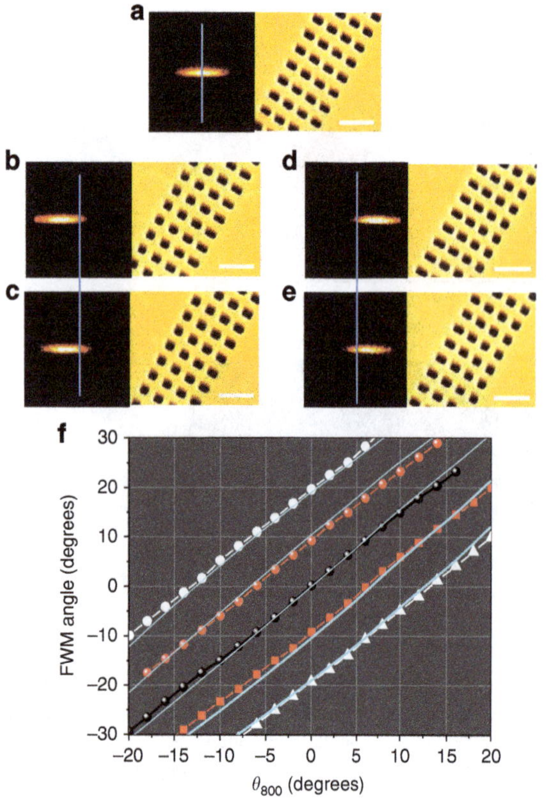

Figure 3 | Beam steering angle of the FWM signal for phase-gradient metasurfaces. (**a–e**) Experimentally measured phase-gradient structures and the generated FWM at normal incidence. (**a**) Uniform structures; (**b,c**) AR increasing to the right; (**d,e**) AR increasing to the left. Scale bar, 1 μm (in all s.e.m.'s). (**f**) Calculated FWM angle dependence for phase gradient structures, where the phase gradient was taken from data such as in Fig. 1 and the calculated blue lines are derived from the anomalous phase-matching condition equation (4). The phase gradients $d\Phi_{FWM}/dx$ 0 (black circles), − 1.1 (white circles), − 0.58 (red circles), 0.58 (red squares), and 1.1 (white triangles).

the FWM signal is generated gradually over the length of propagation through the metasurface, so that the phase accumulated by this wave as it is building up should also be included, and can be incorporated as an effective dielectric constant $\boldsymbol{\varepsilon}_{\text{eff}}$ that appears in the nonlinear wave equation

$$\nabla^2 \mathbf{E}_{\text{FWM}} - \frac{\boldsymbol{\varepsilon}_{\text{eff}}(\omega_{\text{FWM}})}{c^2}\frac{\partial^2}{\partial t^2}\mathbf{E}_{\text{FWM}} = \frac{1}{\varepsilon_0 c^2}\frac{\partial^2}{\partial t^2}\mathbf{P}^{(3)} \quad (3)$$

As mentioned, our direct NL-FDTD calculation provides the phase Φ_{FWM}, and therefore we will use results such as of Fig. 1d for the design of our metasurface.

Experimental arrangement and observations. The experimental arrangement is the standard, forward propagating FWM configuration described in an earlier publication[27]. Two 60-fsec beams, one from a Ti:Sapphire and one from an Optical Parametric Amplifier serve as our inputs—their input angles and time delays are individually controllable. In Fig. 2a, the two ultrashort pulses, with wavevectors $\mathbf{k}_1$ and $\mathbf{k}_2$, respectively, are now spatially and temporally overlapped and focused on the phase-gradient metasurface to generate a FWM signal at $\omega_{\text{FWM}} = 633$ nm and $\mathbf{k}_{\text{FWM}}$. After the sample, the fundamental beams are filtered and the FWM signal is imaged on a CCD camera which records its k-space information. We measure the FWM from two different metasurfaces—each consisting of four rectangles 450 nm apart, in the uniform case all rectangles are with AR = 1.9, and in the phase gradient case the AR covers the range of AR = 1.1–2.9.

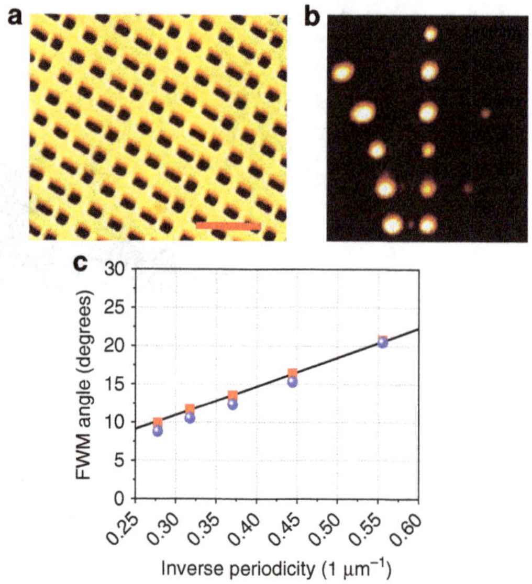

Figure 4 | Nonlinear blazed gratings based on periodic arrangement of phase gradient elements. (**a**) Blazed grating with unitary cell consisting of four elements with AR = 1.1, 1.5, 1.9, 2.9. Scale bar, 1 μm. (**b**) CCD images for the zeroth and first diffraction orders for different periodicities. A much weaker first negative order may also be seen. (**c**) Measured (blue circles) FWM emission angle as a function of the grating periodicity. The red squares are the results from NL-FDTD simulations and the black line is the prediction of the Raman-Nath NL diffraction theory.

In both cases we measure the FWM output angle as a function of the input angle θ_{800} of the $\omega_1 = 800$ nm beam for normal incidence of the ω_2 beam. There is an ~10° difference in the output angle from the two different metasurfaces, indicating a different phase-matching condition. This new phase-matching condition stems from the additional momentum provided by the metasurface, along the gradient direction: $\mathbf{k}_{\text{FWM}}^{\text{new}} = \mathbf{k}_{\text{FWM}} + \Delta\mathbf{k}_x$, where $\mathbf{k}_{\text{FWM}} = 2\mathbf{k}_1 - \mathbf{k}_2$ is the conventional phase-matching condition. The net momentum provided by the metasurface to the FWM signal, $\Delta\mathbf{k}_x = (d\Phi_{\text{FWM}}/dx)\mathbf{u}_x$, is transferred by the metasurface to the beams participating in the nonlinear conversion process.

Thus, the new anomalous phase-matching condition for FWM assumes the form:

$$\mathbf{k}_{FWM}^{new} = \mathbf{k}_{FWM} + \frac{d\Phi_{FWM}}{dx}\mathbf{u}_x \quad (4)$$

Equation (4) combined with the NL-FDTD calculation for Φ_{FWM} provides a framework for the design of phase-gradient nonlinear metasurfaces. In Fig. 2e, the orange curve represents a nonlinear fit to equation (4), from which we extract a value for the phase gradient provided by the metasurface over a unit cell, in this case $\Delta\Phi_{\text{FWM}} = -0.55\pi$.

With the proper choice of $d\Phi_{\text{FWM}}/dx$, one can control the beam steering of the FWM emission. In Fig. 3, we show a NL-FDTD calculation, using parameters similar to the experimental, of the angle dependence of the FWM signal for different phase-gradient metasurfaces. The line fits to equation (4) using the nonlinear phase gradient shown in Fig. 1 (blue lines) are also plotted and are in good agreement to the NL-FDTD calculations.

Beyond the phase gradient unit cell, if several (many) cells are arranged in a periodic manner, they form a blazed grating. The analysis of the anomalous phase matching from such a grating should include the general theory of diffraction[24] in the linear case, or the Raman–Nath diffraction[30] in the nonlinear one. These blazed gratings, unlike gratings resulting from alternating positive- and negative-phased antennas in uniform unit cells[19,31], are not symmetric in terms of 'positive and 'negative' transverse directions, resulting in much lower diffraction efficiency of the negatively diffracted orders ($m = -1, -2...$). In Fig. 4, the scattering from blazed gratings is depicted. The spots seen on the CCD images for collinear, normal excitation are explained by the different orders of diffraction of the blazed grating. The angle of diffraction of the different diffraction orders is determined by the grating period, and agrees with the Raman–Nath diffraction formula $\sin\theta_m = m\lambda_{FWM}/\Lambda$, where Λ is the period of the grating. High-order diffraction modes are also seen, but with weaker intensity compared with the zeroth and first order, also in accordance with the Raman–Nath diffraction theory.

FWM metasurface lenses. To further illustrate the power and flexibility of these nonlinear phase-gradient metasurfaces we designed nonlinear metalenses that focus the wavelength of choice in a specific FWM configuration. The ultrathin lenses operate by imposing a radially dependent relative phase shift at the FWM wavelength $\omega_{FWM} = 2\omega_1 - \omega_2 = 633$ nm:

$$\Phi(r) = \frac{2\pi}{\lambda_0}\left(\sqrt{r^2 + f^2}\right) \quad (5)$$

Here f is the desired focal distance of the lens and λ_0 is the free-space wavelength. Metalenses with different focal lengths are shown and discussed in Fig. 5. They are made of concentric rings of phased gratings consisting of rectangular nanoantennas, and the FWM signal is a tightly focused, nearly diffraction limited Gaussian spot. Note that the experimentally observed focal spots are not as tight as the calculated ones. This discrepancy may be attributed to less than perfect fabrication accuracy, and input beams which are not fully collimated. Both factors will give rise to different focal parameters.

Discussion

In the present work, we demonstrate full control over the nonlinear phase on the subwavelength scale in phase-gradient metasurfaces. As in the linear regime, where Snell's law had to be modified, the phase control over the nonlinear nanoantennas leads to a modified manifestation of the laws governing nonlinear phenomena, such as NL scattering, NL refraction and frequency conversion. Related phase control had been previously reported for the linear case and analysed in terms of the Berry phase[9], and more recently for nonlinear harmonic generation[19,20], while others have discussed shape resonances and their effect on SHG[17,32,33]. For a recent review of nonlinear plasmonics see ref. 11 and references therein. The present implementation of the

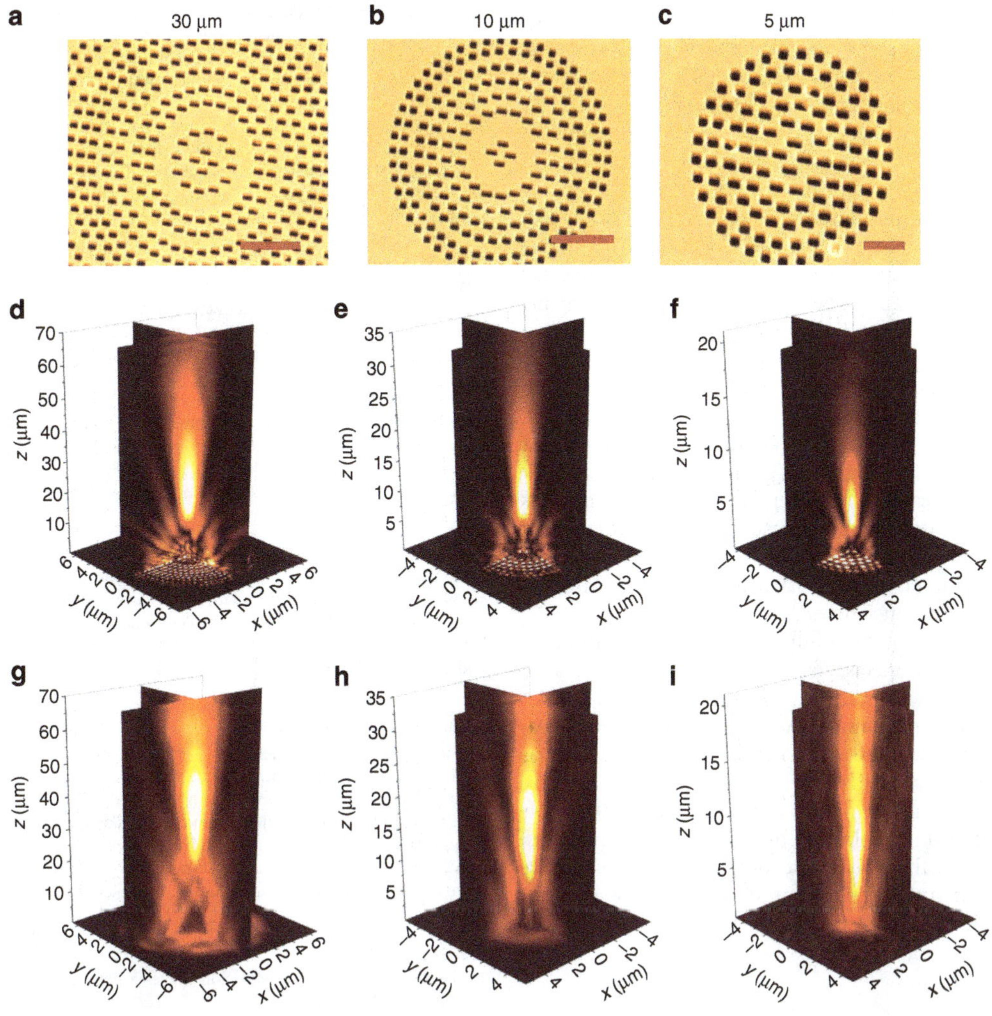

Figure 5 | Nonlinear metalenses of focal lengths of 5, 10, 30 μm based on FWM operating at $\omega_{FWM} = 633$ nm. (**a–c**) SEM images of the fabricated ultrathin lenses; (**d–f**) NL-3D-FDTD simulated images of the focal region; (**g–i**) metalens experimental measurement of the focal region. Scale bars, 5 μm (**a**); 2 μm (**b**); 1 μm (**c**).

subwavelength phase gradients enables treatment of any polarizations and for any input frequency.

Quite a few ultrathin optical components[6,7,34–36] have been proposed and used for beam steering applications. These are based on the abrupt phase changes experienced by light on propagation through a properly designed ultrathin layer of metamaterial. The beam bending (at a specific angle) results from scattering by phase gratings generated by the proper design of the nanoantennas. These nanoantennas are generally uniform across a unit cell, and their phase changes abruptly (typically by π) at a predesigned periodicity. We have carried this concept one step further, designed and fabricated blazed metasurfaces, where the scattering is from the phase gradient across the unit cell, and carefully analysed their nonlinear optical properties. We show that FWM from such metasurfaces reveals a new feature: the scattering from a phase gradient unit cell enables anomalous phase-matching condition for FWM from such metasurfaces. This phase-matched FWM is efficient; its phase-matched direction of propagation may be controlled by proper design of the phase gradients, it enables beam bending at any angle and thus FWM focusing, and it differs from the scattering by phase gratings, as discussed in ref. 24. In all these cases, the experimental results were compared with numerical solutions (NL-FDTD, Lumerical solutions package[37]) of the wave equations with nonlinear terms added to them. The agreement is generally very good, and whenever relevant, comparison with analytical expressions is added.

We designed and fabricated ultrathin FWM lenses, and demonstrated tight focusing with focal lengths of several microns. These FWM lenses do not have any restrictions on the symmetry of the design which is characteristic to elements based on SHG, and can be integrated in light detectors based on frequency conversion to provide more sensitive detection.

For pedagogical reasons we have limited our design to rectangles and the tuning parameter to the AR, but more generally parameters such as area of the hole, or shapes presenting multiple resonances such as V-shaped antennas can and will be used. Interestingly, even though the linear rectangular nanoatennas do not cover the full range of 2π phase shift, it is possible to obtain a 2π phase shift in the nonlinear case due to multiples or combinations of linear phases imprinted on the fundamental beams.

As is well known, metal structures are lossy, especially in the visible range, and thus the search is going on for nonmetallic replacements. This endeavour, however, is hampered by the lack of the electromagnetic enhancement readily available for metals, and presently combined structures consisting of nonlinear materials with plasmonic metallic structures seem to offer a pathway towards higher efficiencies[21].

In conclusion, we demonstrate full control over the nonlinear phase in phase-gradient metasurfaces. We show that in such metasurfaces a new phase-matching condition applies, which differs from the phase-matching schemes known in nonlinear optics, and have designed and built ultrathin elements such as blazed elements and lenses that are based on nonlinear wave mixing. The phase control of nonlinear nanoantennas will enable the design of ultrathin nonlinear metamaterials, which can generate and control the wavefront of light, and may have implications for the next generation of efficient devices for spectroscopic (CARS or SERS) measurements.

Methods

Sample fabrication. The samples were fabricated by focused ion beam (FIB) milling on a high quality free-standing gold film. The procedure for fabrication of free-standing gold films is described in details in ref. 27 Briefly, using an e-beam evaporator, we deposited a 10-nm thick Cr adhesion layer and a 250-nm thick gold layer on a polished silicon wafer. On the opposite side of the wafer, we had previously grown a circular Si_3N_4 mask by plasma-enhanced chemical vapour deposition. The mask was chemically etched using KOH and the remaining free-standing metallic area was etched with HCl to remove the adhesion layer.

Linear FDTD simulations. The transmittance spectrum and the relative spectral phase response of the rectangular metallic nanocavities were calculated using the commercial software Lumerical FDTD solutions. The values of the dielectric constants were taken from the data table of Gold from Palik[38].

Nonlinear FDTD simulations. The nonlinear phase was calculated using the nonlinear material implementation of Lumerical. The base material is Palik gold, which is assumed to have instantaneous (non-dispersive) third-order nonlinearity $\chi^{(3)} = 10^{-18}\,m^2\,V^{-2}$. As input light sources we used two temporally overlapped transform-limited plane wave sources centred at $\omega_1 = 800$ nm and $\omega_2 = 1{,}088$ nm, with pulse duration 60 fs, propagating parallel to the z direction. The polarization of both pulses is perpendicular to the long axis of the rectangles. The y component of the real and imaginary parts of the electric field (that is, the propagating waves) of the FWM signal is recorded on a y normal plane spanning the whole simulation area. The dimensions of the mesh were set to $dx = dy = dz = 5$ nm and perfectly matched layers were added in all dimensions.

k-Space analysis. For the k-space analysis of the FWM signal from phase-gradient antennas, the same simulation parameters used in the phase response were kept, except now the $\omega_1 = 800$ nm source propagates in the x-z plane with a variable incidence angle θ_{800} with respect to the normal. The exit fields on the opposite side of the metasurface are recorded in a z-normal plane and projected to the far field, where the angle θ_{FWM} of the FWM signal at $\omega_{FWM} = 633$ nm is calculated as a function of θ_{800}.

K-space measurements. In the FWM experiments, we used the set-up described in details in ref. 27 An optical parametric amplifier, pumped by a 1-kHz amplified Ti:Sapphire laser, was used as the light source for the $\omega_2 = 1{,}088$ nm pulses, while the pulses of the Ti:Sapphire laser that pumped the optical parametric amplifier were used as the fundamental $\omega_1 = 800$ nm beam. Both ω_1 and ω_2 pulses have the same pulse duration of 60 fs. The beams travel two distinct optical paths, where the intensity and polarization of each individual beam could be controlled by a set of half-wave plate and polarizer to avoid optical damage to the samples. Both beams are focused and overlapped in the sample, by an objective lens of numerical aperture (N.A.) = 0.42 (Mitutoyo M Plan Apo 50X Infinity-Corrected). The incident angle θ_{800} of the ω_1 beam can be varied by controlling the lateral displacement of the beam with computer-controlled translation stage supporting a beam splitter whose primary role was to merge the optical path of two beams. The ω_2 beam at 1,088 nm was always kept normal to the surface. The temporal overlap between the two beams as the input angle was varied, was monitored and controlled by properly delaying the ω_1 beam. The FWM signal centred at $\omega_{FWM} = 633$ nm is collected by an objective lens with N.A. = 0.42 (Mitutoyo M Plan Apo SL50X Infinity-Corrected) and focused by spherical lens of $f = 75$ mm onto an EMCCD camera (Andor iXon DV885) The fundamental beams are filtered by a pair of shortpass filters (Thorlabs FES0700 and FESH0750). In the angular measurements in Figs 3 and 4, the focusing objective with N.A. = 0.42 was replaced by a lens of focal length $f = 50$ mm to illuminate a larger area.

Nonlinear lenses. For the design of our lenses we kept the same parameters used in the calculations of the NL phase response. However, to decrease the simulation time, we used symmetric boundary condition in the x dimension and anti-symmetric in the y dimension. The dimensions of the fine mesh around the lenses were set to $dx = dy = 10$ nm and $dz = 5$ nm for the $f = 5$ and 10 μm lenses and $dx = dy = 15$ nm and $dz = 5$ nm for the $f = 30$ and 60 μm lenses.

Nonlinear lenses measurements. The experimental set-up for measurements of the nonlinear lenses is a modification of the k-space set-up described above. Both ω_1 and ω_2 beams are normally incident. The focusing objective is replaced by a spherical lens with $f = 100$ mm. The FWM signal out of the lenses is collected by the imaging (NA = 0.42) objective and is imaged directly onto the EMCCD in real (physical) space. The imaging objective is supported on a computer-controlled translation stage that can vary the focal plane of the objective and record three-dimensional tomographic images of the focal region.

References

1. Pendry, J. B. Negative refraction makes a perfect lens. *Phys. Rev. Lett.* **85,** 3966–3969 (2000).
2. Veselago, V. G. Electrodynamics of substances with simultaneously negative values of sigma and mu. *Sov. Phys. Uspekhi* **10,** 509–50 (1968).
3. Lapine, M., Shadrivov, I. V. & Kivshar, Y. S. Colloquium: nonlinear metamaterials. *Rev. Mod. Phys.* **86,** 1093–1123 (2014).

4. Yu, N. F. *et al.* Light propagation with phase discontinuities: generalized laws of reflection and refraction. *Science* **334,** 333–337 (2011).
5. Ni, X. J., Emani, N. K., Kildishev, A. V., Boltasseva, A. & Shalaev, V. M. Broadband light bending with plasmonic nanoantennas. *Science* **335,** 427–427 (2012).
6. Kildishev, A. V., Boltasseva, A. & Shalaev, V. M. Planar photonics with metasurfaces. *Science* **339,** 1232009 (2013).
7. Yu, N. F. & Capasso, F. Flat optics with designer metasurfaces. *Nat. Mater.* **13,** 139–150 (2014).
8. Lin, D. M., Fan, P. Y., Hasman, E. & Brongersma, M. L. Dielectric gradient metasurface optical elements. *Science* **345,** 298–302 (2014).
9. Bomzon, Z., Kleiner, V. & Hasman, E. Pancharatnam-Berry phase in space-variant polarization-state manipulations with subwavelength gratings. *Opt. Lett.* **26,** 1424–1426 (2001).
10. Aieta, F., Kats, M. A., Genevet, P. & Capasso, F. Multiwavelength achromatic metasurfaces by dispersive phase compensation. *Science* **347,** 1342–1345 (2015).
11. Kauranen, M. & Zayats, A. V. Nonlinear plasmonics. *Nat. Photonics* **6,** 737–748 (2012).
12. Palomba, S. *et al.* Optical negative refraction by four-wave mixing in thin metallic nanostructures. *Nat. Mater.* **11,** 34–38 (2012).
13. Suchowski, H. *et al.* Phase mismatch-free nonlinear propagation in optical zero-index materials. *Science* **342,** 1223–1226 (2013).
14. Zhang, Y., Wen, F., Zhen, Y. R., Nordlander, P. & Halas, N. J. Coherent Fano resonances in a plasmonic nanocluster enhance optical four-wave mixing. *Proc. Natl Acad. Sci. USA* **110,** 9215–9219 (2013).
15. Simkhovich, B. & Bartal, G. Plasmon-enhanced four-wave mixing for superresolution applications. *Phys. Rev. Lett.* **112,** 056802 (2014).
16. Zhang, Y., Grady, N. K., Ayala-Orozco, C. & Halas, N. J. Three-dimensional nanostructures as highly efficient generators of second harmonic light. *Nano Lett.* **11,** 5519–5523 (2011).
17. Salomon, A., Zielinski, A., Kolkowski, R., Zyss, J. & Prior, Y. Shape and size resonances in second harmonic generation from plasmonic nano-cavities. *J. Phys. Chem. C* **117,** 22377–22382 (2013).
18. Salomon, A. *et al.* Plasmonic coupling between metallic nanocavities. *J. Opt.* **16,** 114012 (2014).
19. Segal, N., Keren-Zur, S., Hendler, N. & Ellenbogen, T. Controlling light with metamaterial-based nonlinear photonic crystals. *Nat. Photonics* **9,** 180–184 (2015).
20. Li, G. X. *et al.* Continuous control of the nonlinearity phase for harmonic generations. *Nat. Mater.* **14,** 607–612 (2015).
21. Lee, J. *et al.* Giant nonlinear response from plasmonic metasurfaces coupled to intersubband transitions. *Nature* **511,** 65–U389 (2014).
22. Wolf, O. *et al.* Phased-array sources based on nonlinear metamaterial nanocavities. *Nat. Commun.* **6,** 7667 (2015).
23. Genevet, P. *et al.* Large enhancement of nonlinear optical phenomena by plasmonic nanocavity gratings. *Nano Lett* **10,** 4880–4883 (2010).
24. Larouche, S. & Smith, D. R. Reconciliation of generalized refraction with diffraction theory. *Opt. Lett.* **37,** 2391–2393 (2012).
25. Armstrong, J. A., Bloembergen, N., Ducuing, J. & Pershan, P. S. Interactions between light waves in a nonlinear dielectric. *Phys. Rev.* **127,** 1918–1939 (1962).
26. Myers, L. E. *et al.* Quasi-phase-matched optical parametric oscillators in bulk periodically poled LiNbO3. *J. Opt. Soc. Am. B* **12,** 2102–2116 (1995).
27. Almeida, E. & Prior, Y. Rational design of metallic nanocavities for resonantly enhanced four-wave mixing. *Sci. Rep.* **5,** 10033 (2015).
28. Mukamel, S. *Principles of Nonlinear Optical Spectroscopy* (Oxford University Press, 1998).
29. Accanto, N., Piatkowski, L., Renger, J. & van Hulst, N. F. Capturing the optical phase response of nanoantennas by coherent second-harmonic microscopy. *Nano Lett.* **14,** 4078–4082 (2014).
30. Raman, C. V. & Nath, N. S. The diffraction of light by high frequency sound waves: part V. *Proc. Indian Acad. Sci.* **3A,** 459 (1936).
31. Saltiel, S. M. *et al.* Multiorder nonlinear diffraction in frequency doubling processes. *Opt. Lett.* **34,** 848–850 (2009).
32. Prangsma, J. C., van Oosten, D., Moerland, R. J. & Kuipers, L. Increase of group delay and nonlinear effects with hole shape in subwavelength hole arrays. *New J. Phys.* **12** (2010).
33. Wang, B. L. *et al.* Origin of shape resonance in second-harmonic generation from metallic nanohole arrays. *Sci. Rep.* **3,** 2358 (2013).
34. Aieta, F. *et al.* Aberration-free ultrathin flat lenses and axicons at telecom wavelengths based on plasmonic metasurfaces. *Nano Lett.* **12,** 4932–4936 (2012).
35. Ni, X. J., Ishii, S., Kildishev, A. V. & Shalaev, V. M. Ultra-thin, planar, Babinet-inverted plasmonic metalenses. *Light Sci. Appl.* **2,** e72 (2013).
36. Walther, B. *et al.* Spatial and spectral light shaping with metamaterials. *Adv. Mater.* **24,** 6300–6304 (2012).
37. Lumerical. Lumerical Solutions, Inc. Available at http://www.lumerical.com/tcad-products/fdtd (2014).
38. Palik, E. D. *Handbook of Optical Constants of Solids* (Academic Press, 1998).

Acknowledgements

This work was funded, in part, by the Israel Science Foundation, by the ICORE program, by an FTA grant from the Israel National Nano Initiative, and by a grant from the Leona M. and Harry B. Helmsley Charitable Trust. Discussions with Roy Kaner, Yaara Bondy and Yael Blechman are gratefully acknowledged.

Author contributions

All authors conceived the idea. E.A. fabricated the samples and performed most of the experimental work. E.A. and G.S. performed the numerical simulations. All authors jointly wrote the paper and contributed to the physical understanding of the phenomena described.

Additional information

Competing financial interests: The authors declare no competing financial interests.

17

Experimental perfect state transfer of an entangled photonic qubit

Robert J. Chapman[1,2], Matteo Santandrea[3,4], Zixin Huang[1,2], Giacomo Corrielli[3,4], Andrea Crespi[3,4], Man-Hong Yung[5], Roberto Osellame[3,4] & Alberto Peruzzo[1,2]

The transfer of data is a fundamental task in information systems. Microprocessors contain dedicated data buses that transmit bits across different locations and implement sophisticated routing protocols. Transferring quantum information with high fidelity is a challenging task, due to the intrinsic fragility of quantum states. Here we report on the implementation of the perfect state transfer protocol applied to a photonic qubit entangled with another qubit at a different location. On a single device we perform three routing procedures on entangled states, preserving the encoded quantum state with an average fidelity of 97.1%, measuring in the coincidence basis. Our protocol extends the regular perfect state transfer by maintaining quantum information encoded in the polarization state of the photonic qubit. Our results demonstrate the key principle of perfect state transfer, opening a route towards data transfer for quantum computing systems.

[1] Quantum Photonics Laboratory, School of Engineering, RMIT University, Melbourne, Victoria 3000, Australia. [2] School of Physics, The University of Sydney, Sydney, New South Wales 2006, Australia. [3] Istituto di Fotonica e Nanotecnologie, Consiglio Nazionale delle Ricerche, Piazza Leonardo da Vinci 32, Milano I-20133, Italy. [4] Dipartimento di Fisica, Politecnico di Milano, Piazza Leonardo da Vinci 32, Milano I-20133, Italy. [5] Department of Physics, South University of Science and Technology of China, Shenzhen 518055, China. Correspondence and requests for materials should be addressed to A.P. (email: alberto.peruzzo@rmit.edu.au).

Transferring quantum information between locations without disrupting the encoded information *en route* is crucial for future quantum technologies[1–8]. Routing quantum information is necessary for communication between quantum processors, addressing single qubits in topological surface architectures, and for quantum memories as well as many other applications.

Coupling between stationary qubits and mobile qubits via cavity and circuit quantum electrodynamics has been an active area of research with promise for long-distance quantum communication[9–12]; however, coupling between different quantum information platforms is challenging as unwanted degrees of freedom lead to increased decoherence[13]. Quantum teleportation between distant qubits allows long-distance quantum communication via shared entangled states[14–17]; however, in most quantum information platforms this would again require coupling between stationary and mobile qubits. Physically relocating trapped ion qubits has also been demonstrated[18,19], however, with additional decoherence incurred during transport.

By taking advantage of coupling between neighbouring qubits, it is possible to transport quantum information across a stationary lattice[2]. This has the benefits that one physical platform is being used and the lattice sites remain at fixed locations. The most basic method is to apply a series of SWAP operations between neighbouring sites such that, with enough iterations, the state of the first qubit is relocated to the last. This method requires a high level of active control on the coupling and is inherently weak as individual errors accumulate after each operation, leading to an exponential decay in fidelity as the number of operations increases[20].

The perfect state transfer (PST) protocol utilizes an engineered but fixed coupled lattice. Quantum states are transferred between sites through Hamiltonian evolution for a specified time[2–7]. For a one-dimensional system with N sites, the state intially at site n is transferred to site $N-n+1$ with 100% probability without need for active control on the coupling[21]. PST can be performed on any quantum computing architecture where coupling between sites can be engineered, such as ion traps[18] and quantum dots[22]. Figure 1 presents an illustration of the PST protocol. The encoded quantum state, initially at the first site, is recovered at the final site after a specific time. In the intermediate stages, the qubit is in a superposition across the lattice. Aside from qubit relocation, the PST framework can be applied to entangled W-state preparation[23], state amplification[24] and even quantum computation[25–29].

To date, most research on PST has been theoretical[2–7,20,21,23,24,28,30–42], with experiments[43,44] being limited to demonstrations where no quantum information is transferred, and do not incorporate entanglement, often considered the defining feature of quantum mechanics[45]. Here, we present the implementation of a protocol that extends PST for relocating a polarization-encoded photonic qubit across a one-dimensional lattice, realized as an array of 11 evanescently coupled waveguides[46–48]. We show that the entanglement between a photon propagating through the PST waveguide array and another photon at a different location is preserved.

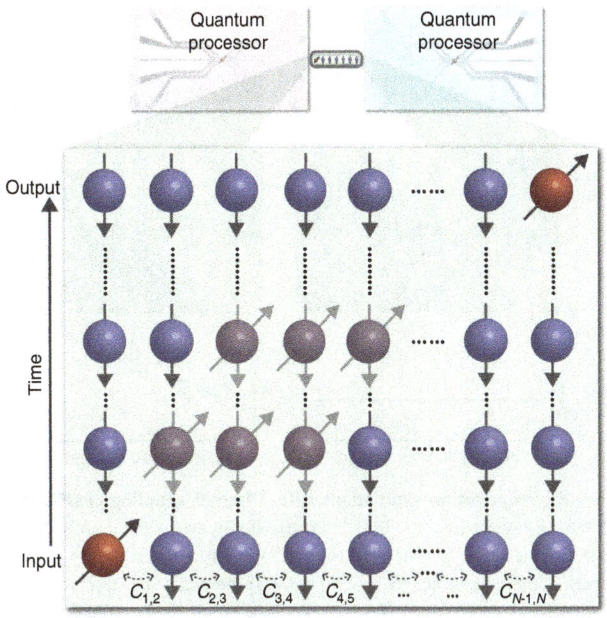

Figure 1 | Illustration of a one-dimensional perfect state transfer lattice connecting two quantum processors. By engineering the Hamiltonian of a lattice, the state at the first site is transferred to the last site after a specific time. This Hamiltonian defines the perfect state transfer protocol[3], which can be used for routing quantum information inside a quantum processor.

Results

PST Hamiltonian. The Hamiltonian for our system in the nearest-neighbour approximation is given by the tight-binding formalism

$$\hat{H} = \sum_{\sigma\in\{H,V\}} \sum_{n=1}^{N-1} C_{n,n+1}\left(\hat{a}^{\dagger}_{n+1,\sigma}\hat{a}_{n,\sigma} + \hat{a}^{\dagger}_{n,\sigma}\hat{a}_{n+1,\sigma}\right), \quad (1)$$

where $C_{n,n+1}$ is the coupling coefficient between waveguides n and $n+1$, and $\hat{a}_{n,\sigma}$ ($\hat{a}^{\dagger}_{n,\sigma}$) is the annihilation (creation) operator applied to waveguide n and polarization σ (horizontal or vertical). Hamiltonian evolution of a state $|\psi_0\rangle$ for a time t is calculated via the Schrödinger equation, giving the final state $|\psi(t)\rangle = \exp(\frac{-i\hat{H}t}{\hbar})|\psi(0)\rangle$ (ref. 49). Equation (1) is constructed of independent tight-binding Hamiltonians acting on each orthogonal polarization. This requires there to be no cross-talk terms $\hat{a}^{\dagger}_{n,H}\hat{a}_{m,V}$ or $\hat{a}^{\dagger}_{n,V}\hat{a}_{m,H}$ $\forall\, m,n$. The spectrum of coupling coefficients $C_{n,n+1}$ is crucial for successful PST. Evolution of this Hamiltonian with a uniform coupling coefficient spectrum, equivalent to equally spaced waveguides, is not sufficient for PST with over three lattice sites as simulated in Fig. 2a. PST requires the coupling coefficient spectrum to follow the function

$$C_{n,n+1} = C_0\sqrt{n(N-n)}, \quad (2)$$

where C_0 is a constant, N is the total number of lattice sites and evolution is for a specific time $t_{\mathrm{PST}}=\frac{\pi}{2C_0}$ (refs 3, 4). This enables arbitrary-length PST as simulated in Fig. 2b for 11 sites. The coupling coefficient spectrum for each polarization must be equal and follow equation (2) for the qubit to be faithfully relocated and the polarization-encoded quantum information to be preserved. The distance between waveguides dictates the coupling coefficient; however, for planar systems, the coupling coefficient of each polarization will in general be unequal due to the waveguide birefringence. To achieve equal coupling between polarizations, the waveguide array is fabricated along a tilted plane in the substrate[50]. This is made possible by the unique three-dimensional capabilities of the femtosecond laser-writing technique (see Supplementary Note 1 for further fabrication and device details). We measure a total propagation loss of 1.8 ± 0.2 dB; however, our figure of merit is how well preserved the polarization quantum state is after the transfer protocol. Therefore we calculate fidelity without loss. Ideally the PST protocol exhibits unit fidelity and efficiency, where the quantum state is reliably transferred and the encoded state is preserved. Due to loss in our experiment, we have less than unit efficiency; however, this loss is largely unrelated to the PST Hamiltonian in equation (1). Further optimizing the fabrication process could

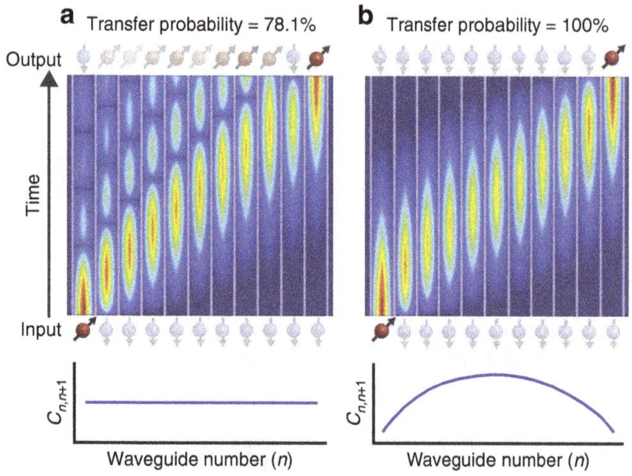

Figure 2 | Propagation simulations with different coupling coefficient spectra. (**a**) A photon is injected into the first waveguide of an array of eleven coupled waveguides with the Hamiltonian in equation (1) and a uniform coupling coefficient spectrum. With the constraint that reflections off boundaries are not allowed, we calculate a maximum probability of transferring the photon to waveguide 11 of 78.1% (ref. 2). (**b**) A photon is injected into the first waveguide of an array of eleven coupled waveguides, this time with the coupling coefficient spectrum of equation (2). After evolution for a pre-determined time, the photon is received at waveguide 11 with 100% probability[3-7].

reduce the level of propagation loss (see Methods and Supplementary Note 2 for further details on loss).

We inject photons into waveguides 1, 6 and 10 of the array, which after time t_{PST} transfer to waveguides 11, 6 and 2, respectively. Figure 3a–c presents propagation simulations for each transfer. Input waveguides extend to the end of the device to allow selective injection.

Transfer characterization. To characterize the coupling coefficient spectra, we inject horizontally and vertically polarized laser light at 808 nm into each input waveguide. Laser light is more robust to noise than single photons and we can monitor the output with a CCD camera to fast gather results. Using laser light at the same wavelength as our single photons will give an output intensity distribution equivalent to the output probability distribution for detecting single photons[46]. Ideally light injected into waveguide n will output the device only in waveguide $N-n+1$; however, this assumes an approximate model of nearest-neighbour coupling only. Taking into account coupling between further separated waveguides reduces the transfer probability. This decrease is greater for light injected closer to the centre of the array (see Supplementary Note 1). Figure 3d–f presents our measured output probability distribution for horizontally $\left(P_n^H\right)$ and vertically $\left(P_n^V\right)$ polarized laser light injected into each input waveguide, where n is the output waveguide number. Fidelity between the probability distributions for each polarization is given by $F_{\text{distribution}} = \sum_n \sqrt{P_n^H P_n^V}$. This fidelity is closely related to how similar the two coupling coefficient spectra are. We measure an average probability distribution fidelity for all transfers of 0.976 ± 0.006 (see Supplementary Table 1 for all fidelity values). We encode quantum information in the polarization state of the photon and are interested in reliably relocating this qubit. We use a single optical fibre to capture photons from the designed output waveguide, which, in all cases, is the waveguide with the greatest output probability.

Quantum process tomography. We perform quantum process tomography to understand the operation performed on the single-photon polarization state during each PST transfer. We inject single-photon states $|\psi_{\text{in}}\rangle = (\alpha\hat{a}^\dagger_{S,H} + \beta\hat{a}^\dagger_{S,V})|0\rangle$ into each input waveguide $S \in \{1,6,10\}$, where α (β) is the probability amplitude of the horizontal (vertical) component of the photon and $|\alpha|^2 + |\beta|^2 = 1$. From quantum process tomography on the output polarization states, we can generate a process matrix χ_{pol} for each transfer[1,51]. We aim to perform the identity operation so that the quantum information encoded in the polarization can be recovered after relocation. We measure a polarization phase shift associated with each transfer. This phase shift can be compensated for with a local polarization rotation applied before injection. Figure 3g–i presents our measured process matrix for each transfer. Across all transfers we demonstrate an average fidelity of the polarization process including compensation to an identity of 0.982 ± 0.003 (see Supplementary Note 3 for details of the compensation scheme and Supplementary Table 2 for all fidelities). Process fidelity is calculated as $F_{\text{process}} = \text{Tr}\{\chi_{\text{I}}\chi_{\text{pol+comp}}\}$ (ref. 52), where χ_{I} is the process matrix for the identity operation and $\chi_{\text{pol+comp}}$ is the combined polarization operation and compensation process matrix.

Ideally the output state for each transfer is $|\psi_{\text{out}}\rangle = (\alpha\hat{a}^\dagger_{T,H} + \beta\hat{a}^\dagger_{T,V})|0\rangle$, where $T \in \{11,6,2\}$ and the probability amplitude of each polarization component remains equal to the input state. Our high-fidelity measurements on single-photon relocation demonstrate that we can route a polarization-encoded photonic qubit across our device and faithfully recover the encoded quantum information.

Entangled state transfer. Entanglement is likely to be a defining feature of quantum computing, and preserving entanglement is therefore critical to the success of any qubit relocation protocol. We prepare the Bell state $\frac{1}{\sqrt{2}}(|H_1V_2\rangle + |V_1H_2\rangle)$ using the spontaneous parametric downconversion process. The polarization is controlled using rotatable half and quarter waveplates (HWPs and QWPs), and polarizing beam splitters (PBSs) as shown in Fig. 4 (ref. 53) (see Methods for details). This set-up prepares a general state $\alpha|H_1V_2\rangle + \beta|V_1H_2\rangle$ when measuring in coincidence, where $|\alpha|^2 + |\beta|^2 = 1$. Photon 1 is injected into the waveguide array, while photon 2 propagates through polarization-maintaining fibre (PMF). In terms of waveguide occupancy, our input state is $|\psi_{\text{in}}\rangle = \frac{1}{\sqrt{2}}(\hat{a}^\dagger_{S,H}\hat{a}^\dagger_{0,V} + \hat{a}^\dagger_{S,V}\hat{a}^\dagger_{0,H})|00\rangle$ for each input waveguide $S \in \{1,6,10\}$, where $\hat{a}^\dagger_{0,\sigma}$ denotes the creation operator acting on polarization σ in PMF. Full two-qubit polarization tomography[54] is performed on the output and the fidelity calculated as

$$F_{\text{quantum}} = \left(\text{Tr}\left\{\sqrt{\sqrt{\rho_{\text{input}}}\rho_{\text{output}}\sqrt{\rho_{\text{input}}}}\right\}\right)^2, \quad (3)$$

where ρ_{output} is the density matrix after the PST protocol has been applied and ρ_{input} is the density matrix after propagation through a reference straight waveguide[55]. After all qubit relocations we measure an average polarization state fidelity of 0.971 ± 0.014. Fidelity is measured in the two-photon coincidence basis. This value is therefore the fidelity on the quantum state transferred without taking into account the loss (see Methods and Supplementary Note 2 for loss analysis). We can use the results from quantum process tomography to generate a characterized model of our device. We can now use this model to calculate the similarity between the predicted output state and our measured output state as

$$S_{\text{quantum}} = \left(\text{Tr}\left\{\sqrt{\sqrt{\rho_{\text{predicted}}}\rho_{\text{output}}\sqrt{\rho_{\text{predicted}}}}\right\}\right)^2. \quad (4)$$

We calculate an average similarity of 0.987 ± 0.014 across all transfers (see Supplementary Table 3 for all fidelities

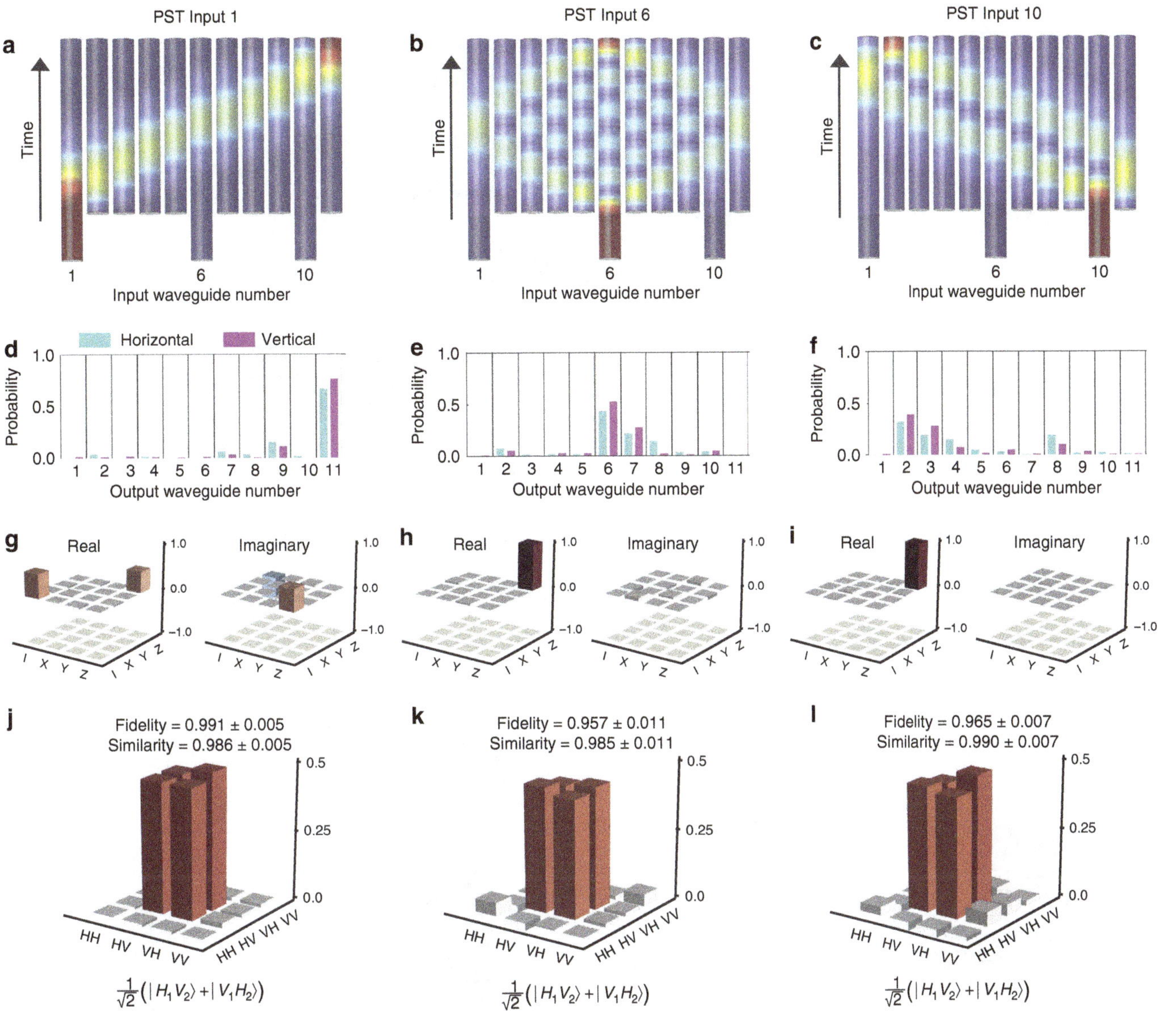

Figure 3 | Experimental data from the characterization and performance of perfect state transfer waveguide array. (a-c) Propagation simulations showing the device implementation to enable specific waveguide input. **(d-f)** Output probability distributions for each input of the PST array for horizontally and vertically polarized laser light. **(g-i)** Quantum process matrix for each transfer in the PST array measured with single-photon quantum process tomography. **(j-l)** Two-photon quantum state tomography is performed after photon 1 of the polarization entangled Bell state $\frac{1}{\sqrt{2}}(|H_1V_2\rangle + |V_1H_2\rangle)$ has been relocated. Results have had the small imaginary components removed for brevity.

and similarities). Figure 3j–l presents our measured density matrix after each entangled state transfer.

Ideally, the output state for each transfer is $|\Psi_{\text{out}}\rangle = \frac{1}{\sqrt{2}}(\hat{a}^\dagger_{T,H}\hat{a}^\dagger_{0,V} + \hat{a}^\dagger_{T,V}\hat{a}^\dagger_{0,H})|00\rangle$, where $T\in\{11,6,2\}$. With high fidelity the probability amplitude of each component is preserved and the state remains almost pure. This result demonstrates that with our device we can relocate a polarization qubit between distant sites and preserve entanglement with another qubit at a different location. In principle our device could route qubits from any waveguide n to waveguide $N-n+1$. Quantum error correction protocols require sophisticated interconnection to access individual qubits for control and measurement within large, highly entangled surface code geometries[56]. PST is a clear gateway towards accessing qubits in such systems without disrupting quantum states and entanglement throughout the surface code.

Decohered state transfer. Decoherence has applications in quantum simulation to emulate systems in nature[57], and it is therefore important to note that this approach for relocating quantum information can be applied to states of any purity[3]. We prepare decohered states by introducing a time delay between the horizontal and vertical components of the polarization qubit. We implement this delay by extending one arm of the source, which reduces the overlap of the photons after they are both incident on the PBS, as shown in Fig. 4. This delay extends the state into a time-bin basis, which we trace over on measurement, leading to a mixed state. The purity of the state can be calculated as the convolution of the horizontal and vertical components with a time delay τ:

$$\text{Purity}(\tau) \equiv \int_{-\infty}^{\infty} H(t)V(\tau - t)\mathrm{d}t, \qquad (5)$$

where τ is controlled by altering the path length of the vertical

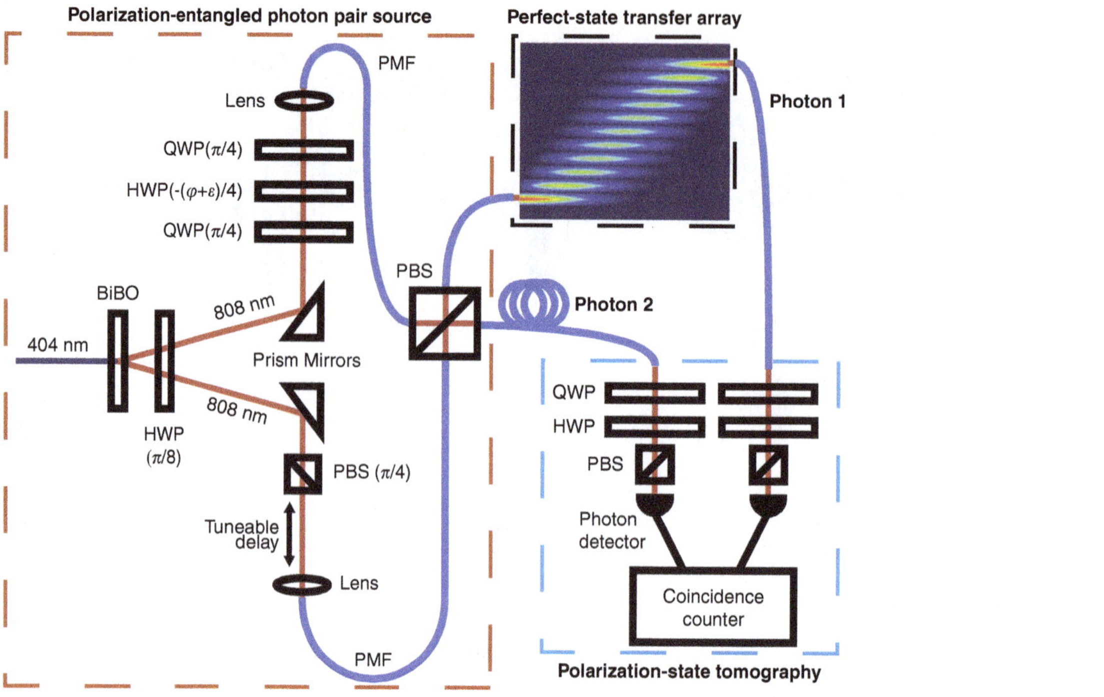

Figure 4 | Experimental set-up. Polarization entangled photons are generated in free space before coupling into PMF. Photon 1 is injected into the perfect state transfer array, while photon 2 travels through PMF. Full two-qubit polarization tomography is performed on the output. See Methods for experimental set-up details.

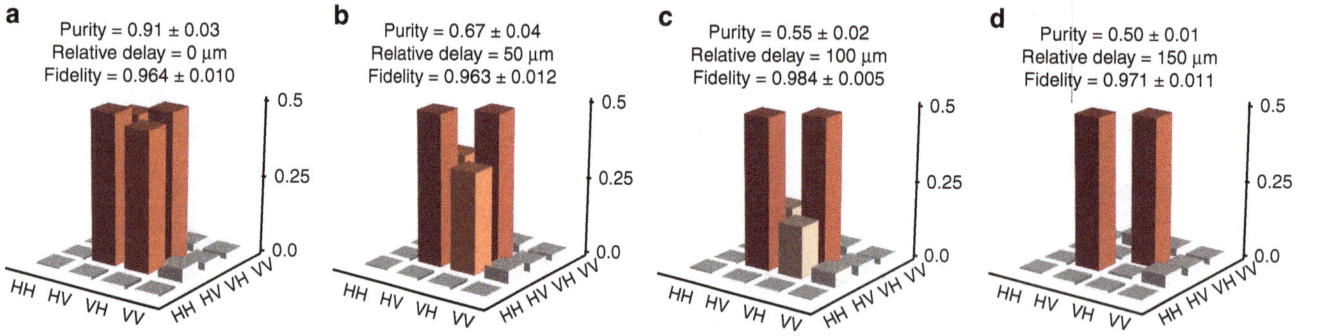

Figure 5 | Perfect state transfer of entangled states with varying purity. Photon 1 of the state $\frac{1}{\sqrt{2}}(|H_1V_2\rangle + |V_1H_2\rangle)$ is injected into waveguide 1 of the PST array. A delay is applied to the vertical component to control the purity of the state. (**a**) Relative delay of 0 μm, (**b**) 50 μm, (**c**) 100 μm and (**d**) 150 μm. Results have had the small imaginary components removed for brevity.

component of the state. $H(V)$ is the horizontal (vertical) component of the photon. Figure 5 presents density matrices for PST from waveguide 1 to waveguide 11 applied to entangled states of varying purity. The injected states are recovered with an average fidelity of 0.971 ± 0.019 and an average similarity of 0.978 ± 0.019 (see Supplementary Table 4 for all values).

Discussion

We have proposed and experimentally demonstrated a protocol for relocating a photonic qubit across eleven discrete sites, maintaining the quantum state with high fidelity and preserving entanglement with another qubit at a different location. We can aim to improve our fidelity by reducing next-nearest-neighbour coupling by further separating the waveguides and having a longer device. This would increase the contrast between nearest- and next-nearest-neighbour coupling to better fit the Hamiltonian in equation (1). A by-product of longer devices, however, is an increase in propagation loss. Depth-dependent spherical aberrations in the laser irradiation process may also affect the homogeneity of the three-dimensional waveguide array. Additional optics in the laser writing set-up could be employed to reduce this effect. Protocols for relocating quantum information across discrete sites are essential for future quantum technologies. Our protocol builds on the PST with extension to include an additional degree of freedom for encoding quantum information. This demonstration opens pathways towards faithful quantum state relocation in quantum computing systems.

Methods

Experimental set-up. Horizontally polarized photon pairs at 807.5 nm are generated via type 1 spontaneous parametric downconversion in a 1-mm-thick BiBO crystal, pumped by an 80-mW, 403.75-nm CW diode laser. Both photons are rotated into a diagonal state $\frac{1}{\sqrt{2}}(|H\rangle + |V\rangle)$ by a half waveplate (HWP) with fast axis at 22.5° from vertical. One photon has a phase applied by two 45° quarter waveplates (QWP) on either side of a HWP at θ°. The second photon has its diagonal state optimized with a PBS at ~45°.

Each photon is collected in PMF and are incident on the two input faces of a fibre pigtailed PBS. When measuring in the coincidence basis, this post-selects the entangled state $\frac{1}{\sqrt{2}}(|H_1V_2\rangle + e^{i\phi}|V_1H_2\rangle)$, where $\phi = 4(\theta + \epsilon)$ and ϵ is the intrinsic phase applied by the whole system. The experimental set-up is illustrated in Fig. 4.

PMF is highly birefringent, resulting in full decoherence of the polarization state after ~1 m of fibre giving a mixed state. To maintain polarization superposition over several metres of fibre, we use 90° connections to ensure that both polarizations propagate through equal proportions of fast- and slow-axis fibre. Slight length differences between fibres and temperature variations mean the whole system applies a residual phase ϵ to the state, which can be compensated for in the source using the phase-controlling HWP.

Polarization state tomography combines statistics from projection measurements to generate the density matrix of a state. Single-photon rotations are applied by a QWP and HWP before a PBS. Single-qubit tomography requires four measurements and two-qubit tomography requires 16. Accidental counts are removed by taking each reading with and without an electronic delay. This helps reduce noise in our measurements.

Photon count rate. In our experiment, we prepare polarization Bell states with a count rate of $\sim 2 \times 10^3$ s^{-1}. After the PST array we measure a count rate of $\sim 10^2$ s^{-1}. The propagation loss of the array is only 1.8 dB. Most of the total loss ($\sim$13 dB) is indeed due to mode mismatch between the waveguides and fibres, imperfect coupling, reflections at interfaces, and non-unit relocation efficiency. We integrate our measurements for 30 s to reduce the statistical noise due to the Poisson distribution of the photon count rate.

References

1. Nielsen, M. A. & Chuang, I. L. *Quantum Computation and Quantum Information* 10th anniversary edition (Cambridge University Press, 2000).
2. Bose, S. Quantum communication through an unmodulated spin chain. *Phys. Rev. Lett.* **91,** 207901 (2003).
3. Christandl, M., Datta, N., Ekert, A. & Landahl, A. J. Perfect state transfer in quantum spin networks. *Phys. Rev. Lett.* **92,** 187902 (2004).
4. Plenio, M. B., Hartley, J. & Eisert, J. Dynamics and manipulation of entanglement in coupled harmonic systems with many degrees of freedom. *New J. Phys.* **6,** 36 (2004).
5. Gordon, R. Harmonic oscillation in a spatially finite array waveguide. *Opt. Lett.* **29,** 2752 (2004).
6. Nikolopoulos, G. M., Petrosyan, D. & Lambropoulos, P. Electron wavepacket propagation in a chain of coupled quantum dots. *J. Phys. Condens. Matter* **16,** 4991 (2004).
7. Nikolopoulos, G. M., Petrosyan, D. & Lambropoulos, P. Coherent electron wavepacket propagation and entanglement in array of coupled quantum dots. *Europhys. Lett.* **65,** 297–303 (2004).
8. DiVincenzo, D. P. The physical implementation of quantum computation. *Fortschr. Phys.* **48,** 771–783 (2000).
9. Wallraff, A. *et al.* Strong coupling of a single photon to a superconducting qubit using circuit quantum electrodynamics. *Nature* **431,** 162–167 (2004).
10. Majer, J. *et al.* Coupling superconducting qubits via a cavity bus. *Nature* **449,** 443–447 (2007).
11. Herskind, P. F., Dantan, A., Marler, J. P., Albert, M. & Drewsen, M. Realization of collective strong coupling with ion Coulomb crystals in an optical cavity. *Nat. Phys.* **5,** 494–498 (2009).
12. Paik, H. *et al.* Observation of high coherence in Josephson junction qubits measured in a three-dimensional circuit QED architecture. *Phys. Rev. Lett.* **107,** 240501 (2011).
13. Schoelkopf, R. J. & Girvin, S. M. Wiring up quantum systems. *Nature* **451,** 664–669 (2008).
14. Bennett, C. H. *et al.* Teleporting an unknown quantum state via dual classical and Einstein-Podolsky-Rosen channels. *Phys. Rev. Lett.* **70,** 1895–1899 (1993).
15. Bouwmeester, D. *et al.* Experimental quantum teleportation. *Nature* **390,** 575–579 (1997).
16. Furusawa, A. *et al.* Unconditional quantum teleportation. *Science* **282,** 706–709 (1998).
17. Braunstein, S. L. & Kimble, H. J. Teleportation of continuous quantum variables. *Phys. Rev. Lett.* **80,** 869–872 (1998).
18. Kielpinski, D., Monroe, C. & Wineland, D. J. Architecture for a large-scale ion-trap quantum computer. *Nature* **417,** 709–711 (2002).
19. Seidelin, S. *et al.* Microfabricated surface-electrode ion trap for scalable quantum information processing. *Phys. Rev. Lett.* **96,** 253003 (2006).
20. Yung, M.-H. Quantum speed limit for perfect state transfer in one dimension. *Phys. Rev. A* **74,** 030303 (2006).
21. Christandl, M. *et al.* Perfect transfer of arbitrary states in quantum spin networks. *Phys. Rev. A* **71,** 032312 (2005).
22. Loss, D. & DiVincenzo, D. P. Quantum computation with quantum dots. *Phys. Rev. A* **57,** 120–126 (1998).
23. Kay, A. Perfect efficient, state transfer and its application as a constructive tool. *Int. J. Quantum Inf.* **08,** 641–676 (2010).
24. Kay, A. Unifying quantum state transfer and state amplification. *Phys. Rev. Lett.* **98,** 010501 (2007).
25. Raussendorf, R. & Briegel, H. J. A one-way quantum computer. *Phys. Rev. Lett.* **86,** 5188–5191 (2001).
26. Zhou, X., Zhou, Z.-W., Guo, G.-C. & Feldman, M. J. Quantum computation with untunable couplings. *Phys. Rev. Lett.* **89,** 197903 (2002).
27. Benjamin, S. C. & Bose, S. Quantum computing in arrays coupled by "always-on" interactions. *Phys. Rev. A* **70,** 032314 (2004).
28. Kay, A. Computational power of symmetric Hamiltonians. *Phys. Rev. A* **78,** 012346 (2008).
29. Mkrtchian, G. F. Universal quantum logic gates in a scalable Ising spin quantum computer. *Phys. Lett. A* **372,** 5270–5273 (2008).
30. Cook, R. J. & Shore, B. W. Coherent dynamics of N-level atoms and molecules. III. An analytically soluble periodic case. *Phys. Rev. A* **20,** 539–544 (1979).
31. Burgarth, D. & Bose, S. Conclusive and arbitrarily perfect quantum-state transfer using parallel spin-chain channels. *Phys. Rev. A* **71,** 052315 (2005).
32. Burgarth, D., Giovannetti, V. & Bose, S. Efficient and perfect state transfer in quantum chains. *J. Phys. A Math. Gen.* **38,** 6793 (2005).
33. Yung, M.-H. & Bose, S. Perfect state transfer, effective gates, and entanglement generation in engineered bosonic and fermionic networks. *Phys. Rev. A* **71,** 032310 (2005).
34. Plenio, M. B. & Semio, F. L. High efficiency transfer of quantum information and multiparticle entanglement generation in translation-invariant quantum chains. *New J. Phys.* **7,** 73 (2005).
35. Zhang, J. *et al.* Simulation of Heisenberg XY interactions and realization of a perfect state transfer in spin chains using liquid nuclear magnetic resonance. *Phys. Rev. A* **72,** 012331 (2005).
36. Kay, A. Perfect state transfer: beyond nearest-neighbor couplings. *Phys. Rev. A* **73,** 032306 (2006).
37. Bose, S. Quantum communication through spin chain dynamics: an introductory overview. *Contemp. Phys.* **48,** 13–30 (2007).
38. Kostak, V., Nikolopoulos, G. M. & Jex, I. Perfect state transfer in networks of arbitrary topology and coupling configuration. *Phys. Rev. A* **75,** 042319 (2007).
39. Di Franco, C., Paternostro, M. & Kim, M. S. Perfect state transfer on a spin chain without state initialization. *Phys. Rev. Lett.* **101,** 230502 (2008).
40. Gualdi, G., Kostak, V., Marzoli, I. & Tombesi, P. Perfect state transfer in long-range interacting spin chains. *Phys. Rev. A* **78,** 022325 (2008).
41. Paz-Silva, G. A., Rebic, S., Twamley, J. & Duty, T. Perfect mirror transport protocol with higher dimensional quantum chains. *Phys. Rev. Lett.* **102,** 020503 (2009).
42. Perez-Leija, A., Keil, R., Moya-Cessa, H., Szameit, A. & Christodoulides, D. N. Perfect transfer of path-entangled photons in Jx photonic lattices. *Phys. Rev. A* **87,** 022303 (2013).
43. Bellec, M., Nikolopoulos, G. M. & Tzortzakis, S. Faithful communication Hamiltonian in photonic lattices. *Opt. Lett.* **37,** 4504 (2012).
44. Perez-Leija, A. *et al.* Coherent quantum transport in photonic lattices. *Phys. Rev. A* **87,** 012309 (2013).
45. Einstein, A., Podolsky, B. & Rosen, N. Can quantum-mechanical description of physical reality be considered complete? *Phys. Rev.* **47,** 777–780 (1935).
46. Perets, H. B. *et al.* Realization of quantum walks with negligible decoherence in waveguide lattices. *Phys. Rev. Lett.* **100,** 170506 (2008).
47. Rai, A., Agarwal, G. S. & Perk, J. H. H. Transport and quantum walk of nonclassical light in coupled waveguides. *Phys. Rev. A* **78,** 042304 (2008).
48. Peruzzo, A. *et al.* Quantum walks of correlated photons. *Science* **329,** 1500–1503 (2010).
49. Bromberg, Y., Lahini, Y., Morandotti, R. & Silberberg, Y. Quantum and classical correlations in waveguide lattices. *Phys. Rev. Lett.* **102,** 253904 (2009).
50. Sansoni, L. *et al.* Two-particle bosonic-fermionic quantum walk via integrated photonics. *Phys. Rev. Lett.* **108,** 010502 (2012).
51. O'Brien, J. L. *et al.* Quantum process tomography of a controlled-NOT gate. *Phys. Rev. Lett.* **93,** 080502 (2004).
52. Gilchrist, A., Langford, N. K. & Nielsen, M. A. Distance measures to compare real and ideal quantum processes. *Phys. Rev. A* **71,** 062310 (2005).
53. Matthews, J. C. F. *et al.* Observing fermionic statistics with photons in arbitrary processes. *Sci. Rep.* **3,** 1539 (2013).
54. James, D. F. V., Kwiat, P. G., Munro, W. J. & White, A. G. Measurement of qubits. *Phys. Rev. A* **64,** 052312 (2001).
55. Jozsa, R. Fidelity for mixed quantum states. *J. Mod. Opt.* **41,** 2315–2323 (1994).
56. Devitt, S. J., Munro, W. J. & Nemoto, K. Quantum error correction for beginners. *Rep. Prog. Phys.* **76,** 076001 (2013).
57. Lloyd, S. Universal quantum simulators. *Science* **273,** 1073–1078 (1996).

Acknowledgements

G.C., A.C. and R.O. acknowledge support from the European Union through the projects FP7-ICT-2011-9-600838 (QWAD-Quantum Waveguides Application and Development; www.qwad-project.eu) and H2020-FETPROACT-2014-641039 (QUCHIP-Quantum Simulation on a Photonic Chip; www.quchip.eu). M.-H.Y. acknowledges support by the National Natural Science Foundation of China under Grants No. 11405093. A.P. acknowledges an Australian Research Council Discovery Early Career Researcher Award under project number DE140101700 and an RMIT University Vice-Chancellor's Senior Research Fellowship.

Author contributions

All authors contributed to all aspects of this work.

Additional information

Competing financial interests: The authors declare no competing financial interests.

18

Implementation of quantum and classical discrete fractional Fourier transforms

Steffen Weimann[1,*], Armando Perez-Leija[1,*], Maxime Lebugle[1,*], Robert Keil[2], Malte Tichy[3], Markus Gräfe[1], René Heilmann[1], Stefan Nolte[1], Hector Moya-Cessa[4], Gregor Weihs[2], Demetrios N. Christodoulides[5] & Alexander Szameit[1]

Fourier transforms, integer and fractional, are ubiquitous mathematical tools in basic and applied science. Certainly, since the ordinary Fourier transform is merely a particular case of a continuous set of fractional Fourier domains, every property and application of the ordinary Fourier transform becomes a special case of the fractional Fourier transform. Despite the great practical importance of the discrete Fourier transform, implementation of fractional orders of the corresponding discrete operation has been elusive. Here we report classical and quantum optical realizations of the discrete fractional Fourier transform. In the context of classical optics, we implement discrete fractional Fourier transforms of exemplary wave functions and experimentally demonstrate the shift theorem. Moreover, we apply this approach in the quantum realm to Fourier transform separable and path-entangled biphoton wave functions. The proposed approach is versatile and could find applications in various fields where Fourier transforms are essential tools.

[1] Institute of Applied Physics, Abbe School of Photonics, Friedrich-Schiller-Universität Jena, Max-Wien Platz 1, 07743 Jena, Germany. [2] Institut für Experimentalphysik, Universität Innsbruck, Technikerstraße 25, 6020 Innsbruck, Austria. [3] Department of Physics and Astronomy, University of Aarhus, 8000 Aarhus, Denmark. [4] INAOE, Coordinacion de Optica, Luis Enrique Erro No. 1, Tonantzintla, Puebla 72840, Mexico. [5] CREOL, The College of Optics & Photonics, University of Central Florida, Orlando, Florida 32816, USA. * These authors contributed equally to this work. Correspondence and requests for materials should be addressed to A.S. (email: alexander.szameit@uni-jena.de).

Two hundred years ago, Joseph Fourier introduced a major concept in mathematics, the so-called Fourier transform (FT). It was not until 1965, when Cooley and Tukey developed the 'fast Fourier transform' algorithm, that Fourier analysis became a standard tool in contemporary sciences[1]. Two crucial requirements in this algorithm are the discretization and truncation of the domain, where the signals to be transformed are defined. These requirements are always satisfiable, since observable quantities in physics must be well behaved and finite in extension and magnitude.

In 1980, Namias made another significant leap with the introduction of the fractional Fourier transform (FrFT), which contains the FT as a special case[2]. Several investigations quickly followed, leading to a more general theory of joint time-frequency signal representations[3] and fractional Fourier optics[4]. The vast scope of the FrFT has been demonstrated in areas such as wave propagation, signal processing and differential equations[3,5–7]. So far, the FT of fractional order was realized only by single-lens systems[8,9], although other theoretical suggestions, including multi-lens systems[10] or graded index fibres exist[11]. The aim to discretize this generalized FT led to the introduction of the discrete fractional Fourier transform (DFrFT) operating on a finite grid in a way similar to that of a discrete FT[12]. Along those lines, several versions of the DFrFT have been introduced[5], however, without any experimental realization, so far. In this work we focus on the optical implementation of the so-called Fourier–Kravchuk transform[12] that can be equally applied to the classical and quantum states. Throughout our paper, we simply refer to this transform as DFrFT whose application reaches from the demonstration of the Fourier suppression law[13], N00N-state generation[14,15] and qubit storage[16] to the realization of perfect discrete lenses for non-uniform input distributions.

In this work, we report on the realization of DFrFTs of one-dimensional optical signals based on an integrated lattice of evanescently coupled waveguides. In these photonic arrangements, the inter-channel couplings are designed in such a way that the system readily performs the DFrFT of any incoming signal. The signal evolution is governed by the Schrödinger equation and the associated Hamilton operator is known as the J_x-operator in the quantum theory of angular momentum or likewise as the Heisenberg *XY* model from the quantum theory of ferromagnetism. We foresee that the inherent versatility of this approach will make other realizations of the DFrFT, the FT and the fast Fourier transform recognizably simple and thus may open the door to many interesting applications in integrated quantum computation[17].

Results

Theoretical approach. Similarly to its continuous counterpart, the DFrFT can be interpreted physically as a continuous rotation of the associated wave functions through an angle Z in phase space (see Fig. 1a)[18]. The idea is thus to construct finite circuits that are capable of imprinting such rotations to any light field. In quantum mechanics, three-dimensional spatial rotations of complex state vectors are generated via operations of the angular momentum operators J_k ($k = x, y, z$) on the Hilbert space of the associated system[19]. In particular, the rotation imprinted by the J_x-operator turns out to be an elaborated definition of the DFrFT (see Methods section for discussion). These concepts can be readily translated to the optical domain by mapping the matrix elements of the J_x-operator over the inter-channel couplings of engineered waveguide arrays (Fig. 1b–e)[20]. The coupling matrix of such waveguide arrays is thus given by $(J_x)_{m,n} = \kappa_0\left(\sqrt{(j-m)(j+m+1)}\delta_{m+1,n} + \sqrt{(j+m)(j-m+1)}\delta_{m-1,n}\right)/2$ (ref. 19). Here, κ_0 is a scaling factor introduced for experimental reasons. The indices m and n range from $-j$ to j in unit steps.

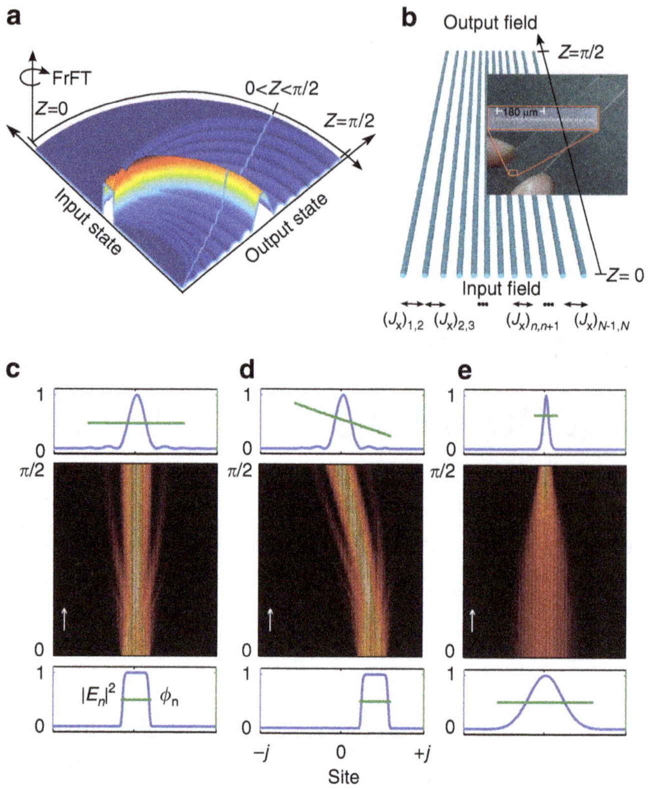

Figure 1 | Discretization of the FrFT. (**a**) Pictorial view of actual fractional Fourier transforms exemplified as continuous rotations in phase space. (**b**) Schematic representation of a pre-engineered J_x-array. (**c–e**) Top views of continuous 'rotations' of a rectangular (**c**), displaced rectangular (**d**) and Gaussian (**e**) optical wave functions in a J_x-array with $N = 151$. The bottom and top plots show the intensities $|E_n|^2$ and phases ϕ_n of the ingoing and outgoing wave packets, respectively. The green lines describe the magnitude of phase distributions of the optical fields, that is, the phase jumps of π due to a change of sign of the signal's amplitude are not shown.

Meanwhile, j represents an arbitrary positive integer or half-integer that determines the total number of waveguides via $N = 2j + 1$ (Fig. 1b).

Coupled mode theory states that the evolution of light in the J_x-waveguide array is governed by the following set of equations[20]

$$i\frac{\mathrm{d}}{\mathrm{d}Z}E_m(Z) = \frac{1}{\kappa_0}\sum_{n=-j}^{j}(J_x)_{m,n}E_n(Z). \qquad (1)$$

Here, $E_n(Z)$ denotes the complex electric field amplitude at site n. In the quantum optics regime, single photons traversing such devices are governed by a set of Heisenberg equations that are isomorphic to equation (1). The only difference is that in the quantum case $E_n(Z)$ must be replaced by the photon creation operator $a_n^\dagger(Z)$. In a spintronic context, the evolution parameter Z is associated with time, whereas in the framework of integrated quantum optics, Z represents the propagation distance, see Fig. 1b–e. A spectral decomposition of the J_x-matrix yields the eigenvectors $u_n^{(m)} = 2^n\sqrt{\frac{(j+n)!(j-n)!}{(j+m)!(j-m)!}}P_{j+n}^{(m-n,-m-n)}(0)$[14,19], which in combination with the eigenvalues, $\beta_m = -j, \ldots, j$, render the closed-form point-spread function

$$G_{p,q}(Z) = i^{p-q}\sqrt{\frac{(j+p)!(j-p)!}{(j+q)!(j-q)!}}\left[\sin\left(\frac{Z}{2}\right)\right]^{q-p}\left[\cos\left(\frac{Z}{2}\right)\right]^{-q-p}P_{j+p}^{(q-p,-q-p)}(\cos(Z)). \qquad (2)$$

Note that q and p represent the excited and observed sites, respectively, and $P_n^{(A,B)}(x)$ are the Jacobi polynomials of order n (see Methods section for discussion). Using equation (2), we can compute the response of the system to any input signal, which in turn gives the DFrFT[12]. Accordingly, DFrFT of any particular order arises at one specific propagation distance Z lying between 0 and $\pi/2$. In the limit $N\rightarrow\infty$, the eigenvectors of J_x, $u^{(m)}_{x\sqrt{2/N}}$, become the continuous Hermite-Gauss polynomials $H_m(x)$, which are known to be the eigenfunctions of the fractional Fourier operator[12]. As a result, in the continuous limit, the DFrFT described by equation (2) converges to the continuous FrFT[5,12]; and the standard FT is recovered at $Z=\pi/2$ (Fig. 1c–e). Note that in general, the DFrFT obtained in our devices and the usual DFT become equal only in the continuous limit $N\rightarrow\infty$ and at $Z=\pi/2$.

Experiments with classical light. To experimentally demonstrate the functionality of the suggested waveguide system, we use $N=21$ waveguides to perform FTs of simple wave packets. We first consider a Gaussian wave packet with a full-width at half-maximum (FWHM) covering the five central sites (Fig. 2a). The input signal is prepared by focusing a Gaussian beam from a HeNe laser onto the front facet of the sample. By exploiting the fluorescence from colour centres within the waveguides[21], we monitor the full intensity evolution from the input to the output plane. The fluorescence image, Fig. 2a, shows a gradual transition from an initially narrow Gaussian distribution at the input to a broader one at the Fourier plane (left and right panels Fig. 2a), demonstrating that narrow signals in space correspond to broad signals in Fourier space. For intermediate propagation distances ($Z\in[0, \pi/2]$) we extract other orders of the DFrFT, simultaneously. For comparison, we plot the continuous FrFT produced by the corresponding continuous Gaussian profile (red curves Fig. 2a). The agreement between the computed FrFT and the experimental DFrFT proves that for the considered Gaussian input signal, $N=21$ is sufficient to achieve the continuous limit. We now shift the input Gaussian beam by six channels towards the edge. Since the separations between adjacent waveguides at the edges are bigger than the separations between adjacent waveguides in the centre, the discretization grid is not perfectly homogeneous. Strictly speaking, the discretized shifted Gaussian just at the input plane covers slightly less than five waveguides FWHM. We observe that the well-approximated off-centre Gaussian travels to the centre at $Z=\pi/2$ (Fig. 2b), hereby showing the famous shift theorem. In additional experiments, extended signals, for example, a shifted top-hat function, are found to be well transformed according to equation (2) as well. However, we find that for this type of excitation $N>21$ would be required to discuss the continuous limit (see Supplementary Note 1 with Supplementary Fig. 1).

An unequivocal criterion, for the functionality of devices that perform the DFrFT, equation (2), can be formulated by evaluating $G_{p,q}(\frac{\pi}{2})$. At this particular distance, point-like excitations will give rise to signal magnitudes that perfectly resemble the magnitudes of one of the eigenstates of the transform. More specifically, for transforms such as equation (2), one finds that an excitation of the qth site excites the qth system eigenstate up to local phases ($G_{p,q}(\frac{\pi}{2})=i^{p-q}u_p^{(q)}$) (see the Methods section for explanations). The experimental demonstration of this intriguing

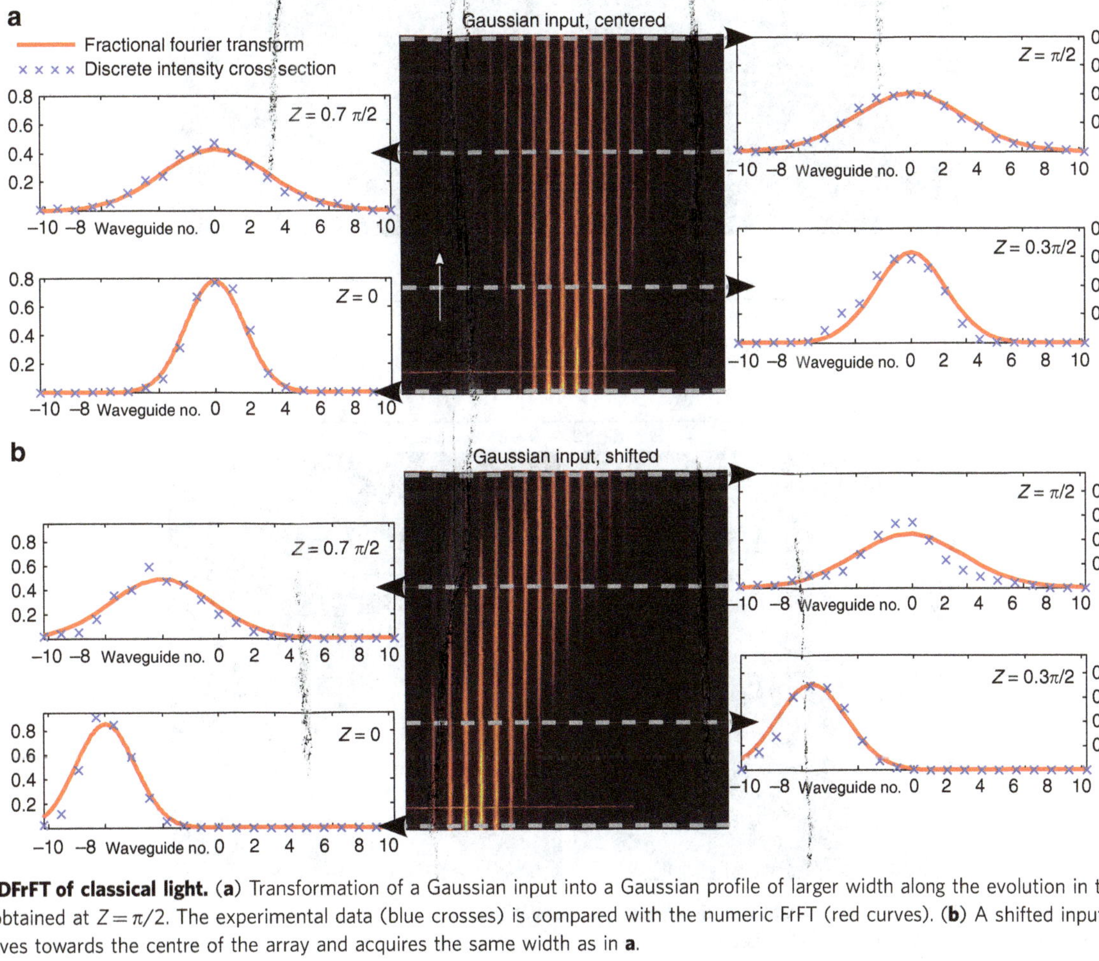

Figure 2 | DFrFT of classical light. (**a**) Transformation of a Gaussian input into a Gaussian profile of larger width along the evolution in the J_x-array. The FT is obtained at $Z=\pi/2$. The experimental data (blue crosses) is compared with the numeric FrFT (red curves). (**b**) A shifted input Gaussian profile evolves towards the centre of the array and acquires the same width as in **a**.

effect is shown in the subpanels of Fig. 3a–d along with the theoretical predictions. It can be argued that for any point-like excitation, the continuous limit cannot be met experimentally (see the Methods for discussion). Instead, equation (2) creates a non-uniform amplitude distribution with a phase difference of $\pi/2$ between adjacent sites. Nevertheless, in the continuous limit, $G_{p,q}\left(\frac{\pi}{2}\right)$ tends to the usual FT kernel[12]. At this point, it is worth emphasizing the formal equivalence to the quantum Heisenberg XY model in condensed matter physics[22,23]. In this respect, our observations demonstrate the capability of the here-presented systems to store quantum information in XY Hamiltonians by converting specific inputs into eigenstates of the system[16]. To our knowledge, this rather rare property has never been thoroughly investigated before.

Quantum experiments. To demonstrate the applicability of our approach in the quantum domain, we now analyse intensity correlations of separable and path-entangled photon pairs propagating through these Fourier transformers. To do so, we fabricated J_x-arrays involving $N=8$ channels. The importance of exploring FTs of such states has been highlighted in several investigations, demonstrating interesting effects such as suppression of states and portraying biphoton spatial correlations[24–26].

In this discrete quantum optical context, pure separable two-photon states are readily produced by coupling pairs of indistinguishable photons into two distinct lattice sites (m, n), this state is mathematically described by $|\Psi(0)\rangle = a_m^\dagger a_n^\dagger|0\rangle$. Conversely, path-entangled two-photon states are created by simultaneously launching both photons at either site m or n with exactly the same probability, that is, $|\Psi(0)\rangle = [(a_m^\dagger)^2 + (a_n^\dagger)^2]|0\rangle/2$. Furthermore, the probability of observing one of the photons at site k and its twin at site l is given by the intensity correlation matrix $\Gamma_{k,l}(Z) = \langle a_k^\dagger a_l^\dagger a_l a_k\rangle$ (ref. 27). An intriguing and unique property of the J_x-systems is that at $Z=\pi/2$ the correlation matrices are

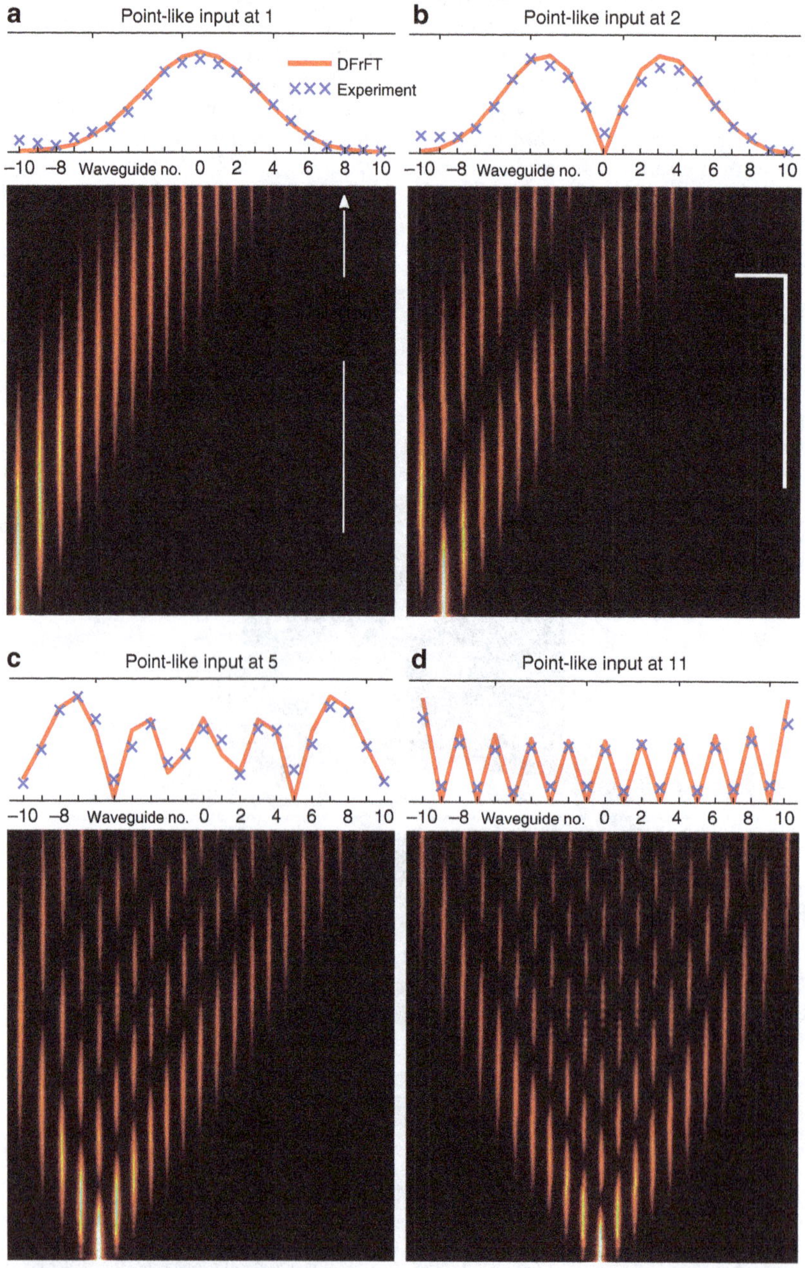

Figure 3 | Experimental visualization of the discrete Hermite-Gauss polynomials. (a–d) Evolution of single-site inputs into the magnitudes of the respective eigensolutions, as predicted theoretically (methods). The experimental data (blue crosses) is compared with the analytic DFrFT (red curves).

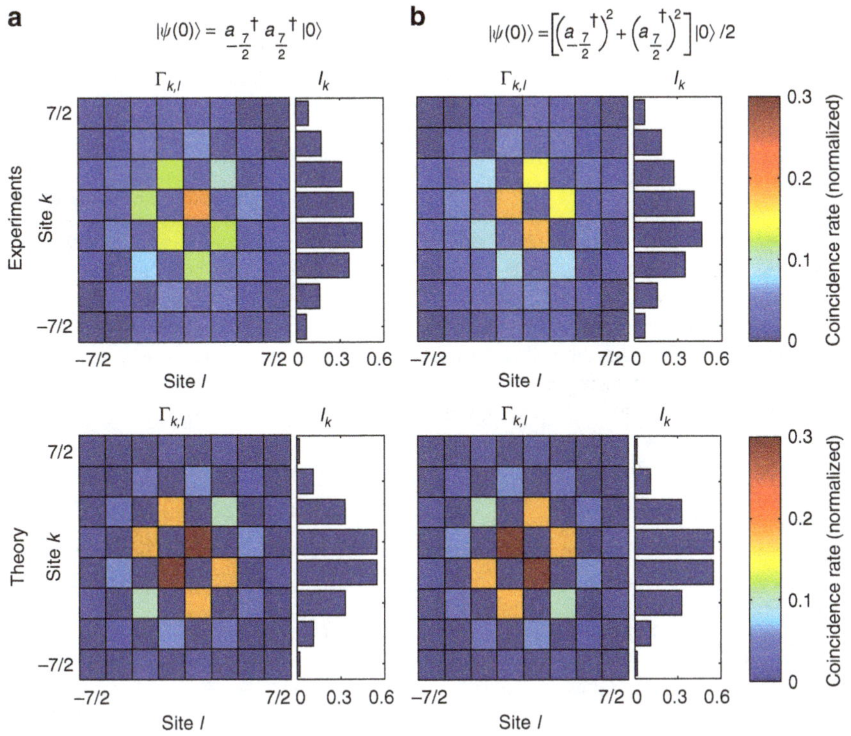

Figure 4 | DFrFT of quantum light. Correlation maps $\Gamma_{k,l}$ of a two-photon state either prepared (**a**) in a product state or (**b**) in a path-entangled state after propagating through a J_x-array. The photon density I_k at the output is shown on the right side of each map. The evaluation of the s.d. of the coincidence rates is presented in the Supplementary Note 2.

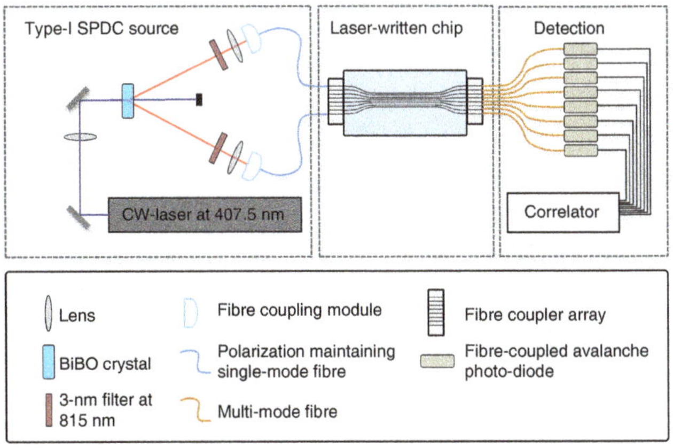

Figure 5 | Set-up used to carry out spatial correlation measurements of photonic quantum states. The set-up consists of three parts—the state preparation, the execution of the DFrFT and the correlation measurement.

given in terms of the eigenstates, as noticed above. Hence, for the separable case, $|\Psi(0)\rangle = a_m^\dagger a_n^\dagger|0\rangle$, the correlation matrices are given by $\Gamma_{k,l} = |u_k^{(m)} u_l^{(n)} + u_k^{(n)} u_l^{(m)}|^2$, whereas for the path-entangled state, $|\Psi(0)\rangle = [(a_m^\dagger)^2 + (a_n^\dagger)^2]|0\rangle/2$, we have $\Gamma_{k,l} = |(-i)^{2m} u_k^{(m)} u_l^{(m)} + (-i)^{2n} u_k^{(n)} u_l^{(n)}|^2$. Of particular interest is the separable case, where the photons are symmetrically coupled into the outermost waveguides, $|\Psi(0)\rangle = a_j^\dagger a_{-j}^\dagger|0\rangle$. In this scenario, only the correlation matrix elements for which $(k+l) = \text{odd}$ are nonzero, and are given by $\Gamma_{k,l} = 4^{k+l+1}(j+k)!(j-k)!(j+l)!(j-l)!(P_{j+k}^{(j-k,-j-k)}(0)P_{j+l}^{(-j-l,j-l)}(0))/[(N-1)!]^2$. These effects are demonstrated for the initial state $|\Psi(0)\rangle = a_{-\frac{7}{2}}^\dagger a_{\frac{7}{2}}^\dagger|0\rangle$ in Fig. 4a, where concentration and absence of probability in the correlation matrix clearly show that some states are completely suppressed—a hallmark of any Fourier unitary process[13]. An estimation of the statistical significance of the data set, along with a short discussion on incoherence effects can be found in the Supplementary Note 2 involving Supplementary Figs 2 and 3.

As a second case, we consider a fully symmetric path-entangled two-photon state of the form $|\Psi(0)\rangle = [(a_j^\dagger)^2 + (a_{-j}^\dagger)^2]|0\rangle/2$. Physically, both photons are entering together into the array at either site j or $-j$ with equal probability[28–30]. The correlations are determined by $\Gamma_{k,l} = |u_k^{(j)} u_l^{(j)} + u_k^{(-j)} u_l^{(-j)}|^2$, from which we infer that the probability of measuring photon coincidences at coordinates (k, l) vanishes at sites where the sum $(k+l)$ is odd. In contrast, at coordinates where $(k+l)$ is even, the correlation function collapses to the expression $\Gamma_{k,l} = 4^{k+l+1}(j+k)!(j-k)!(j+l)!(j-l)!(P_{j+k}^{(j-k,-j-k)}(0)P_{j+l}^{(-j-l,j-l)}(0))^2/[(N-1)!]^2$. This indicates that in this path-entangled case the correlation map appears rotated by 90° with respect to the matrix obtained with separable two-photon states. We performed an experiment to demonstrate these predictions using states of the type $|\Psi(0)\rangle = [(a_{\frac{-7}{2}}^\dagger)^2 + (a_{\frac{7}{2}}^\dagger)^2]|0\rangle/2$, which were prepared using a 50:50 directional coupler acting as a beam splitter[29]. The whole experiment is achieved using a single chip containing both the state preparation stage followed by a J_x-system, yielding high interferometric control over the field dynamics (Fig. 5). The experimental measurements are presented in Fig. 4b. Similarly, suppression of states occurs as a result of destructive quantum interference. As predicted, a closer look into the correlation pattern reveals that indeed the correlation map appears rotated by 90° with respect to the matrix obtained with separable two-photon states.

Discussion

We emphasize that our quantum measurements feature interference fringes akin to the ones observed in quantum Young's two-slit experiments of biphoton wave functions in free space as demonstrated in ref. 24. In such free-space experiments, however, far-field observations were carried out using lenses and the two slits were emulated by optical fibres[24]. Along those lines, we have created a fully integrated quantum interferometer to observe fundamental quantum mechanical features[25]. This additionally suggests an effective way to generate quantum states containing only even (odd) non-vanishing inter-particle distance probabilities for the separable input state (symmetric path-entangled state). In addition, the eigenfunctions associated with the Hamiltonian system explored in our work are specific Jacobi polynomials, which are well known as the optimal basis for quantum phase-retrieval algorithms[31] and these eigenstates can be retrieved by limited phase operations. Knowing a quantum wavefunction and its FT, a phase-retrieval algorithm for signals that are a superposition of a finite number of Hermite–Gauss polynomials has been introduced[32]. This phase-retrieval algorithm might be implemented employing our system, since its discrete character automatically possesses a finite number of polynomials that are closely connected to the Hermite–Gauss polynomials. Another potential application is the realization of the Radon–Wigner transform given by the squared modulus of the FrFT[6,33]. The Radon–Wigner function is a basic tool for the reconstruction of Wigner quasi-probability distributions in quantum optics[34,35]. Also, FrFTs appear naturally in optics as free-space propagation between two spherical reference planes in general[4]. Like in our system, the order of the FrFT is proportional to the propagation distance. On that basis, complex spatial filtering involving several fractional Fourier planes was suggested[36]. In this description, the optical signal is discretized and thus described by a vector of field amplitudes at certain sites. In our approach this is inherently realized.

In conclusion, we have successfully demonstrated a universal discrete optical device capable of performing classical and quantum DFrFTs. Our studies might find applications in developing a more general quantum suppression law[13] and perhaps in the development of new quantum algorithms.

Methods

The fractional Fourier transform in quantum harmonic oscillations. In this section, we briefly describe the relation between the continuous FrFT operator and the Hamiltonian of the quantum harmonic oscillator[2].

The FrFT operator $\hat{F}_Z$ is defined by the following eigenvalue equation involving the Hermite–Gauss polynomials of order n and the eigenvalues $\lambda_n = \exp(inZ)$

$$\hat{F}_Z\{\exp(-x^2/2)H_n(x)\} = \exp(inZ)\exp(-x^2/2)H_n(x), \quad (3)$$

where $Z \in \mathbb{R}$. Concurrently, one can interpret equation (3) as quantum time evolution from time $t=0$ to $t=Z$. We write $\hat{F}_Z$ as $\exp(iZ\hat{A})$.

$$\exp(iZ\hat{A})\{\exp(-x^2/2)H_n(x)\} = \exp(inZ)\exp(-x^2/2)H_n(x). \quad (4)$$

To show that $\hat{A}$ is the Hamilton operator of the harmonic oscillator, we differentiate both sides of equation (4) with respect to Z and evaluate the result at $Z=0$. We obtain

$$\hat{A}\exp(-x^2/2)H_n(x) = n\exp(-x^2/2)H_n(x). \quad (5)$$

To find the spatial representation of $\hat{A}$, consider the differential equation

$$\left(-\frac{1}{2}\frac{d^2}{dx^2} + x\frac{d}{dx}\right)H_n(x) = nH_n(x), \quad (6)$$

for the Hermite polynomials $H_n(x)$ of order n. Using the identities $1=\exp(x^2/2)\exp(-x^2/2)$ and $\exp(\xi x^2/2)\left(\frac{d^n}{dx^n}\right)\exp(-\xi x^2/2)=\left(\frac{d}{dx}-\xi x\right)^n$, one can show that equation (6) can be written as

$$\left(-\frac{1}{2}\frac{d^2}{dx^2} + \frac{1}{2}x^2 - \frac{1}{2}\right)\exp(-x^2/2)H_n(x) = n\exp(-x^2/2)H_n(x), \quad (7)$$

where we have used the commutator $\left[x, \frac{d}{dx}\right] = -1$. Comparing equation (5) and equation (7), one can see that $\hat{A}$ becomes

$$\hat{A} = \left(-\frac{1}{2}\frac{d^2}{dx^2} + \frac{1}{2}x^2 - \frac{1}{2}\right). \quad (8)$$

In summary, the FrFT operator $\hat{F}_z$ can be written as

$$\hat{F}_z = \exp\left(\frac{iz}{2}\left(-\frac{d^2}{dx^2} + x^2 - 1\right)\right). \quad (9)$$

Because of this one-to-one correspondence between the dynamics of the quantum harmonic oscillator and the fractional Fourier operator, the implementation of such transform is immediate using harmonic oscillator systems[37].

J_x-photonic lattices as discrete harmonic oscillators. Our aim in this section is to show that in the continuous limit $N \to \infty$, the eigenvalue equation for J_x-arrays becomes the eigenvalue equation of the quantum harmonic oscillator. Consider the matrix representation of the J_x-operator again. For convenience we take, in this section only, $\kappa_0 = 1$.

$$\begin{aligned}(J_x)_{m,n} &= \tfrac{1}{2}\left(\sqrt{(j-m)(j+m+1)}\delta_{n,m+1} + \sqrt{(j+m)(j-m+1)}\delta_{n,m-1}\right)\\ &= \tfrac{1}{2}\left(\sqrt{j(j+1)-m(m+1)}\delta_{n,m+1} + \sqrt{j(j+1)-m(m-1)}\delta_{n,m-1}\right).\end{aligned} \quad (10)$$

The indices m and n range from $-j$ to j in unit steps and j is an arbitrary positive integer or half-integer. The dimension of the J_x-matrix is $N=2j+1$. We now introduce the variable $\gamma = j(j+1) = (N^2-1)/4$, which implies that $(J_x)_{m,n}$ can be written as

$$(J_x)_{m,n} = \frac{\sqrt{\gamma}}{2}\left(\sqrt{1-\frac{1}{\gamma}m(m+1)}\delta_{n,m+1} + \sqrt{1-\frac{1}{\gamma}m(m-1)}\delta_{n,m-1}\right). \quad (11)$$

Let us consider the eigenvalue equation for this matrix

$$\frac{\sqrt{\gamma}}{2}\left(\sqrt{1-\frac{1}{\gamma}m(m+1)}\psi_{m+1} + \sqrt{1-\frac{1}{\gamma}m(m-1)}\psi_{m-1}\right) = \beta_m\psi_m. \quad (12)$$

Considering the region $m \ll j$, since $\gamma \propto N^2$, in the limit $N \to \infty$, the terms $m(m \pm 1)/\gamma \ll 1$. Hence, in the domain far from the edge of the array a Taylor expansion yields

$$\sqrt{\gamma}\sqrt{1-\frac{m}{\gamma}(m \pm 1)} \approx \sqrt{\gamma}\left(1 - \frac{m}{2\gamma}(m \pm 1) - \frac{m^2}{8\gamma^2}(m \pm 1)^2\right). \quad (13)$$

By defining $m = x\gamma^{1/4}$ (or $x = m/\gamma^{1/4}$), we obtain

$$\sqrt{\gamma}\sqrt{1-\frac{m}{\gamma}(m \pm 1)} \approx \gamma^{1/2} - \frac{x^2}{2} \mp \frac{x}{2\gamma^{1/4}} - \frac{x^4}{8\gamma^{1/2}} \quad (14)$$

Plugging this expression into equation (12)

$$\begin{aligned}&\left(\gamma^{1/2} - \frac{x^2}{2} - \frac{x}{2\gamma^{1/4}} - \frac{x^4}{8\gamma^{1/2}}\right)\psi_{m+1} + \left(\gamma^{1/2} - \frac{x^2}{2} + \frac{x}{2\gamma^{1/4}} - \frac{x^4}{8\gamma^{1/2}}\right)\psi_{m-1}\\ &\approx 2\beta_m\psi_m\end{aligned} \quad (15)$$

We redefine the functions $\psi_m = \psi(x) = \psi\left(\frac{m}{\gamma^{1/4}}\right)$ and $\psi_{m+1} = \psi\left(x + \frac{1}{\gamma^{1/4}}\right) = \psi\left(\frac{m}{\gamma^{1/4}} + \frac{1}{\gamma^{1/4}}\right)$ such that we can introduce the Taylor series

$$\psi_{m \pm 1} = \psi\left(x \pm \frac{1}{\gamma^{1/4}}\right) = \psi(x) \pm \frac{1}{\gamma^{1/4}}\psi'(x) + \frac{1}{2\gamma^{1/2}}\psi''(x), \quad (16)$$

where, again, we have kept only terms up to second order in $1/\gamma^{1/4}$. Substituting equation (16) into equation (15), and using the limit

$$\lim_{N\to\infty}\sqrt[4]{\left(\frac{4}{N^2}\right)\left(\frac{1}{1-\frac{1}{N^2}}\right)} = 0.$$

We obtain the time-independent Schrödinger equation for the harmonic oscillator

$$\left(-\frac{1}{2}\frac{d^2}{dx^2} + \frac{1}{2}x^2\right)\psi(x) = (\sqrt{\gamma} - \beta)\psi(x). \quad (17)$$

Therefore, in the continuous limit $N \to \infty$, the difference equation describing J_x-photonic lattices becomes the time-independent Schrödinger equation for the quantum harmonic oscillator. Note, however, that due to the importance of the condition $m \ll j$ in this derivation, this statement is only valid when dealing with signals that are square integrable in the continuous limit. Thus, the operator equation (11) can be used to define the discrete version of the quantum harmonic oscillator and thus the DFrFT.

The point-spread function for J_x-photonic lattices. In this section, it is shown that at $Z=\pi/2$, the green function of J_x-systems becomes proportional to the amplitude of one of the eigenstates. The evolution of light in J_x-arrays is governed by the set of N coupled differential equations (equation (1)).

The normalized propagation coordinate Z is given by $Z = \kappa_0 z$, where z is the actual propagation distance and κ_0 is an arbitrary scale factor. The quantity $E_n(Z)$ denotes the mode field amplitude at site n. A spectral decomposition of the J_x-matrix yields the eigensolutions

$$u_n^{(m)} = (2)^n\sqrt{\frac{(j+n)!(j-n)!}{(j+m)!(j-m)!}}P_{j+n}^{(m-n,-m-n)}(0). \quad (18)$$

$P_n^{(A,B)}(x)$ are the Jacobi polynomials of order n. And the corresponding eigenvalues are integers or half-integers, $\beta_m = -j, \ldots, j$, depending on the parity of N[14,19]. Using the eigenvectors and eigenvalues we obtain the point-spread function

$$G_{p,q}(Z) = \sum_{r=-j}^{j} u_q^{(r)} u_p^{(r)} \exp(irZ). \tag{19}$$

$G_{p,q}(Z)$ represents the amplitude at site p after an excitation of site q. Using equation (18) and the properties of the Jacobi polynomials one can show that equation (19) reduces to the closed-form expression

$$G_{p,q}(Z) = (-i)^{q-p} \sqrt{\frac{(j+p)!(j-p)!}{(j+q)!(j-q)!}} \left[\sin\left(\frac{Z}{2}\right)\right]^{q-p} \left[\cos\left(\frac{Z}{2}\right)\right]^{-q-p} P_{j+p}^{(q-p,-q-p)}(\cos(Z)). \tag{20}$$

Evaluation of equation (20) at $Z = \pi/2$ yields

$$G_{p,q}\left(\frac{\pi}{2}\right) = (-i)^{q-p}(2)^p \sqrt{\frac{(j+p)!(j-p)!}{(j+q)!(j-q)!}} P_{j+p}^{(q-p,-q-p)}(0) = (-i)^{q-p} u_p^{(q)}. \tag{21}$$

Equation (21) shows that at $Z = \pi/2$ the point-spread function becomes proportional to the amplitude of the corresponding eigenstates depending on the excited site. In other words, there is a one-to-one correspondence between the excited site number and the eigenstates of the system: excitation of the qth site excites the qth eigenstate up to well-defined local phases.

Devices fabrication and specifications. Our devices are fabricated in bulk-fused silica samples (Corning 7980, ArF grade) using the femtosecond laser direct-write approach[21]. The transparent material is modified within the focal region due to nonlinear absorption resulting in a local increase of the refractive index. Effectively, the waveguides possess only the fundamental mode. The coupling between neighbouring waveguides depends on their separation within the glass chip. Regarding the theoretical description of the waveguide array, the validity of equation (1) is only given if the fundamental modes of neighbouring waveguides have a negligible overlap. For a given N and a maximum length of the glass wafers of Z/κ_0, this can only be ensured for a certain range of κ_0. Outside this range of κ_0, coupled mode theory will break down and errors are introduced to the implementation of the DFrFT. In our experiments these errors are kept small but impossibly perfectly zero.

For the fabrication of the devices used to transform classical light, we employed an Yb-doped fibre laser (Amplitude Systèmes) operating at a wavelength of 532 nm, a repetition rate of 200 kHz and a pulse length of 300 fs. Waveguides were written with 300 nJ pulses focused by a 20x objective. The sample was moved at a velocity of 200 mm min^{-1} by high-precision positioning stages (ALS 130, Aerotech Inc.) with a positioning error of ± 0.1 µm. From this random positioning error, the realized inter-channel couplings inherit a relative error of 2%. The mode field diameters of the guided mode were 4 µm × 7 µm at 632 nm. In the classical experiments, the Fourier plane lies at $Z/\kappa_0 = 7.48$ cm, that is, $\kappa_0 = 0.21$ cm^{-1}. The desired nearest-neighbor couplings $(J_x)_{n\pm1,n}$ determine the separations of the waveguides n and $n \pm 1$. The largest separation of 22.8 µm between adjacent waveguides occurs at the edges. In the centre, the separation is 17.6 µm.

For the samples illuminated with single-photon states of light, we used a RegA 9,000 seeded by a Mira Ti:Sa femtosecond laser oscillator. The amplifier produced 150 fs pulses centred at 800 nm at a repetition rate of 100 kHz, with energy of 450 nJ. The structures were permanently inscribed with a 20x objective while moving the sample at a constant speed of 60 mm min^{-1}, using the positioning system described above. The mode field diameters of the guided mode were 18 µm × 20 µm at 815 nm. All structures were designed with fan-in and fan-out sections arranged in a three-dimensional geometry, and located prior and after the J_x-lattice, as illustrated in Fig. 5. This effectively suppresses any unwanted crosstalk between the guides and permits easy coupling to fibre arrays with a standard spacing of 127 µm. In the presented device used to transform quantum states, we have $\kappa_0 = 0.6$ cm^{-1}, that is, the Fourier plane is located 2.62 cm after the beginning of the J_x-array.

Experiment on the characterization of two-photon correlations. A BiB_3O_6 nonlinear crystal was pumped with a 70 mW continuous wave pump laser emitting at 407.5 nm, which provided pairs of indistinguishable photons due to type-I spontaneous parametric down-conversion, see Fig. 5. Photon pairs with a central wavelength of 815 nm were filtered by 3 nm (FWHM) interference filters. They were further coupled to the chip via fibre arrays through polarization maintaining fibres, and subsequently fed into single-photon detectors (avalanche photodiodes). The two-photon correlation function was determined by analysing the twofold coincidences recorded between all output channels with the help of an electronic correlator card (Becker & Hickl: DPC230). The spatial correlation results presented in Fig. 4 were extracted from a data set with total integration time of 5 min. The coincidences were then analysed with a time window set at 5 ns and are corrected for detector efficiencies. To assess the statistical consistency of the results in Fig. 4, we discuss in the Supplementary Note 2 the data set presented in terms of correlation event numbers before normalization. For both measurements, the detector clicks data set initially consisted of $\sim 1.5 \times 10^6$ events in total, which after post selection was reduced to a set of $\sim 5 \times 10^4$ coincidence events in total, for both input states. Accidental coincidences due to simultaneous detection of two photons not coming from the same pair are estimated to occur with a negligible rate of $< 2 \times 10^{-6}$ s^{-1}. Non-deterministic number-resolved photon detection was achieved using fibre beam splitters. This set-up thus allows the determination of all 36 two-photon coincidence events occurring in photonic lattices consisting of eight channels. Furthermore, at the wavelength of interest, propagation losses and birefringence are estimated to be 0.3 dB cm^{-1} and in the order of 10^{-7}, respectively. Additional discrepancies in between the measured correlation matrices and the theoretical ones may appear due to imperfect excitations, asymmetric output coupling losses (or detector efficiencies) or limited indistinguishability of the photons.

References

1. Cooley, J. & Tukey, J. W. An algorithm for the machine calculation of complex Fourier series. *Math. Comput.* **19,** 297–301 (1965).
2. Namias, V. The fractional order Fourier transform and its application to quantum mechanics. *J. Inst. Maths. Applics* **25,** 241–265 (1980).
3. Almeida, L. B. The fractional Fourier transform and time-frequency representations. *IEEE Trans. Signal Process.* **42,** 3084–3091 (1994).
4. Ozaktas, H. M. & Mendlovic, D. Fractional Fourier Optics. *J. Opt. Soc. Am. A* **12,** 743–751 (1995).
5. Ozaktas, H. M., Zalevsky, Z. & Kutay, M. A. *The Fractional Fourier Transform with Applications in Optics and Signal Processing* (Wiley, 2001).
6. Lohmann, A. W. & Soffer, B. H. Relationship between the Radon–Wigner and fractional Fourier transforms. *J. Opt. Soc. Am. A* **11,** 1798–1801 (1994).
7. Man'ko, M. A. Fractional Fourier transform in information processing, tomography of optical signal, and green function of harmonic oscillator. *J. of Russ. Laser Research* **20,** 226–228 (1999).
8. Dorsch, R. G., Lohmann, A. W., Bitran, Y., Mendlovic, D. & Ozaktas, H. M. Chirp filtering in the fractional Fourier domain. *Appl. Opt.* **33,** 7599–7602 (1994).
9. Lohmann, A. W. & Mendlovic, D. Fractional Fourier transform: photonic implementation. *Appl. Opt.* **33,** 7661–7664 (1994).
10. Lohmann, A. W. Image rotation, Wigner rotation, and the Fractional Fourier transform. *J. Opt. Soc. Am. A* **10,** 2181–2186 (1993).
11. Mendlovic, D. & Ozaktas, H. Fractional Fourier transforms and their optical implementation: I. *J. Opt. Soc. Am. A* **10,** 1875–1881 (1993).
12. Atakishiyev, N. M. & Wolf, K. B. Fractional Fourier-Kravchuk transform. *J. Opt. Soc. Am. A* **14,** 1467–1477 (1997).
13. Tichy, M. C., Mayer, K., Buchleitner, A. & Molmer, K. Stringent and efficient assessment of Boson Sampling Devices. *Phys. Rev Lett.* **113,** 020502–020506 (2014).
14. Perez-Leija, A., Keil, R. & Moya-Cessa, H. Perfect transfer of path-entangled photons in J_x-photonic lattices, A. Szameit, and D. N. Christodoulides. *Phys. Rev. A* **87,** 022303 (2013).
15. Humphreys, P. C., Barbieri, M., Datta, A. & Walmsley, I. A. Quantum enhanced multiple phase estimation. *Phys. Rev. Lett.* **111,** 070403 (2013).
16. Perez-Leija, A. *et al.* Eigenstate-assisted longitudinal quantum state transfer and qubit-storage in photonic and spin lattices, 45-th annual meeting of the APS Division of Atomic. *Molecular and optical Physics* **59,** J4.010 (2014).
17. Aspuru-Guzik, A., Dutoi, A. D., Love, P. J. & Head-Gordon, M. Simulated Quantum computation of molecular energies. *Science* **309,** 1704–1707 (2005).
18. Wolf, K. B. *Integral transforms in science and engineering* Vol. 11 (Mathematical Concepts and Methods for Science and Engineering, 1979).
19. Narducci, L. M. & Orzag, M. Eigenvalues and Eigenvectors of angular momentum operator J_x without theory of rotations. *Am. J. of Phys* **40,** 1811–1814 (1972).
20. Christodoulides, D. N., Lederer, F. & Silberberg, Y. Discretizing light behavior in linear and nonlinear waveguide lattices. *Nature* **424,** 817–823 (2003).
21. Szameit, A. & Nolte, S. Discrete optics in femtosecond-laser-written photonic structures. *J. Phys. B* **43,** 163001 (2010).
22. Lieb, E., Schultz, T. & Matts, D. Two soluble models of an antiferromagnetic chain. *Ann. Phys.* **16,** 407–466 (1961).
23. Cormick, C., Bermudez, A., Huelga, S. F. & Plenio, M. Preparation of the ground state of a spin chain by dissipation in a structured environment. *New J. Phys.* **15,** 073027 (2013).
24. Bobrov, I. B., Kalashnikov, D. A. & Krivitsky, L. A. Imaging of spatial correlations of two-photon states. *Phys. Rev. A* **89,** 043814 (2014).
25. Peeters, W. H., Renema, J. J. & van Exter, M. P. Engineering of two-photon spatial quantum correlations behind a double slit. *Phys. Rev. A* **79,** 043817 (2009).
26. Poem, E., Gilead, Y., Lahini, Y. & Silberberg, Y. Fourier processing of quantum light. *Phys. Rev. A* **86,** 023836 (2012).
27. Bromberg, Y., Lahini, Y., Morandotti, R. & Silberberg, Y. Quantum and classical correlations in waveguide lattices. *Phys. Rev. Lett.* **102,** 253904 (2009).

28. Marshall, G. D. *et al.* Laser written waveguide photonic quantum circuits. *Opt. Exp.* **17,** 12546–12554 (2009).
29. Lebugle, M. *et al.* Experimental observation of N00N state Bloch oscillations. *Nat. Commun.* **6,** 8273 (2015).
30. Hong, C. K., Ou, Z. Y. & Mandel, L. Measurement of subpicosecond time intervals between two photons by interference. *Phys. Rev. Lett.* **59,** 2044 (1987).
31. Wiseman, H. M. & Milburn, G. J. *Quantum Measurement and control* (Cambridge University Press, 2010).
32. Orlowski, A. & Paul, H. Phase retrieval in quantum mechanics. *Phys. Rev. A* **50,** R921–R924 (1994).
33. Wood, J. & Barry, D. T. Radon transformation of time-frequency distributions for analysis of multicomponent signals. *IEEE Transac. Signal Process.* **42,** 3166–3177 (1994).
34. Vogel, K. & Risken, H. Determination of quasiprobability distributions in terms of probability distributions for the rotated quadrature phase. *Phys. Rev. A* **40,** 2847 (1989).
35. Raymer, M. G., Beck, M. & McAlister, D. F. *Quantum Optics VI.* (eds Harvey, J. D. & Wall, D. F.) (Springer-Verlag, 1994).
36. Ozaktas, H. M. & Mendlovic, D. Fractional Fourier transforms and their optical implementations. *J. Opt. Soc. Am. A* **10,** 2522–2531 (1993).
37. Marhic, M. E. Roots of the identity operator and optics. *J. Opt. Soc. Am. A* **12,** 1448 (1995).

Acknowledgements

We gratefully acknowledge financial support by the German Ministry of Education and Research (Center for Innovation Competence programme, grant no. 03Z1HN31) and the Deutsche Forschungsgemeinschaft (grant no. NO462/6-1 and SZ276/7-1). M.L. thanks the Initial Training Network PICQUE (grant no. 608062) within the Seventh Framework Programme for Research of the European Commission for funding.

Author contributions

S.W. and A.P.-L. conceived the idea. S.W. and M.L. designed the samples and performed the measurements. A.P.-L. and S.W. developed the theory. S.W., M.L. and A.P.-L. analysed the data. A.S. supervised the project. All authors discussed the results and co-wrote the manuscript.

Additional information

Competing financial interests: The authors declare no competing financial interests.

19

Generalized Brewster effect in dielectric metasurfaces

Ramón Paniagua-Domínguez[1,*], Ye Feng Yu[1,*], Andrey E. Miroshnichenko[2], Leonid A. Krivitsky[1], Yuan Hsing Fu[1], Vytautas Valuckas[1,3], Leonard Gonzaga[1], Yeow Teck Toh[1], Anthony Yew Seng Kay[1], Boris Luk'yanchuk[1] & Arseniy I. Kuznetsov[1]

Polarization is a key property defining the state of light. It was discovered by Brewster, while studying light reflected from materials at different angles. This led to the first polarizers, based on Brewster's effect. Now, one of the trends in photonics is the study of miniaturized devices exhibiting similar, or improved, functionalities compared with bulk optical elements. In this work, it is theoretically predicted that a properly designed all-dielectric metasurface exhibits a generalized Brewster's effect potentially for any angle, wavelength and polarization of choice. The effect is experimentally demonstrated for an array of silicon nanodisks at visible wavelengths. The underlying physics is related to the suppressed scattering at certain angles due to the interference between the electric and magnetic dipole resonances excited in the nanoparticles. These findings open doors for Brewster phenomenon to new applications in photonics, which are not bonded to a specific polarization or angle of incidence.

[1] Data Storage Institute, A*STAR (Agency for Science, Technology and Research), 2 Fusionopolis Way, #08-01, Innovis 138634, Singapore. [2] Nonlinear Physics Centre, Research School of Science and Engineering, The Australian National University, Acton, Australian Capital Territory 2601, Australia. [3] Department of Electrical and Computer Engineering, National University of Singapore, 1 Engineering Drive 2, Singapore 117576, Singapore. * These authors contributed equally to this work. Correspondence and requests for materials should be addressed to R.P.-D. (email: ramon-paniag@dsi.a-star.edu.sg) or to A.I.K. (email: arseniy_k@dsi.a-star.edu.sg).

The oldest, and probably simplest, way to obtain linearly polarized light starting from unpolarized one is impinging it on a dielectric interface at the so-called Brewster's angle. In this way, the reflected light will only have electric field component parallel to the interface. Well-understood since the 1820s after the pioneering work of Fresnel, and experimentally known since the early 1810s from works of Malus and Brewster[1] (see also ref. 2 for a succinct historical perspective), the Brewster's angle for homogeneous isotropic non-magnetic media can be defined as the angle for which Fresnel's reflection coefficient for p-polarized light (that is, with the electric field parallel to the plane of incidence) vanishes, $\Re_p = 0$. An alternative definition states that Brewster's angle is the one at which the reflected and refracted waves are orthogonal, thus fulfilling the condition $\theta_i + \theta_t = \pi/2$, where θ_i is the angle of incidence and θ_t is the angle of refraction/transmission. The common microscopic interpretation of this effect is illustrated in Fig. 1a. In response to the driving electromagnetic wave, electric dipoles are induced within the material. These dipoles oscillate along the direction of the electric field, that is, perpendicular to the propagation direction. As the far-field power radiated by a dipole vanishes along its oscillation axis, whenever the dipoles and the reflection direction are parallel, no radiation is emitted into that direction and reflection is inhibited. In all other directions apart from that of refraction, radiation is compensated by the rest of the dipoles within the medium. If polarization is switched, as shown in Fig. 1b, due to the non-zero radiation in the plane perpendicular to the dipole, it is clear that such effect cannot be achieved.

The situation becomes more interesting when one considers a material that has both electric and magnetic dipoles excited in response to the electric and magnetic components of the incident wave. Such materials should have both electric permittivity and magnetic permeability different from unity ($\varepsilon \neq 1$, $\mu \neq 1$). In this case, the radiation pattern is no longer zero in the direction of oscillation of any of the orthogonal electric or magnetic dipoles (due to the non-zero contribution of the orthogonal dipole) and thus the classical Brewster effect can no longer be observed. Instead, there can be other particular directions at which the collective radiation of both dipoles vanishes due to their destructive interference, as predicted by Kerker *et al.*[3] in early 80s. These directions are determined by the relative amplitudes and phases of the dipoles. In the macroscopic picture, this interference may lead to the appearance of an analogue to Brewster's angle defined by both electric and magnetic properties of the material, as depicted in Fig. 1c. This is, the ratio ε/μ determines the angle at which the condition $\Re_{s,p} = 0$ is satisfied[4]. More importantly, inhibition of radiation from a pair of dipoles can happen at any angle and in any of the two oscillation planes depending on their relative amplitudes and phases. Thus, for such a material Brewster's angle may exist, potentially, for any of the two polarizations and at any angle of incidence (even below 45° without leading to total internal reflection at some higher angles). Both polarizations cannot, however, simultaneously have zero reflection for a given angle, except for the very particular case of $\varepsilon = \mu$ (impedance matched) at normal incidence[5]. In this case, each polarizable portion of matter will have induced electric and magnetic dipoles having the same amplitude and phase leading to inhibition of backscattered radiation, that is, fulfilling the so-called first Kerker's condition, originally derived for small magnetic particles[3]. In case of purely magnetic media, $\mu \neq 1$ and $\varepsilon \approx 1$, one can find a situation when the analogue to Brewster's angle appears for s-polarized light, having the magnetic field vector parallel to the plane of incidence, which is orthogonal to the conventional Brewster effect in dielectric media.

All this findings remained a mere theoretical curiosity for almost 20 years, since for natural materials the magnetic response is typically very weak at optical frequencies ($\mu \approx 1$). Nevertheless, since the advent of metamaterials new ways to produce optical magnetic response have been explored[6–8]. As a result some attempts have been done towards finding Brewster's angle in s-polarization in bulk magnetic metamaterials, both in microwaves[9] in arrays of split ring resonators and at optical frequencies in strongly anisotropic media[10]. Recently, polarization rotation in reflection from meta-films of bi-anisotropic split rings has been theoretically studied at microwave frequencies in connection to Brewster effect[11].

In this paper, it is demonstrated both theoretically and experimentally that the generalized Brewster effect can be observed, potentially, in any two-dimensional (2D) sub-diffractive arrangement of high-index dielectric nanoparticles, or in any other system where strong electric and magnetic resonances can be efficiently excited. It is shown that this effect is a direct consequence of the angle-suppressed radiation/scattering due to interference between the electric and magnetic dipoles excited in the particles within the array, thus connecting two apparently unrelated phenomena such as Brewster's angle and general Kerker's conditions. Silicon (Si) nanoparticles are specifically considered, for which such resonances have been broadly studied both theoretically[12–14] and experimentally[15–17]. They have attracted particular attention within the field of artificial magnetism at optical frequencies[12–17] due to their low intrinsic losses and CMOS compatibility, which holds promise for finding real world applications. Their exciting properties regarding magnetic near-field enhancement[18–22] and directional scattering[23–28], together with their low dissipation, makes them ideal nanoantennas for visible and near-infrared light[29]. The possibility to realize the first Kerker's condition[23–27] has also inspired studies on using them as ideal Huygens' sources in highly efficient transmissive metasurfaces[30–32]. Also their strong interaction with

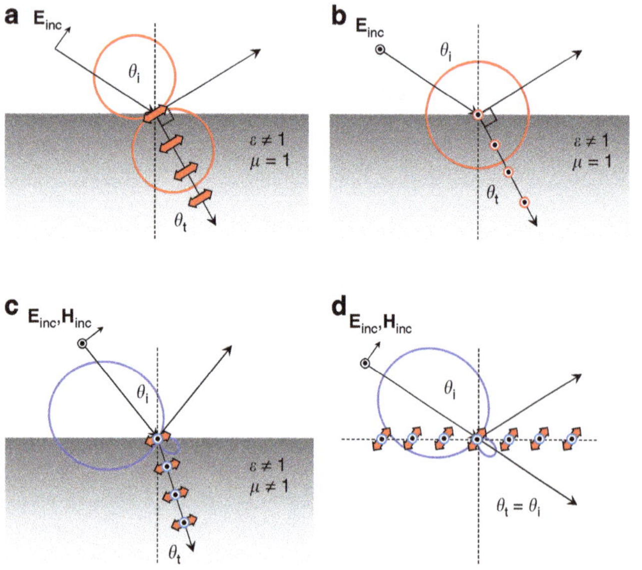

Figure 1 | Microscopic interpretation of Brewster effect and proposed metasurface. (**a**) p-polarized light impinging on a dielectric medium, $\varepsilon \neq 1$ and $\mu = 1$, under usual Brewster's condition. Red line shows 2D emission diagram of electric dipoles excited inside the material by the refracted wave. (**b**) Same as in **a** but for s-polarized incidence, for which no Brewster effect can be observed. (**c**) Generalized Brewster effect for a dielectric medium with electric and magnetic response, $\varepsilon \neq 1$ and $\mu \neq 1$. Blue line shows 2D emission diagram of interfering electric and magnetic dipoles excited inside the material by the refracted wave. (**d**) Generalized Brewster effect in the proposed array of high refractive index nanoparticles.

light, leading to high reflection and phase accumulation, makes them ideal candidates to act as efficient reflectors or phase-controlled mirrors[12,33–35]. The present study comes to extend this already broad realm with new fascinating properties. Moreover, novel generalized Brewster phenomenon giving great degree of freedom in polarization and incident angles may open doors to multiple new applications in photonics, which could not be achieved with standard Brewster effect in conventional dielectric media.

Results

Generalized Brewster effect in arrays of silicon spheres. This section aims to show that the important phenomenology associated to the generalized Brewster effect (see Supplementary Note 1 and associated Supplementary Figs 1, 2 and 3 for a concise review of it) can be found in sub-diffractive, 2D arrays of silicon spheres. We will demonstrate that, in a similar way as in continuous medium with both electric and magnetic response, the suppression of reflection originates from the interference between electric and magnetic dipoles induced in the particles in the array, as schematically depicted in Fig. 1d. For this purpose, let us first start considering a single silicon nanosphere under plane wave illumination (Fig. 2a,b), for which the required electric and magnetic dipole modes can be efficiently excited. The scattering cross section (C_{sca}) for a sphere with diameter $D = 180$ nm, as computed analytically with Mie theory[36], is depicted in Fig. 2c. Partial scattering cross-sections by the first excited resonant modes, namely the electric and magnetic dipoles and the electric and magnetic quadrupoles are also shown. As can be seen, the usual hierarchy of resonances in high-contrast dielectric nanoparticles starts with the lowest-energy magnetic dipole followed by the electric dipole mode[12–16]. Thus, whenever higher-order modes are negligible each sphere can be accurately described by a pair of these dipoles.

Kerker *et al.*[3] showed that, in such systems, the scattered far-field can be completely polarized parallel or perpendicular to the scattering plane in some particular observation direction, and this direction depends on the relative strength of the induced electric (**p**) and magnetic (**m**) dipoles. Originally derived for magnetic spheres, this result relates to interference in the electric far-field radiated by a pair of such dipoles[37], which reads:

$$\mathbf{E}_{ff} = \mathbf{E}_{ff}^{p} + \mathbf{E}_{ff}^{m} = \frac{k_0^2}{4\pi\epsilon_0}\left[\hat{\mathbf{n}}\times(\mathbf{p}\times\hat{\mathbf{n}}) + \frac{1}{c}\mathbf{m}\times\hat{\mathbf{n}}\right] \qquad (1)$$

with $k_0 = 2\pi/\lambda$ the wavenumber and ϵ_0 and c the permittivity and speed of light in vacuum, respectively, and $\hat{\mathbf{n}}$ the unit vector in the observation direction.

Consider now the particular situations depicted in Fig. 2a,b. It also follows from (1), see Supplementary Note 2, that in the plane containing the incident wave-vector and the induced electric dipole (highlighted in Fig. 2a), the radiated electric field vanishes in the observation direction defined by the angle θ if:

$$\cos(\theta - \theta_i) = m/p \qquad (2)$$

in which p and m are the complex amplitudes of electric and magnetic dipoles. In the orthogonal plane, which contains the incident wave-vector and the induced magnetic dipole (case depicted in Fig. 2b), the field vanishes when:

$$\cos(\theta - \theta_i) = p/m \qquad (3)$$

Note that the backscattering direction is defined by $\theta = \theta_i$. In this direction, the field vanishes when $p = m$ (first Kerker's condition[3]). Note also that equations (2) and (3) are, in general, complex and become real only when the dipoles are in phase or anti-phase. From these equations it can be seen that radiation can be totally suppressed for angles in backward directions $|\theta - \theta_i| \leq \pi/2$ exclusively if the dipoles are in phase (p and m having the same sign), and in forward directions ($|\theta - \theta_i| \geq \pi/2$) if they are in anti-phase (p and m having opposite sign). The spectral regions in which the induced dipoles are approximately in phase or anti-phase for the silicon sphere are highlighted in Fig. 2c by yellow and green shading colours, respectively. They indicate the spectral ranges for which scattering cancellation in forward and backward directions may happen.

The partial scattering cross-sections[36] associated with the electric (C_{sca}^{ED}) and magnetic (C_{sca}^{MD}) dipoles are proportional to the squared modulus of the dipole moments ($C_{sca}^{ED} \propto |\mathbf{p}|^2$ and $C_{sca}^{MD} \propto |\mathbf{m}|^2$), and this allows to recast equations (2) and (3) as:

$$C_{sca}^{MD}/C_{sca}^{ED} = |\cos(\theta - \theta_i)|^2 \qquad (4)$$

$$C_{sca}^{ED}/C_{sca}^{MD} = |\cos(\theta - \theta_i)|^2. \qquad (5)$$

It immediately follows from (4) that the electric dipole scattering must dominate $(C_{sca}^{MD}/C_{sca}^{ED} < 1)$ to achieve cancellation in the plane containing the electric dipole. Similarly, it follows from (5) that the magnetic dipole scattering should be dominant $(C_{sca}^{ED}/C_{sca}^{MD} < 1)$ to achieve the scattering cancellation in the plane

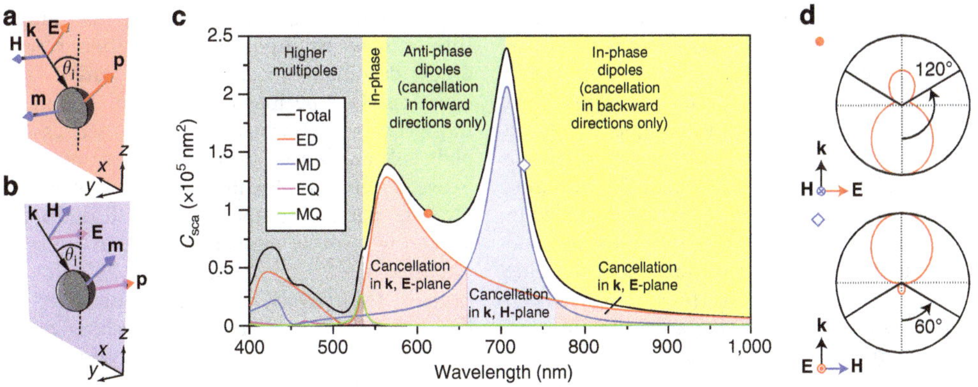

Figure 2 | Optical properties of a silicon spheres in air under plane wave illumination. (**a**,**b**) Schematic representation of the two situations studied. (**c**) Total scattering cross section (black curve) and contributions from the electric dipole (red curve), magnetic dipole (blue curve), electric quadrupole (magenta curve) and magnetic quadrupole (green curve) from a silicon sphere with diameter $D = 180$ nm. The different shaded regions indicate those wavelength windows for which the induced electric and magnetic dipoles are approximately in phase (yellow) or in anti-phase (green), and those for which the electric dipole (light red) or the magnetic dipole (light blue) dominate over the other. (**d**) Far-field radiation patterns for two wavelengths leading to inhibition of radiation at 60° with respect to the forward- ($\lambda_1 = 614$ nm, red solid circle) and backward- ($\lambda_2 = 728$ nm, blue hollow diamond) scattering directions. In the first case zero radiation is only possible in the plane parallel to the electric field whereas in the second it is only possible in the perpendicular one.

containing the magnetic dipole. The regions of dominant electric and magnetic dipoles are highlighted in Fig. 2c by red and purple shading colours, respectively.

In Fig. 2d the 2D scattering pattern of the Si sphere computed from Mie theory is plotted for two selected wavelengths, $\lambda_1 = 614$ nm and $\lambda_2 = 728$ nm, in the plane containing the incident wavevector and the electric or magnetic dipole, respectively. Vanishing scattering intensity angles predicted by equations (4) and (5), respectively, are also shown. At λ_1 the electric dipole dominates and the dipoles are in anti-phase leading to scattering cancellation at an angle $|\theta - \theta_i| \geq \pi/2$ in the plane containing the incident electric field.

At λ_2 the magnetic dipole contribution is dominant and dipoles are in phase leading to scattering cancellation at an angle $|\theta - \theta_i| \leq \pi/2$ in the plane containing the incident magnetic field. Relative amplitudes and phases are such that interference suppresses radiation at 60° with respect to the forward- and back-scattering directions, respectively.

Let us now consider the case of similar spheres arranged in an infinite 2D sub-diffractive array in the *xy*-plane (Fig. 3a) under plane wave oblique incidence. It is clear that for *p*-polarized light the plane of incidence coincides with the plane that contains the incident electric field and the induced electric dipoles (as in Fig. 2a). Correspondingly, for *s*-polarization the incidence plane contains the magnetic field and the induced magnetic dipoles (as in Fig. 2b). Although the effective dipoles induced in the particles in the array are different from the single particle case owing to the lattice interactions[38], they still radiate according to equations (1)–(3) in the plane of incidence. Note that interference from different sites in the infinite array makes radiation of the whole system allowed only as plane waves along the diffraction directions. In the case of sub-diffractive arrays, this implies radiation in the reflection and transmission directions only. Therefore, if the induced dipoles do not radiate along the direction of reflection, no reflection at all will occur in the system, leading to the Brewster's condition (see Supplementary Note 3, and the associated Supplementary Fig. 4, for a demonstration in the context of phased arrays). Following the discussion above, in such systems this may happen for both *s*- and *p*-polarized incident waves.

In the following, we consider an infinite square lattice of silicon spheres with diameter $D = 180$ nm and period $P = 300$ nm, as depicted in Fig. 3a, and study its reflection properties as a function of the wavelength and angle of incidence. We start with *p*-polarized light. Simulated results obtained by means of finite element method (see Methods section for details) are shown in Fig. 3b. At normal incidence, the electric and magnetic resonances of the particles lead to the appearance of well-known bands of high reflectivity[12,33–34]. However, oblique incidence strongly changes this behaviour. At high angles of incidence one can observe three regions of extremely low reflection. Light-white, dashed lines, with numbering ranging from 1 to 3 are included in the figure as guides to the eye to ease their location and referencing. The first one is a narrow region located at the blue side of the resonances (~515 nm). It is present at normal incidence and slightly redshifts for increasing angles (from 0 up to ~40°). The second one, located in the red side of the resonances (~790 nm), is also present at normal incidence and strongly redshifts with increasing angles. Finally, a broad region, both in bandwidth (~150 nm) and angles of incidence (from ~40 to ~80°), appears at higher angles, spectrally located between the positions of the electric and magnetic dipole resonances observed at normal incidence. Importantly, the angle of minimum reflection strongly depends on the wavelength and varies in the wide range. Figure 3c shows the angular dependence of reflection at some selected wavelengths to better illustrate this effect. One can observe that reflection of *p*-polarized light (solid lines) turns into zero at some particular angle of incidence, resembling the conventional Brewster effect in dielectric media, while no special features are observed for *s*-polarized light at this angle (corresponding dashed lines). However, there are two major peculiarities of this system, which should be highlighted. First, the range of angles at which the reflection minimum is observed covers almost the whole 0–90° span, not being restricted to angles above 45° (opposite to the conventional Brewster effect). Second, as will be shown next, the effect is not restricted to *p*-polarization, thus gathering the main features of generalized Brewster phenomenon. It is important to mention that this effect is not related to diffraction. The first non-zero diffraction order, indicated as a dashed white line in Fig. 3b and as shaded regions in Fig. 3c, appears out of the range of wavelengths and angles for which the effect is observed.

Let us now focus on the spectral region between the electric and magnetic dipole resonances. In Fig. 3d–g the case of normal incidence is shown, together with some cases with zero reflection in that region, namely 45, 60 and 75° incidence. For normal incidence both reflection maxima spectrally coincide with the excitation of dipolar resonances inside the particles. Zero reflection is observed at 775 nm, where the induced electric and magnetic dipoles have the same amplitudes and phases meeting the first Kerker's condition[3,23–27], and at 515 nm, which is close to the Kerker's condition but also affected by higher-order contributions. In the cases of oblique incidence, zeros in reflection are observed at 566 nm for 45°, 657 nm for 60°, and 686 nm for 75° (indicated by arrow heads in Fig. 3d–g) showing the strong wavelength dependence.

One explanation of the emergence of the reflection minimum at higher angles could be associated with disappearance of the dipolar resonances at off-normal incidence. However, this is not the case. Both dipolar modes are still efficiently excited, and it is their mutual interference which results in the radiation inhibition in the reflection direction. Figure 3d–g show by red and blue curves (and corresponding shaded areas), the electric dipole and magnetic dipole contributions to the total scattering cross-section (C_{sca}) from each single particle in the array, computed using the multipole decomposition technique, as explained in Methods section. As readily seen, both dipole modes are present for those angles and wavelengths for which the reflection vanishes. The dipolar contributions are dominant and higher-order modes only appear at shorter wavelengths (the complete map can be found in Supplementary Fig. 5, described in the Supplementary Note 4). Interestingly, multipole decomposition reveals that the electric dipole dominates in all the above cases. This is expected from equation (4) to be able to cancel radiation in the plane containing the electric field, which in *p*-polarized case coincides with the plain of incidence. In a simplified case with no interaction between the particles in the array, the induced dipoles should oscillate parallel to the incident field. In that case, to cancel scattering at the reflection angle $\theta = \theta_r = -\theta_i$, equation (4) imposes $C_{sca}^{MD}/C_{sca}^{ED} = \{0, 0.25, 0.75\}$ for $\theta_i = \{45, 60, 75\}$ degrees, respectively. Thus, usual Brewster at 45° is covered in this description and requires vanishing of magnetic dipole for *p*-polarization, as expected. Actual values retrieved from simulations, in which interparticle interaction is taken into account, become $C_{sca}^{MD}/C_{sca}^{ED} = \{0.0078, 0.24, 0.63\}$, which are quite close to the interaction-free case. The second zero in reflection observed at 560 nm for $\theta_i = 60°$ is related to the onset of the diffractive regime (indicated as grey-shaded areas).

As a test of consistency, far-field radiation patterns from each single particle in the infinite array were computed using Stratton–Chu formulas[5] from the fields on the surface of the sphere and plotted in the plane of incidence in Fig. 3h–k

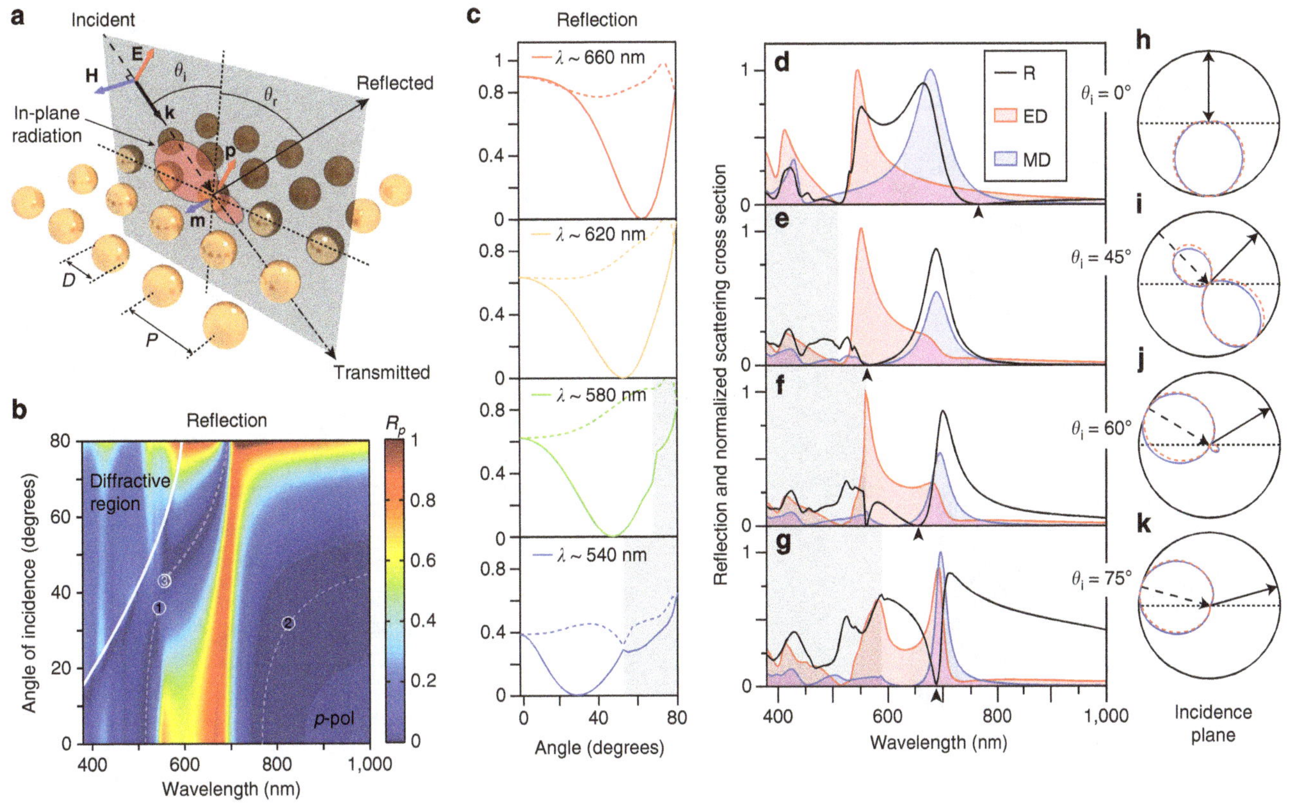

Figure 3 | Simulated optical response of arrays of silicon spheres under *p*-polarized oblique incidence. (**a**) Scheme of the simulated system. (**b**) Numerically calculated reflection versus wavelength and angle of incidence for a square lattice of silicon spheres with diameter $D = 180$ nm and pitch $P = 300$ nm. The diffractive region is indicated and delimited by a thick, solid, white line. The light-white, dashed lines are guides to the eye to help identifying the low reflectivity regions of interest. (**c**) Reflection versus angle of incidence for selected wavelengths showing the strong dependence of Brewster's angle on wavelength as well as the possibility of achieving values below 45°. Solid lines correspond to *p*-polarization while dashed lines are the corresponding curves for *s*-polarization. Grey-shaded areas mark the spectral regions affected by diffraction. (**d,g**) Reflection (black curve), together with electric dipole (red curve and corresponding shaded area) and magnetic dipole (blue curve and corresponding shaded area) contributions to scattering (normalized to their common maximum) as a function of wavelength for the cases of normal incidence and oblique incidence with $\theta_i = 45$, 60 and 70°, respectively. Diffractive region is indicated by a shaded grey area. (**h-k**) Radiation patterns in the plane of incidence numerically computed via Stratton-Chu formulas (blue solid curve) and from the induced electric and magnetic dipoles only (red dashed curve) at the wavelength of minimum reflection (arrow heads in **d-g** respectively). Incidence and reflection direction are shown by arrows.

(blue solid lines). Also shown (red dashed lines) are the patterns radiated by the pair of electric **p** and magnetic **m** dipoles given by the multipole decomposition, computed through equation (1). Both patterns closely coincide and show zero radiation in the direction of the reflected wave (indicated, together with the incident one by arrows), thus confirming the dipole interference origin of the vanishing reflection regions.

Let us now switch to the case of *s*-polarized incidence (as depicted in Fig. 4a) to show that similar effects can be obtained. The change in polarization makes the plane of incidence coincide with that containing the magnetic field in the analysis for a single sphere, thus obeying equations (3) and (5). The simulated reflection versus wavelength and angle of incidence for the same array of spheres in *s*-polarized case is shown in Fig. 4b. Two narrow band frequency windows of vanishing reflection, shifting very weakly with the angle of incidence, can be observed starting at ~515 and 770 nm for normal incidence, indicated by the light-white, dashed lines with numbers 1 and 2, respectively. Also, an omni-directional, high-reflectivity region is observed in between, analogous to that reported for high-index infinite cylinders[39]. Brewster effect in this polarization is evidenced by plotting, as in Fig. 4c, the reflection against the angle of incidence for several wavelengths. We focus on the narrow band 2, observed between 700 and 750 nm, for which no higher-order multipoles are present. For *s*-polarized light (shown as solid lines) emergence of Brewster's angle is apparent, while no special features are observed for *p*-polarization (dashed lines). As readily observed, strong dependence on wavelength and span over the whole 0–80° simulated range are also observed for *s*-polarization.

Now we show that the origin of Brewster's angle in this polarization is totally analogous to that of *p*-polarization. To this end, particular angles are plotted in Fig. 4d together with the electric dipole and magnetic dipole partial scattering cross-sections (normalized to their common maximum). For normal incidence spectral position of the dip corresponds to the first Kerker's condition at which electric and magnetic dipoles have similar amplitude and phases[3,23–27]. The observed weak blue-shift of this dip with increased angle of incidence is a consequence of the particular shape of the resonances excited in the particles and their mutual interplay, which allows fulfilling equation (5) for every angle in a narrow spectral region. Note that within the whole range of wavelengths and angles with vanishing reflection, the magnetic dipole contribution is higher than the electric dipole one, as predicted by equation (5). Similar to the case of *p*-polarized incidence, the radiation patterns of each single particle in the array associated with zero-reflection wavelengths show no radiation in the reflection direction, thus confirming the interference origin of the effect also in *s*-polarization as depicted in Fig. 4e.

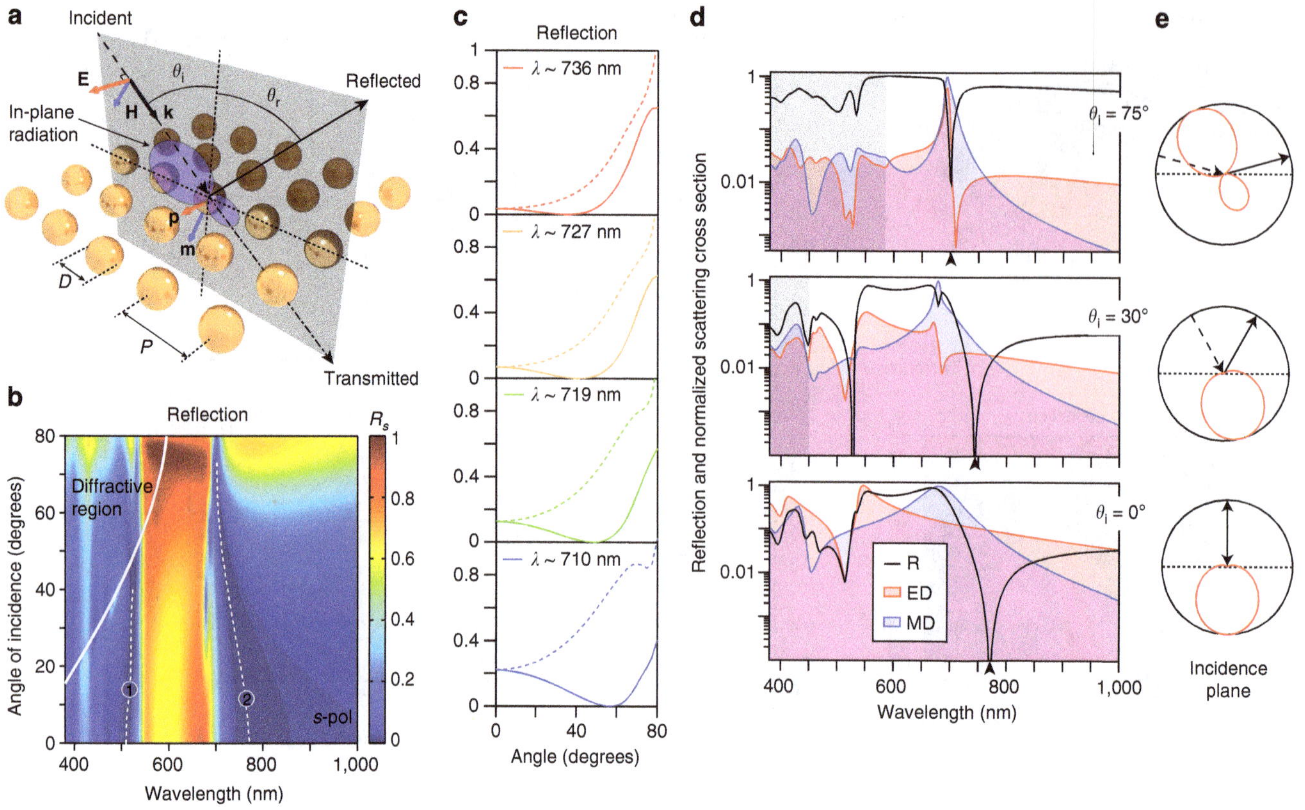

Figure 4 | Simulated optical response of arrays of silicon spheres under *s*-polarized oblique incidence. (**a**) Scheme of the simulated system. (**b**) Numerically calculated reflection versus wavelength and angle of incidence for a square array of silicon spheres with diameter $D = 180$ nm and pitch $P = 300$ nm. The region for which diffraction appears is indicated and delimited by the thick, solid white line. The white dashed lines are guides to the eye to help identifying the low reflectivity regions of interest. (**c**) Reflection versus angle of incidence for selected wavelengths showing Brewster's angle for *s*-polarization. Solid lines represent *s*-polarization while dashed lines are the corresponding curves for *p*-polarization. Dependence on wavelength and the possibility to achieve values below 45° are observed. (**d**) Reflection (black curve) and electric dipole (red curve and corresponding shaded area, electric dipole (ED)) and magnetic dipole (blue curve and corresponding shaded area, magnetic dipole (MD)) contributions to the scattering (normalized to their common maximum) from a single sphere in the array for several angles of incidence. Log_{10} scale is used for better visualization of the minima. Diffractive regions are indicated by the shaded grey area. (**e**) Associated radiation patterns of each single particle in the array in the plane of incidence at the wavelengths of minimum reflection (arrow heads in **d**).

It is important to stress that the observed spectral and angular behaviour of the zero-reflection regions in the metasurface (Figs 3b and 4b) can be directly related to the scattering properties of the single building-blocks through amplitudes and phases of the induced dipoles, as described in detail in Supplementary Note 5 (and associated Supplementary Figs 6 and 7). Thus, engineering these parameters, for example, through the geometry of the inclusions, could lead to the generalized Brewster effect, potentially, at any desired angle, frequency and polarization of interest.

Experimental verification with arrays of silicon nanodisks. To experimentally demonstrate the generalized Brewster effect, an array of silicon nanodisks was fabricated on a fused silica substrate (as described in Methods section) through silicon film deposition, electron-beam lithography and etching. Disks are chosen for ease of fabrication and, for aspect ratios close to unity, they are expected to have similar optical properties to spheres. The actual diameter is around $D = 180$ nm, height $H = 150$ nm and array pitch $P = 300$ nm (see SEM images of the fabricated array in the insets to Fig. 5a). Angular-dependent reflection measurements were performed using a home-built free-space microscopy set-up (see Methods section for details). The measured reflection and transmission spectra under normal incidence are plotted in Fig. 5a as blue and red lines, respectively.

Reflection measurements as a function of the angle of incidence for several wavelengths in the spectral region covering both electric and magnetic dipole resonances are presented as solid circles in Fig. 5b for *p*- (red) and *s*-polarized light (blue), together with results of numerical simulations (corresponding solid lines). The best agreement with the experiment was achieved for simulated diameter $D = 170$ nm, height $H = 160$ nm, pitch $P = 300$ nm and substrate refractive index of 1.45. The origin of the small discrepancy between the experiment and simulations is due to a difference between the refractive index of the fabricated silicon and the tabulated data for α-silicon[40] used in the simulations, suggesting that the fabricated silicon has less dissipation than that commonly found in literature (see Supplementary Note 6 for details). For *p*-polarization, it is clearly observed the appearance of a zero-reflection angle showing strong wavelength dependence and ranging from ~25 to nearly 70° in the studied frequency range, that is, going well below 45°. For those values below 45° no sign of total internal reflection is found. These results are strongly different from conventional Brewster's angle behaviour and represent the first experimental demonstration of the generalized Brewster effect in arrays of particles with both electric and magnetic responses. Numerical simulations, shown as solid lines in Fig. 5c, closely reproduce the experimental values and demonstrate excellent agreement in the position of the minima. The slight differences in

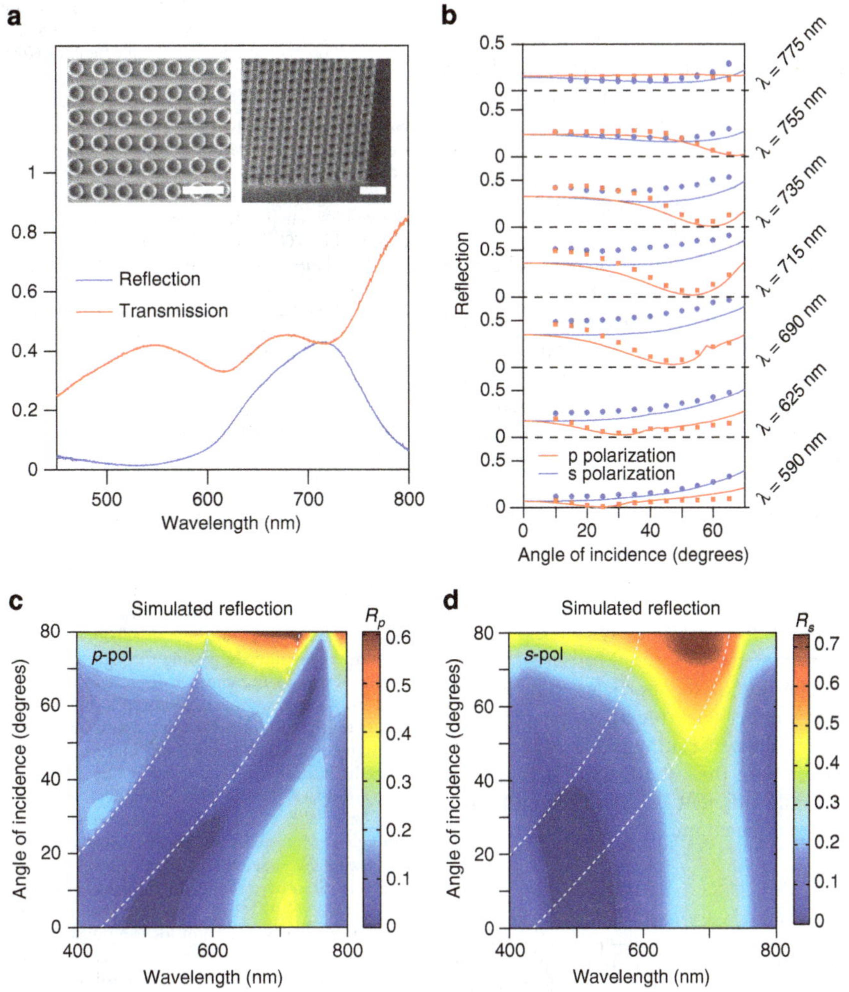

Figure 5 | Angular reflection of light from arrays of silicon nanodisks over a glass substrate. (**a**) Experimentally measured reflection (blue solid curve) and transmission (red solid curve) of a square lattice of silicon disks with diameter $D=180$ nm, height $H=150$ nm and pitch $P=300$ nm under normal incidence. The insets show the top (left) and tilted (right) SEM images of the measured sample. Scale bar, 500 nm. (**b**) Reflection versus angle of incidence measured for different wavelengths under *p*-polarized (red circles) and *s*-polarized (blue circles) illumination. The corresponding simulated data, obtained for $D=170$ nm and $H=160$ nm with the same pitch are shown as solid curves. (**c**,**d**) Maps of simulated values of reflection as a function of angle of incidence and wavelength for *p*-polarized and *s*-polarized incidence, respectively. White dashed lines indicate the onset of the different diffraction orders.

the reflection intensity, as mentioned, are due to the smaller absorption of the deposited silicon compared with the common amorphous silicon data from literature used in simulations[40]. Taking in simulations slightly lower values of the imaginary part of refractive index than in Palik[40] (as given in Supplementary Table 1) leads to almost perfect agreement to the experiment (Supplementary Fig. 8).

To complete the picture, the full simulated maps of reflection of *p*- and *s*-polarized light as a function of angle of incidence and wavelength are shown in Fig. 5c,d. Although with some differences, the general trend observed in the simulated region strongly resembles that shown in Figs 3b and 4a for spheres. For *p*-polarized light, the minimum in reflection strongly varies both with wavelength and angle of incidence, starting in the blue side of the resonances and moving into the region between them for increasing angles. As in the case of spheres both electric dipole and magnetic dipole modes retrieved from multipole decomposition for single disk in the array are strongly excited in the regions of zero reflection (see Supplementary Note 7 and Supplementary Fig. 9). Radiation patterns of these interfering dipoles computed for two of the zero-reflection cases in *p*-polarization are shown in Supplementary Fig. 10 and discussed in Supplementary Note 8. They demonstrate vanishing intensities of the radiation in the direction of the reflected wave at the operation wavelength, thus confirming the interference origin of the observed effect.

For *s*-polarization a shallow minimum in reflection at 735, 755 and 775 nm can be observed both in simulations and experiment (Fig. 5b). These minima correspond to the tail of the vanishing reflection region (Fig. 5d) and provide further experimental evidence of the generalized Brewster effect. Discussion of the experimental plot focused on the cases of 755 and 775 nm can be found in Supplementary Fig. 9, clearly showing a minimum in reflection for angles below and above 45°. It is worth mentioning that for this particular system the complete vanishing of reflection under *s*-polarized incidence can be obtained in the spectral region ~850 nm. However, at these wavelengths the array has very low reflectivity even at normal incidence.

Remarkably, even for the realistic system described above the generalized Brewster effect is very robust and can easily be detected in experiment, the only true requirement being the efficient excitation of electric and magnetic dipole resonances in the particles forming a sub-diffractive array.

Discussion

According to the results shown, sub-diffractive arrays of high-permittivity dielectric nanoparticles supporting both electric and magnetic dipole resonances present a form of generalized Brewster effect leading to vanishing reflection at particular wavelengths and angles both under *p*- and *s*-polarized incidence. The phenomenon can be explained in terms of radiation interference between the electric and magnetic dipoles induced in each particle in the array and connects the angle-suppressed scattering from magneto-electric particles (usually studied in relation to first Kerker's condition) with the zero-reflection (Brewster effect) observed in 2D arrangements of such particles. As a consequence of this interference the range of zero-reflection angles spans almost over the entire 0–90° without implying total internal reflection. It shows a strong dependence on the incident wavelength and is present for both *p* and *s* polarizations. The effect has been experimentally demonstrated in dense arrays of silicon disks over a fused silica substrate, with measured zero-reflection angles ranging from 20 to 70° for wavelengths varying from 590 to 775 nm in the visible spectrum. These results represent the first experimental demonstration of the generalized Brewster's effect at optical frequencies in particle arrays with both electric and magnetic response to incident light.

Since this effect is a universal phenomenon related to the directional interference of resonances excited in the particles, it is foreseen that it will be observed in a variety of systems, provided they present electric and magnetic responses. Moreover, tuning the shape and material properties of the particles may lead to almost-on-demand Brewster's effect with regard to polarization, wavelength and angle of incidence. Taking advantage of the strongly resonant character of the structures may bring opportunities for design of efficient sub-wavelength-thick polarizers with a great degree of freedom.

Methods

Numerical simulations. Finite element method was used to compute the reflection, transmission and absorption of light from infinite square arrays of silicon spheres (commercial COMSOL Multiphysics software was used). Experimentally measured values of the refractive index of crystalline silicon, taken from ref. 40, were used in the simulations. The simulation domain consisted of a single unit cell with Bloch boundary conditions applied in the periodicity directions (x and y axes) to simulate an infinite lattice. The so-called scattered field formulation of the problem was used. The exciting field was defined as a plane wave with the electric field in the incidence plane for *p*-polarized light and perpendicular to it for *s*-polarized light. Perfectly matched layers were applied in the top and bottom directions to absorb all scattered fields from the system. Additionally two planes, $\Sigma\pm$, perpendicular to the z axis at $z = \pm 450$ nm were used as monitors to compute the reflected and transmitted power. Reflection was computed as the flux of the Poynting vector of the scattered fields in the Σ_- plane normalized to the power of the plane wave in the same area. Total fields instead of scattered ones were considered in Σ_+ to compute transmission. Absorption was computed as the volume integration of the Ohmic losses inside the sphere and normalized in the same way. Conservation of energy leads to $R+T+A=1$, condition that allows internal check of consistency. These results were also checked by performing the same calculation in CST Microwave Studio, showing excellent agreement.

Simulations of silicon disk arrays over substrate (with interface in $z=0$ and refractive index $n=1.45$) were carried out using the same approach as described for spheres. The main difference is that, in this case, Fresnel equations were used to explicitly write the excitation fields in the upper ($z>0$) and lower ($z<0$) half spaces. While transmission and absorption are computed in exactly the same way, for reflection calculations one needs to consider the Poynting vector of the scattered fields plus the reflected fields from the substrate. In these simulations the refractive index of amorphous silicon[40] was used to approximate the deposited amorphous silicon in the experiment.

Multipole decomposition. Multipole decomposition technique was employed to analyse the different modes being excited in the particles. For particles in an array embedded in air, multipoles can be computed through the polarization currents induced within them:

$$\mathbf{J} = -i\omega\varepsilon_0(\varepsilon - 1)\mathbf{E}, \tag{6}$$

where ε is the permittivity of the particle and $\mathbf{E}=\mathbf{E}(\mathbf{r})$ is the electric field inside it.

This approach fully takes into account mutual interactions in the lattice[41] as well as the possible presence of a substrate. In particular, a Cartesian basis with origin in the centre of the particles was used in the present work. An accurate description of the radiative properties in this basis involves the introduction of the family of toroidal moments[42–44] and the mean-square radii corrections. The explicit expressions of the multipoles as well as the associated partial scattering cross section can be found in Supplementary Note 9.

Nanodisk array fabrication. Thin films of amorphous silicon of desired thickness were deposited on fused silica substrates via electron-beam evaporation (Angstrom Engineering Evovac). The samples were then patterned by single-step electron-beam lithography: by spin-coating HSQ resist (Dow Corning, XR-1541-006) and a charge-dissipation layer (Espacer 300AX01), e-beam patterning of the resist (Elionix ELS-7000), and subsequent etching via reactive-ion-etching in inductively coupled plasma system (Oxford Plasmalab 100). The remaining HSQ resist (~50 nm after etching) on the top of the nanodisks was not removed since its optical properties after e-beam exposure are close to that of silicon dioxide. To reduce losses the fabricated sample was annealed in vacuum at 600 °C for 40 min by using Rapid Thermal Process system (Model: JetFirst200).

Optical measurements. Transmission and reflection measurements of the nanodisk arrays at normal incidence were conducted using an inverted microscopy set-up (Nikon Ti-U). For transmission measurements, light from a broadband halogen lamp was normally incident onto the sample from the substrate side before being collected by a ×5 objective (Nikon, NA 0.15) and routed to a spectrometer (Andor SR-303i) with a 400 × 1,600 pixel EMCCD detector (Andor Newton), as described in detail elsewhere[15]. Transmitted light through the array was normalized to the transmitted power through the substrate only, after accounting for photodetector noise effects (dark current subtraction). For reflection measurements, light from the broadband halogen lamp was incident into the nanodisk array directly passing through the ×5 objective. The reflected light was then collected by the same objective and routed into the spectrometer. Reflected light from the array was normalized to the incident power, which is characterized by the reflection of a silver mirror with known spectral response.

Angular transmission and reflection measurements were performed using a home-built free-space microscopy set-up. Light originating from a supercontinuum source (SuperK Power, NKT Photonics) was transmitted through a variable band-pass filter for wavelength selection (SuperK Varia, NKT Photonics) and then through a broadband polarizing beam-splitter cube (Thorlabs, PBS252). The linear polarized light passed then through a quarter wave plate (Thorlabs, WPQ10M-808) to obtain circularly polarized light, which was sent to a rotating linear polarizer (Thorlabs, LPNIRE100-B) to obtain linearly polarized light of selected direction. A biconvex lens with 75 mm focusing distance (Thorlabs) was used to focus the light onto the sample surface with silicon nanoparticle arrays. The sample was mounted on a rotation stage for adjusting the angle of incidence. The beam spot size at the sample at normal incidence had a diameter of ~50 µm being smaller than the size of the fabricated arrays (100 × 100 µm). A white-light lamp source was also coupled into the beam path through the same broadband polarizing beam-splitter cube for sample imaging. Both the incident beam power and the transmitted/reflected beam power were measured by a pixel photodetector attached to a digital handheld laser power/energy metre console (Thorlabs, PM100D). A scheme of the experimental set-up is included in Supplementary Fig. 11 and described in Supplementary Note 10.

References

1. Brewster, D. On the laws which regulate the polarisation of light by reflexion from transparent bodies. *Philos. Trans. R. Soc. Lond.* **105,** 125–159 (1815).
2. Lakhtakia, A. Would Brewster recognize today's Brewster angle? *Opt. News* **15,** 14–18 (1989).
3. Kerker, M., Wang, D.-S. & Giles, C. L. Electromagnetic scattering by magnetic spheres. *J. Opt. Soc. Am.* **73,** 765–767 (1983).
4. Giles, C. L. & Wild, W. J. Brewster angles for magnetic media. *Int. J. Infrared Millim. Waves* **6,** 187–197 (1985).
5. Stratton, J. A. *Electromagnetic Theory* (McGraw-Hill, 1941).
6. Shalaev, V. M. Optical negative-index metamaterials. *Nat. Photon.* **1,** 41–48 (2007).
7. Zheludev, N. I. The road ahead for metamaterials. *Science* **328,** 582–583 (2009).
8. Soukoulis, C. M. & Wegener, M. Past achievements and future challenges in the development of three-dimensional photonic metamaterials. *Nat. Photon.* **5,** 523–530 (2011).
9. Tamayama, Y., Nakanishi, T., Sugiyama, K. & Kitano, M. Observation of Brewster's effect for transverse-electric electromagnetic waves in metamaterials: Experiment and theory. *Phys. Rev. B* **73,** 193104 (2006).
10. Watanabe, R., Iwanaga, M. & Ishihara, T. s-polarization Brewster's angle of stratified metal–dielectric metamaterial in optical regime. *Phys. Stat. Sol. b* **245,** 2696–2701 (2008).
11. Tamayama, Y. Brewster effect in metafilms composed of bi-anisotropic split-ring resonators. *Opt. Lett.* **40,** 1382–1385 (2015).
12. Evlyukhin, A. B., Reinhardt, C., Seidel, A., Luk'yanchuk, B. S. & Chichkov, B. N. Optical response features of Si-nanoparticle arrays. *Phys. Rev. B* **82,** 045404 (2010).

13. Garcia-Etxarri, A. *et al.* Strong magnetic response of submicron silicon particles in the infrared. *Opt. Express* **19,** 4815–4826 (2011).
14. Luk'yanchuk, B. S., Voshchinnikov, N. V., Paniagua-Dominguez, R. & Kuznetsov, A. I. Optimum forward light scattering by spherical and spheroidal dielectric nanoparticles with high refractive index. *ACS Photon.* **2,** 993–999 (2015).
15. Kuznetsov, A. I., Miroshnichenko, A. E., Fu, Y. H., Zhang, J. & Luk'yanchuk, B. Magnetic light. *Sci. Rep.* **2,** 492 (2012).
16. Evlyukhin, A. B. *et al.* Demonstration of magnetic dipole resonances of dielectric nanospheres in the visible region. *Nano Lett.* **12,** 3749–3755 (2012).
17. Shi, L., Tuzer, T. U., Fenollosa, R. & Meseguer, F. A new dielectric metamaterial building block with a strong magnetic response in the sub-1.5-micrometer region: silicon colloid nanocavities. *Adv. Mater.* **24,** 5934–5938 (2012).
18. Schmidt, M. K. *et al.* Dielectric antennas-a suitable platform for controlling magnetic dipolar emission. *Opt. Express* **20,** 13636–13650 (2012).
19. Albella, P. *et al.* Low-loss electric and magnetic field-enhanced spectroscopy with subwavelength silicon dimers. *J. Phys. Chem. C* **117,** 13573–13584 (2013).
20. Rolly, B., Stout, B. & Bonod, N. Boosting the directivity of optical antennas with magnetic and electric dipolar resonant particles. *Opt. Express* **20,** 20376–20386 (2012).
21. Bakker, R. M. *et al.* Magnetic and electric hotspots with silicon nanodimers. *Nano Lett.* **15,** 2137–2142 (2015).
22. Albella, P., Alcaraz de la Osa, R., Moreno, F. & Maier, S. A. Electric and magnetic field enhancement with ultra-low heat radiation dielectric nanoantennas: considerations for surface enhanced spectroscopies. *ACS Photon.* **1,** 524–529 (2014).
23. Nieto-Vesperinas, M., Gomez-Medina, R. & Saenz, J. J. Angle-suppressed scattering and optical forces on submicrometer dielectric particles. *J. Opt. Soc. Am. A* **28,** 54–60 (2011).
24. Gomez-Medina, R. *et al.* Electric and magnetic dipolar response of germanium nanospheres: interference effects, scattering anisotropy, and optical forces. *J. Nanophoton.* **5,** 053512 (2011).
25. Geffrin, J.-M. *et al.* Magnetic and electric coherence in forward-and back-scattered electromagnetic waves by a single dielectric subwavelength sphere. *Nat. Commun.* **3,** 1171 (2012).
26. Fu, Y. H., Kuznetsov, A. I., Miroshnichenko, A. E., Yu, Y. F. & Luk'yanchuk, B. Directional visible light scattering by silicon nanoparticles. *Nat. Commun.* **4,** 1527 (2013).
27. Person, S. *et al.* Demonstration of zero optical backscattering from single nanoparticles. *Nano Lett.* **13,** 1806–1809 (2013).
28. Krasnok, A. E., Simovski, C. R., Belov, P. A. & Kivshar, Y. S. Superdirective dielectric nanoantennas. *Nanoscale* **6,** 7354–7361 (2014).
29. Krasnok, A. E. *et al.* All-dielectric optical nanoantennas. in *Progress in Compact Antennas.* (ed Huitema, L.) 143–172 (InTech, 2014).
30. Staude, I. *et al.* Tailoring directional scattering through magnetic and electric resonances in subwavelength silicon nanodisks. *ACS Nano* **7,** 7824–7832 (2013).
31. Decker, M. *et al.* High-efficiency dielectric Huygens' surfaces. *Adv. Opt. Mater.* **3,** 813–820 (2015).
32. Yu, Y. F. *et al.* High-transmission dielectric metasurface with 2π phase control at visible wavelengths. *Laser Photon. Rev.* **9,** 412–418 (2015).
33. Moitra, P., Slovick, B. A., Yu, Z. G., Krishnamurthy, S. & Valentine, J. Experimental demonstration of a broadband all-dielectric metamaterial perfect reflector. *Appl. Phys. Lett.* **104,** 171102 (2014).
34. Moitra, P. *et al.* Large-scale all-dielectric metamaterial perfect reflectors. *ACS Photon.* **2,** 692–698 (2015).
35. Liu, S. *et al.* Optical magnetic mirrors without metals. *Optica* **1,** 250–256 (2014).
36. Bohren, C. F. & Huffmann, D. R. *Absorption and Scattering of Light by Small Particles* (A Wiley-Interscience Publication, 1983).
37. Jackson, J. D. *Classical Electrodynamics* 3rd edn (John Wiley & Sons, 1998).
38. Garcia de Abajo, F. J. Colloquium: Light scattering by particle and hole arrays. *Rev. Mod. Phys.* **79,** 4 (2007).
39. Du, J., Lin, Z., Chui, S. T., Dong, G. & Zhang, W. Nearly Total omnidirectional reflection by a single layer of nanorods. *Phys. Rev. Lett.* **110,** 163902 (2013).
40. Palik, E. D. & Bennett, J. M. (eds) *Handbook of Optical Constants of Solids* (The Optical Society of America, 1995).
41. Grahn, P., Shevchenko, A. & Kaivola, M. Electromagnetic multipole theory for optical nanomaterials. *New J. Phys.* **14,** 093033 (2012).
42. Radescu, E. E. & Vaman, G. Exact calculation of the angular momentum loss, recoil force, and radiation intensity for an arbitrary source in terms of the electric, magnetic, and toroid multipoles. *Phys. Rev. E* **65,** 046609 (2002).
43. Kaelberer, T., Fedotov, V. A., Papasimakis, N., Tsai, D. P. & Zheludev, N. I. Toroidal dipolar response in a metamaterial. *Science* **330,** 1510 (2010).
44. Miroshnichenko, A. E. *et al.* Nonradiating anapole modes in dielectric nanoparticles. *Nat. Commun.* **6,** 8069 (2015).

Acknowledgements

Authors from DSI were supported by DSI core funds and A*STAR SERC Pharos program, Grant No. 152 73 00025 (Singapore). Fabrication and Scanning Electron Microscope imaging works were carried out at the SnFPC cleanroom facility at DSI (SERC Grant No. 092 160 0139). The authors are grateful to Yi Yang (DSI) and Seng Kai Wong (DSI) for help with SEM imaging. The work of AEM was supported by the Australian Research Council via Future Fellowship program (FT110100037).

Author contributions

R.P.-D. proposed the explanation of the experimental results, performed the theoretical analysis and numerical simulations. Y.F.Y. performed the sample nanofabrication and the reflection and transmission angular measurements. A.I.K. and L.A.K. built the angular optical measurement set-up and performed early stage measurements. A.E.M. performed early stage simulations and developed the phased array model. Y.H.F. helped to perform angular and normal incidence measurements. V.V. performed SEM imaging. L.G., Y.T.T., and A.Y.S.K. contributed to development of various nanofabrication procedures. B.L. performed the analysis of the homogeneous slab. A.I.K. proposed the initial idea, and supervised the work. All authors contributed to the manuscript preparation and reviewed the final version of the manuscript.

Additional information

Competing financial interests: The authors declare no competing financial interests.

20

Intensifying the response of distributed optical fibre sensors using 2D and 3D image restoration

Marcelo A. Soto[1], Jaime A. Ramírez[1] & Luc Thévenaz[1]

Distributed optical fibre sensors possess the unique capability of measuring the spatial and temporal map of environmental quantities that can be of great interest for several field applications. Although existing methods for performance enhancement have enabled important progresses in the field, they do not take full advantage of all information present in the measured data, still giving room for substantial improvement over the state-of-the-art. Here we propose and experimentally demonstrate an approach for performance enhancement that exploits the high level of similitude and redundancy contained on the multidimensional information measured by distributed fibre sensors. Exploiting conventional image and video processing, an unprecedented boost in signal-to-noise ratio and measurement contrast is experimentally demonstrated. The method can be applied to any white-noise-limited distributed fibre sensor and can remarkably provide a 100-fold improvement in the sensor performance with no hardware modification.

[1] EPFL Swiss Federal Institute of Technology, Group for Fibre Optics,, SCI-STI-LT, Station 11, CH-1015 Lausanne, Switzerland. Correspondence and requests for materials should be addressed to M.A.S. (email: marcelo.soto@epfl.ch).

Distributed fibre sensors[1,2] exploit specific optical effects activated along optical fibres, to obtain a spatially distributed profile of environmental quantities such as temperature, strain, pressure, electromagnetic fields and so on. This feature offers unique attributes and capabilities compared with conventional discrete sensing methods[1,2]. Conventional distributed fibre sensors can be classified in a wide range of types[2] based on the nature of the exploited optical effect. There exist types based on optical absorption, fluorescence, evanescent field or interferometers, among others[1,2]; however, most of the distributed sensors use natural scattering processes[3,4] present in optical fibres along with interrogating methods based on time-domain[5] or frequency-domain[6] reflectometry. These are sensors essentially based on Rayleigh scattering[7–11], spontaneous Raman scattering[12–14] and spontaneous or stimulated Brillouin scattering[15–24]. The spatial resolution of distributed sensors is primarily determined by the bandwidth of the interrogating signal, which is typically traded off with the power contrast in the detected signal. Under an optimized configuration, the best performance attained by any distributed sensor is ultimately determined by the signal-to-noise ratio (SNR) of the measurements[16,24]. Therefore, to ensure a given measurement quality (defined by the measurand resolution), a minimum SNR has to be secured in the system[24,25]. This imposes an important tradeoff between spatial resolution and sensing range, which affects all kinds of distributed fibre sensors[1,2,25–27]. Sensors with high (that is, sharp) spatial resolution are inherently limited by the short interaction time and low energy involved in the scattering process (locally activated at each fibre position), enabling measurements only along short ranges[19,20] (typically below 1 km). On the other hand, sensors with metric resolution can typically operate over many tens of kilometres of optical fibre. In this case, the SNR and quality of the measurements are basically limited by the onset of nonlinear effects[28–32], which constraint the maximum optical power launched into the sensing fibre, and by the fibre attenuation, which leads to an exponentially decaying sensor response and poor measurement contrast at the end of long optical fibres[24,25].

The clear understanding reached in recent years on the factors ultimately limiting the sensor capabilities has motivated researchers to propose and demonstrate specially designed methods for performance enhancement[33–54]. Among several advanced techniques, methods such as distributed Raman amplification[34–37], optical pulse coding[38–43] or different signal processing methods[44–51] have resulted in implementations outperforming classical standard configurations. Whereas each of these methods can individually provide up to about 10–12 dB SNR enhancement, a higher improvement can only be reached by a proper combination of several of those techniques[52–54], at the cost of complex and expensive implementations.

Among several existing methods, signal processing techniques, such as optical pulse coding[38–43], wavelet transform[44–48] and Fourier transform[49] have demonstrated to be very efficient tools to remove noise. However, their exploitation for distributed sensing has been restricted so far only to one-dimensional (1D) arrays of data. Those techniques can be readily applied, for instance, to Raman-distributed fibre sensors[38,44], owing to the 1D nature of the acquired data (corresponding to 1D traces of the anti-Stokes, Rayleigh and Stokes backscattered light[12–14]). In the case of Brillouin- and Rayleigh-based distributed sensors, in which time and frequency are scanned, signal processing has been used to denoise individual longitudinal traces (that is, at a fixed scanned frequency) independently from each other[39–41,45], the measured local spectrum at each fibre location[47] or the retrieved measurand profile[48]. Although methods such as time–frequency coding[42,43] take advantage of the double scanning (fibre position and pump–probe frequency detuning) required in Brillouin sensing, the provided SNR enhancement is basically given by the ability of the code to reduce noise in a 1D array of data. Indeed, none of the existing methods for performance enhancement exploit the redundancies and correlations contained in the multidimensional domain of the measured information. This is so far a feature of distributed fibre sensors that has been completely unexplored in the state-of-the-art; however, as such measurements contain repeated structures of information in a multidimensional domain (time, frequency and position), they can be smartly and efficiently exploited to improve the SNR of the measurements.

In the following, we propose and experimentally demonstrate an approach that exploits correlated patterns of information and their high degree of redundancy for enhancing the measurement quality and performance of distributed optical fibre sensors. In particular, this approach makes use of image and video enhancement processing for removing noise and increasing the contrast of noisy measurements obtained by any kind of distributed sensor. To the best of our knowledge, this is the first time that such a multidimensional approach is used to restore information and enhance the capabilities of distributed fibre sensors. Here we demonstrate an unprecedented boost in SNR, which can reach two orders of magnitude (that is, ~20 dB SNR enhancement), being equivalent or even superior to the use of extensively complex hardware sophistications but at a minor fraction of the cost. Any dB gained in SNR can be used to improve the sensor performance, that is, to extend the range, to sharpen the spatial resolution, to reduce the measurement time (reducing the number of averages), or simply to improve the measurand accuracy[24]. The technique can be applied to any conventional or advanced sensor in which the acquired data can be arranged in a two-dimensional (2D) or three-dimensional (3D) data structure. This includes any possible configuration for distributed fibre sensing based on, for example, faint long gratings[55], Rayleigh[7–11], Raman[12–14] or Brillouin[15–24] scattering (or any combination of them); however, the use of the method can also be extended to any reflectometry-based technique for fibre characterization[5,6] as well as for quasi-distributed or multiplexed sensors, such as arrays of fibre Bragg gratings[56], in which the measured information can be arranged in a 2D or 3D data structure.

Results

2D image processing for sensor data restoration. We first tested the proposed method on measurements obtained by a standard Brillouin optical time-domain analyser[15,16] (BOTDA), using the proof-of-concept experimental setup shown in Fig. 1. For a 2-m spatial resolution over a 50-km-long sensing fibre, the SNR of the implemented system turns out to be optimized using a 125-MHz high-transimpedance photodetector[57]. Although this high-transimpendance makes the system thermal-noise dominated, this leads to an optimized electrical SNR[57]. To provide a reliable demonstration of the technique proposed in this study, noisy distributed measurements are intentionally acquired using only four time-averaged traces per scanned frequency (two averaged traces for each orthogonal polarization). It is noteworthy that the acquisition procedure in Brillouin sensors makes inherently use of a 2D data structure $\mathbf{M}(z,\Delta f)$ in the position z and frequency Δf domains, from which the environmental information is retrieved by detecting spectral shifts of the peak gain frequency. Figure 2a shows the noisy 3D map of the Brillouin gain spectrum (BGS) measured along the sensing fibre for different pump–probe frequency offsets Δf. In this case we use a sampling rate of 5 ns per digital point in the

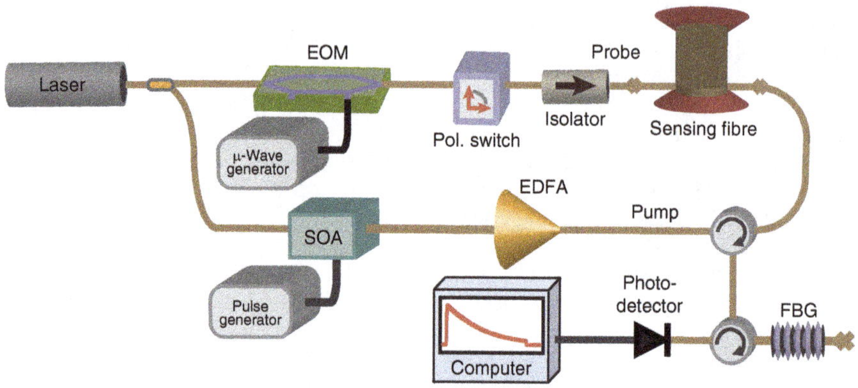

Figure 1 | Experimental setup of a basic BOTDA sensor. The light of a conventional distributed-feedback laser operating at 1,550 nm is split into distinct branches to generate the pump and probe signals. In the upper branch, a high-extinction ratio (>40 dB) electro-optic modulator (EOM), driven by a microwave signal and operating in carrier-suppression mode, generates a two-sideband probe signal. A polarization switch is used to compensate for the polarization-induced fading affecting the Brillouin gain along the fibre. The probe power launched into the fibre is set to −6 dBm to avoid unwanted spectral distortions resulting from non-local effects as reported in ref. 23. In the lower branch, a high on-off ratio (> 50 dB) pump pulse of 20 ns is generated by a semiconductor optical amplifier (SOA) and then boosted by an erbium-doped fibre amplifier (EDFA) up to 100 mW (limit imposed by modulation instability, as indicated in ref. 31). The sensing fibre is a 50-km-long standard single-mode fibre. On the receiver side, a narrowband (10 GHz) fibre Bragg grating (FBG) is inserted to select the lower-frequency probe sideband, which is detected by a 125-MHz photoreceiver. Time-domain traces are then collected using an acquisition card connected to a computer.

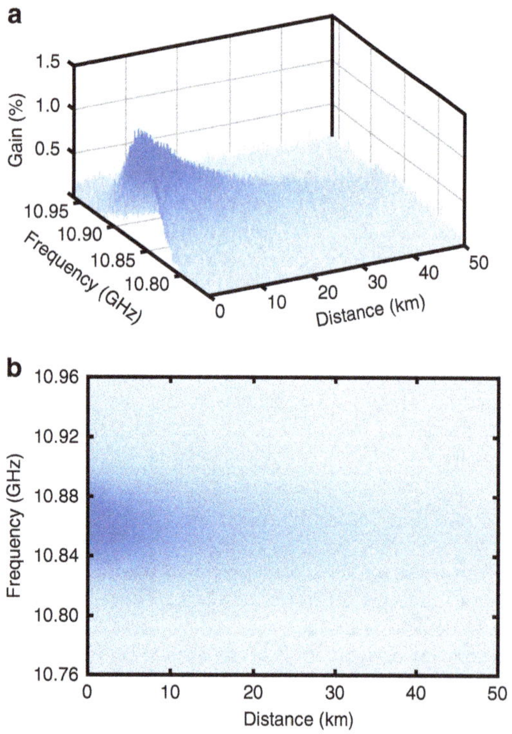

Figure 2 | Measured BGS versus distance. (**a**) Three-dimensional map of the measured BGS as a function of distance. Measurements are obtained along a 50-km-long sensing fibre using a spatial resolution of 2 m, 4 temporal averages (2 averages per each orthogonal polarization state), a sampling interval of 0.5 m, a spectral scanning range of 200 MHz and a frequency step of 1 MHz. (**b**) Top view of the measured BGS, where pixels with darker blue tones represent position-frequency pairs with higher Brillouin gain. This image provides a visual representation of the noisy measured data and depicts the 'image' to be enhanced by image processing.

analogue-to-digital converter, corresponding to a sampling interval of 0.5 m per point (using for convenience the typical time-to-position conversion based on the group velocity of light in the fibre), whereas a spectral range of 200 MHz is scanned with steps of 1 MHz. The measurement process therefore results in a 2D matrix $\mathbf{M}(z,\Delta f)$ of 100,000 × 200 data points containing the local BGS at each sampled fibre location. Image processing[58,59] here associates each acquired position–frequency pair $(z,\Delta f)$ to a pixel (x,y) of a noisy digital image (illustrated in Fig. 2b), where x and y are the spatial coordinates of the image. This image can be represented by a two-variable function $f(x,y)$ with values belonging to a 1D space, such as in a greyscale image[58,59] and mapping the local Brillouin gain measured at a given position z and frequency offset Δf. For a better visual perception of the data, a blue scale is chosen in Fig. 2, where pixels with darker tones represent position–frequency pairs having higher Brillouin gain (it is noteworthy that the scale is inverted when compared with the traditional representation of monochrome images[58,59]). A high level of redundancy in the signal amplitude associated to given position–frequency pairs can be found along the entire 2D data structure $\mathbf{M}(z,\Delta f)$; this actually becomes evident if we consider that the same BGS (having a known spectral shape) is repeatedly measured along the fibre, being only spectrally shifted at positions where the local environmental conditions change[16,17]. Although the fibre attenuation can alter the BGS peak amplitude with distance, Fig. 2 shows that the matrix $\mathbf{M}(z,\Delta f)$ can still be decomposed into small 2D patches containing several closely located longitudinal points having essentially the same average amplitude.

To exploit the high level of similitude and redundancy present in the 2D domain of the data, we here use two of the best-known image denoising methods as proof-of-concept, to evaluate the effectiveness of the proposed approach: the so-called non-local means (NLM)[59–63] and wavelet denoising (WD)[64–67] (see Methods for details). Whereas the former method operates in the spatial domain of the image, that is, making direct use of the measured data points (raw data), the later method converts the raw data into the wavelet domain (corresponding to a particular representation of the spectral domain of the image), where the components associated with noise are filtered out by wavelet shrinkage using a hard thresholding function[64–66]. Supplementary Fig. 1 shows the 'denoised images' resulting after processing the raw measured data points. Although a simple visual inspection of these images indicates a clear improvement in the data quality (compared with Fig. 2b), we evaluate the effectiveness of the denoising using an objective metric by calculating the SNR of the time-domain trace obtained at the peak Brillouin gain frequency[24]. To allow a fair comparison with existing techniques, it is worth mentioning that we use here the definition of SNR calculated to be proportional to the trace amplitude, which means the ratio between the mean amplitude of the measured local response and its standard deviation[24,25,51–53], contrarily to some works where the SNR is defined to be proportional to the electrical power[16,45,50].

Figure 3a,b compare the SNR of the raw noisy traces (blue lines) and the ones obtained after denoising (red lines) with the NLM and WD methods, respectively. Black dashed lines correspond to the respective linear fitting (in dB scale) of the SNR versus distance. Experimental results point out that the SNR of 1.4 dB obtained at 50 km distance on the raw data can be substantially boosted up to 15.2 and 15.6 dB by applying the NLM and WD methods, respectively. It is noteworthy that those values represent a remarkable SNR enhancement of 13.8 and 14.2 dB using each of the respective methods. Although the SNR improvement provided by these two denoising methods is fairly equivalent, we should mention that the computational complexity of the NLM is normally much larger, especially in our implementation (see Discussion section), giving a crucial advantage to the WD, unless dedicated programming and implementation are used for the NLM processing[63]. It is also important to mention that although the Brillouin frequency shift (BFS) of the sensing fibre is in this case longitudinally quite uniform, the reported SNR improvement is obtained by the NLM and WD methods using parameters that secure a spatial resolution of 2 m, as described in the Methods and demonstrated hereafter in Fig. 4.

The SNR enhancement demonstrated above has actually a massive impact on the quality of the obtained BGS. Figure 3c,d highlight the huge contrast enhancement in the BGS measurements and the considerable noise reduction provided by image processing. No relevant distortion can be observed in the BGS obtained after processing. An accurate distributed BFS profile is then obtained by fitting a quadratic curve to the BGS obtained at each fibre location[24]. The ultimate impact of image processing on

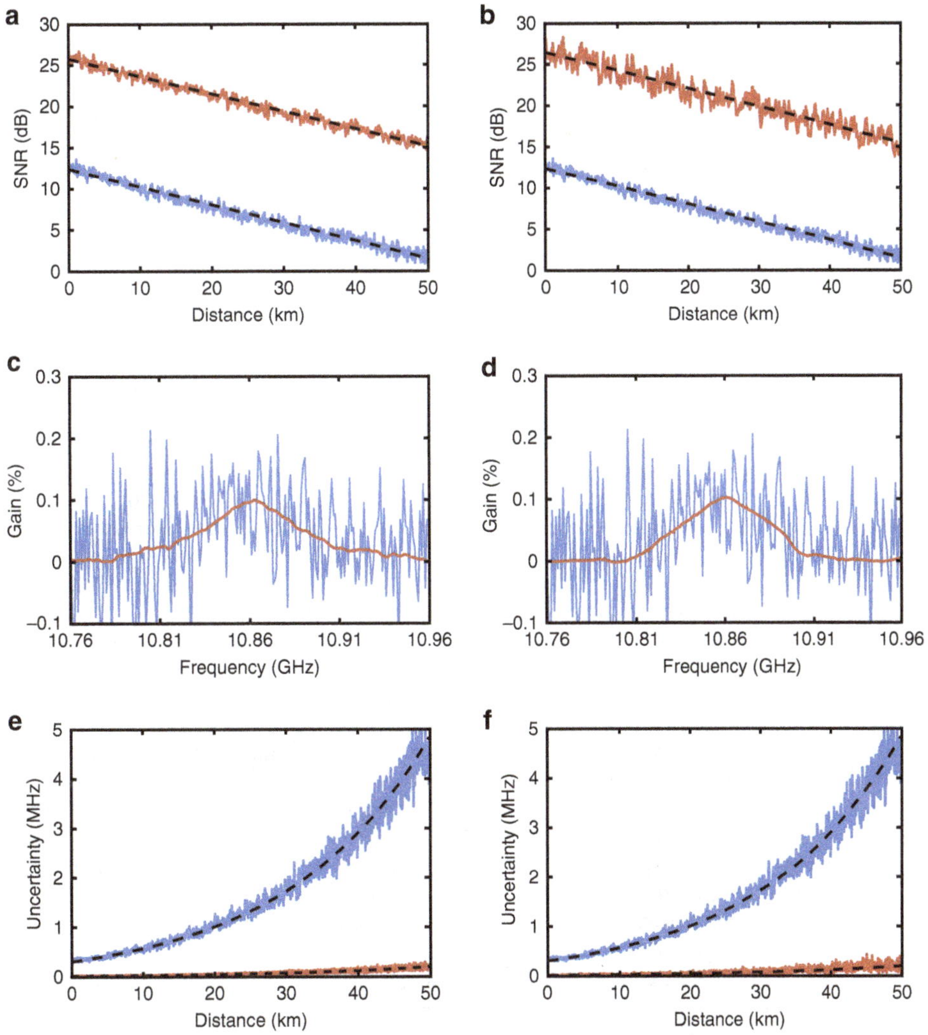

Figure 3 | Impact of 2D image processing on BOTDA measurements. The sensing performance enhancement provided by the 2D NLM (left) and 2D WD (right) methods is evaluated by comparing the raw (blue curves) and processed (red curves) data. (**a**,**b**) SNR versus distance, showing that the raw SNR of 1.4 dB (obtained at 50 km distance with 2 m spatial resolution and 4 averaged traces) can be improved up to 15.2 and 15.6 dB with the NLM and 2D WD methods, respectively. The black dashed lines show a linear fitting (in dB scale) of the SNR curves versus distance. (**c**,**d**) BGS measured near the far fibre end, pointing out that a significant increase in the measurement contrast can be obtained by image processing. (**e**,**f**) Uncertainty on the determination of the peak gain frequency versus distance, showing that the uncertainty of 4.8 MHz obtained in the raw data at 50 km can be significantly reduced down to 0.20 and 0.19 MHz with the NLM and 2D WD methods, respectively.

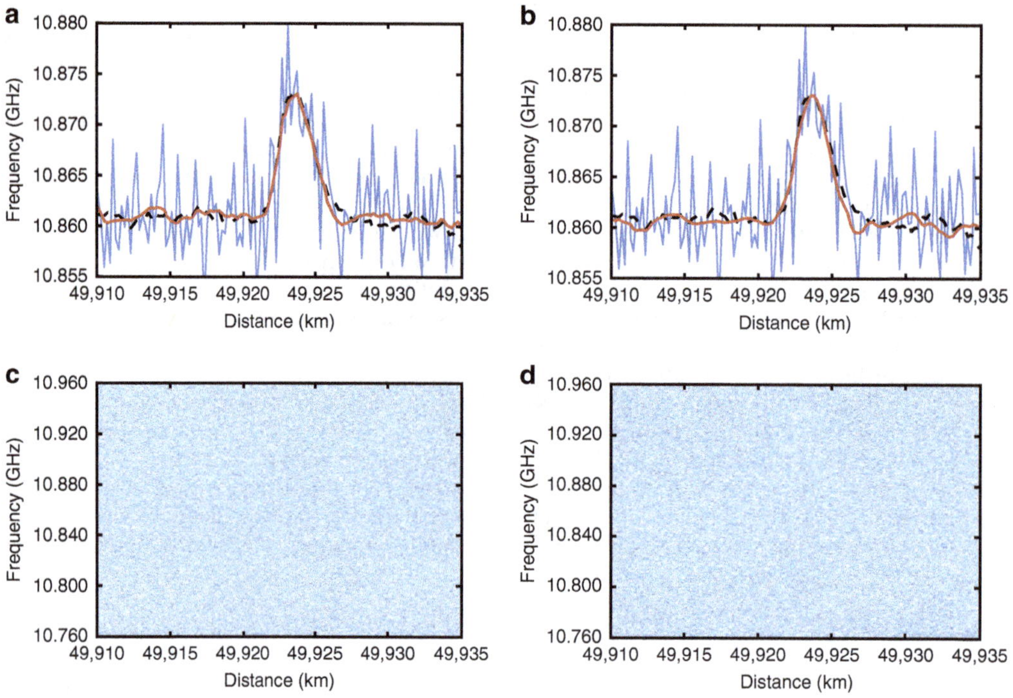

Figure 4 | Validation of the maintained spatial resolution. Impact of the 2D image processing technique on the spatial resolution of the implemented BOTDA sensor, for the 2D NLM (left) and 2D WD (right) methods. (**a**,**b**) BFS profiles around a 2-m-long hotspot located near the far end of the sensing fibre, for the NLM and WD methods, respectively. The black dashed lines show a reference BFS profile retrieved directly from measurements (that is, with no image processing) acquired with a much larger number (4,000) of averaged traces. Results demonstrate that the applied image processing methods have only an imperceptible impact on the spatial resolution of the sensor. (**c**,**d**) Two-dimensional representation of the noise component eliminated by 2D NLM and 2D WD methods. These figures are obtained subtracting the 'denoised images' from the 'original image' (2D data matrix with the raw BOTDA traces), in the case of a hotspot measurement. Results show only the fibre section where the hotspot is located, indicating that the eliminated noise components do not contain any evidence of visual patterns or relevant information that could affect the proper hotspot detection.

the measurement quality turns evident when calculating the standard deviation of the retrieved BFS. Figure 3e,f show that the frequency uncertainty of 4.8 MHz obtained with the raw data at 50 km distance can be remarkably improved down to 0.20 and 0.19 MHz using the NLM and WD methods, respectively. These values are actually in perfect agreement with the uncertainty expected from the experimental SNR[24] obtained after denoising.

To corroborate that the proposed technique does not actually lead to a penalizing loss of information (for example, loss of spatial resolution capabilities) and eliminates mostly the uncorrelated noise present in the measurements, it is essential to demonstrate that the applied 2D processing does not excessively blur the 'denoised images'. For this purpose, a 2-m section of fibre at 50 km distance is heated up to 40 °C, while the rest of the fibre is kept at room temperature (27 °C). Figure 4a,b show the BFS profiles around the hotspot location obtained from the raw and denoised data for the NLM and WD methods. To accurately assess the impact of image processing on the spatial resolution, the sampling interval in this figure is reduced down to 0.2 m, while a reference BFS profile of the hotspot (black dashed line in the figures) is obtained using 4,000 averages and no processing, thus providing a reliable reference profile with comparable SNR. Results highlight the correct detection of the 2-m-long hotspot, demonstrating that the applied processing has an imperceptible impact on the spatial resolution even under low SNR conditions. We should emphasize that this verification of the spatial resolution is obtained with the same denoising parameters used to process the raw data when estimating the SNR improvement reported in Fig. 3a,b, and therefore they fully represent the spatial resolution capabilities reached by the NLM and WD techniques when enhancing the SNR by $\sim$14 dB. This clearly demonstrates the benefits of the proposed method when, for instance, compared with classical low-pass filtering techniques, which are highly affected by the tradeoff between noise reduction and spatial resolution. In contrast to low-pass filtering, the here-proposed technique uses a nonlinear approach, in which no explicit bandwidth notion is relevant; the amount of removed noise depends exclusively on the frequency components of the useful signal and their level of redundancy. The redundancy of information increases the amplitude of the frequency components inherent to the useful signal, while low-amplitude components are assumed to be noise and removed by the processing. This nonlinear denoising approach enables us to measure events with a sharp spatial resolution (that is, maintaining high-frequency components), while reducing significantly the noise (that is, random and low-amplitude components present in the low and/or high frequency range). Thus, fundamentally there is no discernible bandwidth change between raw and processed data, as proved by the hotspot measurements in Fig. 4a,b. More quantitatively, if we consider the spatial resolution of 2 m and the sampling interval of 0.5 m, a digital low-pass filter of maximum four points could be used (equivalent to a four-point moving average window). This leads to a 3-dB SNR improvement, which is much lower than the 14-dB improvement here demonstrated. Considering that the electrical bandwidth in the system is 125 MHz, an electrical low-pass filter of 50 MHz bandwidth could still be used to secure a spatial resolution of 2 m, but this has also minor impact on the SNR. Figure 4c,d actually confirm that the eliminated 2D component (obtained subtracting the 'denoised image' from the original 'noisy image') is essentially white noise, showing no specific features or patterns that could indicate the elimination of relevant information.

Extending the concept to video processing. The principle of distributed fibre sensing assumes that the temporal evolution of the measurand changes slowly compared with the acquisition time. In the case of Brillouin and Rayleigh sensors, this leads to consecutive 2D measurements containing highly correlated information. Based on this feature, the concept proposed in this study can be extended to the 3D case[60–62], in which each measurement (in the position–frequency domain) is assimilated to a frame of a video sequence. This approach exploits not only the redundancy found in the 2D domain of the measurements but also in the temporal dimension, thus leading to a very powerful tool for a better data restoration and noise removal in distributed fibre sensing. Using the same setup reported in Fig. 1, we acquire consecutive 2D data matrices $\mathbf{M}(z,\Delta f)$ every $\sim$42 s. Although this measurement time can be still reduced due to the small averaging number (four averaged traces per scanned frequency), the response time of the equipment used in this lab demonstration eventually limited our minimum acquisition time.

Considering that the temporal evolution of the measurand might reduce the correlation existing between consecutive measurements, the effectiveness of video processing[60–62] is experimentally verified under conditions in which the fibre temperature changes during the measurement process. It is important to bear in mind that if the measurands were completely static (that is, showing no temporal variations), an excellent and trivial solution to increase the SNR would be the use of a simple (linear) temporal averaging of consecutive measurements. However, owing to the dynamical nature of the environment, such a kind of averaging leads to 'blurred images', resulting in loss of information and details, as depicted in Supplementary Fig. 2 (see description in next paragraphs). Video processing here takes into account the non-stationary characteristics of the data in the temporal domain[60–62].

Measurements are processed using a 3D NLM algorithm[60–62] (see Methods). As in the 2D case, the 3D spatio-temporal NLM method uses the self-similarity of an image, but the search for repeated patterns is extended to several consecutive frames[60–62]. Figure 5a shows the SNR of the trace measured at the peak gain frequency for the raw (blue curve) and denoised (red curve) data using the 3D NLM method[60–62] based on ten consecutive measurements (frames). The figure indicates that the raw SNR of 1.4 dB obtained at 50 km distance can be increased up to 22.1 dB using 3D NLM processing. This is >6.5 dB improvement with respect to the 2D image processing reported above and corresponds to a remarkable absolute SNR enhancement of 20.7 dB. The benefits provided by this SNR enhancement can be clearly observed in Fig. 5b, which shows the BGS measured at 50 km distance obtained from the raw and denoised data. Figure 5c highlights that the BFS uncertainty of 4.85 MHz, obtained from the raw data, can be significantly reduced down to 0.055 MHz with this 3D NLM processing, being in good agreement with the attained SNR enhancement[24].

Supplementary Movie 1 shows the BFS profile along the last metres of fibre measured when the temperature of a 2-m section changes from 10 °C up to 40 °C (the rest of the fibre is at 27 °C). As also shown in Fig. 5d and Supplementary Fig. 2, results demonstrate that the retrieved BFS profiles present no observable distortion of the hotspot, while negligible delay is observed when compared with the temperature evolution retrieved from the raw measurements. This represents a key advantage of video processing when compared, for example, with linear temporal averaging, which generally leads to delays in the temporal evolution of the measurand, as demonstrated in Supplementary Fig. 2.

Denoising 1D data using 2D image processing. In sensors measuring only 1D data, such as Raman-distributed temperature sensors, a 2D image can be formed considering time as a second dimension when stacking successive sequential measurements. To validate the proposed method in this case, we use a generic Raman-distributed sensor scheme[12–14], as depicted in Fig. 6. Using a 9-km-long sensing fibre, a spatial resolution of 2 m and a sampling interval of 0.5 m, we measure 1D time-domain traces of the spontaneous Raman anti-Stokes and Stokes backscattered components using 4,096 averages, within a measurement time of $\sim$25 s. Traces are consecutively measured, while the temperature of $\sim$10 m of fibre (at the end of the sensing range) is changed from room temperature ($\sim$25 °C) up to 40 °C.

As suggested, we form two 2D images (one for the Stokes and another for the anti-Stokes signal) stacking together consecutive time-domain traces. Figure 7a,b provide 2D image representations of the evolution of consecutive Raman anti-Stokes and Stokes traces within the last 200 m of fibre. The temperature evolution affecting the anti-Stokes trace amplitude around 8.86 km is evident in Fig. 7a. Although the entire set of data contained in those images could be used for denoising, a sliding temporal window of 21 consecutive measurements is chosen here. Thanks to this restricted temporal window, the processing time is highly reduced without significantly compromising the denoising capabilities of the method[63]. Thus, to remove noise from a 'current' measurement, matrices $\mathbf{M}_{\mathbf{aS}}(z,t)$ and $\mathbf{M}_{\mathbf{S}}(z,t)$ of $18{,}000 \times 21$ points (including the raw measurements of the 'current' and previous 20 anti-Stokes and Stokes traces) are processed by the 2D NLM method. This procedure is continuously repeated for each new measured trace and independently for the Stokes and anti-Stokes signals. After applying image denoising to the raw measurements, the ratio anti-Stokes to Stokes is calculated at each fibre location, and by following a standard calibration procedure the actual distributed temperature profile is finally obtained[12,13].

Figure 8a shows the temporal evolution of the measured temperature within the hotspot section, whereas Fig. 8b shows the measured temperature profile around the hotspot after the fibre temperature reaches a stable temperature of 40 °C. Results demonstrate that no loss of spatial resolution and no distortion or delay in the temperature evolution are induced by the processing. Figure 8c shows that the SNR of 26.2 dB obtained at 8.86 km distance with the raw data can be significantly improved up to 39.8 dB after 2D NLM denoising. This corresponds to an SNR enhancement of 13.6 dB, which leads to a temperature resolution improvement from 0.5 °C (obtained from raw data at the fibre end) down to 0.022 °C after NLM processing, as shown in Fig. 8d.

Discussion

In this study, we have proposed and demonstrated the use of image/video denoising techniques[58–67] as an efficient approach to enhance the SNR of distributed fibre sensors. To the best of our knowledge, this is the first time that the high levels of correlation and redundancy contained in the multidimensional domain of the measurements obtained by distributed fibre sensors are exploited for performance improvement. Compared with state-of-the-art methods, the multidimensional processing approach here proposed turns much more efficient than applying known (1D) denoising algorithms[44–51] simply replicated in the different dimensions of interest. For instance, the independent use of 1D processing to denoise time-domain traces and then applied to the processed data in frequency domain leads to denoised data points that do not benefit from the similitude and redundancy that can only be found in a 2D or 3D

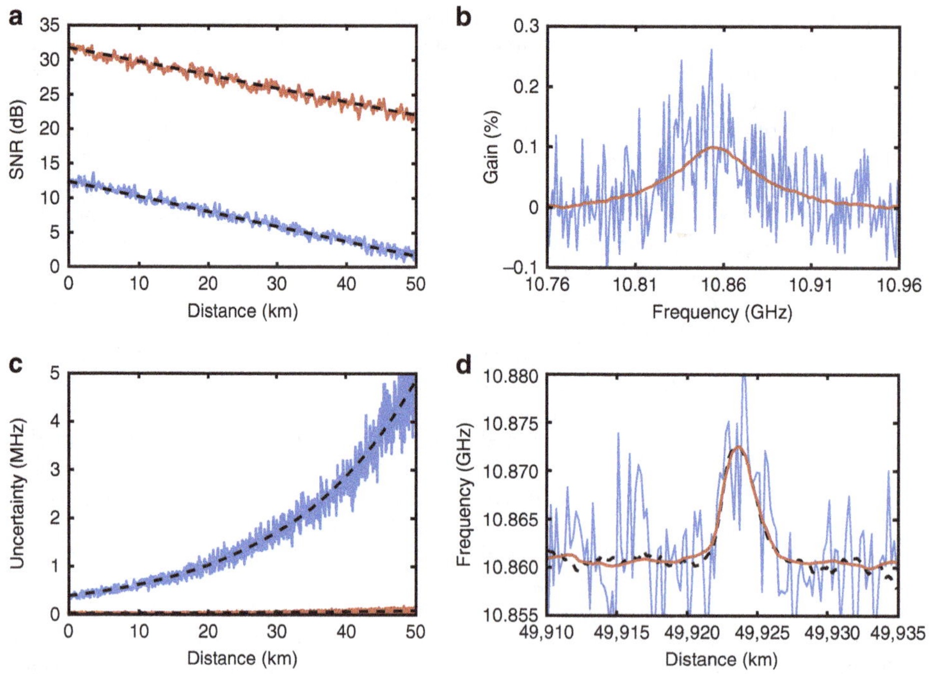

Figure 5 | Impact of video processing on BOTDA measurements. The sensing performance enhancement provided by the 3D NLM methods is evaluated by comparing the raw (blue curves) and processed (red curves) data for different parameters. (**a**) SNR versus distance, showing that the raw SNR of 1.4 dB can be improved up to 22.1 dB when 10 consecutive measurements are used for denoising. (**b**) BGS measured near 50 km distance. (**c**) Uncertainty on the BGS peak frequency versus distance, showing that the uncertainty of 4.85 MHz obtained in the raw data at 50 km distance can be significantly reduced down to 0.055 MHz. (**d**) BFS profile around a 2-m-long hotspot located near the far end of the sensing fibre. Negligible impact on the spatial resolution of the sensor can be observed. The black dashed line in **d** shows a reference BFS profile retrieved directly from a measurement obtained with 4,000 averaged traces.

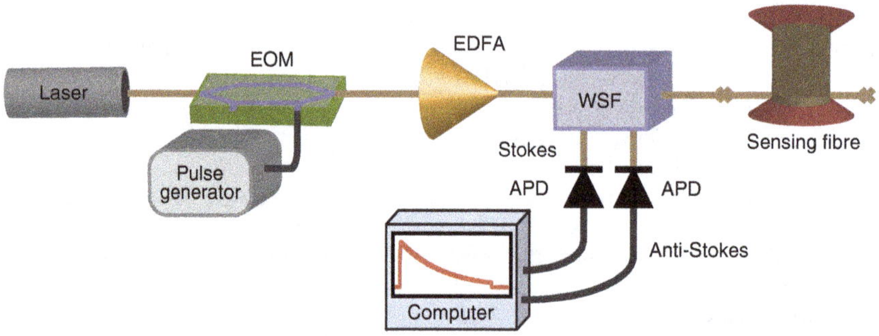

Figure 6 | Experimental setup of a basic Raman-distributed sensor. The system operates using a conventional distributed-feedback laser operating at 1,552 nm, followed by an electro-optic modulator (EOM) to generate pulses of 10 ns and an erbium-doped fibre amplifier (EDFA) to boost the pulse power. Optical pulses of about 4W peak power are launched into the sensing fibre through a wavelength-selective filter (WSF), which also separates the spontaneous Stokes and anti-Stokes Raman components backscattered from the sensing fibre into two branches. These two spectral components are sent into two parallel 50 MHz avalanche photodetectors (APD), followed by an acquisition system connected to a computer. The sensing fibre corresponds to a 50/125-µm graded-index multimode fibre 9 km long. As a result of intermodal dispersion, the spatial resolution at the end of the sensing fibre is ~2 m.

data structure containing the entire measured information. For this reason, the method here proposed offers exceptional denoising capabilities when compared with state-of-the-art techniques[33–54], enabling a remarkable and unprecedented SNR enhancement, boosting the sensor performance[24] up to about two orders of magnitude with no loss of relevant information at minor added cost. This translates, for instance, into an unmatched 100-fold improvement in the measurand accuracy of conventional distributed sensors. Furthermore, this multidimensional processing can be freely implemented on top of existing sophisticated methods for performance improvement[33–43] such as optical pulse coding and/or distributed Raman amplification, thus resulting in a fully additional SNR enhancement.

Although the Brillouin system used in this study as a proof-of-concept is thermal-noise dominated, it is worth mentioning that the efficiency of the proposed technique would be similar if shot noise, spontaneous-signal beat noise (when using an erbium-doped fibre amplifier in the receiver) or relative intensity noise (RIN) had an impact on the measurements[57]. In fact, the largest contribution to noise in those cases is given basically by the continuous-wave probe power reaching the receiver[25,57], whereas the response of the BOTDA trace only corresponds to a very small fraction of this continuous-wave level (typically 1% for 1 m spatial resolution), having a negligible impact on the signal noise. Such sources of noise can therefore as well be considered additive and uniform along the fibre[57]. When RIN dominates, traces can

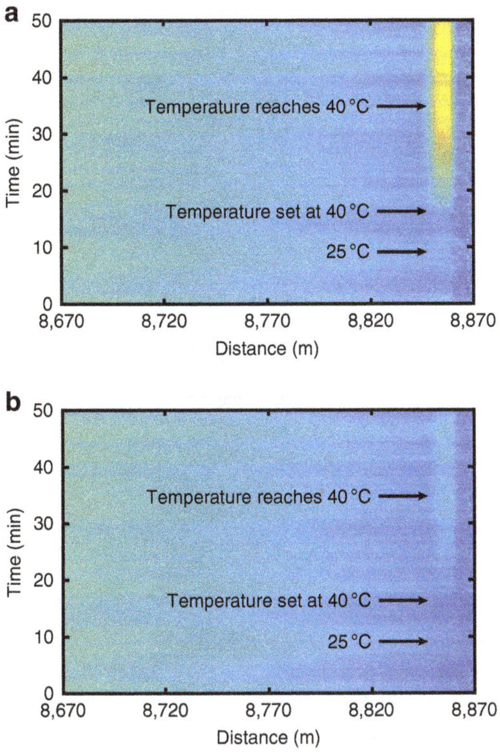

Figure 7 | Two-dimensional images formed from stacked 1D Raman time-domain traces. The images show excerpts from the time-domain traces of the spontaneous Raman (**a**) anti-Stokes and (**b**) Stokes components, measured using a 9-km-long sensing fibre. During the measurement process, the temperature of 10 m of fibre has been increased from room temperature (25 °C) up to 40 °C. For the sake of clarity, only the last 200 m are shown in the figure.

be affected by noisy patterns that could partially impact on the gained SNR of the non-local means and WD methods used in this study; however, we should mention that there exists a vast variety of image processing algorithms and some of them could be more efficient for removing RIN. Further research on this subject is still necessary and goes beyond the scope of this study. If Brillouin time-domain traces are affected by stationary patterns, for instance, when polarization fading are imperfectly compensated, then image processing will consider those patterns as real signal, leading to a non-uniform noise removal along the fibre. However, as fading are proportional to the Brillouin gain, they would eventually affect the noise removal along the first kilometres of fibre, where the SNR is still high, and therefore under normal operation conditions there should be no real detrimental effects at long distances (low SNR region dominated by white noise) where the benefit from image denoising is more crucial. Furthermore, if slow-varying temporal instabilities affect the sensor operation, for instance, due to thermal fluctuations, the effectiveness of the denoising might also be reduced when the temporal dimension is considered in the processing. This is because slow-varying fluctuations and instabilities could be interpreted by image/video processing as a correlated signal containing relevant (that is, non-random) information, thus imposing an ultimate limit to the estimated SNR improvement.

Although the concept proposed in this study has only been demonstrated for BOTDA and Raman sensors, it is envisaged that image/video processing can also be applied to improve SNR of other sensors, such as Brillouin optical-time domain reflectometers[21], as well as schemes based on frequency domain[18] and correlation domain[19,20]. Furthermore, the concept can also be extended to Rayleigh-based sensors[7–11], such as phase-sensitive optical time-domain or frequency-domain reflectometers for quasi-static strain and temperature sensing, where image/video processing can be applied to the 2D data matrix containing the calculated correlation spectrum versus distance[9–11]. Although correlation noise could dominate the correlation spectrum in that case, some dedicated image/video processing algorithm could still provide an important enhancement in the contrast of the main correlation peak. This certainly requires a further study and goes beyond the main scope of this study. In addition, the approach followed for denoising 1D traces can also be applied to standard reflectometry measurements[5,6], including even more sophisticated methods for fibre characterization.

An interesting aspect that should be highlighted is that the 3D approach here proposed has proved to be very efficient when dealing with the dynamical character of the measurements. This is because video sequences are also inherently non-stationary[60–62] and therefore conventional video processing can easily deal with the motion of pixels among different frames. This, if properly tackled and implemented for instance in a dedicated graphic processing unit, could have interesting applications in dynamic distributed sensing[22]. Nevertheless, we should clarify that video denoising typically follow two different approaches: (i) the use of methods with no motion compensation[60–62] and (ii) the use of methods with explicit motion compensation[68–70]. Whereas the 3D NLM[60–62] used here as a proof-of-concept belongs to the first group, there exist also several movie denoising techniques belonging to the second category, such as the 3D WD[68–70]. Further investigation can be still carried out to verify the effectiveness of methods based on explicit motion compensation[68–70]. However, we should mention that the concept of calculating the motion of pixels, that is, the trajectory of pixels through consecutive frames, is already inherently incorporated in the 3D NLM method[60,61] and therefore already demonstrated in this study.

It is also worth mentioning that the experimental results shown in this study have been obtained by simple image/video denoising implementations, where parameters have been selected by following standard recommendations and fine empirical adjustments to maximize the noise removal avoiding any loss of relevant information. However, more dedicated strategies for adjusting parameters could be still developed and further investigated. For instance, the possibility for auto-tuning parameters and adaptive window sizes to avoid oversmoothing fast spatial and temporal changes of the measurand could also be investigated.

The processing time of the method could also be further investigated and improved. No special hardware has been here used to optimize the processing time. Considering the large number of acquired points in the demonstration sensors, the implementation of a standard NLM algorithm turns out to be computationally very demanding. Using a conventional computer with a 3.5-GHz processor and 8 GB RAM, the 2D NLM processing time of each matrix of $18,000 \times 21$ points in the implemented Raman sensor turns out to be about 1 s. On the other hand, denoising the 2D data matrix of $100,000 \times 200$ points in the BOTDA sensor takes about 30 s: this is about one order of magnitude larger than the time required by the WD, which requires <1 s for processing the same data. Furthermore, the 3D NLM denoising of the same matrix size, but considering ten consecutive frames, is about 4 min. This time can however be highly reduced following specific strategies[62,63]: for example, using parallel computation to distribute the operations on several processors. In fact, the NLM method is intrinsically well suited for parallelization and multithreading[63], making this kind of

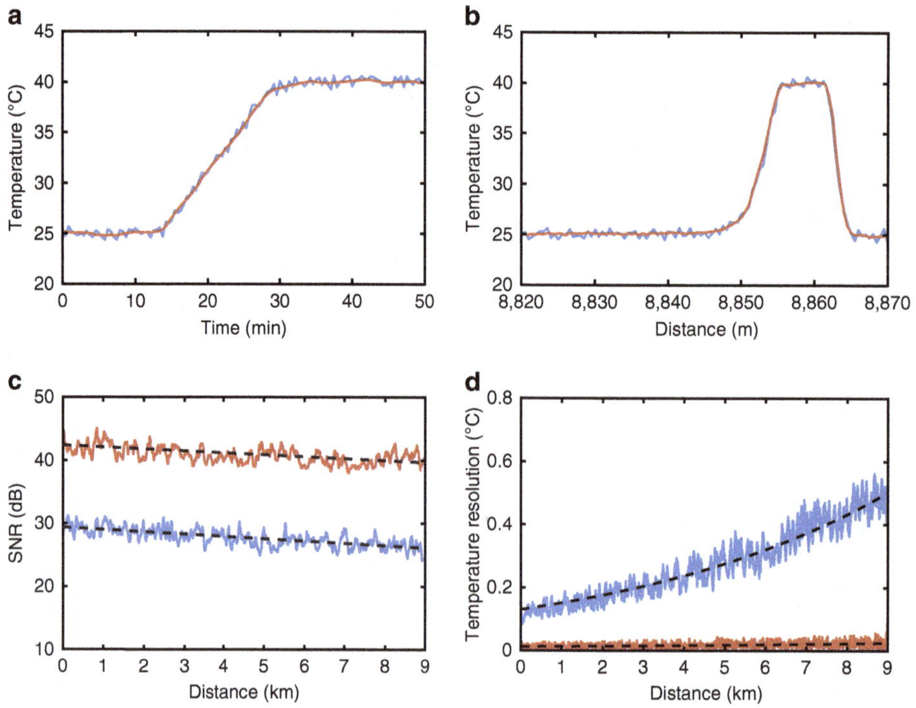

Figure 8 | Impact of 2D image processing on 1D Raman traces. The performance of the sensor and the impact of 2D image denoising is evaluated while the temperature of 10 m at the end of the sensing fibre is slowly increased in time. Results obtained from the raw data (blue) are compared with the ones obtained from the denoised data using the 2D NLM method (red curve), in which a moving window of 21 consecutive traces is considered. **(a)** Temporal evolution of the measured temperature at the hotspot location; this result shows that image processing induces no delay or distortion in the measured hotspot temperature. **(b)** Distributed temperature profile near the hotspot location, demonstrating that the denoising process produces no perceptible loss of spatial resolution. **(c)** SNR versus distance, validating an SNR enhancement of 13.6 dB at the end of the sensing range. **(d)** Temperature resolution versus distance, showing that the use of 2D NLM method can improve the temperature uncertainty from 0.5 °C down to 0.022 °C.

approach very efficient and straightforward. Much further improvement can also be obtained if dedicated algorithms and programming are used to implement image/video denoising in a dedicated graphic processing unit installed in a sensor system. This shows that we are just at the early stages of research, only unveiling a part of the promising potential of these techniques.

Although in this study we demonstrate only the effectiveness of image denoising methods, other techniques for image enhancement[58] (different from denoising) can also be employed to increase the measurement quality of distributed fibre sensors in some particular conditions or to spot particular events in a massive data flow. This can be obtained using dedicated algorithms[58], for instance, to sharpen image details, increase the dynamic range of particular features, restore blurring effects, enhance contrast and edges, and several other approaches. Many of those methods actually offer the possibility to recognize objects or patterns, which could be very helpful for the detection of special features that could be present in the measurand. Blur-removing strategies could also be investigated to sharpen the detection of small events, comparable to or potentially shorter than the spatial resolution of the sensor.

Methods

Non-local means. This technique[59–63] proposes an original paradigm in noise reduction by taking advantage of the high degree of redundancy contained in the 2D or 3D data measured by a distributed fibre sensor. The method is based on the use of sliding neighbourhoods or 2D patches, which correspond to sets of all pixels $j=(x,y)$ that surround a certain pixel at $i=(x',y')$ within a window of a predefined size. The similarity between two pixels i and j is performed by comparing, not only the values $f(i)=f(x',y')$ and $f(j)=f(x,y)$ assigned to the pixels but the entire 2D patches or neighbourhoods surrounding the pixels of interest[59,60]. For this, a so-called similarity neighbourhood $\boldsymbol{\eta}_i$ surrounding the pixel i is defined and then compared with all other similarity neighbourhoods $\boldsymbol{\eta}_j$ of the same size existing in the entire 2D matrix containing the data provided by the sensor.

The degree of similitude and redundancy in the data is evaluated in a patch-by-patch basis by calculating the Euclidean distance[59,60] $||f(\boldsymbol{\eta}_i)\text{-}f(\boldsymbol{\eta}_j)||$ between all values $f(\boldsymbol{\eta}_i)$ and $f(\boldsymbol{\eta}_j)$ within neighbourhoods $\boldsymbol{\eta}_i$ and $\boldsymbol{\eta}_j$. It turns out that the Euclidean distance is small when a high level of similitude exist between both compared windows $\boldsymbol{\eta}_i$ and $\boldsymbol{\eta}_j$, and therefore the highly similar values $f(i)$ and $f(j)$ associated to both pixels i and j can be averaged to reduce noise[59,60]. More specifically, to eliminate noise from a pixel i in the 2D data matrix, the value $f(i)$ associated to such a pixel is processed by the NLM method calculating the following weighted average[59,60]:

$$\mathrm{NLM}\{f(i)\} = \sum_{\forall j \in I} w(i,j)\cdot f(j), \tag{1}$$

where I is the entire domain of the image, $f(j)$ corresponds to the value of the image associated to the pixel j and $w(i,j)$ are the weighting factors calculated as[59,60]

$$w(i,j) = \frac{1}{Z(i)} e^{\frac{-||f(\boldsymbol{\eta}_i)-f(\boldsymbol{\eta}_j)||^2}{h^2}}, \tag{2}$$

where h is a smoothing control parameter and $Z(i)$ is a normalization factor defined so that the sum of all values of $w(i,j)$, for a given pixel i, results to be equal to one. It should be noted that the weighting factors $w(i,j)$ are independent of the geometry and only depend on the similarity of the data around pixels i and j. This feature characterizes the method as non-local, as pixels j whose surroundings are similar to the pixel i are associated to a higher weight $w(i,j)$, regardless of the relative (spatial) distance between the two pixels.

A strict condition for the processing is that the similarity window size must be comparable to the smallest details in the image[59,60], which in the context of distributed fibre sensing is associated to the spatial resolution capabilities of the sensor. Considering that all implemented and analysed systems in this study (Brillouin and Raman sensors) have a spatial resolution of 2 m and a sampling interval of 0.5 m, the similarity window has been chosen of size 3 × 3. This size, corresponding to three longitudinal data points (that is, 1.5 m long), is smaller than the spatial resolution of the system, ensuring that the processing does not have any detrimental impact on the real spatial resolution of the sensor.

On the other hand, the parameter h controls the level of blurring of the method and its optimum value depends on the noise level of the data. Following the recommendation given in ref. 59, h has been set to ten times the noise standard deviation σ. In the Brillouin gain data shown in Fig. 2, the noise standard deviation is $\sigma = 7.2 \times 10^{-4}$; thus h has been set to 7.2×10^{-3}. In the case of the Raman sensor implemented in this work, the noise standard deviation of data in Fig. 7 is evaluated to be $\sigma = 6.9 \times 10^{-4}$ and therefore h has been set to 6.9×10^{-3}.

Furthermore, we should consider the large amount of data obtained by the implemented sensors. For instance, in the case of the implemented BOTDA sensor 20 M points (that is, 20 Mpix) are acquired to cover 50 km of sensing fibre, sampled every 0.5 m and scanning 200 frequencies, so that the calculation of equations (1) and (2) turns out to be extremely demanding. To reduce the processing time, the strategy proposed in ref. 59 is applied, in which a spatially constrained search window is defined. This window is larger than the similarity window, but smaller than the entire 2D matrix (image) so that the search for repeated patterns is only restricted to the area covered by this search window. In practice, using a similarity window of 3 × 3 and $h = 7.2 \times 10^{-3}$, it is empirically observed that a satisfactory noise removal is obtained at a reasonable computational cost when defining a search window of 13 × 13. Increasing this search area has only led to a marginal SNR improvement, at the cost of a substantial increase in the processing time.

As in the 2D case, the 3D spatio-temporal NLM method[60–62] uses the self-similarity of an image to reduce the noise of a pixel (x', y') by averaging weighed image sections at coordinates (x,y) that have high level of similarity; however, in this case the search for repeated patterns is also extended to several consecutive frames[60–62]. As the spatial features of the acquired BOTDA data are maintained, the same parameters as in the 2D case are used for video denoising, but with the difference that the search for repeated patterns is extended to ten consecutive measurements.

Finally, in the case of processing 1D data provided by the implemented Raman distributed sensor, two 2D images (one for the Stokes and another for the anti-Stokes component) are formed by stacking together consecutive time-domain traces. A sliding search window of size 23 × 23 is used for 2D NLM denoising.

Wavelet denoising. The method is based on the 2D discrete wavelet transform (DWT) and a wavelet shrinkage strategy[64–66], and has been applied only to BOTDA measurements as a proof-of-concept. To eliminate noise using a 2D WD approach, data provided by the BOTDA sensor are decomposed using the Mallat algorithm[67] into versions of sub-images containing different levels of details. After testing many mother wavelets, the wavelet sym7 has been chosen because of the better denoising capabilities obtained in this case, while the number of levels of decomposition has been set to 5. By comparing the wavelet coefficients obtained in the 2D DWT with a predefined threshold, wavelet shrinkage is then applied to the wavelet coefficients using a hard thresholding strategy[66]. This means that all wavelet coefficients having an amplitude below a given threshold are associated to noise and set to zero, whereas high-amplitude coefficients are associated to useful information provided by the sensor. Following the recommendation given in ref. 59, in this case the threshold level has been set to three times the noise standard deviation from which a small adjustment has been performed to optimize the amount of removed noise. Thus, considering that the noise standard deviation is $\sigma = 7.2 \times 10^{-4}$ for the BOTDA measurements, the threshold value is set to 2.6×10^{-3} ($\approx 3.6\sigma$). The output 2D data matrix (considered as the 'output image') is reconstructed from the result of this thresholding stage, using an inverse 2D DWT procedure, which converts the data back to the spatial domain of the image.

References

1. Rogers, A. J. Distributed optical-fibre sensing. *Meas. Sci. Technol.* **10,** R75–R99 (1999).
2. Hartog, A. H. *Optical Fiber Sensor Technology* Vol. 1 347–382 (Springer, 1995).
3. Agrawal, G. P. *Nonlinear Fiber Optics* 4th edn (Academic Press, 2007).
4. Boyd, R. W. *Nonlinear Optical* 2nd edn (Academic Press, 2003).
5. Aoyama, K.-I., Nakagawa, K. & Itoh, T. Optical time domain reflectometry in a single-mode fiber. *IEEE J. Quantum Electron.* **QE-17,** 862–868 (1981).
6. Eickhoff, W. & Ulrich, R. Optical frequency domain reflectometry in single-mode fiber. *Appl. Phys. Lett.* **39,** 693–695 (1981).
7. Rathod, R., Pechstedt, R. D., Jackson, D. A. & Webb, D. J. Distributed temperature-change sensor based on Rayleigh backscattering in an optical fiber. *Opt. Lett.* **19,** 593–595 (1994).
8. Shatalin, S. V., Treschikov, V. N. & Rogers, A. J. Interferometric optical time-domain reflectometry for distributed optical-fiber sensing. *Appl. Opt.* **37,** 5600 5604 (1998).
9. Froggatt, M. & Moore, J. High-spatial-resolution distributed strain measurement in optical fiber with Rayleigh scatter. *Appl. Opt.* **37,** 1735–1740 (1998).
10. Juarez, J. C., Maier, E. W., Choi, K. N. & Taylor, H. F. Distributed fiber-optic intrusion sensor system. *J. Lightwave Technol.* **23,** 2081–2087 (2005).
11. Koyamada, Y., Imahama, M., Kubota, K. & Hogari, K. Fiber-optic distributed strain and temperature sensing with very high measurand resolution over long range using coherent OTDR. *J. Lightwave Technol.* **27,** 1142–1146 (2009).
12. Dakin, J. P., Pratt, D. J., Bibby, G. W. & Ross, J. N. Distributed antistokes ratio thermometry. In Optical Fiber Sensors, (Optical Society of America, 1985), paper PDS3; doi:10.1364/OFS.1985.PDS3. *Electron. Lett.* **21,** 569–570 (1985).
13. Dakin, J. P., Pratt, D. J., Bibby, G. W. & Ross, J. N. Distributed optical fibre Raman temperature sensor using a semiconductor light source and detector. *Electron. Lett.* **21,** 569–570 (1985).
14. Hartog, A. H. & Leach, A. P. Distributed temperature sensing in solid-core fibres. *Electron. Lett.* **21,** 1061–1062 (1985).
15. Horiguchi, T. & Tateda, M. BOTDA-nondestructive measurement of single-mode optical fiber attenuation characteristics using Brillouin interaction: theory. *J. Lightwave Technol.* **7,** 1170–1176 (1989).
16. Horiguchi, T., Shimizu, K., Kurashima, T., Tateda, M. & Koyamada, Y. Development of a distributed sensing technique using Brillouin scattering. *J. Lightwave Technol.* **13,** 1296–1302 (1995).
17. Niklès, M., Thévenaz, L. & Robert, P. A Simple distributed fiber sensor based on Brillouin gain spectrum analysis. *Opt. Lett.* **21,** 758–760 (1996).
18. Garus, D., Gogolla, T., Krebber, K. & Schliep, F. Distributed sensing technique based on Brillouin optical-fiber frequency-domain analysis. *Opt. Lett.* **21,** 1402–1404 (1996).
19. Hotate, K. Measurement of Brillouin gain spectrum distribution along an optical fiber using a correlation-based technique-proposal, experiment and simulation. *IEICE Transact. Electron.* **E83-C,** 405–411 (2000).
20. Zadok, A. *et al.* Random-access distributed fiber sensing. *Laser Photon Rev.* **6,** L1–L5 (2012).
21. Shimizu, K., Horiguchi, T., Koyamada, Y. & Kurashima, T. Coherent self-heterodyne Brillouin OTDR for measurement of Brillouin frequency shift distribution in optical fibers. *J. Lightwave Technol.* **12,** 730–736 (1994).
22. Peled, Y., Motil, A. & Tur, M. Fast Brillouin optical time domain analysis for dynamic sensing. *Opt. Express* **20,** 8584–8591 (2012).
23. Dominguez-Lopez, A. *et al.* Reaching the ultimate performance limit given by non-local effects in BOTDA sensors. Proc. SPIE 9634, 24th International Conference on Optical Fibre Sensors, 96342E (September 28, 2015); doi:10.1117/12.2205440).
24. Soto, M. A. & Thévenaz, L. Towards 1 000 000 resolved points in a distributed optical fibre sensor. Proc. SPIE 9157, 23rd International Conference on Optical Fibre Sensors, 9157C3 (June 2, 2014); doi:10.1117/12.2072358.
25. Soto, M. A. & Thévenaz, L. Towards 1 000 000 resolved points in a distributed optical fibre sensor. Proc. SPIE 9157, 23rd International Conference on Optical Fibre Sensors, 9157C3 (June 2, 2014); doi:10.1117/12.2072358.
26. Thévenaz, L. Brillouin distributed time-domain sensing in optical fibers: state of the art and perspectives. *Front. Optoelectron. China* **3,** 13–21 (2010).
27. Bao, X. & Chen, L. Recent progress in distributed fiber optic sensors. *Sensors* **12,** 8601–8639 (2012).
28. Wait, P. C., De Souza, K. & Newson, T. P. A theoretical comparison of spontaneous Raman and Brillouin based fibre optic distributed temperature sensors. *Optics Commun.* **144,** 17–23 (1997).
29. Foaleng, S. M. & Thévenaz, L. Impact of Raman scattering and modulation instability on the performances of Brillouin sensors. Proc. SPIE 7753, 21st International Conference on Optical Fiber Sensors, 77539V (May 17, 2011); doi:10.1117/12.885105.
30. Thévenaz, L., Foaleng, S. M. & Lin, J. Effect of pulse depletion in a Brillouin optical time-domain analysis system. *Opt. Express* **21,** 14017–14035 (2013).
31. Alem, M., Soto, M. A. & Thévenaz, L. Analytical model and experimental verification of the critical power for modulation instability in optical fibers. *Opt. Express* **23,** 29514–29532 (2015).
32. Martins, H. F. *et al.* Modulation instability-induced fading in phase-sensitive optical time-domain reflectometry. *Opt. Lett.* **38,** 872–874 (2013).
33. Angulo-Vinuesa, X. *et al.* Limits of BOTDA range extension techniques. *Sensors J. IEEE.* doi: 10.1109/JSEN.2015.2424293 2015).
34. Rodriguez-Barrios, F. *et al.* Distributed Brillouin fiber sensor assisted by first-order Raman amplification. *J. Lightwave Technol.* **28,** 2162–2172 (2010).
35. Martin-Lopez, S. *et al.* Brillouin optical time-domain analysis assisted by second-order Raman amplification. *Opt. Express* **18,** 18769–18778 (2010).
36. Soto, M. A., Bolognini, G. & Di Pasquale, F. Optimization of long-range BOTDA sensors with high resolution using first-order bi-directional Raman amplification. *Opt. Express* **19,** 4444–4457 (2011).
37. Martins, H. F. *et al.* Phase-sensitive optical time domain reflectometer assisted by first-order Raman amplification for distributed vibration sensing over > 100 km. *J. Lightwave Technol.* **32,** 1510–1518 (2014).
38. Bolognini, G., Park, J., Soto, M. A., Park, N. & Di Pasquale, F. Analysis of distributed temperature sensing based on Raman scattering using OTDR coding and discrete Raman amplification. *Meas. Sci. Technol.* **18,** 3211–3218 (2007).
39. Soto, M. A., Bolognini, G. & Di Pasquale, F. Analysis of optical pulse coding in spontaneous Brillouin-based distributed temperature sensors. *Opt. Express* **16,** 19097–19111 (2008).
40. Soto, M. A., Bolognini, G., Di Pasquale, F. & Thévenaz, L. Simplex-coded BOTDA fiber sensor with 1 m spatial resolution over a 50 km range. *Opt. Lett.* **35,** 259–261 (2010).

41. Liang, H., Li, W., Linze, N., Chen, L. & Bao, X. High-resolution DPP-BOTDA over 50 km LEAF using return-to-zero coded pulses. *Opt. Lett.* **35,** 1503–1505 (2010).
42. Le Floch, S., Sauser, F., Soto, M. A. & Thévenaz, L. Time/frequency coding for Brillouin distributed sensors. Proc. SPIE 8421, OFS-2012 22nd International Conference on Optical Fiber Sensors, 84211J (October 4, 2012); doi:10.1117/12.975001.
43. Le Floch, S., Sauser, F., Llera, M., Soto, M. A. & Thévenaz, L. Colour simplex coding for Brillouin distributed sensors. Proc. SPIE 8794, Fifth European Workshop on Optical Fibre Sensors, 879437 (May 20, 2013); doi:10.1117/12.2025795.
44. Saxena, M. K. *et al.* Raman optical fiber distributed temperature sensor using wavelet transform based simplified signal processing of Raman backscattered signals. *Optics Laser Technol.* **65,** 14–24 (2015).
45. Farahani, M., Wylie, M., Castillo-Guerra, E. & Colpitts, B. Reduction in the number of averages required in BOTDA sensors using wavelet denoising techniques. *J. Lightwave Technol.* **30,** 1134–1142 (2012).
46. Qin, Z., Chen, L. & Bao, X. Continuous wavelet transform for non-stationary vibration detection with phase-OTDR. *Opt. Express* **20,** 20459–20465 (2012).
47. Zhang, Z.-H., Hu, W.-L., Yan, J.-S. & Zhang, P. The research of optical fiber Brillouin spectrum denoising based on wavelet transform and neural network. Proc. SPIE 8914, International Symposium on Photoelectronic Detection and Imaging 2013: Fiber Optic Sensors and Optical Coherence Tomography, 891408 (August 29, 2013); doi:10.1117/12.2032008.
48. Xu, H. -Z. & Zhang, D. Wavelet-Based Data Processing for Distributed Fiber Optic Sensors. in Machine Learning and Cybernetics, 2006 International Conference on, pp. 4040-4045, 13-16 August 2006; doi: 10.1109/ICMLC.2006.258858.
49. Saxena, M. K. *et al.* Optical fiber distributed temperature sensor using short term Fourier transform based simplified signal processing of Raman signals. *Measurement* **47,** 345–355 (2014).
50. Farahani, M. A., Castillo-Guerra, E. & Colpitts, B. G. Acceleration of measurements in BOTDA sensors using adaptive linear prediction. *IEEE Sensors J.* **13,** 263–272 (2013).
51. Muanenda, Y., Taki, M. & Di Pasquale, F. Long-range accelerated BOTDA sensor using adaptive linear prediction and cyclic coding. *Opt. Lett.* **39,** 5411–5414 (2014).
52. Soto, M. A., Bolognini, G. & Di Pasquale, F. Simplex-coded BOTDA sensor over 120 km SMF with 1 m Spatial resolution assisted by optimized bidirectional Raman amplification. *IEEE Photon. Technol. Lett.* **24,** 1823–1826 (2012).
53. Soto, M. A *et al.* Extending the real remoteness of long-range Brillouin optical time-domain fiber analyzers. *J. Lightwave Technol.* **32,** 152–162 (2014).
54. Jia, X.-H. *et al.* Experimental demonstration on 2.5-m spatial resolution and 1°C temperature uncertainty over long-distance BOTDA with combined Raman amplification and optical pulse coding. *IEEE Photon. Technol. Lett.* **23,** 435–437 (2011).
55. Thévenaz, L., Chin, S., Sancho, J. & Sales, S. Novel technique for distributed fibre sensing based on faint long gratings (FLOGs). Proc. SPIE 9157, 23rd International Conference on Optical Fibre Sensors, 91576W (June 2, 2014); doi:10.1117/12.2059668.
56. Rao, Y.-J. In-fibre Bragg grating sensors. *Meas. Sci. Technol.* **8,** 355–375 (1997).
57. Urricelqui, J, Soto, M. A. & Thévenaz, L. . Sources of noise in Brillouin optical time-domain analyzers. Proc. SPIE 9634, 24th International Conference on Optical Fibre Sensors, 963434 (September 28, 2015); doi:10.1117/12.2195298.
58. Szeliski, R. *Computer vision: algorithms and applications* (Springer Science & Business Media, 2010).
59. Buades, A., Coll, B. & Morel, J. M. A review of image denoising methods, with a new one. *Multiscale Model. Simul.* **4,** 490–530 (2005).
60. Buades, A., Coll, B. & Morel, J. M. Nonlocal image and movie denoising. *Int. J. Comput. Vis.* **76,** 123–139 (2008).
61. Buades, A., Coll, B. & Morel, J. M. Denoising image sequences does not require motion estimation. in Advanced Video and Signal Based Surveillance, 2005. AVSS 2005. IEEE Conference on, pp.70-74, 15-16 Sept. 2005; doi: 10.1109/AVSS.2005.1577245.
62. Mahmoudi, M. & Sapiro, G. Fast image and video denoising via nonlocal means of similar neighborhoods. *IEEE Signal Process. Lett.* **12,** 839–842 (2005).
63. Coupe, P. *et al.* An optimized blockwise nonlocal means denoising filter for 3-D magnetic resonance images. *IEEE Trans. Med. Imag.* **27,** 425–441 (2008).
64. Donoho, D. L. & Johnstone, I. M. Ideal spatial adaptation via wavelet shrinkage. *Biometrika* **81,** 425–455 (1994).
65. Donoho, D. L. De-noising by soft-thresholding. *IEEE Trans. Inform. Theory* **41,** 613–627 (1995).
66. Jansen, M. *Noise Reduction by Wavelet Thresholding* (Springer-Verlag, 2001).
67. Mallat, S. G. A theory for multiresolution signal decomposition: the wavelet representation. *IEEE Trans. Pattern Anal. Mach. Intell.* **11,** 674–693 (1989).
68. Zlokolica, V., Pizurica, A. & Philips, W. Wavelet-domain video denoising based on reliability measures. *IEEE Trans. Circuits Syst. Video Technol.* **16,** 993–1007 (2006).
69. Shigong, Y., Ahmad, M. O. & Swamy, M. N. S. Video denoising using motion compensated 3-d wavelet transform with integrated recursive temporal filtering. *IEEE Trans. Circuits Syst. Video Technol.* **20,** 780–791 (2010).
70. Balster, E. J., Zheng, Y. F. & Ewing, R. L. Combined spatial and temporal domain wavelet shrinkage algorithm for video denoising. *IEEE Trans. Circuits Syst. Video Technol.* **16,** 220–230 (2006).

Acknowledgements

We are thankful to Omnisens SA for their active interest. This research receives financial support from the Swiss Commission for Technology and Innovation (Project 18337.2 PFNM-NM).

Author contributions

M.A.S. and J.A.R. invented the presented concept of using image and video processing to enhance the performance of distributed fibre sensors. M.A.S. implemented the experiments and carried out data analysis. J.A.R. implemented and optimized the denoising algorithms. L.T. contributed to the methodology and supervised the entire work at the Group for Fibre Optics. All authors contributed to the writing of the paper.

Additional information

Competing financial interests: The authors declare no competing financial interests.

21

Geometry-invariant resonant cavities

I. Liberal[1,2], A.M. Mahmoud[2] & N. Engheta[2]

Resonant cavities are one of the basic building blocks in various disciplines of science and technology, with numerous applications ranging from abstract theoretical modelling to everyday life devices. The eigenfrequencies of conventional cavities are a function of their geometry, and, thus, the size and shape of a resonant cavity is selected to operate at a specific frequency. Here we demonstrate theoretically the existence of geometry-invariant resonant cavities, that is, resonators whose eigenfrequencies are invariant with respect to geometrical deformations of their external boundaries. This effect is obtained by exploiting the unusual properties of zero-index metamaterials, such as epsilon-near-zero media, which enable decoupling of the temporal and spatial field variations in the lossless limit. This new class of resonators may inspire alternative design concepts, and it might lead to the first generation of deformable resonant devices.

[1] Department of Electrical and Electronic Engineering, Universidad Pública de Navarra, E31006 Pamplona, Spain. [2] Department of Electrical and Systems Engineering, University of Pennsylvania, Philadelphia, Pennsylvania 19104, USA. Correspondence and requests for materials should be addressed to N.E. (email: engheta@ee.upenn.edu).

The dynamics of many physical systems are usually described in terms of wave equations subject to certain boundary conditions. This is the case, for example, in classical and quantum mechanics, electromagnetics, acoustics and fluid dynamics. Specifically, when considering source-free time-harmonic $\exp(-i\omega t)$ fields, one finds that the solutions to these equations, subject to specific boundary conditions, often take place at specific discrete ω-frequency values, usually labelled as eigenfrequencies, or resonance frequencies[1]. In general, wave equations interrelate both spatial and temporal variations of the fields (for example, consider the vector wave equation of the electric field in classical electromagnetics: $\nabla\times\nabla\times\boldsymbol{\mathcal{E}}(\mathbf{r},t)+c^{-2}\partial_t^2\boldsymbol{\mathcal{E}}(\mathbf{r},t)=\mathbf{0}$ (ref. 2). Consequently, eigenfrequencies are determined by the geometry at hand.

This fundamental principle shapes the way we address various phenomena and develop technology. In fact, one of the main conceptual challenges that researchers, engineers and designers face across multiple disciplines is to come up with the appropriate geometry to operate at a specific frequency. On the other hand, fabrication imperfections degrade the performance of the devices, as well as hinder the application of thrilling physical concepts that may unfortunately require too stringent fabrication tolerances. Therefore, we could wonder if, as it is symbolically sketched in Fig. 1, it could be possible to find scenarios in which the eigenfrequencies of a resonant cavity are invariant with respect to geometrical transformations. If so, this would propose a complete change in the mindset behind design processes, and in turn open up the possibility for developing resonant devices that still be functional even under severe geometrical deformations with interesting applications, for example, in flexible photonics, as well as in tailoring light–matter interaction and quantum emission in such deformable structures.

Naturally, the idea of a geometry-invariant resonator challenges our intuition on how waves usually behave. However, the fields of topological insulators[3,4] and topological photonics[5–10] have revealed that certain physical quantities are preserved under continuous deformations. Moreover, during the past several years metamaterials have demonstrated that waves can be manipulated in unconventional manners[11–20]. For instance, metamaterials featuring extreme parameters, such as epsilon-and-mu-near-zero and zero refractive index structures, have been found to support fields with static spatial distributions, while maintaining their temporally dynamic properties[21–26]. This apparent decoupling between spatial and temporal domains encouraged us to believe that, indeed, resonators whose eigenfrequencies are invariant under geometrical transformations could be possible.

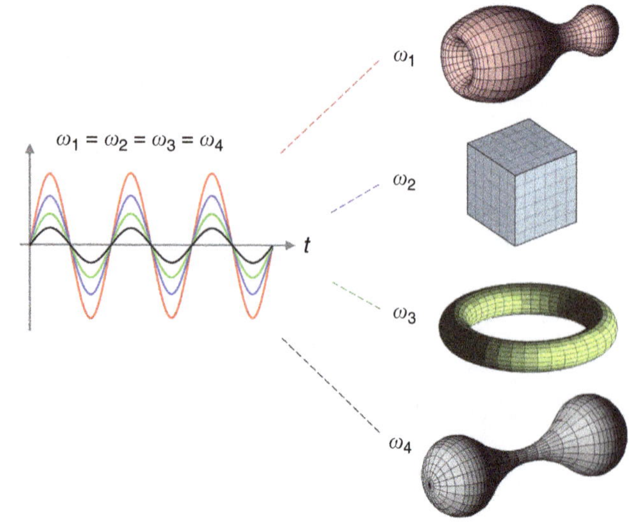

Figure 1 | Geometry-invariant resonant cavities. Conceptual sketch of resonant cavities that, despite their very distinct geometry (shape, size, topology), support an eigenmode at the same resonance frequency.

In the following, we will concentrate on the classical source-free time-harmonic wave equation for the electric field $\mathbf{E}$ in nonmagnetic media: $\nabla\times\nabla\times\mathbf{E}-\varepsilon(\omega/c)^2\mathbf{E}=\mathbf{0}$ (ref. 2), with c being the speed of light in vacuum and ε the relative permittivity of the medium at hand. However, this must be considered only as a specific example of a more general concept that, as many other metamaterial paradigms, can be extrapolated to other forms of waves such as acoustic, elastic, mechanical and matter waves. Moreover, electromagnetic systems also represent an excellent test bench for future experimental verifications of the concepts introduced in this work. In fact, different experimental realizations of zero-index electromagnetic metamaterials have already been reported in the form of naturally available materials[27,28], dispersion engineering in waveguides[29,30], photonic crystals[25] and artificial electromagnetic materials[31,32].

In this work we analytically and numerically demonstrate that there are at least three distinct physical mechanisms in which zero-index metamaterials, and, in particular, epsilon-near-zero (ENZ) media, enable the development of cavities supporting eigenmodes whose eigenfrequency is invariant with respect to geometrical deformations of their external boundary.

Results

2D cavities invariant under equi-areal transformations. One key property of ideal zero-index metamaterials is their ability to 'stop' the spatial variations of the phase, and for some cases also the magnitude, of electromagnetic fields[21–24]. For instance, in ENZ media—that is, media whose relative permittivity is approximately zero, $\varepsilon\approx 0$—the magnetic field parallel with the axis of a two-dimensional (2D) system must be uniform, $H_z(\boldsymbol{\rho})=H_z^{\mathrm{h}}$, to avoid a singularity of the electric field $\mathbf{E}=i/(\omega\varepsilon)\nabla H_z(\boldsymbol{\rho})\times\hat{\mathbf{z}}$ (refs 22–24). One could anticipate that the influence of geometry is lessened in the presence of spatially uniform fields, since effectively the apparent wavelength in such media is very large. This intuition is indeed correct, and we show that uniform magnetic field distributions can be associated with 2D cavities whose eigenfrequency is invariant with respect to equi-areal transformations.

To this end, let us consider, for example, a 2D cavity composed of a 2D dielectric particle of relative permittivity ε_{i}, cross-sectional area A_{i} and perimeter L_{i}, immersed in a 2D ENZ host of arbitrary cross-sectional shape but area A_{h} (Fig. 2a), bounded by perfectly electric conducting (PEC) walls. As demonstrated in Supplementary Note 1, the eigenfrequencies obtained as solutions to the source-free electric field time-harmonic wave equation subject to the boundary condition $\hat{\mathbf{n}}\times\mathbf{E}=\mathbf{0}$ on the PEC wall are determined by the solutions to the following characteristic equation: $\omega=\frac{i}{\mu_0}\frac{L_{\mathrm{i}}}{A_{\mathrm{h}}}Z_{\mathrm{S}}$, where $Z_{\mathrm{S}}=\oint_{\partial A_{\mathrm{i}}}\mathbf{E}\cdot\mathbf{dl}/(L_{\mathrm{i}}H_z^{\mathrm{h}})$ is the surface impedance of the particle embedded in ENZ. It is thus clear that the existence of an eigenmode at the ENZ frequency is completely determined by the overall cross-sectional area of the ENZ host, A_{h}, and the properties of the internal 2D particle, encapsulated in Z_{S}. Therefore, if the internal particle is designed so that the characteristic equation has a solution at the plasma frequency of the host, then the cavity will have an eigenmode at such eigenfrequency, independently of the shape of its external boundary, as long as its cross-sectional area remains the same. As if it were an incompressible fluid, the 2D cavity can be exposed to any equi-areal geometrical deformation while keeping the same eigenfrequency. Note that the set of allowed geometrical deformations also includes piercing (making 2D

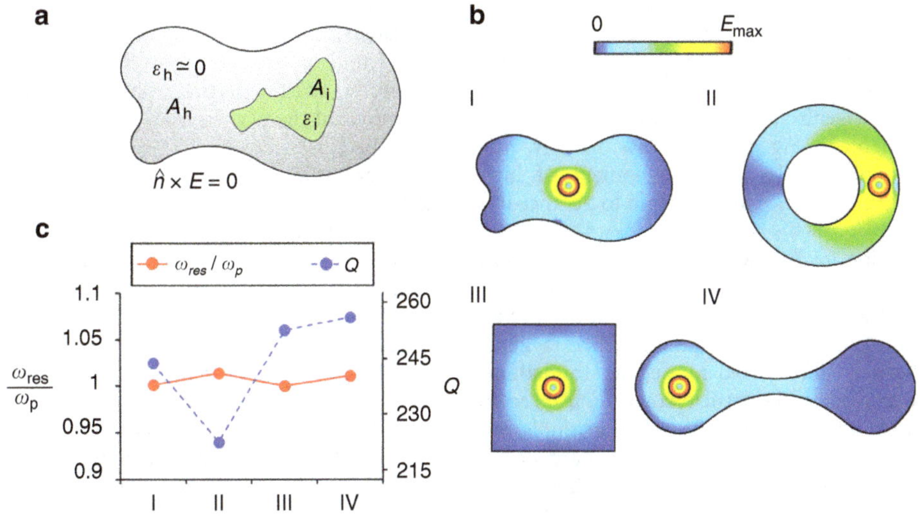

Figure 2 | Two-dimensional (2D) cavities with eigenfrequencies invariant under equi-areal geometry deformations. (**a**) Sketch of a generic 2D cavity composed of a 2D ENZ host of cross-sectional area A_h containing a 2D dielectric (nonmagnetic) particle of cross-sectional area A_i, perimeter L_i and relative permittivity ε_i. The entire 2D cavity is bounded by a PEC wall on which the tangential component of **E** field must vanish. (**b**) Colourmaps of the electric field magnitude distributions of resonant eigenmodes, obtained using numerical simulation, in four cavities consisting of a Si (ε_i=11.7) cylinder of radius $r_i = 1.165\,\mu$m, immersed in a 2D SiC host of different shapes but equal cross-sectional area $A_i + A_h = 49\pi\,\mu m^2$. (**c**) Linear graph portraying the resonance frequency (normalized to the SiC plasma frequency) and the quality factor Q as a function of cavity number. Here the imaginary part of relative permittivity of SiC is assumed to be 0.1 at the SiC plasma frequency. In Supplementary Fig. 5, we show how these quantities vary with different level of loss (as represented by the different value of the imaginary part of permittivity of SiC).

holes in) the cavity. For example, a simply connected cavity can be transformed into a 2D annulus-like cavity. If the total area as well as the geometry and the material of the internal 2D particle are kept the same, the eigenfrequency will be immune even to evident changes in the topology of the cavity.

To illustrate this fact, we can, for example, set the internal particle as an infinitely long cylinder of relative permittivity ε_i and radius r_i. In such a case, the surface impedance of the particle can be written in closed form: $Z_S = -i\eta_i J'_0(\sqrt{\varepsilon_i}\frac{\omega}{c}r_i)/J_0(\sqrt{\varepsilon_i}\frac{\omega}{c}r_i)$, where $\eta_i = \sqrt{\mu_0/(\varepsilon_0\varepsilon_i)}$ is the intrinsic medium impedance in the particle, $J_0(x)$ is the cylindrical Bessel function of the first kind and order zero, and $J'_0(x) = \partial_x J_0(x)$. Consequently, the characteristic equation can be explicitly written as follows: $\omega = c2\pi r_i/(A_h\sqrt{\varepsilon_i})J'_0(\sqrt{\varepsilon_i}\frac{\omega}{c}r_i)/J_0(\sqrt{\varepsilon_i}\frac{\omega}{c}r_i)$, where it is again evident that the existence of an eigenmode with uniform magnetic field in the ENZ host medium only depends on the characteristics of the internal particle (ε_i and r_i), and the cross-sectional area of the ENZ host A_h. Therefore, if the radius of the cylinder is set so that the characteristic equation has a solution at the plasma frequency of the host, then the cavity will have an eigenmode whose eigenfrequency is independent with respect to equi-areal transformations of the ENZ host.

To validate this property, we have numerically studied a few examples of 2D cavities by carrying out an eigenfrequency analysis with a full-wave electromagnetic solver (Fig. 2b). These specific examples have been selected to illustrate the high degree of arbitrariness in the geometry of the 2D cavities, including non-canonical shapes (Fig. 2b, I), different topologies (Fig. 2b, II), sharp corners (Fig. 2b, III) and high-aspect ratios (Fig. 2b, IV). A more detailed description of the geometry of these cavities is gathered in Supplementary Figs 1-4. Anticipating future experimental verifications of the presented results, the ENZ host has been modelled using silicon carbide (SiC)[27], whereas the internal dielectric cylinder is assumed to be silicon (Si), with relative permittivity $\varepsilon_i = 11.7$ (ref. 33). In this manner, our analysis includes the effect of the relatively high losses of SiC with relative value of the imaginary part of its permittivity to be 0.1 ($\varepsilon'' \approx 0.1$) in the vicinity of the SiC plasma frequency, $\omega_p = 2\pi \times 29.08 \times 10^{12}$ rad s^{-1} (corresponding to a free-space wavelength of around 10.3 µm), where the real part of the relative permittivity is near zero ($\varepsilon' \approx 0$)[27]. The radius of the cylinder ($r_i = 1.165\,\mu$m) and the area of the host ($A_h = 49\pi\,\mu m^2$) have been selected such that the characteristic equation is satisfied at the SiC plasma frequency using a cylinder that is subwavelength in its cross-section.

The eigenfrequencies of these 2D resonators were computed numerically and are depicted in Fig. 2c. It is apparent from the figure that despite their very distinct geometry, and despite the fact that realistic losses have been taken into account, the eigenfrequencies of these resonators deviate by $<1.5\%$ from the plasma frequency of SiC. These small disagreements are mainly caused by the losses of SiC, which slightly deviate the response of the host from that of a pure ENZ medium. As a matter of fact, the eigenfrequencies converge even more closely to the SiC plasma frequency as losses decrease (Supplementary Fig. 5).

It is worth remarking that, unlike its eigenfrequency, not all properties of the resonator are invariant with respect to geometrical deformations. For example, the quality factor Q—i.e., the ratio between the stored and dissipated energies per cycle—strongly depends on the field intensity distributions in the resonator[2]. Subsequently, as it is illustrated in Fig. 2c, deforming the cavities results in changes in the quality factor in an excess of 10%, while their resonance frequencies stay effectively unchanged. This exotic feature could be potentially exploited, for instance, in designing flexible resonators in which the strength of the coupling with a quantum emitter embedded in them can be modified by deforming their external boundary, while keeping a constant resonance frequency, hence enabling a fine tuning of the decay dynamics of the quantum emitter. This is the subject of our ongoing study and will be reported in a future publication.

Moreover, Fig. 2c and Supplementary Fig. 5 also serve to illustrate a unique property of the proposed geometry-invariant resonators that, to the best of our knowledge, has no counterpart in conventional resonators. Specifically, in a conventional resonator, the quality factor increases as losses decrease, and the eigenfrequency becomes more sensitive to geometrical deformations. On the contrary, in our proposed geometry-invariant resonators, the smaller the losses the larger the quality factor, but also the more robust the eigenfrequency is towards geometrical deformations (Supplementary Fig. 5). In this manner, the proposed idea in principle enables the development of high Q resonators whose eigenfrequencies are immune to geometrical deformations.

To finalize the discussion for this set of modes, we emphasize that the choice of a 2D cylindrical internal particle with circular cross-section was made for the sake of simplicity, and to have an analytical solution to the characteristic equation, hence facilitating the comparison between theory and numerical simulations. However, in principle any other 2D particle could be utilized to induce an eigenmode in the 2D ENZ host. For example, an analogous analysis by using a 2D particle with square cross-section is reported in Supplementary Figs 6 and 7, leading to the same conclusions.

3D cavities supporting spatially 'electrostatic' eigenmodes. Next, even a more general invariance with respect to geometrical deformations may be found by noting that, as demonstrated in Supplementary Note 2, ENZ media may also support other modes as $\exp(-i\omega t)$ time-varying spatially 'electrostatic' fields, which are different from what we discussed above. In other words, as the medium relative permittivity goes to zero, the Maxwell curl equation $\nabla \times \mathbf{H} = -i\omega\varepsilon_0\varepsilon\mathbf{E}$ may also support solutions with zero magnetic field $\mathbf{H} = \mathbf{0}$, but a non-zero and time-varying electric field, $i\omega\mathbf{E} \neq \mathbf{0}$. Naturally, the other Maxwell curl equation imposes that the associated electric field is irrotational $\nabla \times \mathbf{E} = i\omega\mu_0\mathbf{H} = \mathbf{0}$, since for this mode $\mathbf{H}$ is zero in the ENZ region. Thus, interestingly, ENZ media may support solutions to the wave equation in the form of spatially 'electrostatic' distributions that are dynamically varying in time. We note that the existence of time-varying electrostatic field distributions had already been discovered in the field of plasma physics, mostly in the form of longitudinal waves[34]. Here we remark that ENZ media support generic electrostatic field distributions, which can be excited in a wide set of cavities.

The field distributions of these spatially electrostatic eigenmodes correspond to the solution of Laplace's equation ($\nabla^2\varphi(\mathbf{r}) = 0$, $\mathbf{E} = -\nabla\varphi(\mathbf{r})$) in the ENZ host, subject to the appropriate boundary conditions. Interestingly, the solution to the Dirichlet problem of Laplace's equation is known to exist and be unique if the boundary is sufficiently smooth and the potential prescribed at the boundary is continuous[35]. Therefore, if the boundary conditions on the ENZ host enable the existence of spatially electrostatic modes, then such a cavity has an eigenfrequency at the ENZ frequency, no matter what its geometry is.

To illustrate this phenomenon with a specific example, let us consider a three-dimensional (3D) scenario in which a resonator is composed by a 3D dielectric particle immersed in a 3D ENZ host. A detailed theoretical derivation of the conditions under which this composite cavity supports an eigenmode is included in Supplementary Note 3 and Supplementary Fig. 8. However, it is actually sufficient to simply note that to excite a spatially 'electrostatic' eigenmode with zero magnetic field in the ENZ region, the continuity of the fields imposes that the magnetic field (normal and tangential) at the boundary of the dielectric particle must be zero. In addition, at this boundary the normal component of the electric field inside the dielectric particle must also be zero to preserve the continuity of the normal displacement vector. (Note that the electric field inside the ENZ host might have a normal component to this boundary of dielectric particle, still satisfying the continuity of the normal displacement vector.) If we find a particle satisfying these conditions, then the composite cavity particle plus ENZ host will support an eigenmode at the plasma frequency, no matter what the geometry of the ENZ host is.

For example, if the internal particle is a dielectric sphere of radius r_i and relative permittivity ε_i, then these conditions are met at the solutions of the following characteristic equation: $\hat{J}_n\left(\sqrt{\varepsilon_i}\frac{\omega}{c}r_i\right) = 0$ for $n = 1, 2, \ldots$ (see also Supplementary Notes 3 and 4, as well as Supplementary Fig. 9 for the analysis of a canonical core-shell cavity). That is to say, the eigenfrequencies of the resonator correspond to the zeros of the functions $\hat{J}_n(x) \equiv \sqrt{\frac{\pi x}{2}}J_{n+\frac{1}{2}}(x)$, representing the Schelkunoff form of the spherical Bessel functions of the first kind and order n, where $J_n(x)$ is the cylindrical Bessel function of the first kind and order n (ref. 2). Note that in this case there is not only one, but an infinite number of possible eigenmodes $n = 1, 2, 3\ldots$ with geometry-invariant properties. Moreover, due to the spherical symmetry of the internal particle, there are $2n+1$ degenerate modes for each n-th eigenmode. We emphasize that the solutions to this characteristic equation only depend on the properties on the internal particle (ε_i and r_i), and are independent of the geometry of the main cavity. Therefore, once the internal particle has been correctly designed, then the ideal ENZ host, and hence the external boundaries of the cavity, can in principle be of any size and shape. What is more, the cavity could even be 'polluted' with other particles made of different dielectric materials sharing the same ENZ host medium. In all these cases, the cavity will support an eigenmode at the ENZ frequency. As shown in Supplementary Note 5 and Supplementary Fig. 10, the invariance of the eigenfrequency in the presence of time-harmonic spatially electrostatic fields can also be justified by using perturbational techniques. These modes can be excited in both 2D and 3D systems.

The geometry-invariant properties of these eigenmodes are numerically validated in Fig. 3a, which shows four cavities with very distinct geometries, but that nevertheless support eigenmodes at the same eigenfrequency. Again, these specific cavities have been chosen to illustrate the high degree of arbitrariness in the geometry of the cavities (shape, topology and in this case also size). A more detailed description of their geometry can be found in Supplementary Figs 11–14. All cavities are composed of a SiC host containing a Si particle. In this case, a spherical particle of radius r_i=2.155 μm has been selected to satisfy the characteristic equation, $\hat{J}_1\left(\sqrt{\varepsilon_i}\frac{\omega}{c}r_i\right)$=0 (that is, n=1), at the SiC plasma frequency. For the sake of brevity, Fig. 3a only depicts the electric field magnitude distribution of one of the three degenerate modes that can be excited in the vicinity of the SiC plasma frequency (each eigenmode corresponding to a different orientation of the electric dipolar mode within the Si spherical particle). The electric and magnetic field magnitude distributions of all degenerate modes are depicted in Supplementary Figs 15–18. The fact that the magnetic field vanishes in the ENZ host can also be more clearly appreciated in those figures. The resonance frequencies and quality factors of these modes have been numerically computed and are depicted in Fig. 3b. Despite the use of the realistic losses of SiC ($\varepsilon'' \approx 0.1$), the numerical computation of the resonance frequencies reveals that all degenerate modes in all four cavities deviate $<0.3\%$ from the SiC plasma frequency. Furthermore, the fact that degenerate modes exhibit different quality factors allows us to envision the design of a new class of resonators, in which the fields excited by quantum emitters immersed within them exhibit a different Purcell factor and decay dynamics as a function of their polarization, while maintaining the same resonance frequency.

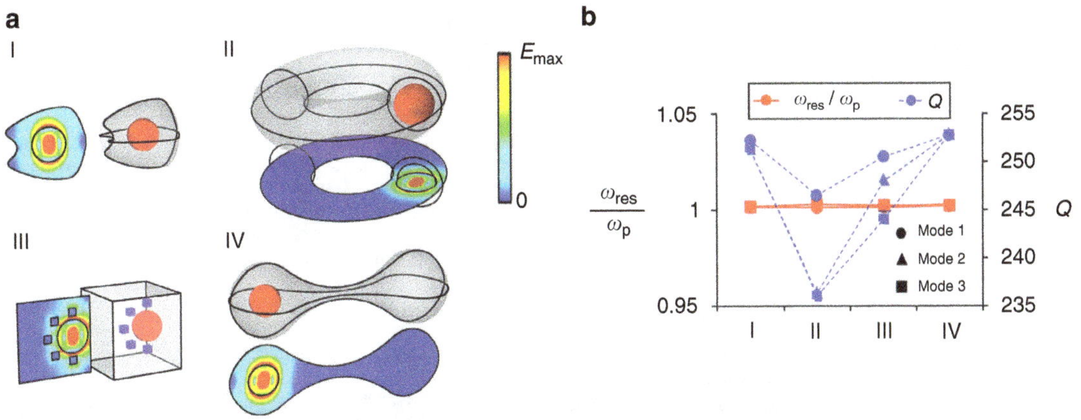

Figure 3 | Three-dimensional (3D) cavities supporting spatially 'electrostatic' modes with the same resonance eigenfrequency. (**a**) Four 3D cavities with different geometries supporting the same resonance eigenfrequency. Each cavity consists of a Si sphere (ε_i=11.7) (shown as a red sphere) of radius r_i=2.155 µm immersed in a 3D SiC host (shown as grey background), bounded by a PEC wall. Cavity III also contains several additional cubic dielectric particles (shown in blue) with permittivity $\varepsilon_p = 2$, and side l_p=1 µm inserted in the ENZ host. Next to each we show the colourmaps of the electric field magnitude distributions obtained using numerical simulation of one of the three supported degenerate eigenmodes (the other eigenmodes can be found in Supplementary Figs 15-18). (**b**) Linear graph portraying the resonance frequency (normalized to the SiC plasma frequency) and quality factors for these four cavities, demonstrating that the resonance eigenfrequencies are the same, while the quality factors of these eigenmodes are different. Here the imaginary part of relative permittivity of SiC is assumed to be 0.1 at the SiC plasma frequency.

Again, we remark that while the internal particle must be designed to enable the excitation of an eigemode at the ENZ frequency, this particle must not necessarily be the sphere used in the current example. In essence, any particle supporting a solution to the wave equation in which the magnetic field and the normal electric field are zero at its boundary can trigger the excitation of a spatially 'electrostatic' mode in an arbitrarily shaped ENZ region. For instance, Supplementary Figs 19–23 present an equivalent analysis for the case in which the cavities contain a cylindrical dielectric particle whose top and bottom walls have been covered by perfect magnetic conductor layers. The results are very similar to those obtained in Fig. 3.

Surface-avoiding modes. To finalize, there is at least a third set of modes present in ENZ media that are invariant under certain (but not completely arbitrary) geometrical transformations. These modes correspond to the cases where both electric and magnetic fields are neither constant nor zero in the ENZ region. Note that, even if not constant, the magnetic field must always be irrotational $\nabla\times\mathbf{H} = -i\omega\varepsilon_0\varepsilon\mathbf{E} \approx \mathbf{0}$, and thus it features a 'quasi-static' spatial distribution while it is temporally dynamic. However, in this case, the electric field cannot be curl free, $\nabla\times\mathbf{E} = i\omega\mu_0\mathbf{H}$, and it indeed takes the form of a solenoidal field forming closed loops in the cavity. The geometry-invariant properties of this set of modes arise from the fact that the modes can concentrate the fields on the vicinity of the internal dielectric particle, resulting in a negligible field at the outer boundaries of the cavity, which naturally satisfies the $\hat{\mathbf{n}}\times\mathbf{E}=\mathbf{0}$ boundary condition. In essence, the ENZ properties of the medium ensure that the propagation constant vanishes, $k=\omega\sqrt{\mu_0\varepsilon} \approx 0$, and, hence, the fields cannot propagate as in a conventional dielectric through the ENZ host towards the external surface of the cavity.

For instance, let us consider a spherical cavity with two concentric layers, as schematically depicted in Fig. 4a. In particular, we assume that the inner layer (the internal particle) is made of Si, whereas the external layer (the background host) is filled with SiC. In this manner, when the radius of the internal particle, r_i, is much smaller than the external radius of the cavity, r_{out}, that is, $r_i \ll r_{out}$, the field on the surface of the

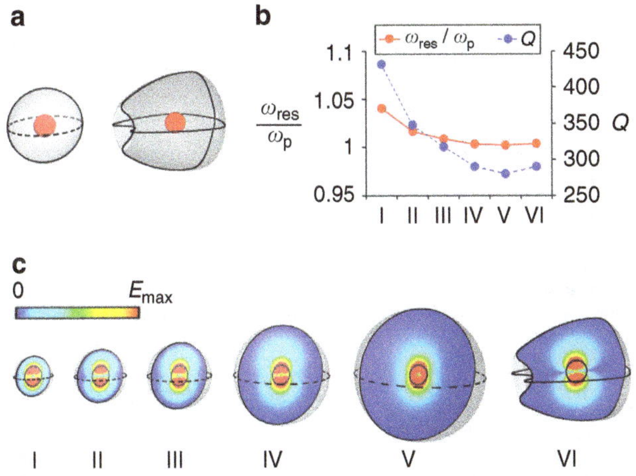

Figure 4 | Three-dimensional (3D) cavities supporting surface-avoiding modes. (**a**) Sketches of two cavities consisting of a Si sphere (shown as a red sphere) with ε_i=11.7 and radius r_i=1.507 µm immersed in a 3D SiC host (shown as a grey background). (**b**) Linear graph portraying the resonance frequency (normalized to the SiC plasma frequency) and quality factor for six cavities, showing small variation in the resonance frequency, while having different quality factors. Here the imaginary part of relative permittivity of SiC is assumed to be 0.1 at the SiC plasma frequency. (**c**) Colourmaps of the electric field magnitude distributions of the surface-avoiding eigenmodes, obtained using numerical simulations, in six different cavities (five of which are obtained from the left cavity but with different sizes of the ENZ host, whereas the sixth is the second cavity, shown in the top left panel).

cavity is negligible, and the eigenfrequency becomes approximately independent of the volume and shape of the resonator. Specifically, and as shown in Supplementary Note 4, the characteristic equation determining the eigenfrequencies of this set of modes in our spherical example can be asymptotically written as follows: $\hat{J}_n(\sqrt{\varepsilon_i}\frac{\omega}{c}r_i) = -(\sqrt{\varepsilon_i}\frac{\omega}{c}r_i)\hat{J}'_n(\sqrt{\varepsilon_i}\frac{\omega}{c}r_i)/n$, for $r_i/r_{out} \to 0$, $n = 1, 2, \ldots$. Fig. 4 gathers a set of examples illustrating how, as the volume of the cavity increases, the eigenfrequencies converge towards the value prescribed by the characteristic equation. It also exemplifies that once the cavity is sufficiently large, the eigenfrequency becomes independent of the shape of the external surface of the cavity. The examples have been chosen to illustrate the impact of progressively increasing the size, and the geometry of the cavities is detailed in Supplementary Figs 24 and 25. In this case, the radius of the internal particle (r_i=1.507 μm) is selected so that the eigenfrequency satisfying the characteristic equation for $n = 1$ equals the SiC plasma frequency.

Discussion

In summary, our theoretical study demonstrates that in the context of ENZ and zero-index metamaterials there are multiple solutions to the wave equation whose eigenfrequency is invariant under geometrical transformations. It was demonstrated that these solutions enable the design of resonant cavities whose eigenfrequencies are invariant with respect to geometrical transformations of their external boundary, and, hence, they inspire new design philosophies in which the geometry of a device is not determined by, and locked to, its frequency of operation. While our analytical and numerical analyses have been focused on closed cavities bounded by PEC walls, we expect that analogous phenomena could be observed in open resonators. A set of preliminary simulations is included in Supplementary Figs 26–31. We believe that these unconventional resonators could give rise to a new generation of deformable resonant devices. Among other applications, the proposed resonators appear to be particularly well suited for flexible photonics and cavity quantum electrodynamics. For instance, the proposed cavities can be locked with their resonances overlapping the atomic transitions of quantum emitters embedded within them, while different aspects of the emitter-cavity interaction can be dynamically tuned by means of deforming the cavity.

Methods

Numerical simulations. The commercially available full-wave electromagnetic simulator software COMSOL Multiphysics, version 5.0, was used to generate all 2D and 3D numerical simulations presented in the figures of the main text and the Supplementary Material. Specifically, we carried out eigenfrequency analyses in which the software makes use of the finite element method to determine the eigenmodes and eigenfrequencies of the wave equation: $\nabla\times\nabla\times\mathbf{E} - \varepsilon(\omega/c)^2\mathbf{E} = \mathbf{0}$, subject to be PEC boundary condition, that is, the tangential electric field vanishes on the external surface of the cavity, $\hat{\mathbf{n}}\times\mathbf{E}=\mathbf{0}$, where $\hat{\mathbf{n}}$ stands for the outward normal vector to the surface of the cavity. The solver was requested to search for eigenfrequencies around the frequency where the real part of the permittivity is near zero ($\varepsilon'(\omega_p)=0$), and the eigenmodes with the closest eigenfrequencies to such a frequency were selected. The ENZ host was modelled as SiC in accordance with ref. 27. Thus, the plasma frequency, at which the real part of the permittivity vanishes, $\varepsilon'(\omega_p)=0$, takes place at $\omega_p = 2\pi \times 29.08\ 10^{12}\ \mathrm{rad\,s^{-1}}$, with losses represented by $\varepsilon''(\omega_p) = 0.1$ unless otherwise stated. The numerical solver provided the field distributions of the eigenmodes and their associated eigenfrequencies. These data were also used to calculate the quality factor Q as the ratio of the energies stored and dissipated per unit cycle: $Q=\omega W_{\mathrm{stored}}/P_{\mathrm{loss}}$, where $P_{\mathrm{loss}}=\frac{\omega}{2}\int \varepsilon_0\, \varepsilon''|\mathbf{E}|^2 dV$ and $W_{\mathrm{stored}}=\frac{1}{4}\int \varepsilon_0\, \partial_\omega\{\omega\varepsilon'\}|\mathbf{E}|^2 + \mu_0|\mathbf{H}|^2 dV$ were computed via integration of the electric and magnetic field intensities. In Supplementary Figs 26–31 the analysis was carried out without any PEC boundary and using the frequency domain solver. The simulation set-up is described in Supplementary Figs 26 and 29.

References

1. Morse, P. M. & Feshbach, H. *Methods of Theoretical Physics* (McGraw-Hill, 1953).
2. Harrington, R. F. *Time-Harmonic Electromagnetic Fields* (McGraw-Hill, 1961).
3. Hasan, M. Z. & Kane, C. L. Colloquium: topological insulators. *Rev. Mod. Phys.* **82,** 3045–3067 (2010).
4. Qi, X.-L. & Zhang, S.-C. Topological insulators and superconductors. *Rev. Mod. Phys.* **83,** 1057–1110 (2011).
5. Lu, L., Joannopoulos, J. D. & Soljačić, M. Topological photonics. *Nat. Photon.* **8,** 821–829 (2014).
6. Haldane, F. & Raghu, S. Possible realization of directional optical waveguides in photonic crystals with broken time-reversal symmetry. *Phys. Rev. Lett.* **100,** 013904 (2008).
7. Raghu, S. & Haldane, F. Analogs of quantum-Hall-effect edge states in photonic crystals. *Phys. Rev. A* **78,** 033834 (2008).
8. Wang, Z., Chong, Y., Joannopoulos, J. D. & Soljacić, M. Observation of unidirectional backscattering-immune topological electromagnetic states. *Nature* **461,** 772–775 (2009).
9. Rechtsman, M. C. *et al.* Floquet topological insulators. *Nature* **196,** 196 (2013).
10. Lumer, Y., Plotnik, Y., Rechtsman, M. C. & Segev, M. Self-localized states in photonic topological insulators. *Phys. Rev. Lett.* **111,** 1–5 (2013).
11. Engheta, N. & Ziolkowski, R. W. *Metamaterials: Physics and Engineering Explorations* (IEEE-Wiley, 2006).
12. Eleftheriades, G. V. & Balmain, K. G. *Negative-Refraction Metamaterials* (IEEE, 2005).
13. Pendry, J. B., Schurig, D. & Smith, D. R. Controlling electromagnetic fields. *Science* **312,** 1780–1782 (2006).
14. Leonhardt, U. Optical conformal mapping. *Science* **312,** 1777–1780 (2006).
15. Cai, W. & Shalaev, V. M. *Optical Metamaterials: Fundamentals and Applications* (Springer, 2010).
16. Smith, D. R., Pendry, J. B. & Wiltshire, M. C. K. Metamaterials and negative refractive index. *Science* **305,** 788–793 (2004).
17. Yu, N. *et al.* Light propagation with phase reflection and refraction: generalized laws of reflection and refraction. *Science* **334,** 333–337 (2011).
18. Vakil, A. & Engheta, N. Transformation optics using graphene. *Science* **332,** 1291–1295 (2011).
19. Ni, X., Emani, N. K., Kildishev, A. V, Boltasseva, A. & Shalaev, V. M. Broadband light bending with plasmonic nanoantennas. *Science* **335,** 427 (2012).
20. Silva, A. *et al.* Performing mathematical operations with metamaterials. *Science* **343,** 160–164 (2014).
21. Ziolkowski, R. W. Propagation in and scattering from a matched metamaterial having a zero index of refraction. *Phys. Rev. E* **70,** 046608 (2004).
22. Silveirinha, M. G. & Engheta, N. Design of matched zero-index metamaterials using nonmagnetic inclusions in epsilon-near-zero media. *Phys. Rev. B* **75,** 075119 (2007).
23. Mahmoud, A. M. & Engheta, N. Wave-matter interactions in epsilon-and-mu-near-zero structures. *Nat. Commun.* **5,** 5638 (2014).
24. Silveirinha, M. G. & Engheta, N. Tunneling of electromagnetic energy through subwavelength channels and bends using ε-near-zero materials. *Phys. Rev. Lett.* **97,** 157403 (2006).
25. Huang, X., Lai, Y., Hang, Z. H., Zheng, H. & Chan, C. T. Dirac cones induced by accidental degeneracy in photonic crystals and zero-refractive-index materials. *Nat. Mater.* **10,** 582–586 (2011).
26. Liberal, I., Ederra, I., Gonzalo, R. & Ziolkowski, R. W. Electromagnetic force density in electrically and magnetically polarizable media. *Phys. Rev. A* **88,** 053808 (2013).
27. Spitzer, W. G., Kleinman, D. & Walsh, D. Infrared properties of hexagonal silicon carbide. *Phys. Rev.* **113,** 127–132 (1959).
28. Naik, G. V., Kim, J. & Boltasseva, A. Oxides and nitrides as alternative plasmonic materials in the optical range. *Opt. Mater. Express* **1,** 1090–1099 (2011).
29. Edwards, B., Alù, A., Young, M. E., Silveirinha, M. G. & Engheta, N. Experimental verification of epsilon-near-zero metamaterial coupling and energy squeezing using a microwave waveguide. *Phys. Rev. Lett.* **100,** 033903 (2008).
30. Vesseur, E. J. R., Coenen, T., Caglayan, H., Engheta, N. & Polman, A. Experimental verification of n = 0 structures for visible light. *Phys. Rev. Lett.* **110,** 1–5 (2013).
31. Maas, R., Parsons, J., Engheta, N. & Polman, A. Experimental realization of an epsilon-near-zero metamaterial at visible wavelengths. *Nat. Photon.* **7,** 907–912 (2013).
32. Rizza, C., Di Falco, A. & Ciattoni, A. Gain assisted nanocomposite multilayers with near zero permittivity modulus at visible frequencies. *Appl. Phys. Lett.* **99,** 221107 (2011).
33. Weber, M. J. *Handbook of Optical Materials* (CRC, 2003).
34. Swanson, D. G. *Plasma Waves* (Academic, 1989).
35. John, F. *Partial Differential Equations* (Springer, 1978).

Acknowledgements

This work is supported in part by the US Air Force Office of Scientific Research (AFOSR) Multidisciplinary University Research Initiative (MURI) on Quantum Metaphotonics and Metamaterials, Award No. FA9550-12-1-0488. I.L. acknowledges financial support from a FPI scholarship from the Public University of Navarre (UPNA).

Author contributions

I.L. carried out the analytical derivations; I.L. and A.M.M. carried out the numerical simulations; N.E. conceived the idea and supervised the project; all authors discussed the theoretical and numerical aspects and interpreted the results, and contributed to the preparation and writing of the manuscript.

Additional information

Competing financial interests: The authors declare no competing financial interests.

22

Cavity-excited Huygens' metasurface antennas for near-unity aperture illumination efficiency from arbitrarily large apertures

Ariel Epstein[1], Joseph P.S. Wong[1] & George V. Eleftheriades[1]

One of the long-standing problems in antenna engineering is the realization of highly directive beams using low-profile devices. In this paper, we provide a solution to this problem by means of Huygens' metasurfaces (HMSs), based on the equivalence principle. This principle states that a given excitation can be transformed to a desirable aperture field by inducing suitable electric and (equivalent) magnetic surface currents. Building on this concept, we propose and demonstrate cavity-excited HMS antennas, where the single-source-fed cavity is designed to optimize aperture illumination, while the HMS facilitates the current distribution that ensures phase purity of aperture fields. The HMS breaks the coupling between the excitation and radiation spectra typical to standard partially reflecting surfaces, allowing tailoring of the aperture properties to produce a desirable radiation pattern, without incurring edge-taper losses. The proposed low-profile design yields near-unity aperture illumination efficiencies from arbitrarily large apertures, offering new capabilities for microwave, terahertz and optical radiators.

[1] The Edward S. Rogers Department of Electrical and Computer Engineering, University of Toronto, Toronto, Ontario, Canada M5S 2E4. Correspondence and requests for materials should be addressed to A.E. (email: ariel.epstein@utoronto.ca) or to G.V.E. (email: gelefth@waves.utoronto.ca).

Achieving high directivity with compact radiators has been a major problem in antenna science since early days[1–3]. Still today, many applications, such as automotive radars and satellite communication, strive for simple and efficient low-profile antennas producing the narrowest beams[4–7]. Increasing the radiating aperture size enhances directivity, but only if the aperture is efficiently excited. To date, uniform illumination of large apertures is achievable with reflectors and lenses; however, these require substantial separation between the source and the aperture, resulting in a large overall antenna size[8,9]. High aperture illumination efficiencies can also be achieved using antenna arrays[10], but the elaborated feed networks increase complexity and cost, and can lead to high losses[11].

Contrarily, leaky-wave antennas (LWAs) can produce directive beams using a low-profile structure fed by a simple single source[12]. In Fabry–Pérot (FP) LWAs, a localized source is sandwiched between a perfect electric conductor (PEC) and a partially reflecting surface (PRS), forming a longitudinal FP cavity[2,13]. By tuning the cavity dimensions and source position, favourable coupling to a single waveguided mode is achieved, forming a leaky wave emanating from the source; typical device thicknesses lie around half of a wavelength. The leaky mode is characterized by a transverse wavenumber whose real-part k_t corresponds to the waveguide dispersion, and is accompanied by a small imaginary part α determined by the PRS. Assuming $|k_t| \gg \alpha$, this leads to a conical directive radiation through the PRS towards $\theta_{out} \approx \pm \arcsin(|k_t|/k)$, $k = \omega\sqrt{\mu\varepsilon}$ being the free-space wavenumber, with a beamwidth proportional to α. Broadside radiation is achieved when $|k_t|$ is small enough such that the splitting condition $|k_t| < \alpha$ is satisfied, and the peaks of the conical beam merge[14]. LWAs based on modulated metasurfaces (MoMetAs) are also compact and probe-fed, but utilize a surface wave $|k_t| > k$ guided on a PEC-backed dielectric sheet covered with metallic patches[15–18]. This mode is coupled to radiation via periodic modulation of the patch geometry; its leakage rate α is determined by the modulation depth[18].

Although FP-LWAs and MoMetAs have compact configurations, they suffer from a fundamental efficiency limitation for finite structures: designing a moderate leakage rate α yields uniform illumination but results in considerable losses from the edges; on the other hand, for large leakage rates only a portion of the aperture is effectively radiating[19–21].

To mitigate edge-taper losses, shielded FP-LWA structures have been recently proposed, using PEC side walls which form a lateral cavity[22–28]. Nevertheless, the tight coupling between the propagation of the leaky mode inside the FP cavity and the angular distribution of the radiated power manifested by $\theta_{out} \approx \arcsin(k_t/k)$ poses serious limitations on the achievable aperture illumination efficiency. This is most prominent for antennas radiating at broadside, in which only low-order lateral modes (satisfying the splitting condition) can be used. Consequently, such antennas are designed to excite exclusively the TE_{10} lateral mode, which inherently limits the aperture illumination efficiency, defined as the relative directivity with respect to the case of uniform illumination, to 81% (ref. 29). In addition, as the dominant spectral components of the cavity fields directly translate to prominent lobes in the radiation pattern, only a single mode should be excited to guarantee high directivity. However, suppression of parasitic cavity modes is a very difficult problem[30], especially for large apertures.

From the so far discussion it follows that it would be very beneficial if the fields inside the cavity and those formed on the aperture could be optimized independently. This would facilitate good aperture illumination without the necessity to meet excitation-related restricting conditions. But how to achieve such a separation? The equivalence principle suggests that for a given field exciting a surface, desirable (arbitrary) aperture fields can be formed by inducing suitable electric and (equivalent) magnetic surface currents[29]. On the basis of this idea, the concept of Huygens' metasurfaces (HMSs) has been recently proposed, where subwavelength electric and magnetic polarizable particles (meta-atoms) are used to generate these surface currents in response to a known incident field[31–39]. In previous work, we have shown that if the reflected and transmitted fields are properly set, the aperture phase can be tailored by a passive and lossless HMS to produce prescribed directive radiation, for any given excitation source[40].

In this paper, we harness the equivalence principle to efficiently convert fields excited in a cavity by a localized source to highly directive radiation using a Huygens' metasurface: cavity-excited HMS antenna. The device structure resembles a typical shielded FP-LWA, with an electric line source surrounded by three PEC walls and a HMS replacing the standard PRS (Fig. 1). For a given aperture length L and a desirable transmission angle θ_{out}, we optimize the cavity thickness and source position to predominantly excite a high-order lateral mode, thus guaranteeing good aperture illumination. Once the source configuration is established, we stipulate the aperture fields to follow the power profile of the cavity mode, and impose a linear phase to promote radiation towards θ_{out}. With the cavity and aperture fields in hand, we invoke the equivalence principle and evaluate the (purely reactive[40]) electric surface impedance and magnetic surface admittance required to support the resultant field discontinuity[31,32,41,42]. As the power profile of the chosen high-order mode creates hot spots of radiating surface currents approximately half a wavelength apart, a uniform virtual phased array is formed on the HMS aperture; such excitation profile is expected to yield very high directivity with no grating lobes regardless of θ_{out} (ref. 10). Furthermore, in contrast to LWAs, the antenna directivity does not deteriorate significantly even if other modes are partially excited, as these would merely vary the amplitude of the virtual array elements, without affecting the phase purity. This semianalytical design procedure can be applied to arbitrarily large apertures, yielding near-unity aperture illumination efficiencies. With the PEC side walls, no power is lost via the edges, offering an effective way to overcome the efficiency tradeoff inherent to FP-LWAs and MoMetAs, while preserving the advantages of a single-feed low-profile antenna.

Results

Cavity-excited Huygens' metasurface antennas. To design the HMS-based antenna, we apply the general methodology developed in ref. 40 to the source configuration of Fig. 1; for

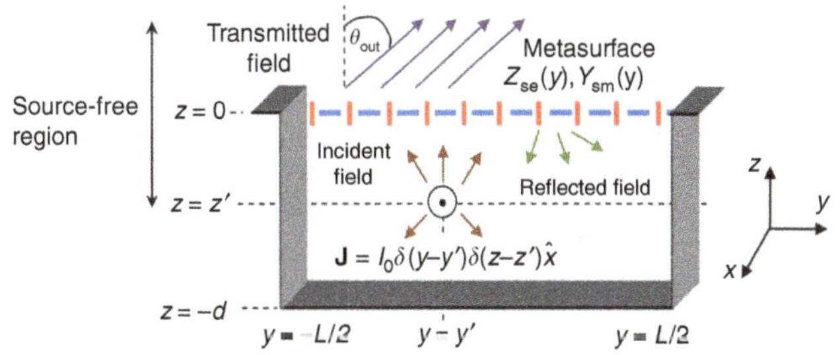

Figure 1 | Physical configuration of a cavity-excited Huygens' metasurface antenna. An electric line source is positioned at (y',z'), surrounded by three perfect-electric-conductor (PEC) walls at $z = -d, y = \pm L/2$, forming a lateral cavity. The cavity is covered by a Huygens' metasurface of aperture length L situated at $z = 0$, facilitating directive radiation towards θ_{out}.

completeness, we recall briefly its main steps. We consider a two-dimensional (2D) scenario ($\partial/\partial x = 0$) with the HMS at $z = 0$ and a given excitation geometry at $z \le z' < 0$ embedded in a homogeneous medium ($k = \omega\sqrt{\epsilon\mu}$, $\eta = \sqrt{\mu/\varepsilon}$). Under these circumstances, the incident, reflected and transmitted fields in the vicinity of the HMS can be expressed via their plane-wave spectrum[43]

$$\begin{cases} E_x^{\text{inc}}(y,z) = k\eta I_0 \mathcal{F}^{-1}\left\{\frac{1}{2\beta} f(k_t) e^{-j\beta z}\right\} \\ E_x^{\text{ref}}(y,z) = -k\eta I_0 \mathcal{F}^{-1}\left\{\frac{1}{2\beta}\Gamma(k_t) f(k_t) e^{j\beta z}\right\} \\ E_x^{\text{trans}}(y,z) = k\eta I_0 \mathcal{F}^{-1}\left\{\frac{1}{2\beta}\overline{T}(k_t) e^{-j\beta z}\right\}, \end{cases} \tag{1}$$

where $\mathcal{F}^{-1}\{g(k_t;z)\} \triangleq \frac{1}{2\pi}\int_{-\infty}^{\infty} dk_t g(k_t;z) e^{jk_t y}$ is the inverse spatial Fourier transform of $g(k_t;z)$ (ref. 44), $f(k_t)$ is the source spectrum, $\Gamma(k_t)$ is the HMS reflection coefficient, and $\overline{T}(k_t) \triangleq T(k_t)[1+\Gamma(k_t)]$ is the transmission spectrum. As before, k_t denotes the transverse wavenumber and the longitudinal wavenumber is $\beta = \sqrt{k^2 - k_t^2}$. For simplicity, we only consider here transverse electric (TE) fields ($E_z = E_y = H_x = 0$); the nonvanishing magnetic field components H_y, H_z can be calculated from E_x via Maxwell's equations.

For a given source spectrum, it is required to determine the reflected and transmitted fields, through the respective degrees of freedom $\Gamma(k_t)$ and $T(k_t)$, that would implement the desirable functionality. Once the tangential fields on the two facets of the HMS are set, the equivalence principle is invoked to evaluate the required electric and magnetic surface currents to induce them[29]. The polarizable particles comprising the HMS are then designed such that the average fields acting on them effectively induce these surface currents[41,42]. Analogously, the HMS can be characterized by its electric surface impedance $Z_{se}(y)$ and magnetic surface admittance $Y_{sm}(y)$, relating the field discontinuity and the average excitation via the generalized sheet transition conditions (GSTCs)[31,32,40,41].

To promote directive radiation towards θ_{out} we require that the aperture (transmitted) fields approximately follow the suitable plane-wave-like relation (Supplementary Note 1)

$$\begin{aligned} E_x(\mathbf{r})|_{z\to 0^+} &\approx Z_{out} H_y(\mathbf{r})|_{z\to 0^+} \\ &\approx k\eta I_0 \mathcal{F}^{-1}\left\{\frac{1}{2\beta} T(k_t)\right\} \\ &\triangleq k\eta I_0 W_0(y) e^{-jky \sin\theta_{out}}, \end{aligned} \tag{2}$$

where $W_0(y)$ is the aperture window (envelope) function (yet to be determined) and $Z_{out} = 1/Y_{out} = \eta/\cos\theta_{out}$ is the TE wave impedance of a plane-wave directed towards θ_{out}.

In previous work[40], we have shown that if the wave impedance and the real power are continuous across the metasurface, then these aperture fields can be supported by a passive lossless HMS (purely reactive Z_{se} and Y_{sm}). The first condition, local impedance equalization, means that the total (incident + reflected) fields on the bottom facet of the metasurface should exhibit the same wave impedance as the aperture fields, that is, $E_x(\mathbf{r})|_{z\to 0^-} = Z_{out} H_y(\mathbf{r})|_{z\to 0^-}$; this is achieved by setting the reflection coefficient to a Fresnel-like form

$$\Gamma(k_t) = \frac{k\cos\theta_{out} - \beta}{k\cos\theta_{out} + \beta}, \tag{3}$$

determining the reflected fields everywhere, fixing our first degree of freedom.

To satisfy the second condition, local power conservation, we require that the aperture window function follows the magnitude of the total (incident + reflected) fields at $z \to 0^-$, namely,

$$\begin{aligned} W_0(y) = |E_x(\mathbf{r})|_{z\to 0^-} &= \left|\mathcal{F}^{-1}\left\{\frac{1}{2\beta}[1-\Gamma(k_t)]f(k_t)\right\}\right| \\ &= \left(\mathcal{F}^{-1}\left\{\left[\frac{1}{2\beta}(1-\Gamma)f\right]\star\left[\frac{1}{2\beta}(1-\Gamma)f\right]\right\}\right)^{1/2}, \end{aligned} \tag{4}$$

where $g^{\star}g$ is the autocorrelation of the spectral-domain function $g(k_t)$ (ref. 44); this determines the transmitted fields everywhere, fixing our second degree of freedom.

The absolute value operator in the last equality is of utmost significance: it indicates that the transmission spectrum of the aperture fields follows, up to a square root, the power spectral density of $E_x(\mathbf{r})|_{z\to 0^-}$, and not the spectral content of the incident and reflected fields. This is directly related to the balanced (plane-wave-like) contribution of the electric and magnetic fields to the power flow that we stipulated in equation (2), and results in a significantly favourable plane-wave spectrum, as will be discussed in detail in the next section.

Finally, we use these semianalytically predicted fields and the equivalence principle, manifested by the GSTCs, to calculate the required HMS surface impedance, yielding the desirable purely reactive modulation given by[40],

$$\frac{Z_{se}(y)}{Z_{out}} = \frac{Y_{sm}(y)}{Y_{out}} = -\frac{j}{2}\cot\left[\frac{\phi_-(y) - \phi_+(y)}{2}\right] \tag{5}$$

where $\phi_\pm(y) \triangleq \angle E_x(y,z)|_{z\to 0^\pm}$ are the phases of the stipulated fields just above and below the metasurface.

Once the general design procedure is established, applying it to the configuration of Fig. 1, which includes an electric line source at (y', z') surrounded by PEC walls at $z = -d$, $y = \pm L/2$, is straightforward: it is reduced to finding the corresponding source spectrum. The latter is quantized due to the lateral cavity, and includes multiple reflections between the HMS at $z = 0$ and the PEC at $z = -d$; explicitly[43],

$$f(k_t) = \frac{\pi}{2L}\sum_{n=-\infty}^{\infty}\left\{\begin{array}{c}\dfrac{e^{-j\beta(d+z')} - e^{j\beta(d+z')}}{e^{j\beta d} - \Gamma(k_t)e^{-j\beta d}} \\ \left[e^{-jk_t y'} + (-1)^{n+1} e^{jk_t y'}\right]\delta\left(k_t - \frac{n\pi}{L}\right)\end{array}\right\}. \tag{6}$$

We refer to the sum of the fields corresponding to the n, $-n$ terms in the summation as the field of the nth mode of the lateral cavity, where $n \ge 0$.

Although this procedure is applicable for any transmission angle, we restrict ourselves from now on to the case of broadside radiation $\theta_{out} = 0$, where the performance of shielded and unshielded FP-LWAs is the most problematic due to the splitting condition[14] (design of oblique-angle radiators is addressed in the Supplementary Methods). For simplicity, we set the lateral position of the source to be $y' = 0$; with this choice, the even modes vanish, and the odd modes follow a cosine profile in the lateral dimension.

Optimizing the cavity excitation. One of the key differences between the cavity-excited HMS antenna and FP-LWAs is that by harnessing the equivalence principle we control the individual contributions of the electric and magnetic fields to the flow of power, expressed by the lateral distribution of the z-component of the Poynting vector on the aperture. More specifically, the resultant (transmitted) aperture fields corresponding to equation (4) actually follow the square root of the power profile dictated by the cavity mode, and not the profile of the cavity fields. This distinction is very important, as the power profile of a standing wave is always positive, whereas the field profile changes signs along the lateral dimension. Hence, the spectral content of

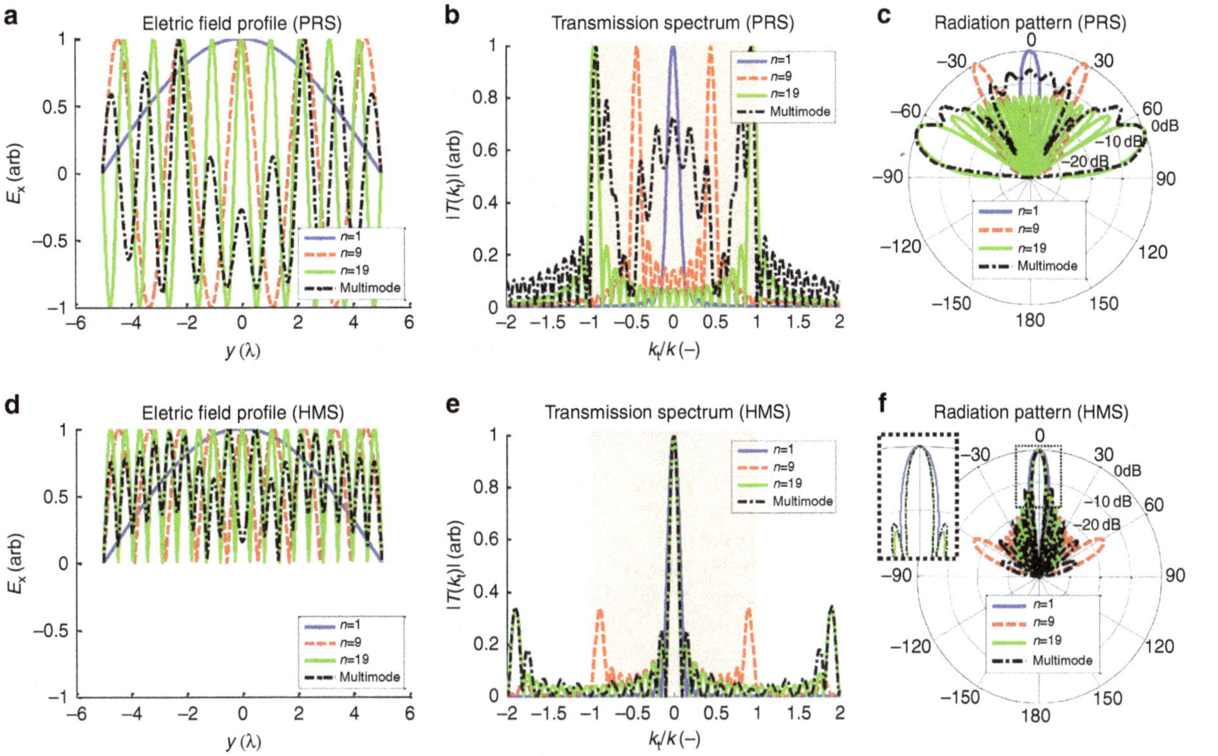

Figure 2 | Comparison between aperture profiles and radiation patterns of cavity-excited PRSs and cavity-excited HMSs. Single-mode excitations of the $n=1$ (blue solid line), $n=9$ (red dashed line), and $n=19$ (green solid line) modes of an aperture of length $L=10\lambda$ are compared to the multimode excitation corresponding to the HMS antenna presented in Fig. 1 with $L=10\lambda$, $z'=-\lambda$, and $d=1.61\lambda$ (black dash-dotted line). (**a**,**d**) Normalized spatial profile of the tangential electric field on the aperture. (**b**,**e**) Normalized spectral content of the aperture field; shaded region correspond to the visible part of the spectrum. (**c**,**f**) Normalized radiation patterns. Inset: close-up of the radiation pattern around $\theta=0$.

the aperture fields, which determines the far-field radiation pattern, is fundamentally different.

To illustrate this point, we compare the fields formed on the device aperture for a shielded FP-LWA, where a standard PRS is used, and for a cavity-excited HMS antenna with the same excitation. Figure 2 presents the spatial profile of the tangential electric field, its spatial Fourier transform, and corresponding radiation patterns (calculated following ref. 29), for single-mode excitation of the $n=1$ (blue solid lines), $n=9$ (red dashed lines) and $n=19$ (green solid lines) modes, for an aperture length of $L=10\lambda$. All plots are normalized to their maximum, as the radiation pattern is sensitive to the variation of the fields, and not to their magnitude.

As follows from equations (4) and (6), the spatial profile of the nth-mode aperture field is proportional to $\cos(n\pi y/L)$ for a standard PRS, but for an HMS it is proportional to $|\cos(n\pi y/L)|$ (Fig. 2a,d). Except for the lowest order mode $n=1$, for which the two functions coincide, the difference in the spatial profile translates into distinctively different features in the spectral content (Fig. 2b,e). For the nth mode, the transmission spectrum of the HMS aperture corresponds to the autocorrelation of the PRS aperture spectrum, leading to formation of peaks centred around the second harmonics ($k_t=\pm 2n\pi/L$) and d.c. ($k_t=0$). As both the right-propagating and left-propagating components of the standing wave coherently contribute to the d.c. peak, the latter dominates the transmission spectrum, and the radiation patterns corresponding to the HMS aperture exhibit highly directive radiation towards broadside (Fig. 2f). In contrast, the PRS-based devices exhibit conical radiation to angles determined by the dominant spectral components of the aperture fields, that is, towards $\theta=\pm\arcsin[n\lambda/(2L)]$ (Fig. 2c)[12,13].

The transverse wavenumber $k_t=\pi/L$ corresponding to the lowest order mode $n=1$ is small enough such that the two symmetric beams merge[14], which enables the PRS aperture to radiate a single beam at broadside. Indeed, small-aperture shielded FP-LWAs utilize this TE_{10} mode to generate broadside radiation. However, as demonstrated by ref. 29, the aperture illumination efficiency of this mode is inherently limited to 81%, due to the non-optimal cosine-shaped aperture illumination[22–28], leading to broadening of the main beam (inset of Fig. 2f). This highlights a key benefit of using an HMS-based antenna, as it is clear from Fig. 2f that we can use high-order mode excitations, which provide a more uniform illumination of the aperture, for generating narrow broadside beams with enhanced directivities.

In fact, as the mode index n increases, the autocorrelation of equation (4) drives the second harmonic peaks outside the visible region of the spectrum (shaded region in Fig. 2b,e), funnelling all the HMS-radiated power to the broadside beam, subsequently increasing the overall directivity. This improvement in radiation properties can be explained using ordinary array theory. As seen from Fig. 2d, the peaks of the field profile generated by the nth mode on the HMS aperture form hot spots of radiating currents separated by a distance of L/n. The radiation from such an aperture profile would resemble the one of a uniform array with the same element separation. As known from established array theory, to avoid grating lobes the element separation should be smaller than a wavelength[10]. For an aperture length of $L=N\lambda$, where N is an integer, the hot-spot separation satisfies this condition for mode indices $n>N$; specifically, for $N=10$ (Fig. 2), grating lobes would not be present in the radiation pattern for mode indices $n>10$. In agreement with this argument, Fig. 2f

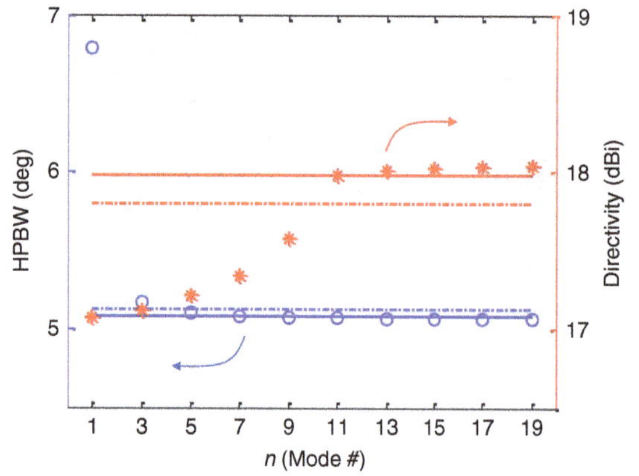

Figure 3 | Radiation characteristics of different lateral cavity modes. Half-power beamwidth (HPBW, blue circles) and 2D directivity[58] (red asterisks) of an HMS aperture of length $L=10\lambda$ excited by a single mode as a function of the mode index n. Solid lines denote the respective radiation characteristics of a uniformly excited aperture[29] and dash-dotted lines mark the HPBW (blue) and directivity (red) of *multimode* excitation corresponding to the HMS antenna of Fig. 1 with $L=10\lambda$, $z'=-\lambda$, and $d=1.61\lambda$.

shows that for $n=9$ grating lobes still exist, while for the highest order fast mode $n=19$ they indeed vanish.

These observations are summarized in Fig. 3, where the radiation characteristics of an HMS aperture of $L=10\lambda$ excited by a single mode are plotted as a function of the mode index n (only fast modes $k_{t,n}=n\pi/L<k$ are considered). Indeed, it can be seen that the lowest order lateral mode exhibits the worst performance, and the performance improves as the mode index increases. While the half-power beamwidth (HPBW) saturates quickly, the directivity D continues to increase with n until the point in which grating lobes disappear $n=N=10$ is crossed; for mode indices $n>10$ the radiation characteristics of the HMS aperture are comparable to those of the optimal uniformly excited aperture (solid lines).

From an array theory point of view, excitation of the highest order fast mode is preferable, as the corresponding equivalent element separation approaches $\lambda/2$, implying that such aperture profile would be suitable for directing the radiation to large oblique angles $\theta_{out}\neq 0$ without generating grating lobes[10]. Furthermore, as the HMS reflection coefficient $\Gamma(k_t)$ grows larger with $k_t=n\pi/L$ (equation (3)), the power carried by the highest order fast-mode $n=2N-1$ is best-trapped in the cavity, guaranteeing uniform illumination even in the case of very large apertures.

Nevertheless, generating a single-mode excitation of a cavity via a localized source can be very problematic[30,45]. Fortunately, the cavity-excited HMS antenna can function very well also with multimode excitation, as long as high-order modes dominate the transmission spectrum. This is demonstrated by the dot-dashed lines in Figs 2 and 3, corresponding to a multimode excitation generated by the configuration depicted in Fig. 1 with $L=10\lambda$, $z'=-\lambda$, and $d=1.61\lambda$. As expected from the expression for the source spectrum (equation (6)), for a given aperture length L, the field just below the aperture due to a line source would be a superposition of lateral modes, the weights of which are determined by the particular source configuration, namely the cavity thickness d and source position z'. The multimode transmission spectrum in Fig. 2b indicates that for the chosen parameter values, high-order modes ($k_t\to\pm k$) predominantly populate the aperture spectrum, however, low-order modes ($k_t\to 0$) are present as well, to a non-negligible extent. Considering that the far-field angular power distribution $S(\theta)$ is proportional to $\cos^2\theta|T(k_t=k\sin\theta)|^2$, the multimode excitation of the PRS aperture results in a radiation pattern resembling the one corresponding to single-mode excitation of the highest order fast-mode ($n=19$) but with significant lobes around broadside (Fig. 2c); consequently, the directivity is significantly deteriorated.

On the other hand, the same multimode excitation does not degrade substantially the performance of the HMS antenna. The autocorrelated spectrum results in merging of all spectral components into a sharp d.c. peak, with most grating lobes pushed to the evanescent region of the spectrum (Fig. 2e). This retains a beamwidth comparable to that resulting from a single-mode excitation of the highest order fast mode, with only slight deterioration of the directivity due to increased side-lobe level (Fig. 2f and inset). Continuing the analogy to array theory, such multimode excitation introduces slight variations to the magnitude of the array elements, forming an equivalent non-uniform array[10]. The corresponding multimode HPBW and directivity values marked by dash-dotted lines in Fig. 3 verify that, indeed, cavity-excited HMS antennas achieve near-unity aperture illumination efficiencies with a practical multimode excitation; this points out another key advantage of the cavity-excited HMS antenna with respect to shielded FP-LWAs.

We utilize these observations to formulate guidelines for optimizing the cavity excitation for maximal directivity. For a given aperture length $L=N\lambda$, with respect to equation (6), we maximize the coupling to the $n=2N-1$ mode (which exhibits the best directivity) by tuning the cavity thickness d as to minimize the denominator of the corresponding coupling coefficient; equally important, we minimize the coupling to the $n=1$ mode (which exhibits the worst directivity) by tuning the source position z' as to minimize the numerator of the corresponding coupling coefficient. To achieve these with minimal device thickness we derive the following design rules

$$d=\frac{\lambda}{2}\frac{2N}{\sqrt{4N-1}}\xrightarrow{N\gg 1}\frac{\lambda}{2}\sqrt{\frac{L}{\lambda}},\quad z'\approx-\left(d-\frac{\lambda}{2}\right). \tag{7}$$

Although this is somewhat analogous to the typical design rules for FP-LWAs[13], the key difference is that for HMS-based antennas we optimize the source configuration regardless of the desirable transmission angle θ_{out}. This difference is directly related to the utilization of the equivalence principle for the design of the proposed device, which provides certain decoupling between its excitation and radiation spectra (cf. Fig. 2b,e). This decoupling becomes very apparent when the HMS antenna is designed to radiate towards oblique angles $\theta_{out}\neq 0$, in which case the same cavity excitation yields optimal directivity as well (see Supplementary Methods, Supplementary Fig. 1, and Supplementary Table 1).

Two important comments are relevant when considering these design rules. First, even though following equation (7) maximizes the coupling coefficient of the highest order fast mode and minimizes the coupling coefficient of the lowest order mode, it does not prohibit coupling to other modes. The particular superposition of lateral modes exhibits a tradeoff between beamwidth and side-lobe level (as for non-uniform arrays[10]). Thus, final semianalytical optimization of the cavity illumination profile is achieved by fine tuning the source position z' for the cavity thickness d derived in equation (7). In fact, the source position z' is another degree of freedom that can be used to optimize the radiation pattern for other desirable performance features, such as minimal side-lobe level; this feature is further discussed in the Supplementary Methods and demonstrated in Supplementary Figs. 2 and 3, and Supplementary Table 2. Second, although the optimal device thickness increases with increasing aperture length, the increase is sublinear. Therefore, applying the proposed concept to very large apertures would still result in a relatively compact device, while efficiently utilizing the aperture for producing highly directive beams.

Physical implementation and radiation characteristics. We follow the design procedure and considerations discussed above to design cavity-excited HMS antennas for broadside radiation with different aperture lengths: $L=10\lambda$, $L=14\lambda$, and $L=25\lambda$. The cavity thickness was determined via equation (7) to be $d=1.61\lambda$, $d=1.89\lambda$ and $d=2.50\lambda$, respectively; the source position was set to $z'=-1.00\lambda$, $z'=-1.33\lambda$, and $z'=-1.94\lambda$, respectively, exhibiting maximal directivity.

The required electric surface impedance and magnetic surface admittance modulations are implemented using the 'spider' unit cells depicted in Fig. 4. At the design frequency $f=20$ GHz ($\lambda\approx 15$ mm), the unit cell transverse dimensions are $\lambda/10\times\lambda/10$ and the longitudinal thickness is 52 mil $\approx\lambda/12$. Each unit cell consists of three layers of metal traces defined on two bonded laminates of high-dielectric-constant substrate (see Methods). The two (identical) external layers provide the magnetic response of the unit cell, corresponding to the magnetic surface susceptance $B_{sm}=\Im\{Y_{sm}\}$, which is tuned by modifying the arm length L_m (affects equivalent magnetic currents induced by tangential magnetic fields H_y). Analogously, the middle layer is responsible for the electric response of the meta-atom, corresponding to the electric surface reactance $X_{se}=\Im\{Z_{se}\}$, which is tuned by modifying the capacitor width W_e (affects electric currents induced by tangential electric fields E_x). By controlling L_m and W_e, these unit cells can be designed to exhibit Huygens source behaviour, with balanced electric and magnetic responses ranging from $B_{sm}\eta=X_{se}/\eta=-3.1$ to $B_{sm}\eta=X_{se}/\eta=0.9$ (see Methods and Supplementary Fig. 4).

To experimentally verify our theory, we have fabricated and characterized the $L=14\lambda$ cavity-excited HMS antenna, based on the simulated spider cell design at $f=20$ GHz (see Methods). The antenna is composed of a one unit-cell-wide metastrip excited by a coaxial-cable-fed short dipole positioned inside an Aluminium cavity, forming the suitable 2D excitation configuration (Fig. 5). The aperture fields were allowed to radiate into (3D) free-space; the far-field radiation measured in the $\widehat{yz}$ plane then corresponds to the theoretically predicted 2D radiation patterns.

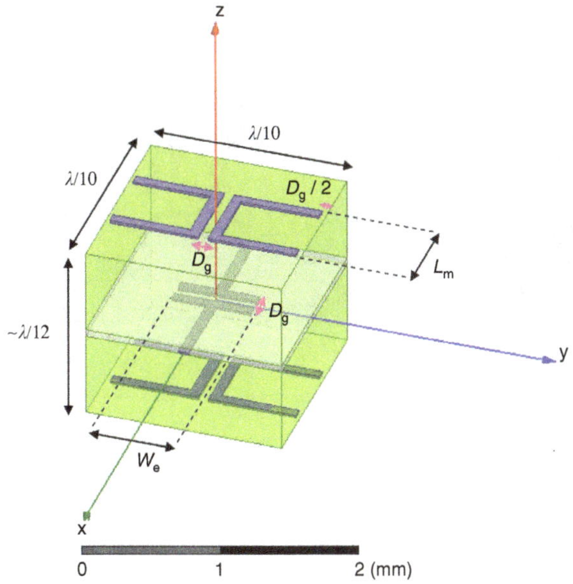

Figure 4 | Spider unit cells. Physical configuration of the meta-atoms used for implementing the HMS at a frequency of $f=20$ GHz ($\lambda\approx 15$ mm). The electric response is controlled by the capacitor width W_e of the electric dipole, while the magnetic response is determined by the magnetic dipole arm length L_m. The gap size (magenta) and copper trace width are fixed to $D_g=3$ mil ≈ 76 μm to comply with standard printed-circuit board fabrication techniques.

Figure 6 presents the design specifications, field distributions, and radiation patterns for the three cavity-excited HMS antennas; Table 1 summarizes the antenna performance parameters (for reference, parameters for uniformly excited apertures[29] are also included). The semianalytical predictions[40] are compared with full-wave simulations conducted with commercially available finite-element solver (ANSYS HFSS), as well as to experimental measurements where applicable (see Methods). As demonstrated by Fig. 6a–c, the realized unit cells are capable of reproducing the required surface impedance modulation, except maybe around large values of $B_{sm}\eta=X_{se}/\eta$; however, such discrepancies usually have little effect on the performance of HMSs[46].

The results in Fig. 6 and Table 1 indicate that the fields and radiation properties predicted by the semianalyical formalism are in excellent agreement with the full-wave simulations for a wide range of aperture lengths. The utilization of realistic (lossy) models for the conductors and dielectrics in the simulated device, as well as other deviations from the assumptions of the design procedure (Supplementary Note 1), result in some discrepancies between the full-wave simulations and predicted performance; however, these mostly affect radiation to large angles (Fig. 6d–f). While this contributes to a minor quantitative difference in the directivity, the properties of the main beam and the side lobes follow accurately the semianalytical results (Table 1), indicating that the theory can reliably predict the dominant contributions to the radiation pattern, as discussed in reference to Fig. 2.

This conclusion is further supported by the experimental results presented for the $L=14\lambda$ antenna at $f=20.04$ GHz, where good agreement between theoretical and measured radiation patterns is observed (Fig. 6e). The experimental values of the HPBW, directivity and side-lobe level and position documented in Table 1 also agree quite well with the simulated ones. The slightly higher side-lobe levels and the broadening of the side lobe at $\theta=-7.2°$ contribute to a smaller measured directivity value, and can be attributed to fabrication errors. Nevertheless, the fact that the main features of the radiation pattern are reproduced well with only negligible deviation of 0.2% from the design frequency, and the fact that the predicted, simulated and measured main beams practically coincide, forms a solid validation of our theory.

The measured frequency response of the antenna presented in Supplementary Fig. 5a also compares very well with the simulated

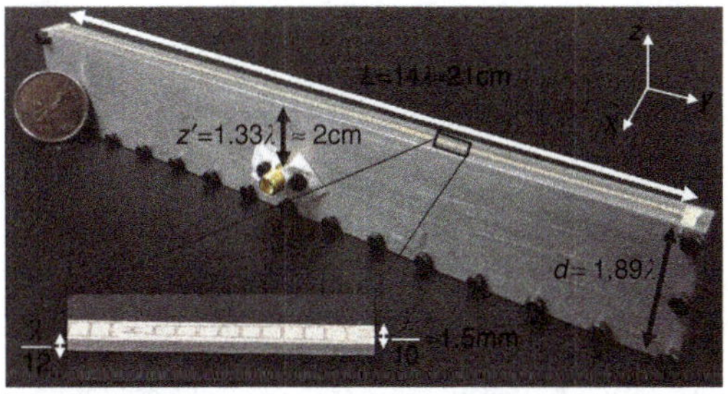

Figure 5 | Fabricated cavity-excited HMS antenna. Probe-fed cavity excites a $\lambda/10$-wide $L=14\lambda$-long metastrip implementing the simulated design corresponding to Fig. 6b, based on the spider unit cells. The two metallic walls parallel to the $\widehat{yz}$ plane form a 2D excitation environment, while the two (shorter) metallic walls parallel to the $\widehat{xz}$ plane form the lateral cavity. Inset: close-up of a section of the metastrip before integration with the cavity.

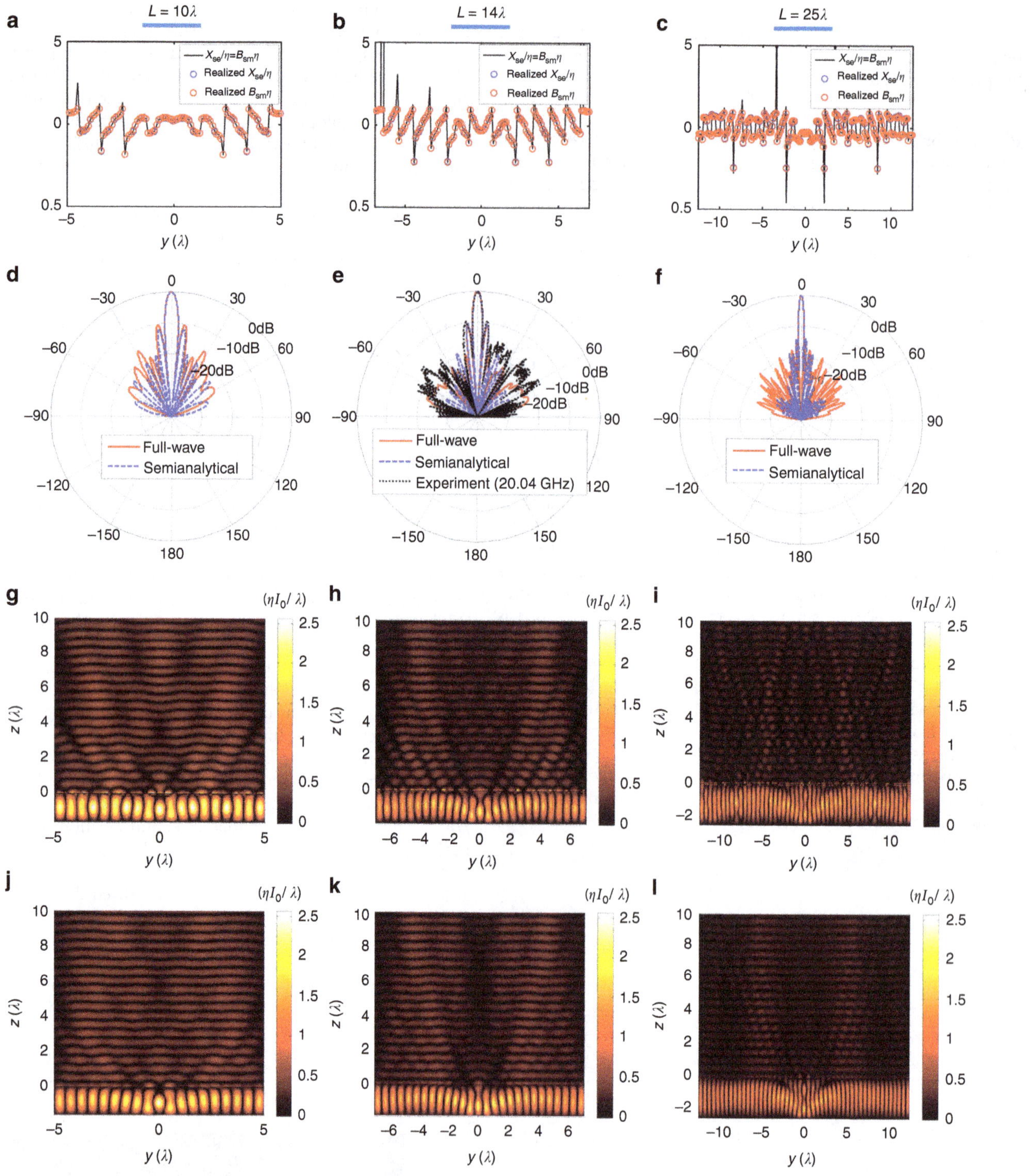

Figure 6 | Performance of cavity-excited HMS antennas. Results are presented for devices with aperture lengths of $L = 10\lambda$, $L = 14\lambda$, and $L = 25\lambda$. (**a–c**) HMS design specifications $X_{se}(y)/\eta = B_{sm}(y)\eta$ (black solid line) derived from equation (5), and the realized electric surface reactance (blue circles) and magnetic surface susceptance (red circles) using the spider unit cells. (**d–f**) Radiation patterns produced by semianalytical formalism (blue dashed line) and full-wave simulations (red solid line). For $L = 14\lambda$, the experimentally obtained radiation pattern is presented in black dotted line. (**g–i**) Field distribution $|\Re\{E_x(y,z)\}|$ produced by full-wave simulations. (**j–l**) Semianalytical prediction of $|\Re\{E_x(y,z)\}|$ (ref. 40).

results, indicating a measured 3- and 1-dB directivity bandwidths of 5.5% and 1.3%, respectively. These values are not very high, but this is expected due to the resonant nature of the cavity and metasurface[24,38]. Bandwidth enhancement can be likely achieved by using stacked HMSs or increasing the number of layers comprising the unit cells[24,47]. The measured 10-dB return-loss bandwidth is 0.2% (Supplementary Fig. 5b), comparable to the values reported for cavity-based antennas[48]; nonetheless, it could be improved by using suitable matching circuitry[49]. The measured (3D) realized gain at $f = 20.04$ GHz is 16.12 ± 1 dBi, corresponding to a gain of 17.21 ± 1 dBi (with 2.7% 3-dB bandwidth). From the 3D directivity estimated from the measurements (Supplementary Methods and Supplementary Fig. 6) we evaluate a radiation efficiency of $75 \pm 17\%$, in a

Table 1 | Radiation characteristics of cavity-excited HMS antennas.

	$L=10\lambda$ ($d=1.61\lambda$, $\|z'\|=1.00\lambda$)			$L=14\lambda$ ($d=1.89\lambda$, $\|z'\|=1.33\lambda$)				$L=25\lambda$ ($d=2.50\lambda$, $\|z'\|=1.94\lambda$)		
	Full-wave	Semianalytic	Uniform	Experiment	Full-wave	Semianalytic	Uniform	Full-wave	Semianalytic	Uniform
HPBW	5.11°	5.13°	5.08°	3.63°	3.83°	3.64°	3.63°	2.09°	2.13°	2.03°
Directivity (2D) (dBi)	17.42	17.84	17.98	18.20	18.79	19.15	19.44	21.33	21.75	21.96
First side-lobe	8.6°	8.6°	8.2°	− 7.2°|6.0°	6.4°	6.3°	5.9°	3.4°	3.4°	3.3°
Side-lobe level (dB)	− 10.4	− 12.0	− 13.5	− 9.1	− 11.1	− 10.4	− 13.5	− 14.6	− 14.0	− 13.5

HMS, Huygens' metasurface; HPBW, half-power beamwidth; 2D, two dimensional.

reasonable agreement with the $\approx 15\%$ conductor and dielectric losses predicted by full-wave simulations (Supplementary Fig. 7). These results indicate that if impedance matching can be achieved via suitable circuitry, then cavity-excited HMS antennas could exhibit reasonably high antenna efficiencies. Lastly, the measured cross-polarization gain was below the noise floor, indicating at least 30-dB cross-polar discrimination at broadside.

For all cases considered, the excitation of the highest order lateral fast mode is clearly visible (Fig. 6g–l), leading to beamwidth and (2D) directivity values comparable to the ones achieved by uniform excitation of the aperture (Table 1). In particular, the simulated (measured) radiation patterns yield aperture illumination efficiencies $\eta_{\mathrm{apt}} \triangleq D/(2\pi L/\lambda)$ of 88%, 86% (75%) and 87% for the $L=10\lambda$, $L=14\lambda$ and $L=25\lambda$ devices, respectively, retained even when the aperture length is very large. In terms of HPBW, often taken as a measure for effective exploitation of the aperture[19], the device performance is even closer to that of a unifromly excited aperture, with pencil-beam HPBWs reaching 99%, 95% (100%) and 97% of the optimal beamwidth, for $L=10\lambda$, $L=14\lambda$ and $L=25\lambda$ devices, respectively. Similar to FP-LWAs[13], the simulated 3-dB bandwidths of the antennas reveal a tradeoff between directivity and bandwidth, reflected by an approximately constant value of 3.1 for the directivity-bandwidth product (Supplementary Fig. 8).

It should be stressed that even though optimized TE_{10} shielded FP-LWAs can reach aperture illumination efficiencies of 81%, their HPBWs are limited to about 75% of the optimal beamwidth[29]; more importantly, their PRS-based design requires single-mode excitation to achieve this performance, thus preventing practical realization of large-aperture devices. Just recently, a low-profile metasurface-based lens antenna with a smaller aperture length (diameter of 6.6λ) operating at a lower frequency (10 GHz), has been reported[50], also exhibiting a high aperture illumination efficiency. Nevertheless, this antenna utilizes a metasurface designed to convert Bessel beams to Gaussian beams; hence, to excite it, a Bessel beam launcher has to be separately designed and fabricated for each application, in addition to the metasurface lens. In contrast, our design approach and rationale can be straightforwardly applied to apertures of any size, and yields an integrated air-filled device requiring only a single simple feed.

While the main purpose of the proposed solution is to maximize antenna directivity, this design goal is suitable in cases where the environment clatter is relatively controlled (for example, point-to-point communications[4]), as uniformly illuminated apertures incur relatively high side-lobe levels; for many applications lower side-lobe levels are required[7]. Nonetheless, as discussed in the Supplementary Methods and demonstrated in Supplementary Fig. 3, a desirable compromise between directivity and side-lobe level can be achieved in the framework of our theory, by simple tuning of the source position z'. Setting the cavity thickness d following equation (7) still forms a virtual antenna array with linear phase, while the different values of z' effectively vary the magnitude of the array elements, facilitating side-lobe control[10]. The example presented in Supplementary Fig. 2 and Supplementary Table 2 harnesses this scheme to design ($L=10\lambda$)-long cavity-excited HMS antennas with side-lobe level of − 20 dB, still retaining a rather high aperture illumination efficiency of 81%.

Discussion

We have introduced a novel design for low-profile single-fed antennas exhibiting beamwidth and directivity values comparable with uniformly excited apertures. To that end, we harness the equivalence principle to devise a cavity-excited Huygens' metasurface, setting the source configuration, HMS reflection coefficient, and aperture fields such that (1) the highest order (lateral) fast mode is predominantly excited, which guarantees that the aperture is well-illuminated; (2) the aperture fields follow the incident power profile and not the incident field profile, which forms an array-like aperture profile with favourable transmission spectrum; and (3) the power flow and wave impedance are continuous across the metasurface, which ensures the design can be implemented by a passive and lossless HMS[40]. The possibility to control the field discontinuities using the electric and (equivalent) magnetic currents induced on the HMS allows us to optimize separately the cavity excitation and the radiated fields, thus overcoming the fundamental tradeoff existing in FP-LWAs between aperture illumination efficiency and edge-taper losses.

It should be emphasized that the general design procedure formulated and demonstrated herein facilitates further optimization of such devices for various applications. The extensive freedom one has in choosing the source configuration, combined with the efficient semianalytical approach, allows explorations of other excitation sources, for example, with different orientations and current distributions, to further tailor the aperture fields (extension to 3D configurations and polarization control are discussed in Supplementary Note 2); once the source spectrum is characterized, the rest of the procedure is straightforward, and the fields and radiation patterns are readily predicted. In addition, considering the recent demonstrations of metasurfaces in general and Huygens' metasurfaces in particular, operating at terahertz and optical frequencies[36,39,51–56], the proposed methodology could be applied to realize compact and efficient pencil beam radiators across the electromagnetic spectrum, extending the range of applications even further.

Methods

Spider unit-cell modelling. The spider unit cells depicted in Fig. 4 were defined in ANSYS Electromagnetic Suite 15.0 (HFSS 2014) with two 25-mil-thick (≈ 0.64 mm) Rogers RT/duroid 6010LM laminates (green boxes in Fig. 4) bonded by 2-mil-thick (≈ 51 μm) Rogers 2929 bondply (white box in Fig. 4). The electromagnetic properties of these products at 20 GHz, namely, permittivity tensor and dielectric loss tangent, as were provided to us by Rogers Corporation, have been inserted to the model. Specifically, a uniaxial permittivity tensor with $\varepsilon_{xx}=\varepsilon_{yy}=13.3\varepsilon_0$, $\varepsilon_{zz}=10.81\varepsilon_0$ and loss tangent of $\tan\delta=0.0023$ were considered for Rogers RT/duroid 6010LM laminates, while an isotropic permittivity of $\varepsilon=2.94\varepsilon_0$ and loss tangent $\tan\delta=0.003$ were considered for Rogers 2929 bondply. The copper traces corresponded to 1/2 oz. cladding, featuring a thickness of 18 μm; the standard value of $\sigma=58\times10^6\,\mathrm{S\,m^{-1}}$ bulk conductivity was used in the model. To comply with standard printed-circuit board manufacturing processes, all copper traces were 3 mil ($\approx 76\,\mu m$) wide, and a minimal distance of 3 mil was kept between adjacent traces (within the cell or between adjacent cells). This implies that the fixed gaps between the capacitor traces (along the x axis) of the electric dipole in the middle layer, as well as between the two arms (along the y axis) of the magnetic dipole in the top and bottom layer (Fig. 4), were fixed to a value of $D_g = 3$ mil ($\approx 76\,\mu m$); the distance from the arm edge to the edge of the unit cell was fixed to $D_g/2 = 1.5$ mil ($\approx 38\,\mu m$).

Unit cells with different values of magnetic dipole arm length L_m and electric dipole capacitor width W_e were simulated using periodic boundary conditions; HFSS Floquet ports were placed at $z = \pm\lambda$ and used to characterize the scattering of a normally incident plane wave off the periodic structure (the interface between the bondply and the bottom laminate was defined as the $z = 0$ plane). For each combination of L_m and W_e, the corresponding magnetic surface susceptance B_{sm} and electric surface reactance X_{se} were extracted from the simulated impedance matrix of this two-port configuration, following the derivation in ref. 57.

The magnetic response B_{sm} was found to be proportional to the magnetic dipole arm length L_m, with almost no dependency in W_e (ref. 37). Thus, to create an adequate lookup table for implementing broadside-radiating HMSs, we varied L_m by constant increments, and for a given L_m, plotted $B_{sm}\eta$ and X_{se}/η as a function of W_e. The value of W_e for which the two curves intersected corresponded to a balanced-impedance point ($Z_{se}/Z_{out} = Y_{sm}/Y_{out}$), where the unit cell acts as a Huygens source, and thus suitable for implementing our metasurface. A lookup table composed of (B_{sm}, X_{se}) pairs and the corresponding unit cell geometries (L_m, W_e) was constructed, and refined through interpolation. The interpolated unit cell geometries were eventually simulated again, to verify the interpolation accuracy and finalize the lookup table entries, as presented in Supplementary Fig. 4. Finally, for a given HMS with prescribed surface impedance modulation ($B_{sm}(y)$, $X_{se}(y)$), a corresponding structure could be defined in HFSS using the unit cells ($L_m(y)$, $W_e(y)$) found via the lookup table in terms of least squares error.

Antenna simulations. To verify our semianalytical design via full-wave simulations, each of the cavity-excited HMS antennas designed in this paper was defined in HFSS using a single strip of unit cells implementing the metasurface, occupying the region $|x| \leq \lambda/20$, $|y| \leq L/2$ (L being the aperture length of the antenna), and $-0.64\,mm \leq z \leq 0.69\,mm$ (in correspondence to the laminate and bondply thicknesses). The simulation domain included $|x| \leq \lambda/20$, $|y| \leq L/2 + 2.5\lambda$, and $-d \leq z \leq 10\lambda$ (d being the cavity thickness), where PEC boundary conditions were applied to the $x = \pm\lambda/20$ planes to form the equivalence of a 2D scenario. PEC boundary conditions were also applied to the $z = -d$ plane, and to two 18-μm-thick side-walls at $y = \pm L/2$, forming the cavity. The line-source excitation was modelled by a $\lambda/20$-wide 1A current sheet at $z = z'$, with the current aligned with the x axis. Radiation boundary conditions were applied to the rest of the simulation space boundaries, namely $z = 10\lambda$, and $y = \pm(L/2 + 2.5\lambda)$, allowing proper numerical evaluation of the fields surrounding the antenna.

To reduce the computational effort required to solve this configuration, we utilized the symmetries of our TE scenario. Specifically, we placed a perfect-magnetic-conductor (PMC) symmetry boundary conditions at the $\hat{x}\hat{z}$ plane, and a PEC symmetry boundary conditions at the $\hat{y}\hat{z}$ plane. We also noticed that adding a thin layer (1 mil $\approx 25\,\mu m$) of copper between the electric dipole edges and the PEC parallel plates at $x = \pm\lambda/20$ enhanced the convergence of the simulation results. With that minor modification, all of the simulated antennas converged within < 40 iterations (maximum refinement of 10% per pass), where the stop conditions was three consecutive iterations in which ΔEnergy < 0.03.

Antenna realization and measurement. The simulated design corresponding to the cavity-excited HMS antenna with $L = 14\lambda$ aperture (Fig. 6b) has been realized and measured to obtain experimental validation of our theory. As in the full-wave simulations, the fabricated metasurface consisted of two 25-mil-thick (≈ 0.64 mm) Rogers RT/duroid 6010LM laminates. The copper traces' geometry used in the simulations to implement the spider unit cells was exported to standard grbr files, later used by Candor Industries Inc. to accordingly etch the electrodeposited 1/2 oz. copper foils covering the laminates. The etched laminates were then bonded using 2-mil-thick ($\approx 51\,\mu m$) Rogers 2929 bondply, forming the desirable three-layer metasurface. Lastly, unit-cell-wide ($\lambda/10 \approx 1.5$ mm) metastrips were achieved via routing (see inset of Fig. 5).

The cavity was realized by replacing the PEC walls utilized in simulations by 4-mm-thick Aluminium. The resultant five-faceted box was split along the $\hat{y}\hat{z}$ plane into two parts to enable accurate fabrication using computerized numerical control (CNC) machines in the University of Toronto; the two parts of the box were attached using multiple metallic screws along the box perimeter (Fig. 5). A subminiature version A (SMA) female connector flange with an exposed pin was mounted using screws on the front facet to feed the current source exciting the structure. The current source was created by extending the exposed pin using a soldered copper wire, forming an electric dipole with an overall length of $\lambda/10$, in accordance to the simulated configuration.

Finally, the metastrip was vertically positioned using two dowel pins inserted at the edges of the cavity aperture, where the electric field is predicted to be negligible. When assembling the antenna, the horizontal alignment of the metastrip was insured by the use of masking tape strips while the box peripheral screws were tightened; the tape was removed before measurements were conducted. To guarantee good coupling between the middle layer copper traces (Fig. 4) and the metallic walls, a thin adhesive copper film was attached to the metallic walls where they made contact with the the metastrip facets.

Antenna measurements were conducted in the far-field measurement chamber at the University of Toronto, calibrated using Quinstar Technology Inc. QWH-KPRS00 standard-gain horn antennas following the gain comparison method. The HMS antenna was then mounted on a stage situated in the far field of a transmitting horn antenna, and radiation patterns were obtained by rotating the stage in steps of 0.2°. The received power was spectrally resolved with a frequency resolution of 0.02 GHz within the frequency range $f \in [18.1, 21.9\,GHz]$. The procedure for evaluating the antenna gain, directivity, radiation efficiency, and aperture illumination efficiency out of the measured radiation patterns is addressed in detail in the Supplementary Methods.

References

1. Kraus, J. The corner-reflector antenna. *Proc. IRE* **28,** 513–519 (1940).
2. von Trentini, G. Partially reflecting sheet arrays. *IRE Trans. Antennas Propag.* **4,** 666–671 (1956).
3. Hansen, R. Communications satellites using arrays. *Proc. IRE* **49,** 1066–1074 (1961).
4. Franson, S. & Ziolkowski, R. Gigabit per second data transfer in high-gain metamaterial structures at 60 GHz. *IEEE Trans. Antennas Propag.* **57,** 2913–2925 (2009).
5. Tichit, P.-H., Burokur, S. N., Germain, D. & de Lustrac, A. Design and experimental demonstration of a high-directive emission with transformation optics. *Phys. Rev. B* **83,** 155108 (2011).
6. Jiang, Z. H., Gregory, M. D. & Werner, D. H. Experimental demonstration of a broadband transformation optics lens for highly directive multibeam emission. *Phys. Rev. B* **84,** 165111 (2011).
7. Lier, E., Werner, D. H., Scarborough, C. P., Wu, Q. & Bossard, J. A. An octave-bandwidth negligible-loss radiofrequency metamaterial. *Nat. Mater.* **10,** 216–222 (2011).
8. Imbriale, W. A. *Modern Antenna Handbook*. Ch. 5 (Wiley, 2008).
9. Hum, S. V. & Perruisseau-Carrier, J. Reconfigurable reflectarrays and array lenses for dynamic antenna beam control: A review. *IEEE Trans. Antennas Propag.* **62,** 183–198 (2014).
10. Tsoilos, G. V. & Christodoulou, C. G. *Modern Antenna Handbook*. Ch. 11 (Wiley, 2008).
11. Haupt, R. & Rahmat-Samii, Y. Antenna array developments: A perspective on the past, present and future. *IEEE Antennas Propag. Mag.* **57,** 86–96 (2015).
12. Jackson, D. R. & Oliner, A. A. *Modern antenna handbook*. Ch. 7 (Wiley, 2008).
13. Jackson, D. R. *et al.* The fundamental physics of directive beaming at microwave and optical frequencies and the role of leaky waves. *Proc. IEEE* **99,** 1780–1805 (2011).
14. Lovat, G., Burghignoli, P. & Jackson, D. R. Fundamental properties and optimization of broadside radiation from uniform leaky-wave antennas. *IEEE Trans. Antennas Propag.* **54,** 1442–1452 (2006).
15. Fong, B., Colburn, J., Ottusch, J., Visher, J. & Sievenpiper, D. Scalar and tensor holographic artificial impedance surfaces. *IEEE Trans. Antennas Propag.* **58,** 3212–3221 (2010).
16. Minatti, G., Caminita, F., Casaletti, M. & Maci, S. Spiral leaky-wave antennas based on modulated surface impedance. *IEEE Trans. Antennas Propag.* **59,** 4436–4444 (2011).
17. Patel, A. M. & Grbic, A. A printed leaky-wave antenna based on a sinusoidally-modulated reactance surface. *IEEE Trans. Antennas Propag.* **59,** 2087–2096 (2011).
18. Minatti, G. *et al.* Modulated metasurface antennas for space: Synthesis, analysis and realizations. *IEEE Trans. Antennas Propag.* **63,** 1288–1300 (2015).
19. Sievenpiper, D. Forward and backward leaky wave radiation with large effective aperture from an electronically tunable textured surface. *IEEE Trans. Antennas Propag.* **53,** 236–247 (2005).
20. Komanduri, V. R., Jackson, D. R. & Long, S. A. in *Proceedings of IEEE International Symposium on Antennas and Propagation (APSURSI)*, 1–4 (Toronto, ON, Canada, 2010).
21. Garca-Vigueras, M., Delara-Guarch, P., Gómez-Tornero, J. L., Guzmán-Quirós, R. & Goussetis, G. in *Proceedings of the 6th European Conference on Antennas and Propagation (EuCAP)*, 247-251 (Prague, Czech Republic, 2012).
22. Feresidis, A. P. & Vardaxoglou, J. C. in *Proceedings of the 1st European Conference on Antennas and Propagation (EUCAP)*, 3–7 (Nice, France, 2006).
23. Ju, J., Kim, D. & Choi, J. Fabry-Pérot cavity antenna with lateral metalic walls for WiBro base station applications. *Electron. Lett.* **45,** 141–142 (2009).
24. Muhammad, S. A., Sauleau, R. & Legay, H. Small-size shielded metallic stacked Fabry-Pérot cavity antennas with large bandwidth for space applications. *IEEE Trans. Antennas Propag.* **60,** 792–802 (2012).
25. Muhammad, S. A. & Sauleau, R. in *Proceedings of the 11th International Bhurban Conference on Applied Sciences and Technology (IBCAST)*, 14–17 (Islamabad, Pakistan, 2014).

26. Kim, D., Ju, J. & Choi, J. A mobile communication base station antenna using a genetic algorithm based Fabry-Pérot resonance optimization. *IEEE Trans. Antennas Propag.* **60,** 1053–1058 (2012).
27. Hosseini, S. A., Capolino, F. & De Flaviis, F. in *Proceedings of the IEEE International Symposium on Antennas and Propagation (APSURSI)*, 746–747 (Orlando, FL, USA, 2013).
28. Haralambiev, L. A. & Hristov, H. D. Radiation characteristics of 3D resonant cavity antenna with grid-oscillator integrated inside. *Int. J. Antennas Propag.* **2014,** 479189 (2014).
29. Balanis, C. *Antenna Theory: Analysis and Design.* Ch. 12 (Wiley, 1997).
30. Muhammad, S. A., Sauleau, R. & Legay, H. in *Proceedings of the 5th European Conference on Antennas and Propagation (EUCAP)*, 1526–1530 (Rome, Italy, 2011).
31. Pfeiffer, C. & Grbic, A. Metamaterial Huygens' surfaces: tailoring wave fronts with reflectionless sheets. *Phys. Rev. Lett.* **110,** 197401 (2013).
32. Selvanayagam, M. & Eleftheriades, G. V. Discontinuous electromagnetic fields using orthogonal electric and magnetic currents for wavefront manipulation. *Opt. Express* **21,** 14409–14429 (2013).
33. Selvanayagam, M. & Eleftheriades, G. V. Polarization control using tensor Huygens surfaces. *IEEE Trans. Antennas Propag.* **62,** 6155–6168 (2014).
34. Ra'di, Y., Asadchy, V. S. & Tretyakov, S. A. One-way transparent sheets. *Phys. Rev. B* **89,** 075109 (2014).
35. Zhu, B. O. *et al.* Dynamic control of electromagnetic wave propagation with the equivalent principle inspired tunable metasurface. *Sci. Rep.* **4,** 4971 (2014).
36. Pfeiffer, C. *et al.* Efficient light bending with isotropic metamaterial Huygens' surfaces. *Nano Lett.* **14,** 2491–2497 (2014).
37. Wong, J. P. S., Selvanayagam, M. & Eleftheriades, G. V. Design of unit cells and demonstration of methods for synthesizing Huygens metasurfaces. *Photon. Nanostruct.* **12,** 360–375 (2014).
38. Wong, J. P. S., Selvanayagam, M. & Eleftheriades, G. V. Polarization considerations for scalar Huygens metasurfaces and characterization for 2-D refraction. *IEEE Trans. Microwave Theor. Technol.* **63,** 913–924 (2015).
39. Decker, M. *et al.* High-efficiency dielectric Huygens' surfaces. *Adv. Opt. Mater.* **3,** 813–820 (2015).
40. Epstein, A. & Eleftheriades, G. V. Passive lossless Huygens metasurfaces for conversion of arbitrary source field to directive radiation. *IEEE Trans. Antennas Propag.* **62,** 5680–5695 (2014).
41. Kuester, E., Mohamed, M., Piket-May, M. & Holloway, C. Averaged transition conditions for electromagnetic fields at a metafilm. *IEEE Trans. Antennas Propag.* **51,** 2641–2651 (2003).
42. Tretyakov, S. *Analytical Modeling in Applied Electromagnetics* (Artech House, 2003).
43. Felsen, L. B. & Marcuvitz, N. *Radiation and Scattering of Waves,* 1st edn (Prentice-Hall, 1973).
44. Howell, K. B. *The Transforms and Applications Handbook.* Ch. 2, 2nd edn (CRC Press, 2000).
45. Chung-Shu, K. S. Cavity-backed spiral antenna with mode suppression. (U.S. Patent 3,555,554, 1971).
46. Epstein, A. & Eleftheriades, G. V. Floquet-Bloch analysis of refracting Huygens metasurfaces. *Phys. Rev. B* **90,** 235127 (2014).
47. Abadi, S. M. A. M. H. & Behdad, N. Design of wideband, FSS-based multibeam antennas using the effective medium approach. *IEEE Trans. Antennas Propag.* **62,** 5557–5564 (2014).
48. Martinis, M., Mahdjoubi, K., Sauleau, R., Collardey, S. & Bernard, L. Bandwidth behavior and improvement of miniature cavity antennas with broadside radiation pattern using a metasurface. *IEEE Trans. Antennas Propag.* **63,** 1899–1908 (2015).
49. Pozar, D. M. *Microwave Engineering.* Ch. 5, 4th edn (John Wiley and Sons, Inc., 2011).
50. Pfeiffer, C. & Grbic, A. Planar lens antennas of subwavelength thickness: collimating leaky-waves with metasurfaces. *IEEE Trans. Antennas Propag.* **63,** 3248–3253 (2015).
51. Monticone, F., Estakhri, N. M. & Alù, A. Full control of nanoscale optical transmission with a composite metascreen. *Phys. Rev. Lett.* **110,** 203903 (2013).
52. Cheng, J. & Mosallaei, H. Optical metasurfaces for beam scanning in space. *Opt. Lett.* **39,** 2719–2722 (2014).
53. Yu, N. & Capasso, F. Flat optics with designer metasurfaces. *Nat. Mater.* **13,** 139–150 (2014).
54. Lin, D., Fan, P., Hasman, E. & Brongersma, M. L. Dielectric gradient metasurface optical elements. *Science* **345,** 298–302 (2014).
55. Campione, S., Basilio, L. I., Warne, L. K. & Sinclair, M. B. Tailoring dielectric resonator geometries for directional scattering and Huygens' metasurfaces. *Opt. Express* **23,** 2293 (2015).
56. Aieta, F., Kats, M. A., Genevet, P. & Capasso, F. Multiwavelength achromatic metasurfaces by dispersive phase compensation. *Science* **347,** 1342–1345 (2015).
57. Selvanayagam, M. & Eleftheriades, G. V. Circuit modelling of Huygens surfaces. *IEEE Antennas Wirel. Propag. Lett* **12,** 1642–1645 (2013).
58. Lovat, G., Burghignoli, P., Capolino, F., Jackson, D. & Wilton, D. Analysis of directive radiation from a line source in a metamaterial slab with low permittivity. *IEEE Trans. Antennas Propag.* **54,** 1017–1030 (2006).

Acknowledgements

Financial support from the Natural Sciences and Engineering Research Council of Canada (NSERC) is gratefully acknowledged. A.E. gratefully acknowledges the support of the Lyon Sachs Postdoctoral Fellowship Foundation as well as the Andrew and Erna Finci Viterbi Postdoctoral Fellowship Foundation of the Technion - Israel Institute of Technology, Haifa, Israel. He thanks M. Selvanayagam, T. Cameron and L. Liang for valuable advice and assistance with the experimental set-up.

Author contributions

A.E. performed formulation, analysis, simulations, physical design, experiments and generation of the results, J.P.S.W. conceived the unit-cell structure and performed simulations, and G.V.E. supervised all stages. A.E. and G.V.E. contributed to conceiving the idea, and writing and editing the manuscript.

Additional information

Competing financial interests: The authors declare no competing financial interests.

23

Dispersion and shape engineered plasmonic nanosensors

Hyeon-Ho Jeong[1,2,*], Andrew G. Mark[1,*], Mariana Alarcón-Correa[1,3], Insook Kim[1,3], Peter Oswald[1], Tung-Chun Lee[1,4] & Peer Fischer[1,3]

Biosensors based on the localized surface plasmon resonance (LSPR) of individual metallic nanoparticles promise to deliver modular, low-cost sensing with high-detection thresholds. However, they continue to suffer from relatively low sensitivity and figures of merit (FOMs). Herein we introduce the idea of sensitivity enhancement of LSPR sensors through engineering of the material dispersion function. Employing dispersion and shape engineering of chiral nanoparticles leads to remarkable refractive index sensitivities (1,091 nm RIU^{-1} at $\lambda = 921$ nm) and FOMs (>2,800 RIU^{-1}). A key feature is that the polarization-dependent extinction of the nanoparticles is now characterized by rich spectral features, including bipolar peaks and nulls, suitable for tracking refractive index changes. This sensing modality offers strong optical contrast even in the presence of highly absorbing media, an important consideration for use in complex biological media with limited transmission. The technique is sensitive to surface-specific binding events which we demonstrate through biotin-avidin surface coupling.

[1] Max Planck Institute for Intelligent Systems, Heisenbergstrasse 3, 70569 Stuttgart, Germany. [2] Institute of Materials, École Polytechnique Fédérale de Lausanne (EPFL), CH-1015 Lausanne, Switzerland. [3] Institute for Physical Chemistry, University of Stuttgart, Pfaffenwaldring 55, 70569 Stuttgart, Germany. [4] UCL Institute for Materials Discovery and Department of Chemistry, University College London, Christopher Ingold Building, 20 Gordon Street, London WC1H 0AJ, UK. * These authors contributed equally to this work. Correspondence and requests for materials should be addressed to P.F. (email: fischer@is.mpg.de).

Devices based on surface plasmon resonance (SPR) phenomena detect shifts of the resonance wavelength in response to changes of the refractive index of the medium surrounding the plasmonic material[1]. This may, for instance, be due to biomolecules that bind to the sensor. SPR offers high sensitivities[2], but requires extended smooth surfaces[3]. In many situations, it would be desirable to have a local sensor for use *in situ* or *in vivo* (for instance within a cell) and here the localized SPR (LSPR) supported by nanostructures offers substantial advantages[4–6]. The short penetration depth of plasmon oscillations into the surrounding fluid makes for small, spatially localized sensors which promise to be effective in a range of biomedical applications[7–9]. However, compared with plasmonic biosensors that utilize extended SPR, which serve as the reference standard for optically addressed sensors, LSPR sensors generally have a reduced sensitivity ($S_n < 1{,}000\,\mathrm{nm\,RIU}^{-1}$) and a lower figure of merit (FOM) $< 100\,\mathrm{RIU}^{-1}$ (refs 2,5).

The typical strategy employed to enhance the sensitivity of LSPR nanosensors is to manipulate the aspect ratio of symmetrical nanoparticles, with more elongated particles yielding higher sensitivities[10,11]. This approach has been demonstrated with nanorods and nanoprisms that show improved sensitivites[12–14]. On the other hand, the materials selected for such particles have, quite reasonably, been restricted to those with strong plasmonic properties: for example, Ag (ref. 15), Au (ref. 9), Al (ref. 16). This has been driven by the desire for a high figure of merit (FOM) that comes from the low-interband damping, and sharp resonance of pure metals. However, it means that the material dispersion function, a key factor in the sensitivity of LSPR nanosensors, has been limited to the dielectric functions inherent to those pure metals.

Here, we show how dispersion-engineering introduces a new material-based parameter for improving the sensitivity of LSPR sensors. When combined with shape engineering, this leads to extremely high LSPR sensitivities and FOMs, which we report herein. We introduce an analytical model of chiral plasmonic sensing that illustrates the roles of chirality and materials properties for the important sensing characteristics. Based on this understanding we present colloidal nanostructures that are, to the best of our knowledge, the most sensitive LSPR sensors reported to date[12,17]. Furthermore, we demonstrate the utility of the engineered particles as surface-sensitive probes for biotin–avidin binding. Our scheme is particularly robust as it is immune to changes in the optical density of the background, and can thus be used in complex environments.

Results

Theoretical concept. Plasmon-based LSPR sensors operate by tracking the shift in the resonance peak of the plasmon absorption in response to changes in n, the refractive index of the local medium[5]. The resonance condition is met when $\varepsilon_r^* = \varepsilon_r(\lambda^*) = -\chi n^2$, where λ^* is the peak wavelength, ε_r is the real part of the dielectric function describing the plasmonic material, and the factor χ describes the shape of the particle (for spheres $\chi = 2$). Ideally, the dispersion of ε_r should be such that even very small changes in n cause appreciable changes in the resonance wavelength λ^*, that is, large resonance shifts. The sensitivity of the system is defined as[11]

$$S_n = \frac{\mathrm{d}\lambda^*}{\mathrm{d}n} = \frac{\frac{\mathrm{d}\varepsilon_r^*}{\mathrm{d}n}}{\left(\frac{\mathrm{d}\varepsilon_r}{\mathrm{d}\lambda}\right)_{\lambda^*}} = \frac{-2\chi n}{\left(\frac{\mathrm{d}\varepsilon_r}{\mathrm{d}\lambda}\right)_{\lambda^*}} \quad (1)$$

so decreasing the wavelength dependence of the real part of the material's dielectric constant (the denominator) leads to an increase in sensitivity. Here, we engineer the dielectric response of the nanoparticles by alloying a plasmonic material with a weakly dispersive one to yield a negative, but flatter ε_r. This dramatically increases the sensitivity of any plasmonic sensor (of any shape).

In conjunction with the sensitivity, it is also of interest to consider the accuracy with which a spectral feature can be resolved. For a spectral peak, the standard measure is its full width at half maximum (FWHM). Engineering the real part of the particle's dielectric function[18] in equation (1) through alloying is generally accompanied by a concomitant increase in the plasmon damping and broadening of the extinction spectrum[19], and thus suggests that the FOM ($= S_n$/FWHM) is also lowered. This can be addressed by noting that traditional plasmon-based sensors are generally polarization independent, because the particles themselves are highly symmetrical, Fig. 1a. Circular dichroism (CD) offers an alternative to extinction-based sensing techniques[20,21], and CD spectra are typically more feature rich than extinction spectra[22,23], which increases the number of spectral signatures that can be tracked when the local environment changes, Fig. 1b. Crucially, CD spectra are also bipolar, and the crossing points where the signal changes sign are ideal features to track, because of the simplicity of identifying the

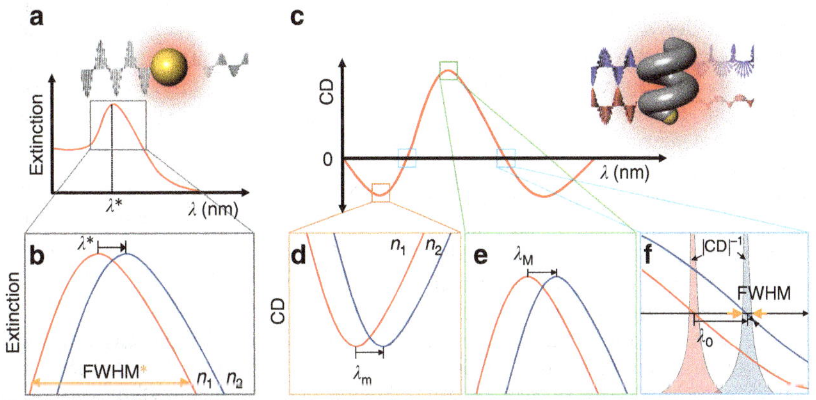

Figure 1 | Chiral plasmonic sensing. (**a**,**b**) Schematic view of conventional plasmonic sensing. (**a**) The plasmonic resonance of the metallic nanoparticle as a function of λ. Inset shows a conventional plasmonic sensing system where a metallic nanoparticle (here a sphere) interacts with light and generates a detectable absorption peak due to plasmon resonance. (**b**) A zoomed-in region near the peak λ^* showing how it shifts in wavelength as the refractive index of the surrounding medium changes $n_1 < n_2$. (**c**-**f**) Schematic view of polarization-dependent chiral plasmonic sensing. (**c**) The CD spectrum of a plasmonic enantiomer as a function of λ. Inset scheme illustrates the interaction of a left-handed nanohelix with circularly polarized light. Three bottom panels indicate the resonance shifts at (**d**) λ_m, (**e**) λ_M and (**f**) λ_0 where the refractive indices of the surrounding media are varied between n_1 and n_2.

null point. Earlier work on CD-based sensing made use of magneto-optical modulation of an achiral nanoantenna LSPR to induce ellipticity in the transmitted beam[24]. In this case, the choice of material was dictated by the requirement to induce a magneto-optical response. However, performing CD-based measurements on chiral particles, that exhibit a natural CD, offers the possibility of tuning the material properties independently of the chirality (shape), and we exploit this approach to maximize the LSPR sensitivity (and FOM).

The exact optical responses for such complex nanoparticles are only calculable using numerical methods[25]. However, we introduce the following chiral form of nanoparticle plasmonic absorption[10,11] that captures the key features of extinction in response to left (−) and right (+) circularly polarized light.

$$E_{\pm} = \frac{2\pi l N V n^3}{\lambda \ln(10)} \left(\frac{\varepsilon_i}{\left(\varepsilon_r + (\bar{\chi} \pm \delta\chi_{D/L}) n^2\right)^2 + \varepsilon_i^2} \right). \quad (2)$$

The shape factor now includes an achiral term $\bar{\chi}$ plus a chiral term $\delta\chi_D = -\delta\chi_L$ specific to right (D) or left (L) handed enantiomers. The CD is the difference between the extinction for ± polarizations, $CD \propto E_- - E_+$. The CD signal is bipolar, and the wavelength at which the extinctions are equal yields a zero-crossing in the spectrum, which serves as a natural point for tracking. Within this formulation the crossover wavelength coincides with the achiral resonance condition $\varepsilon_r^* = -\bar{\chi}n^2$, in the limit of small, slowly-varying ε_i. So the sensitivity of nanohelix sensors follows that of achiral ones. Plotting the reciprocal of the absolute value of the CD (for example, Fig. 2d) gives an intrinsically sharp representation of the crossover point, one whose FWHM is defined by the instrumental resolution σ (ref. 24) and $\frac{dCD}{d\lambda}\big|_{\lambda^*}$, the slope of the CD spectrum at the crossing. This yields a FWHM (see Supplementary Note 1 for details), based on equation (2), of

$$\mathrm{FWHM} = \frac{4\sigma\varepsilon_i^2}{|\delta\chi| n^2 \left|\frac{d\varepsilon_r}{d\lambda}\right|_{\lambda^*}} \quad (3)$$

and a FOM of

$$\mathrm{FOM} = \frac{S_n}{\mathrm{FWHM}} = \frac{2\bar{\chi}|\delta\chi| n^3}{\varepsilon_i^2 \sigma} \quad (4)$$

Thus, higher FOM can be achieved by increasing the magnitude of the chiral shape factor $\delta\chi$.

Fabrication and bulk refractive index sensing. We fabricate a series of hybrid nanohelix sensors using a physical vapour deposition technique that allows precise control of the particles' alloy composition and shape $\delta\chi$ (Supplementary Note 2, Supplementary Fig. 1 and Supplementary Table 1)[25,26]. The former can be used to engineer the material dielectric function to maximize sensitivity, while the latter is used to adjust the achiral and chiral shape factors to affect sensitivity and introduce a chiroptical response. A typical particle, a two-turn left-handed 128-nm-tall helix composed of 97% Ag and 3%Ti, is shown in Fig. 2a. The small amount of Ti alloying agent improves the helix fidelity over pure Ag (ref. 27). The particles are tested in the form of a colloid suspension whose stability was confirmed through dynamic light scattering (DLS, more details in Supplementary Note 3 and Supplementary Fig. 2). Figure 2b shows CD spectra of the suspensions in glycerol–water mixtures varying from 0 to 20% concentration (refractive indices between 1.333 and 1.357 (ref. 28)). As expected, the CD spectra exhibit multiple features, shown in Fig. 2c–f: a maximum λ_M, a minimum λ_m and two crossing points λ_{01} and λ_{02}, all of which can be tracked in

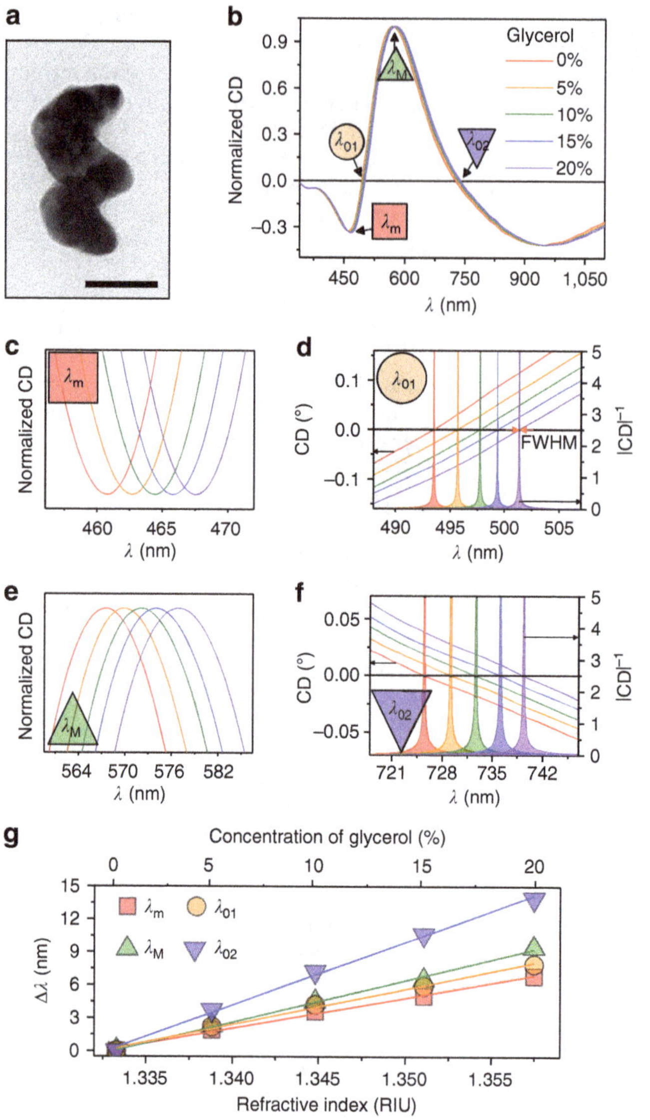

Figure 2 | Bulk refractive index sensing. (**a**) TEM image of a single $Ag_{0.97}Ti_{0.03}{:}L_2$ nanohelix (Scale bar, 50 nm). (**b**) CD spectra of colloidal $Ag_{0.97}Ti_{0.03}{:}L_2$ nanohelices in media of five different refractive indices (red: 0%, orange: 5%, green: 10%, blue: 15% and violet: 20% glycerol-water mixtures) over the full spectral range and detailed plots of the resonance shifts at (**c**) λ_m (red square), (**d**) λ_{01} (orange circle), (**e**) λ_M (green top triangle) and (**f**) λ_{02} (blue bottom triangle). For **d** and **f** the filled curves represent $|CD|^{-1}$. (**g**) Wavelength shift, relative to water, of the four spectral features as functions of the glycerol-water concentration (top *x* axis) and its corresponding *n* (bottom *x* axis).

response to changes in the refractive index of the medium. The wavelength shifts, relative to the pure water reference, are shown in Fig. 2g, and indicate sensitivities S_n of 275, 320, 379 and 571 nm RIU^{-1} for λ_m, λ_{01}, λ_M and λ_{02}, respectively.

Dispersion and shape engineering of chiral nanoparticles. These results demonstrate that features, like λ_M, and λ_{02}, found at longer wavelengths exhibit greater sensitivity, consistent with equation (1). Since larger values of the achiral shape factor increase the wavelength of the resonance condition, it is possible to increase the sensitivity by growing particles that are more elongated (Fig. 3a). This has been the approach most often pursued in the search for higher sensitivities from extinction-based LSPR sensors[10,11,12,17]. Here we apply the same principle to

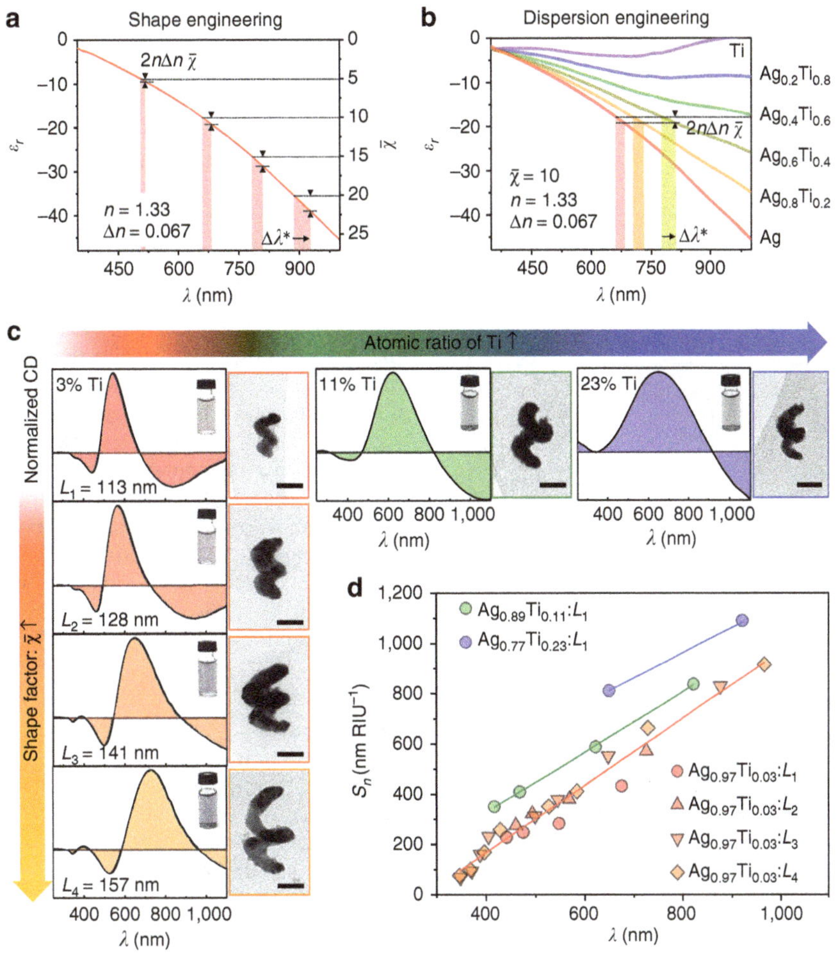

Figure 3 | Shape and dispersion engineering of Ag-Ti nanohelices. (**a**) The effect of the achiral shape factor $\bar{\chi}$ on the wavelength shift $\Delta\lambda$ observed for a given change in refractive index Δn (=0.067). (**b**) The effect of ε_r on the wavelength shift based on calculated dielectric constant of Ag-Ti alloys with varying Ti composition. As the proportion of Ti increases the real part of the dielectric function becomes flatter, and the resulting wavelength shift increases. (**c**) CD spectra (panel left), TEM images (panel right) and colloidal solutions (each inset) of the grown Ag-Ti nanohelices. The achiral shape factor $\bar{\chi}$ increases moving down the rows, and the atomic ratio of Ti increases moving to the right. (**d**) The measured S_n of the Ag-Ti nanohelices as a function of λ at 1.333 RIU (the symbols and colours are different for the $\bar{\chi}$ and the atomic ratio of Ti, respectively).

a chiral sensor, by growing a series of nanohelices with a range of heights, labelled L_1 (113 nm) through L_4 (157 nm). The results, shown in Fig. 3c, confirm that the peak and crossing features of longer structures are red-shifted relative to the smaller ones and that the sensitivity of each feature is directly proportional to its wavelength (Fig. 3d, red points).

However, equation (1) also suggests that the sensitivity can be improved by reducing the wavelength dependence of ε_r, the real part of the material dielectric function. We do this by exercising control over the structures' Ag–Ti stoichiometry during their growth. Figure 3b shows the effective ε_r of Ag–Ti alloys of varying compositions (more details in Supplementary Note 4). As the alloy becomes more Ti-rich, the wavelength shift $\Delta\lambda$ increases for a given change in the medium's refractive index, due to a progressive flattening of the material's dielectric function (see also Supplementary Fig. 3 and Supplementary Table 2). The effect of dispersion engineering in practice is illustrated in Fig. 3c for nanohelices having fixed size, but composed of alloys containing 3, 11 and 23% Ti. Higher Ti composition red-shifts the features, and as shown in Fig. 3d, and increases the sensitivity. Indeed, the sensitivity trends show a systematic enhancement for the 11% (green) and 23% (blue) alloy relative to the minimal Ti samples (red; see Supplementary Note 5 and Supplementary Figs 4–9 for detailed plots).

The results are summarized in Fig. 4. The left column in the plots shows the effect of increased nanohelix length on the crossing point wavelength, sensitivity, FWHM and FOM. The null wavelength and sensitivity, both increase as the nanohelices lengthen. However, the FWHM remains unchanged, so that the FOM increases dramatically. The L_4 helix exhibits a FOM = 2,859 RIU^{-1}, which is larger than what has been previously reported for LSPR-based sensors[13,14,29] including those based on a magnetochiral response[24]. Relative to the latter, the improvement comes from a combination of improved sensitivity ($\sim 4\times$) and decreased FWHM ($\sim 5\times$) thanks to a steeper crossing of the CD signal at the null point (Supplementary Note 6 and Supplementary Fig. 10). The right hand column of Fig. 4 illustrates the effect of engineering the dispersion function through the addition of Ti. In this case the increase in sensitivity that comes from flattening the material dispersion function also leads to an increase in the FWHM due to the ε_i contribution in equation (3). The net effect is a decrease in the FOM relative to the low-alloy sample. Nevertheless, the flatter dispersion curve does have the effect of increasing the sensitivity beyond what is achievable in the pure metal or the low-alloy sample. Here we observe a sensitivity $>$1,000 nm RIU^{-1} at a wavelength of 921 nm. For applications where sensitivity is the primary concern, dispersion engineering provides a powerful technique

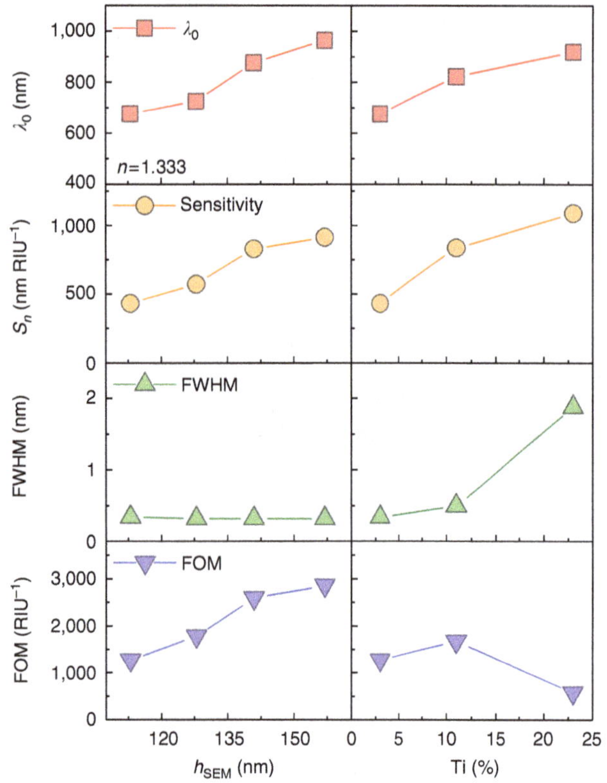

Figure 4 | Summary of shape and dispersion engineering of nanohelices. Left panels show the effect of increasing nanohelix length and the right panels show the effect of increasing Ti composition (red square: zero-crossing wavelengths at $n = 1.333$, orange circle: refractive index sensitivities, green top triangles: FWHMs and blue bottom triangles: FOMs).

for reaching the highest possible refractive index sensitivities (Supplementary Note 7 and Supplementary Fig. 11).

Strong optical response in the presence of absorbers. One appealing characteristic of CD is that it is insensitive to achiral absorbers or scatterers[30]. So unlike traditional extinction measurements, high signal-to-noise CD measurements of a chiral analyte are possible even in the presence of strong achiral absorbers. Figure 5a illustrates such a measurement, where we use a colloidal suspension of $Ag_{0.97}Ti_{0.03}$:L_3 nanohelices in water with the addition of absorbers in the form of blue and yellow filters in the optical path. In both cases the maximum absorbance exceeds an optical density of 2 (<0.1% transmission, right panels of Fig. 5a). This introduces small distortions into the CD signal, however, as we have already seen, CD offers several spectral features for sensing, the majority of which are here unperturbed. Notably, since CD-based sensing offers a clearly distinguishable signal even in the presence of absorbance that would be strong enough to obscure optical extinction-based sensing methodologies, Fig. 5b. The same experiment demonstrated with molecular absorbers rhodamine and indigo yield similar results (Supplementary Note 8 and Supplementary Fig. 12).

Surface-sensitive sensing. Since the evanescent field of the excited plasmon decays rapidly into the medium, LSPR sensors are ideally suited to making surface-sensitive measurements[31]. Here we use our chiral plasmonic nanohelices to sense avidin-binding events. First, the surfaces of our nanohelices ($Ag_{0.97}Ti_{0.03}$:L_1 and $Ag_{0.89}Ti_{0.11}$:L_1) are functionalized with biotin (see Methods section for details), which acts as a complementary and specific-binding factor for avidin in solution (Fig. 5c). The wavelength of the λ_{02} crossing in the CD spectrum (Fig. 5d, lower), and the change in the CD amplitude at the initial crossing wavelength (Fig. 5d, upper) were measured at 1 min intervals. Before the addition of avidin both signals are stable, but they experience a prompt shift on the addition of 1 μg ml^{-1} avidin. For the $Ag_{0.97}Ti_{0.03}$:L_1 nanohelices the crossing point $\Delta\lambda_{02}$ redshifts by ~3 nm (blue), and the CD intensity at the initial crossing wavelength increases by 6 mdeg (green). The rise time is ~5 min. The $Ag_{0.89}Ti_{0.11}$:L_1 nanohelices show an even stronger response with shifts of 5 nm (red) and 10 mdeg (yellow) respectively, thanks to the increased sensitivity that comes from the reduced dispersion. Control experiments (grey) where the nanohelices lack biotin functionalization show no distinct shift on the addition of avidin, confirming that the signal is truly surface sensitive and arises from specific binding (see Supplementary Note 9 and Supplementary Figs 13 and 14 for additional details). The two measurement modes suggest alternative sensing schemes for real-time monitoring of binding events: one based on tracking the crossing point wavelength (Fig. 5e, blue and red), and another based on identifying the change in CD intensity at fixed wavelength (Fig. 5e, green and yellow). Both offer a highly sensitive and rapid measurement of the specific binding at surfaces.

Discussion

We have shown that flattening the dielectric function of plasmonic nanomaterials by alloying results in large LSPR shifts in response to changes in their local environment. If these nanostructures are also chiral, then their polarization-dependent spectra offer additional and sharper spectral features relative to optical extinction-based measurement of achiral particles. These characteristics make chiral plasmonic nanohelices especially suited for sensing applications. Furthermore, the bipolar nature of CD spectra of chiral structures leads to points of zero-crossing, which have very small effective line-widths and thus result in high FOM. Both bulk refractive index changes as well as surface-specific binding events are tracked by the engineered nanocolloidal LSPR sensors with sensitivity (~1,100 nm RIU^{-1}) and FOM (~2,900 RIU^{-1}) higher than any previously reported for plasmonic sensors.

Aside from extremely high-detection sensitivities, these chiral sensors offer unique advantages over existing plasmonic devices. Since they are freely suspended colloids, they naturally suggest themselves as sensors for *in situ* or *in vivo* biological applications, particularly, because of their small size. Since the CD signal is robust against background interference, especially in absorbing media, the sensing scheme reported herein is particularly suited to real biological media, which are typically optically dense. For instance, in the presence of whole blood, our scheme has the potential for real-time measurement of protein corona formation, a vitally important factor for the function of nanoparticles in biological media, but one that remains poorly understood.

Methods

Nano glancing angle deposition (nanoGLAD). The shadow growth technique[32] was used to grow the three-dimensional nanocolloids in the manner previously reported[25]. First, a hexagonal array of 12 nm Au nanoparticles was prepared by block-copolymer micelle nanolithography[33]. Next, Ag–Ti nanohelices were grown on the array of the Au nanoparticles in a GLAD system based on co-deposition from dual electron-beam evaporators at $T = 90$ K with a base pressure of 1×10^{-6} mbar. Their alloy stoichiometry was controlled by the deposition rates measured by quartz crystal microbalance (QCM) for each evaporator independently. Particles were grown with lengths (measured by scanning electron microscopy (SEM)) of $L_1 = 113$ nm, $L_2 = 128$ nm, $L_3 = 141$ nm, $L_4 = 157$ nm and Ti contents of 3, 11 and 23%. Finally, the grown Ag–Ti nanohelices were lifted off from the wafer by sonicating a piece of sample wafer (0.5–1 cm^2) in an aqueous solution of 0.15 mM polyvinylpyrrolidone for ~2 min to prepare a stock solution.

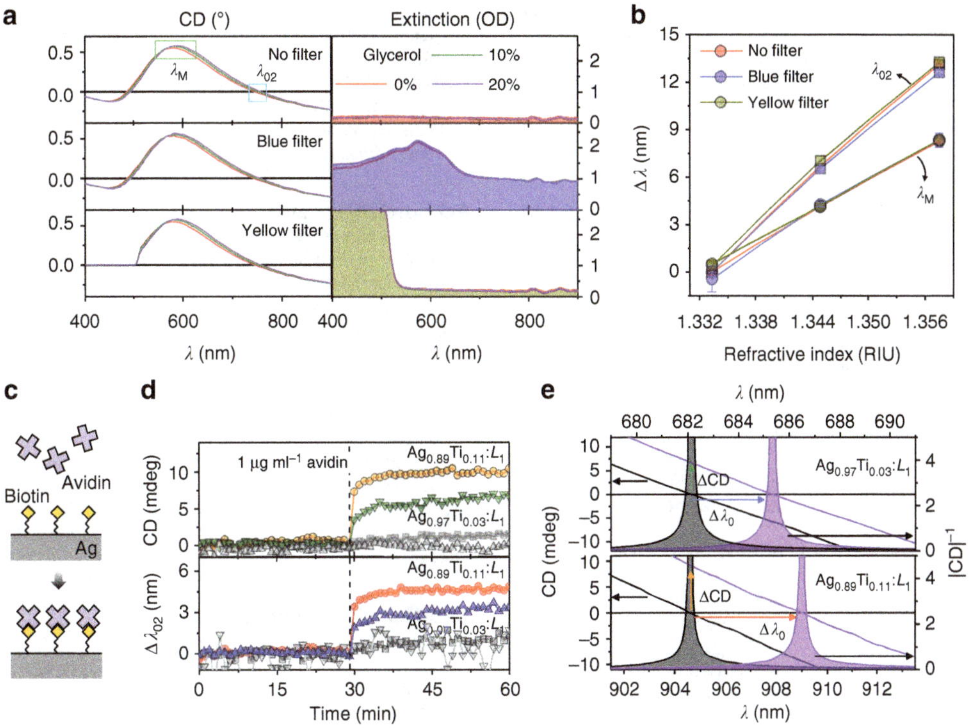

Figure 5 | Sensing in absorbing media and Sensing of specific binding events. CD spectra (left panels) and extinction spectra (right panels) of the colloidal $Ag_{0.97}Ti_{0.03}$:L_3 nanohelices in the presence of complex absorbing environments (top: no filter, middle: a blue filter, bottom: yellow filter), (**b**) their corresponding wavelength shifts, relative to water, of the two spectral features (λ_M and λ_{02}) as functions of the n (error bar: s.d.). (**c**) Schematic view of biotin-avidin interaction on the surface of a Ag-Ti nanohelix. (**d**) *In situ* measurements of the biotin-avidin interaction by monitoring the change of CD at the initial λ_{02} (upper plot) and the wavelength shift of $\Delta\lambda_{02}$ (lower plot) with 1 min intervals. The coloured plots indicate the response of specific binding of biotin-avidin ($Ag_{0.97}Ti_{0.03}$:L_1 blue, green; $Ag_{0.89}Ti_{0.11}$:L_1 red, yellow) and the grey plots indicate non-specific binding of avidin with Ag-Ti nanohelices (without biotin) in the control group. (**e**) Close-up view of the CD spectra for the two biotinylated nanoparticle systems showing the wavelength shift and CD amplitude increase after avidin introduction.

To minimize the effect of possible variations in structure and concentration, samples for each set of tests were always drawn from the same stock solution.

SEM and TEM analysis. The Ag–Ti helices were imaged in SEM (Ultra 55, Zeiss) and transmission electron microscopy (TEM) (CM200, Philips) under the accelerating voltages of 10 and 200 kV, respectively.

Circular dichroism analysis. CD spectra were obtained with a Jasco J-810 CD spectrometer. All the spectra were measured with 500 nm min^{-1} scan rate in the wavelength range of 300–1,100 nm with 0.1 nm intervals. For selected regions of interest smaller 0.025 nm intervals were used.

DLS analysis. Colloidal solution (200 μl) of nanohelices was measured using a zeta potential analyser (Zetasizer Nano ZS, Malvern) repeated ~10 times for 20 min with 2 min intervals. The material property of Ag–Ti nanohelices was fixed to Ag (RI: 0.135 and Absorption: 3.990) and the environmental parameter was matched to literature values for the viscosity, and refractive index of the solution based on the temperature and concentration[28].

ICP-OES analysis. Approximately 1 cm^2 as-grown nanohelices supported on a Si substrate was dissolved into HNO_3/HF etchant and this solution was analysed by the inductively coupled plasma atomic emission spectroscopy (ICP-OES) (Ciros CCD, Spectro). The material composition of the nanohelices was evaluated by repeating the analysis three times with samples cleaved from different areas of the growth wafer.

Absorbers. Blue and yellow filters were separately inserted into the optical patch either before or after the sample cuvette. Similarly, 10 μM rhodamine 6G and 100 μM indigo were added to separate samples after acquiring baseline measurements.

Biotin-avidin binding procedure. The mixture of biotin-Polyethylene glycol (PEG)-SH (25 mg ml^{-1} in 3-(N-morpholino)propanesulfonic acid (MOPS)-DNA buffer of pH7.5) and CH_3O-PEG-SH (250 mg ml^{-1} in MOPS-DNA buffer of pH7.5) in the ratio of 1:1 (v v^{-1}) was drop-cast onto ~0.5 cm^2 of as-grown nanohelix substrate and kept under humid condition overnight. For the control group, the nanohelices were exposed to only CH_3O-PEG-SH. Next, each substrate was immersed in 1 ml of 0.15 mM polyvinylpyrrolidone solution and sonicated to remove the particles from the wafer and suspend them in solution. Finally, for the biotin–avidin interaction, 1 μg ml^{-1} of avidin was injected to the colloidal solution and mixed by pipetting.

References

1. Homola, J. Present and future of surface plasmon resonance biosensors. *Anal. Bioanal. Chem.* **377,** 528–539 (2003).
2. Haes, A. J. & Duyne, R. P. V. Preliminary studies and potential applications of localized surface plasmon resonance spectroscopy in medical diagnostics. *Expert Rev. Mol. Diagn.* **4,** 527–537 (2004).
3. Nagpal, P., Lindquist, N. C., Oh, S.-H. & Norris, D. J. Ultrasmooth patterned metals for plasmonics and metamaterials. *Science* **325,** 594–597 (2009).
4. Brolo, A. G. Plasmonics for future biosensors. *Nat. Photon.* **6,** 709–713 (2012).
5. Mayer, K. M. & Hafner, J. H. Localized surface plasmon resonance sensors. *Chem. Rev.* **111,** 3828–3857 (2011).
6. Stockman, M. I. Nanoplasmonic sensing and detection. *Science* **348,** 287–288 (2015).
7. Anker, J. N. *et al.* Biosensing with plasmonic nanosensors. *Nat. Mater.* **7,** 442–453 (2008).
8. Howes, P. D., Rana, S. & Stevens, M. M. Plasmonic nanomaterials for biodiagnostics. *Chem. Soc. Rev.* **43,** 3835–3853 (2014).
9. Yang, X., Yang, M., Pang, B., Vara, M. & Xia, Y. Gold nanomaterials at work in biomedicine. *Chem. Rev.* **115,** 10410–10488 (2015).
10. Lee, K.-S. & El-Sayed, M. A. Gold and silver nanoparticles in sensing and imaging: sensitivity of plasmon response to size, shape, and metal composition. *J. Phys. Chem. B* **110,** 19220–19225 (2006).
11. Miller, M. M. & Lazarides, A. A. Sensitivity of metal nanoparticle surface plasmon resonance to the dielectric environment. *J. Phys. Chem. B* **109,** 21556–21565 (2005).

12. Charles, D. E. *et al.* Versatile solution phase triangular silver nanoplates for highly sensitive plasmon resonance sensing. *ACS Nano* **4,** 55–64 (2010).
13. Shen, Y. *et al.* Plasmonic gold mushroom arrays with refractive index sensing figures of merit approaching the theoretical limit. *Nat. Commun.* **4,** 2381 (2013).
14. Gartia, M. R. *et al.* Colorimetric plasmon resonance imaging using nano lycurgus cup arrays. *Adv. Opt. Mater.* **1,** 68–76 (2013).
15. Jin, R. *et al.* Photoinduced conversion of silver nanospheres to nanoprisms. *Science* **294,** 1901–1903 (2001).
16. Knight, M. W. *et al.* Aluminum for plasmonics. *ACS Nano* **8,** 834–840 (2014).
17. Chen, H., Kou, X., Yang, Z., Ni, W. & Wang, J. Shape- and size-dependent refractive index sensitivity of gold nanoparticles. *Langmuir* **24,** 5233–5237 (2008).
18. Boltasseva, A. & Atwater, H. A. Low-loss plasmonic metamaterials. *Science* **331,** 290–291 (2011).
19. Chang, W.-S. *et al.* Tuning the acoustic frequency of a gold nanodisk through its adhesion layer. *Nat. Commun.* **6,** 7022 (2015).
20. Hendry, E. *et al.* Ultrasensitive detection and characterization of biomolecules using superchiral fields. *Nat. Nanotechnol.* **5,** 783–787 (2010).
21. Ma, W. *et al.* Attomolar DNA detection with chiral nanorod assemblies. *Nat. Commun.* **4,** 2689 (2013).
22. Valev, V. K., Baumberg, J. J., Sibilia, C. & Verbiest, T. Chirality and chiroptical effects in plasmonic nanostructures: fundamentals, recent progress, and outlook. *Adv. Mater.* **25,** 2517–2534 (2013).
23. Zhang, S. *et al.* Photoinduced handedness switching in terahertz chiral metamolecules. *Nat. Commun.* **3,** 942 (2012).
24. Maccaferri, N. *et al.* Ultrasensitive and label-free molecular-level detection enabled by light phase control in magnetoplasmonic nanoantennas. *Nat. Commun.* **6,** 6150 (2015).
25. Mark, A. G., Gibbs, J. G., Lee, T.-C. & Fischer, P. Hybrid nanocolloids with programmed three-dimensional shape and material composition. *Nat. Mater.* **12,** 802–807 (2013).
26. Gibbs, J. G., Mark, A. G., Eslami, S. & Fischer, P. Plasmonic nanohelix metamaterials with tailorable giant circular dichroism. *Appl. Phys. Lett.* **103,** 213101 (2013).
27. Larsen, G. K., He, Y., Wang, J. & Zhao, Y. Scalable fabrication of composite Ti/Ag plasmonic helices: controlling morphology and optical activity by tailoring material properties. *Adv. Opt. Mater.* **2,** 245–249 (2014).
28. Cheng, N.-S. Formula for the viscosity of a glycerol – water mixture. *Ind. Eng. Chem. Res.* **47,** 3285–3288 (2008).
29. Kabashin, A. V. *et al.* Plasmonic nanorod metamaterials for biosensing. *Nat. Mater.* **8,** 867–871 (2009).
30. Claborn, K., Isborn, C., Kaminsky, W. & Kahr, B. Optical rotation of achiral compounds. *Angew. Chem. Int. Ed.* **47,** 5706–5717 (2008).
31. Liu, N., Tang, M. L., Hentschel, M., Giessen, H. & Alivisatos, A. P. Nanoantenna-enhanced gas sensing in a single tailored nanofocus. *Nat. Mater.* **10,** 631–636 (2011).
32. Robbie, K., Sit, J. C. & Brett, M. J. Advanced techniques for glancing angle deposition. *J. Vac. Sci. Technol. B* **16,** 1115–1122 (1998).
33. Glass, R., Möller, M. & Spatz, J. P. Block copolymer micelle nanolithography. *Nanotechnology* **14,** 1153–1160 (2003).

Acknowledgements

We are grateful to C. Miksch and J.P. Spatz for providing us with micellar nanolithographically patterned substrates and for SEM access. We thank the ZWE Analytical Chemistry for the ICP-OES analysis and the Stuttgart Centre for Electron Microscopy for technical support with the TEM imaging. This work was supported by the Max Planck-EPFL centre for molecular nanoscience and technology, and the European Research Council under the ERC Grant agreement 278213.

Author contributions

P.F. proposed the idea and H.-H.J., A.G.M., T.-C.L and P.F. designed the experiments. H.-H.J. performed the experiments except for TEM, ellipsometry and ICP-OES. A.G.M. developed the analytical theory. M.A.-C. carried out the TEM imaging and the DLS. I.K. conducted thin-film analysis theoretically and experimentally. P.O. designed bio-experiment. H.-H.J. and A.G.M. analysed the data and H.-H.J., A.G.M. and P.F. wrote the paper.

Additional information

Competing financial interests: The authors declare no competing financial interests.

24

Magnetic hyperbolic optical metamaterials

Sergey S. Kruk[1], Zi Jing Wong[2], Ekaterina Pshenay-Severin[1,3], Kevin O'Brien[2], Dragomir N. Neshev[1], Yuri S. Kivshar[1] & Xiang Zhang[2,4,5]

Strongly anisotropic media where the principal components of electric permittivity or magnetic permeability tensors have opposite signs are termed as hyperbolic media. Such media support propagating electromagnetic waves with extremely large wave vectors exhibiting unique optical properties. However, in all artificial and natural optical materials studied to date, the hyperbolic dispersion originates solely from the electric response. This restricts material functionality to one polarization of light and inhibits free-space impedance matching. Such restrictions can be overcome in media having components of opposite signs for both electric and magnetic tensors. Here we present the experimental demonstration of the magnetic hyperbolic dispersion in three-dimensional metamaterials. We measure metamaterial isofrequency contours and reveal the topological phase transition between the elliptic and hyperbolic dispersion. In the hyperbolic regime, we demonstrate the strong enhancement of thermal emission, which becomes directional, coherent and polarized. Our findings show the possibilities for realizing efficient impedance-matched hyperbolic media for unpolarized light.

[1] Nonlinear Physics Center and Center for Ultrahigh Bandwidth Devices for Optical Systems (CUDOS), Research School of Physics and Engineering, The Australian National University, Canberra, Australian Capital Territory 2601, Australia. [2] NSF Nanoscale Science and Engineering Center, University of California, Berkeley, California 94720, USA. [3] Institute of Applied Physics, Abbe Center of Photonics, Friedrich-Schiller-Universität Jena, 07743 Jena, Germany. [4] Materials Sciences Division, Lawrence Berkeley National Laboratory, Berkeley, California 94720, USA. [5] Department of Physics, King Abdulaziz University, Jeddah 21589, Saudi Arabia. Correspondence and requests for materials should be addressed to S.S.K. (email: Sergey.Kruk@anu.edu.au).

The study of hyperbolic media and hyperbolic metamaterials have attracted significant attention in recent years due to their relatively simple geometry and many interesting properties, such as high density of states, all-angle negative refraction and hyperlens imaging beyond the diffraction limit[1–3]. Usually, both artificial[4–9] and natural[10,11] media with hyperbolic dispersion are uniaxial materials whose axial and tangential dielectric permittivities have opposite signs. In general, however, the propagation of electromagnetic waves inside a material is defined by both the dielectric permittivity and magnetic permeability tensors. Specifically, the electric response defines the dispersion for the transverse magnetic (TM) linearly polarized light, and the magnetic response defines the dispersion for transverse electric (TE) polarization (see details in Supplementary Note 1). Therefore, the ability to control both the electric permittivity and magnetic permeability gives a full flexibility for the dispersion engineering for any arbitrary polarization and nonpolarized light. This is of a major importance for the efficient interaction with randomly positioned emitters or thermal radiation. Moreover, a control over both electric and magnetic responses allows one to engineer the material dispersion and impedance independently, and, in particular, to achieve impedance matching between a hyperbolic material and the free space. Impedance matching prevents any light reflections at the interfaces and allows for efficient light coupling and extraction from the hyperbolic materials. Therefore, the development of magnetic hyperbolic materials with a simultaneous control over both dielectric permittivity and magnetic permeability tensors remains an important milestone. In particular, it is of a special interest to realize experimentally a magnetic hyperbolic material with the effective magnetic permeability tensor having principal components of the opposite signs[1,12]. Such a development would open new opportunities for super-resolution imaging, nanoscale optical cavities or control over the density of photon states and, in particular, the magnetic density of states for enhancing brightness of magnetic emitters[13,14].

In recent years, we have seen the development of numerous structures with artificial magnetism. However, these structures are largely limited to planar metasurfaces of deeply subwavelength thickness. Importantly, many properties and functionalities of hyperbolic media rely on wave propagation inside them and therefore require essentially a three-dimensional design. For example, hyperlens super-resolution imaging relies on conversion of evanescent waves propagating in a bulk of hyperbolic media[15,16]. Nowadays, the realization of three-dimensional metamaterial structures[17] is at the edge of technological possibilities associated with extreme fabrication difficulties and material constraints. To date, no photonic structures with hyperbolic dispersion in the magnetic response have been demonstrated, and such types of structures are only known for microwave systems[18,19].

Here we demonstrate experimentally optical magnetic hyperbolic metamaterial with principal components of the magnetic permeability tensor having the opposite signs. We directly observe a topological transition between the elliptic and hyperbolic dispersions in metamaterials. In the hyperbolic regime, the length of wave vectors inside the metamaterial is diverging towards infinity. We reveal the effect of the hyperbolic dispersion on thermal emission, where the magnetic hyperbolic metamaterial demonstrates enhanced, directional, coherent and polarized thermal emission. Our experimental observations are supported by analytical calculations as well as full-wave numerical simulations.

Results

Sample fabrication. To realize a magnetic hyperbolic medium in optics, we employ multilayer fishnet metamaterials, known as the bulk-type metamaterials with negative refractive index at optical frequencies[20]. Multilayer fishnets were predicted theoretically to possess a magnetic hyperbolic dispersion[21], but direct measurements of their dispersion remained out of reach. To test this, we fabricate a fishnet metamaterial by using focused ion beam milling through a stack of 20 alternating films of gold and magnesium fluoride (see details in Methods). The sample is fabricated on a 50-nm-thin silicon nitride membrane. A sketch and scanning electron microscopy image of the fabricated structure are shown in Fig. 1. The fishnets feature high optical transmission in the near-infrared spectral region, exhibiting a transmission maximum of 42% at ~1,320 nm wavelength, as shown in Fig. 1b. We also measure the fishnet refractive index at normal incidence using spectrally and spatially resolved interferometry[22]. The fishnet's refractive index shown in Fig. 1c is constantly decreasing with an increase of the wavelength exhibiting negative values at wavelengths above 1,410 nm.

Angular dispersion measurements. To reconstruct the dispersion isofrequency contours experimentally, we determine the length of the **k**-vectors of light inside the materials for a range of different directions. For this, we measure both amplitude and phase of the transmitted and reflected light, and find the **k**-vectors via the reverted Fresnel equations (see details in Supplementary Note 2).

For the phase measurements, we employ interferometry techniques. Specifically, for measuring a phase in transmission,

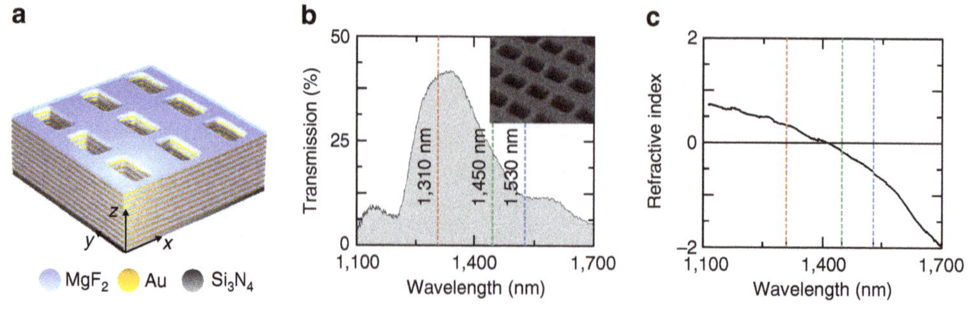

Figure 1 | Multilayer fishnet metamaterial. (**a**) Sketch of the structure. Thicknesses of MgF_2 and Au layers are 45 and 30 nm, respectively. Thickness of Si_3N_4 membrane is 50 nm. Lattice period is 750 × 750 nm. Size of holes is 260 × 530 nm. (**b**) Experimentally measured transmission spectrum of the fishnet metamaterial. Inset shows a scanning electron microscopy image of the fabricated structure. (**c**) Effective refractive index of the fishnet metamaterial extracted for the normal incidence. The marked lines in **b** and **c** represent the wavelengths in the regions of elliptic dispersion (red), crossover optical topological transition (green) and hyperbolic dispersion (blue).

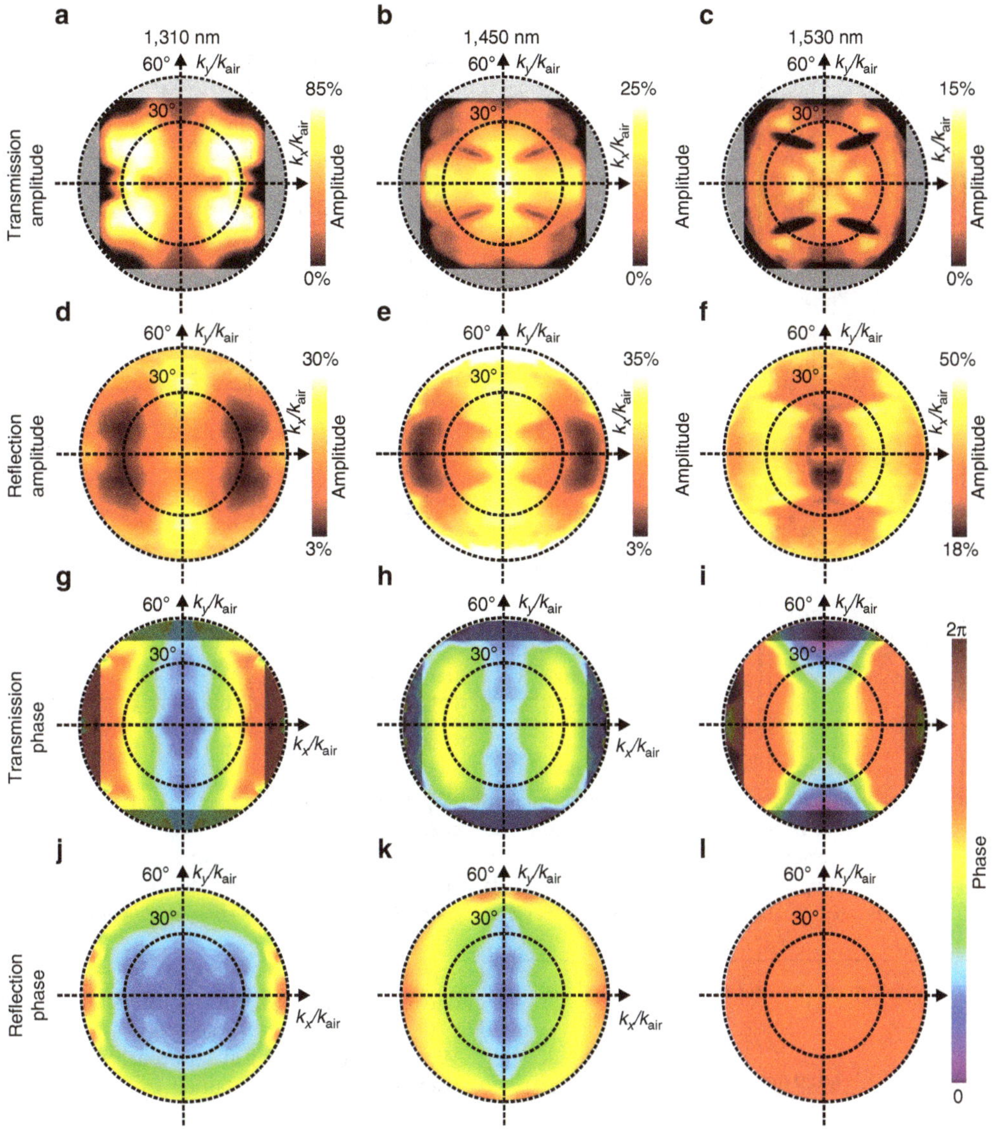

Figure 2 | Experimental results. Measured (**a-c**) transmission and (**d-f**) reflection amplitudes, (**g-i**) transmission and (**j-l**) reflection phase for three different wavelengths: (**a,d,g,j**) 1,310 nm, (**b,e,h,k**) 1,450 nm and (**c,f,i,l**) 1,530 nm. All the measurements are performed for the range of incident angles 0-60°and are plotted versus wave vector components k_x and k_y normalized by the length of the wave vector in air k_{air}. Horizontal axes correspond to TM-polarized illumination. Vertical axes correspond to TE illumination. Square apertures for the transmission amplitude and phase measurements show the numerical aperture limited by the size of the windows in the supporting silicon wafer.

we use a Mach–Zehnder-type interferometer[23], while for the measurements of a phase in reflection we employ a Michelson–Morley interferometer[23]. To resolve transmission and reflection at different angles, we focus and collect the light using objective lens with high numerical aperture (Olympus LCPLN100XIR NA 0.85) and project the objective's back-focal plane image onto an infrared camera (Xenics XS-1.7-320). The resulting image on the camera represents the **k**-space spectrum of the fishnet metamaterial with the central point of the image corresponding to the **k**-vectors normal to the fishnet's surface. The edge of the image corresponds to **k**-vectors oblique to the fishnet at an angle ~58°, limited by the numerical aperture of the objective. We note that for transmission measurements the numerical aperture is limited to 0.7 by the finite size of the silicon nitride membrane window etched into the supporting silicon handle wafer. This restriction also assures insignificant non-paraxial effects due to the sharp focusing. To obtain the phase information, we interfere the back-focal plane image with a reference beam. To reconstruct the phase information from the interference pattern, we employ off-axis digital holography technique[24] (see details in Supplementary Note 3).

We measure the complex transmission and reflection for the following three different wavelengths: 1,310, 1,450 and 1,530 nm, marked in Fig. 1b,c with red, green and blue, respectively. We notice that for the first wavelength, the metamaterial exhibits positive refractive index for normal incidence of light; for the second wavelength, the refractive index is close to zero and for the latter wavelength, the refractive index is negative (Fig. 1c). The results of our angular measurements are presented in Fig. 2. We use a linearly polarized light source with the electric field polarized in the x direction. After the objective lens, the focusing beam has TE polarization along the k_y axis and TM polarization along the k_x axis. As a result, the back-focal plane images (Fig. 2) contain information about the optical response of the sample with respect to both TE- and TM-polarized light along the k_y and k_x axes, respectively. We notice that the material's magnetic dispersion is measured along the k_y axis, and the electric dispersion is measured along the k_x axis.

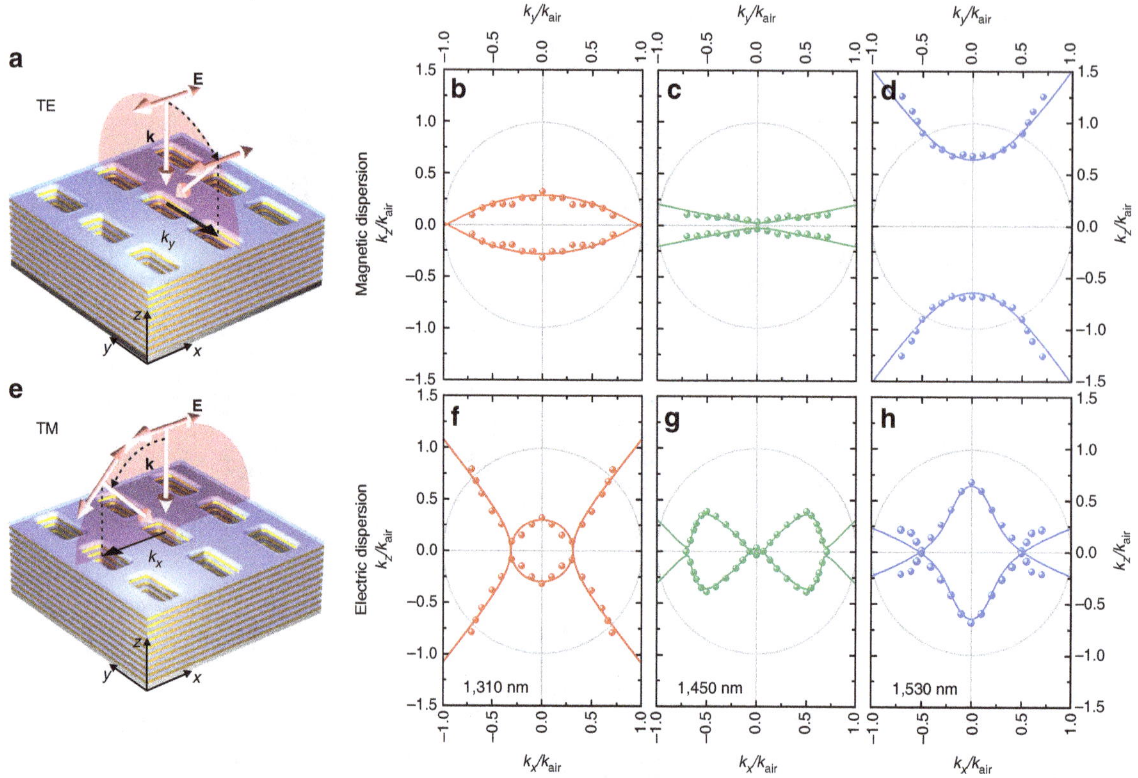

Figure 3 | Experimental observation of a transition from elliptic to hyperbolic dispersion. (**a**) Sketch of the TE illumination geometry showing relative orientations of the sample, wave vector **k** and electric field **E**. (**b–d**) Isofrequency dispersion contours for the TE polarization at wavelengths 1,310, 1,450 and 1,530 nm, respectively. Wave vector components k_y and k_z are normalized by the length of the wave vector in air k_{air}. (**e**) Sketch of the TM illumination geometry. (**f–h**) Isofrequency dispersion contours for the TM polarization at wavelengths 1,310, 1,450 and 1,530 nm, respectively. Dots mark experimental data, and lines correspond to analytical results. Grey circles correspond to the isofrequency contour of light in vacuum.

Table 1 | Effective parameters of the metamaterial dispersion.

Parameter	1,310 nm	1,450 nm*	1,530 nm
ε_x	$0.45+i\,0.8$	$-0.045+i\,0.03$	$-0.14+i\,0.25$
ε_z	$0.2+i\,0.01$	$0.08+i\,0.002$	$-0.02+i\,0.023$
μ_y	$0.18+i\,0.3$	$0.06+i\,0.13$	$-2+i\,0.1$
μ_z	1	1	1

*Nonlocal parameters at 1,450 nm wavelength $\partial^2\varepsilon_x/\partial k_z^2=7$, $\partial^2\varepsilon_z/\partial k_x^2=0.82$ and $\partial^2\varepsilon_z/\partial k_z^2=0$.

From our measurements, we can see that for a specific range of angles of incidence at 1,310 nm and 1,450 nm wavelengths, the reflection becomes $<4\%$, which, for comparison, is lower than the reflection of glass. This is a direct consequence of the impedance matching of the metamaterial to air.

Next, we analyse the phase accumulation of light inside the metamaterial (Fig. 2g–i). At 1,310 nm wavelength (Fig. 2g), the phase accumulation increases from the centre (normal incidence) to the edges (60° oblique incidence). This is similar to the response of a usual dielectric, where the phase accumulation increases with a growth of an optical path inside the material. At 1,450 nm wavelength, the phase accumulation along the k_y axis remains nearly unchanged being close to zero for the entire range of incident angles. This corresponds to the case of ε-near-zero[25] and μ-near-zero[26] materials. Finally, at 1,530 nm wavelength, the phase accumulation is decreasing from the centre to the edges along the k_y axis, while it is increasing along the k_x axis.

Isofrequency analysis. Next, we reconstruct the isofrequency contours out of the measured transmission and reflection data (see details in Supplementary Note 2). For the three wavelengths and two polarizations, these dispersion contours are shown in Fig. 3b–d,f–h.

We use an analytical approach to reveal the shapes of isofrequency surfaces obtained experimentally. We write explicitly a set of two equations for two principal linear polarizations: TE and TM. Without a lack of generality, we assume that for TE polarization the electric field component is pointing in the x direction, and for TM polarization the magnetic field component is pointing in the y direction. The resulting dispersion relations take the form:

$$\text{TE:}\quad \frac{k_y^2}{\varepsilon_x\mu_z}+\frac{k_z^2}{\varepsilon_x\mu_y}=\frac{\omega^2}{c^2};\qquad\qquad (1)$$
$$\text{TM:}\quad \frac{k_x^2}{\varepsilon_z\mu_y}+\frac{k_z^2}{\varepsilon_x\mu_y}=\frac{\omega^2}{c^2}.$$

Thus, the material response is described by a set of parameters ε_x, ε_z, μ_y and μ_z. For the case of purely real values of parameters (for example, for materials with no loss or gain), the dispersion equations describe two types of isofrequency contours: either elliptic or hyperbolic depending on the relative signs of the parameters. In particular, opposite signs of the electric permittivity components ε_x and ε_z lead to hyperbolic isofrequency contours for TM polarization, while the opposite signs of the magnetic permeability components μ_y and μ_z lead to hyperbolic isofrequency contours for the TE polarization. Here we take into account the absorption of light in the metamaterial and consider the parameters ε_x, ε_z and μ_y as complex numbers with the imaginary parts representing losses. We assume $\mu_z=1$ for all the cases, as we do not expect artificial magnetic response

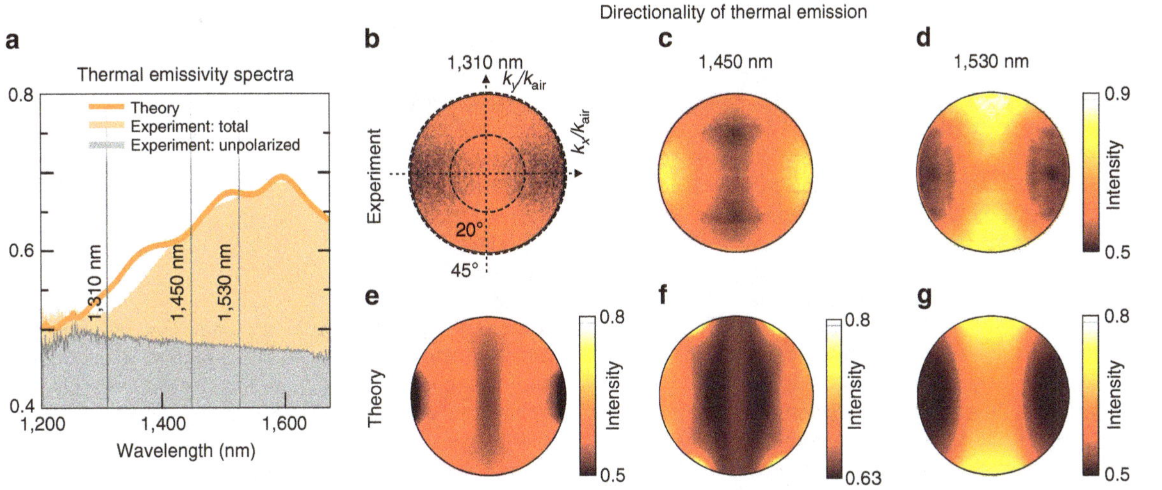

Figure 4 | Thermal emission from bulk magnetic hyperbolic metamaterials. (**a**) Spectrum of thermal emission normalized by the black body spectrum (emissivity). The unpolarized portion of emission is shown in grey, whereas the total emissivity is shown in orange above the unpolarized part. Line represents the corresponding theoretical calculation for the total emissivity. (**b–d**) Experimentally measured directionality of thermal emissivity at wavelengths 1,310, 1,450 and 1,530 nm, respectively, plotted versus wave vector components k_x and k_y normalized by the length of the wave vector in air k_{air}. (**e–g**) Theoretically calculated directionality for the same three wavelengths, respectively. Images (**c–g**) have the same coordinate system as **b**.

from the structure in the z direction. Table 1 summarizes the values of ε_x, ε_z and μ_y used in our analytical model to describe the experimental data.

As we observe, at 1,310 nm wavelength (Fig. 3b,f), all the material parameters have positive real parts, thus representing the cases of the elliptic dispersion. Interestingly, in both cases, the shapes of the isofrequency contours deviate from elliptical. This effect comes from the imaginary parts of ε_x, ε_z and μ_y. It is known that finite material losses lead to a hybridization of propagating and evanescent modes[27]. Importantly, in our case for the TM polarization (see Fig. 3f), the hybridization leads to a new class of topology of isofrequency contours that is different from either elliptic or hyperbolic.

At 1,530 nm wavelength and TE polarization, the permeability coefficient μ_y has a negative real part, which is opposite to the sign of the permeability coefficient $\mu_z = 1$. Therefore, in this spectral region, the material dispersion becomes magnetic hyperbolic (see Fig. 3d). The branches of the hyperbola go beyond the isofrequency contour of light in air. We notice that the **k**-vectors with tangential components larger than $|k_{air}|$ are not accessible experimentally when the metamaterial is illuminated from free space. However, an analytical extrapolation of the experimental curves supports the existence of propagating waves with large wave vectors. This is a key to achieve extraordinary optical properties of hyperbolic media, such as super-resolution imaging, nanoscale optical cavities and control over the density of photon states. For the other TM polarization, however, all three material parameters ε_x, ε_z and μ_y are simultaneously negative resulting in elliptic dispersion with a complex topology of isofrequency contours due to the presence of losses (see Fig. 3h).

At 1,450 nm wavelength, the parameters ε_x, ε_z and μ_y are vanishing simultaneously, representing the regime of optical topological transition[2]. Around the topological transition, ε_x, ε_z and μ_y change their signs due to the resonant nature of the metamaterial's response. This results in an increase of the phase velocity of light towards infinity inside the structure. Importantly, at this wavelength, the structure supports propagating waves with **k**-vectors substantially smaller than the **k**-vectors in air, while all conventional optical materials support **k**-vectors larger than those in air. As local material parameters become close to zero, we expect to see strong contributions from nonlocal response of the metamaterials[28]. This implies that the permittivity coefficients ε_x and ε_z become functions of the wave vector **k** (see details in Supplementary Note 4). We find that for the magnetic (TE) dispersion of the fishnet metamaterials, both local and nonlocal models result in the same dispersion relation. Therefore, we consider the electric (TM) dispersion equation only, which in the nonlocal case takes the form:

$$\text{TM}: \quad \frac{k_x^2}{\varepsilon_z + \frac{\partial^2 \varepsilon_z}{\partial k_x^2} k_x^2 + \frac{\partial^2 \varepsilon_z}{\partial k_z^2} k_z^2} + \frac{k_z^2}{\varepsilon_x + \frac{\partial^2 \varepsilon_x}{\partial k_x^2} k_x^2 + \frac{\partial^2 \varepsilon_x}{\partial k_z^2} k_z^2} = \frac{\omega^2}{c^2} \quad (2)$$

We notice that $\mu_y = \left(1 - \omega^2/c^2 \; \partial^2 \varepsilon_x / \partial k_z^2\right)^{-1}$ (ref. 28). We further neglect the nonlocal parameter $\partial^2 \varepsilon_z / \partial k_z^2$, as we find it to be of a minor importance[28]. Thus, for the case of 1,450 nm wavelength, near the point of the optical topological transition of the metamaterial, we introduce two extra spatially dispersive terms $\partial^2 \varepsilon_x / \partial k_z^2$ and $\partial^2 \varepsilon_z / \partial k_x^2$ to describe the experimental dispersion. The values of the material parameters for the 1,450 nm wavelength are also given in Table 1. Our results suggest that a wide range of nontrivial isofrequency dispersion contours can be realized by an appropriate tuning of material's loss, gain and spatial dispersion. While all possible types of isofrequency contours for local media without loss/gain are limited to the second-order geometrical curves[29] (such as an ellipse or hyperbola), the presence of loss, gain and spatial dispersion extends the possible cases of isofrequency contours to the fourth-order curves. This leads to new topologies of the metamaterial dispersion.

In addition, we use full-wave numerical simulations to calculate material's isofrequency contours (see details in Supplementary Note 5 and Supplementary Fig. 1). We find that numerical results are in a good agreement with both experimental measurements and analytical calculations.

Manipulation of thermal emission from fishnet metamaterials. Next, we study the effect of hyperbolic dispersion on far-field thermal emission. In our experiments, we heat the sample up to 400 °C with a ceramic heater. At this temperature, the metamaterial gives relatively bright thermal emission in the spectral region of interest, and it remains undamaged by heating. We collect the thermal emission of the fishnet metamaterial sample by an objective lens with 10 mm working distance and 0.7 numerical aperture. In our experiments, we ensure that only thermal emission from the metamaterial sample is collected by the objective lens. We then direct it onto an infrared spectrometer and measure the thermal emission spectra. We take a reference measurement of the thermal emission from a silicon sample next to the fishnet metamaterial The reference measurement allows us to find emissivity of fishnets (radiation of fishnets normalized by the black body radiation) using the known emissivity of silicon[30] and, in particular, characterize the degree of polarization of the emitted light (see details in Supplementary Note 6 and Supplementary Figs 2 and 3). We then measure the polarization states of the emissivity by employing Stokes vector formalism (see details in Supplementary Note 6 and Supplementary Figs 4 and 5). We find that in the spectral region with the magnetic elliptic dispersion, the thermal emission remains largely unpolarized. However, the degree of polarized light grows rapidly as we approach the point of the optical topological transition. In the spectral region of the magnetic hyperbolic dispersion, the thermal emission becomes partially linearly polarized. Figure 4a shows the emissivity spectrum of our sample. We notice that the unpolarized fraction of the emissivity remains almost unchanged over the measurement spectral range. The polarized part of the emission, however, increases at around the topological transition region and in the region with hyperbolic dispersion. These phenomena can be explained only by the enhanced density of photon states due to the magnetic hyperbolic dispersion. In addition, we argue that our far-field results suggest that the near-field thermal radiation can be characterized as super-Planckian, that is, exceeding the black body limit[31].

Further, we study directionality of thermal emission at the three wavelengths of 1,310 nm, 1,450 and 1,530 nm. For this, we translate the back-focal plane image of the collecting objective onto the infrared camera through a corresponding band-pass filter. We again employ Stokes formalism to characterize the polarization states of directionality diagrams (details of the back-focal plane polarimetry method can be found in ref. 32). With this method, we retrieve the polarized portion of the thermal emission and plot its directionality diagrams for the three wavelengths in Fig. 4b–d. We notice that the directionality of emission at 1,310 nm (elliptic dispersion) is not pronounced, while the emission at 1,530 nm has a noticeable north–south directionality. Importantly, the directions of high thermal radiation correspond to the directions with large **k**-vectors on the magnetic hyperbolic dispersion curve in Fig. 3d. Emission at the point of topological transition exhibits noticeable directionality as well; in particular, the emission in the direction normal to the sample is suppressed (the centre of the image in Fig. 4c). This corresponds to the region with near-zero **k**-vectors. The fact that in the regime of the magnetic hyperbolic dispersion, the thermal emission is directional implies that it exhibits a high degree of spatial coherence.

We further calculate the spectra and directionalities of thermal emission theoretically (see details in Supplementary Note 7). The results of our calculations are sown in Fig. 4a with a line for the spectral density and Fig. 4e–g for the directionality. The calculated spectra and directionality diagrams show an excellent qualitative agreement with our experimental measurements.

In addition, we compare experimental thermal emission directionalities at 400 °C with experimentally measured absorption directionalities at room temperature (see details in Supplementary Note 6 and Supplementary Fig. 6). The directionalities look similar, while resembling some differences in details associated with the change of material properties with temperature.

Discussion

We have demonstrated experimentally optical magnetic hyperbolic metamaterial with the principal components of the magnetic permeability tensor having the opposite signs. We have developed an experimental method for direct measurements of isofrequency dispersion contours of three-dimensional metamaterials and directly observed a topological transition between the elliptic and hyperbolic dispersions in metamaterials. In the hyperbolic regime, the length of wave vectors inside the metamaterial is diverging towards infinity.

We have applied an analytical theory that takes into account losses and spatial dispersion to describe the measured isofrequency contours, and demonstrated the importance of nonlocal contributions[33] in the regime of optical topological transitions, associated with vanishing local parameters. A control of loss, gain and spatial nonlocalities in metamaterials opens up new opportunities for engineering isofrequency dispersion contours beyond elliptic or hyperbolic, with nontrivial geometry and topology. The magnetic hyperbolic dispersion of metamaterials together with their electric response enables impedance matching between hyperbolic media and air, resulting in an efficient light transfer through interfaces. Our results suggest that other three-dimensional metamaterials assembled from magnetically polarizable or chiral elements[34–36] may possess magnetic hyperbolic dispersion as well.

In addition, we have studied the effect of the hyperbolic dispersion on thermal emission of matematerials, and revealed that the magnetic hyperbolic metamaterial demonstrates enhanced, directional, coherent and polarized thermal emission. These results suggest an advanced thermal management that can find applications in thermophotovoltaics[37], scanning thermal microscopy[38], coherent thermal sources[39] and other thermal devices.

Methods

Nanofabrication. The bulk fishnet metamaterial is fabricated on a suspended 50-nm low-stress silicon nitride (Si_3N_4) membrane made from standard microelectromechanical systems fabrication technologies. The metal–dielectric stack is then deposited onto the Si_3N_4 membrane using layer-by-layer electron beam evaporation technique at pressure $\sim 1 \times 10^{(-6)}$ Torr without vacuum break. The chamber temperature is cooled down on each layer of evaporation to avoid buildup of excessive heating and stress. Essentially 10 repeating layers of gold (Au, 30 nm) and magnesium fluoride (MgF_2, 45 nm) are deposited. Next, the sample is turned upside down and mounted on a special stage holder, which has a matching trench that prevents any mechanical contact with the fragile multilayer structure sitting on the membrane. The nanostructures are milled by gallium (Ga) focused ion beam from the membrane side. Milling from the membrane side prevents the implantation of Ga ions into the metal layers at the unpatterned areas that reduces optical losses and improves the overall quality. This is essential to mask the implantation of Ga ions into the metal layers at the unpatterned areas. The final structure has a slight sidewall angle along the thickness direction, but is found to have minor influence on the optical properties. Another important advantage of focused ion beam fabrication of the structure on a thin membrane compared with conventional bulk substrates is the ability for Ga ions to enter the free space (that is, no substrate for Ga ions to accumulate and cause undesired contamination and absorption).

References

1. Smith, D. & Schurig, D. Electromagnetic wave propagation in media with indefinite permittivity and permeability tensors. *Phys. Rev. Lett.* **90,** 077405 (2003).
2. Krishnamoorthy, H. N., Jacob, Z., Narimanov, E., Kretzschmar, I. & Menon, V. M. Topological transitions in metamaterials. *Science* **336,** 205–209 (2012).
3. Poddubny, A., Iorsh, I., Belov, P. & Kivshar, Y. Hyperbolic metamaterials. *Nat. Photon.* **7,** 948–957 (2013).
4. Yao, J. *et al.* Optical negative refraction in bulk metamaterials of nanowires. *Science* **321,** 930–930 (2008).
5. Noginov, M. *et al.* Bulk photonic metamaterial with hyperbolic dispersion. *Appl. Phys. Lett.* **94,** 151105 (2009).
6. Kabashin, A. *et al.* Plasmonic nanorod metamaterials for biosensing. *Nat. Mater.* **8,** 867–871 (2009).

7. Kanungo, J. & Schilling, J. Experimental determination of the principal dielectric functions in silver nanowire metamaterials. *Appl. Phys. Lett.* **97,** 021903 (2010).
8. Noginov, M. *et al.* Controlling spontaneous emission with metamaterials. *Opt. Lett.* **35,** 1863–1865 (2010).
9. Wurtz, G. A. *et al.* Designed ultrafast optical nonlinearity in a plasmonic nanorod metamaterial enhanced by nonlocality. *Nat. Nanotechnol.* **6,** 107–111 (2011).
10. Sun, J., Litchinitser, N. M. & Zhou, J. Indefinite by nature: from ultraviolet to terahertz. *ACS Photon.* **1,** 293–303 (2014).
11. Dai, S. *et al.* Tunable phonon polaritons in atomically thin van der waals crystals of boron nitride. *Science* **343,** 1125–1129 (2014).
12. Smith, D. R., Kolinko, P. & Schurig, D. Negative refraction in indefinite media. *J. Opt. Soc. Am. B* **21,** 1032–1043 (2004).
13. Hussain, R. *et al.* Enhancing eu3 + magnetic dipole emission by resonant plasmonic nanostructures. *Opt. Lett.* **40,** 1659–1662 (2015).
14. Kasperczyk, M., Person, S., Ananias, D., Carlos, L. D. & Novotny, L. Excitation of magnetic dipole transitions at optical frequencies. *Phys. Rev. Lett.* **114,** 163903 (2015).
15. Jacob, Z., Alekseyev, L. V. & Narimanov, E. Optical hyperlens: far-field imaging beyond the diffraction limit. *Opt. Express* **14,** 8247–8256 (2006).
16. Liu, Z., Lee, H., Xiong, Y., Sun, C. & Zhang, X. Far-field optical hyperlens magnifying sub-diffraction-limited objects. *Science* **315,** 1686–1686 (2007).
17. Soukoulis, C. M. & Wegener, M. Past achievements and future challenges in the development of three-dimensional photonic metamaterials. *Nat. Photon.* **5,** 523–530 (2011).
18. Sun, J. *et al.* Low loss negative refraction metamaterial using a close arrangement of split-ring resonator arrays. *New J. Phys.* **12,** 083020 (2010).
19. Shchelokova, A. V., Filonov, D. S., Kapitanova, P. V. & Belov, P. A. Magnetic topological transition in transmission line metamaterials. *Phys. Rev. B* **90,** 115155 (2014).
20. Valentine, J. *et al.* Three-dimensional optical metamaterial with a negative refractive index. *Nature* **455,** 376–379 (2008).
21. Kruk, S. S., Powell, D. A., Minovich, A., Neshev, D. N. & Kivshar, Y. S. Spatial dispersion of multilayer fishnet metamaterials. *Opt. Express* **20,** 15100–15105 (2012).
22. O'Brien, K. *et al.* Reflective interferometry for optical metamaterial phase measurements. *Opt. Lett.* **37,** 4089–4091 (2012).
23. Saleh, B. & Teich, M. *Wiley Series in Pure and Applied Optics* (Wiley, 2007).
24. Cuche, E., Marquet, P. & Depeursinge, C. Simultaneous amplitude-contrast and quantitative phase-contrast microscopy by numerical reconstruction of fresnel off-axis holograms. *Appl. Opt.* **38,** 6994–7001 (1999).
25. Alù, A., Silveirinha, M. G., Salandrino, A. & Engheta, N. Epsilon-near-zero metamaterials and electromagnetic sources: Tailoring the radiation phase pattern. *Phys. Rev. B* **75,** 155410 (2007).
26. Engheta, N. & Ziolkowski, R. *Metamaterials: Physics and Engineering Explorations* (Wiley, 2006).
27. Davoyan, A. R. *et al.* Mode transformation in waveguiding plasmonic structures. *Photon. Nanostruct. Fundam. Appl.* **9,** 207–212 (2011).
28. Gorlach, M. A. & Belov, P. A. Nonlocality in uniaxially polarizable media. *Phys. Rev. B* **92,** 085107 (2015).
29. Berger, M., Cole, M. & Levy, S. *Geometry II. Universitext* (Springer, 2009).
30. Ravindra, N. *et al.* Emissivity measurements and modeling of silicon-related materials: an overview. *Int. J. Thermophys.* **22,** 1593–1611 (2001).
31. Guo, Y., Cortes, C. L., Molesky, S. & Jacob, Z. Broadband super-planckian thermal emission from hyperbolic metamaterials. *Appl. Phys. Lett.* **101,** 131106 (2012).
32. Kruk, S. S. *et al.* Spin-polarized photon emission by resonant multipolar nanoantennas. *ACS Photon.* **1,** 1218–1223 (2014).
33. Orlov, A. A., Voroshilov, P. M., Belov, P. A. & Kivshar, Y. S. Engineered optical nonlocality in nanostructured metamaterials. *Phys. Rev. B* **84,** 045424 (2011).
34. Liu, N. *et al.* Three-dimensional photonic metamaterials at optical frequencies. *Nat. Mater.* **7,** 31–37 (2008).
35. Pendry, J. A chiral route to negative refraction. *Science* **306,** 1353–1355 (2004).
36. Gansel, J. K. *et al.* Gold helix photonic metamaterial as broadband circular polarizer. *Science* **325,** 1513–1515 (2009).
37. Lenert, A. *et al.* A nanophotonic solar thermophotovoltaic device. *Nat. Nanotechnol.* **9,** 126–130 (2014).
38. De Wilde, Y. *et al.* Thermal radiation scanning tunnelling microscopy. *Nature* **444,** 740–743 (2006).
39. Greffet, J.-J. *et al.* Coherent emission of light by thermal sources. *Nature* **416,** 61–64 (2002).

Acknowledgements

We thank D. Smith and D. Basov for discussions and also acknowledge useful suggestions from S. Fan and C. Simovski. The work was partially supported by the Australian Research Council. Z.J.W., K.O. and X.Z. were funded by the Director, Office of Science, Office of Basic Energy Sciences, Materials Sciences and Engineering Division, of the U.S. Department of Energy under Contract No. DE-AC02-05-CH11231.

Author contributions

S.S.K., D.N.N. and Y.S.K. conceived the idea. S.S.K. conducted the experiments. E.P.-S. assisted with the spatially resolved interferometric measurements. Z.J.W., K.O'.B. and X.Z. designed the samples. Z.J.W. fabricated the samples. Z.J.W. and K.O'.B. conducted sample quality control. S.S.K. performed the analytical modelling and numerical simulations. S.S.K. and Y.S.K. wrote the manuscript. S.S.K., Z.J.W., K.O'.B., D.N.N., Y.S.K. and X.Z. analysed the experimental and theoretical results. All authors contributed to discussions and edited the manuscript.

Additional information

Competing financial interests: The authors declare no competing financial interests.

25

Free-carrier-induced soliton fission unveiled by *in situ* measurements in nanophotonic waveguides

Chad Husko[1,*,†], Matthias Wulf[2,*,†], Simon Lefrancois[1], Sylvain Combrié[3], Gaëlle Lehoucq[3], Alfredo De Rossi[3], Benjamin J. Eggleton[1] & L. Kuipers[2]

Solitons are localized waves formed by a balance of focusing and defocusing effects. These nonlinear waves exist in diverse forms of matter yet exhibit similar properties including stability, periodic recurrence and particle-like trajectories. One important property is soliton fission, a process by which an energetic higher-order soliton breaks apart due to dispersive or nonlinear perturbations. Here we demonstrate through both experiment and theory that nonlinear photocarrier generation can induce soliton fission. Using near-field measurements, we directly observe the nonlinear spatial and temporal evolution of optical pulses *in situ* in a nanophotonic semiconductor waveguide. We develop an analytic formalism describing the free-carrier dispersion (FCD) perturbation and show the experiment exceeds the minimum threshold by an order of magnitude. We confirm these observations with a numerical nonlinear Schrödinger equation model. These results provide a fundamental explanation and physical scaling of optical pulse evolution in free-carrier media and could enable improved supercontinuum sources in gas based and integrated semiconductor waveguides.

[1] Centre for Ultrahigh bandwidth Devices for Optical Systems (CUDOS), Institute of Photonics and Optical Science (IPOS), School of Physics, University of Sydney, Sydney, New South Wales 2006, Australia. [2] Center for Nanophotonics, FOM Institute AMOLF, Science Park 104, 1098 XG, Amsterdam, The Netherlands. [3] Thales Research and Technology, 1 Avenue. A. Fresnel, 91767 Palaiseau, France. * These authors contributed equally to this work. † Present addresses: Center for Nanoscale Materials, Argonne National Laboratory, Argonne, Illinois USA. (C.H.); Institute of Science and Technology (IST) Austria, Klosterneuburg, Austria (M.W.). Correspondence and requests for materials should be addressed to C.H. (email: chusko@anl.gov) or to M.W. (email: matthias.wulf@ist.ac.at).

Soliton fission occurs when a fundamental soliton is ejected and temporally separates from a higher-order soliton due to a sufficiently strong perturbation to the system. This behaviour strongly contrasts with the expected periodic recurrence for ideal higher-order solitons[1]. In the optical domain, soliton fission or 'soliton decay' as it is also known, was first numerically shown to occur due to perturbations of the traditional nonlinear Schrödinger equation including self-steepening (SS), third-order dispersion (TOD) and Raman scattering[2–4] with experimental demonstrations following soon after[5,6]. Since that time, nonlinear optical waveguides have evolved from glass optical fibres to new platforms such as semiconductors[7] and gas-filled microstructured fibres[8] where the dominant perturbation is a plasma effect due to nonlinear photogeneration of free electrons or free carriers (electron–hole pairs) similar to light in bulk ionized gases[9].

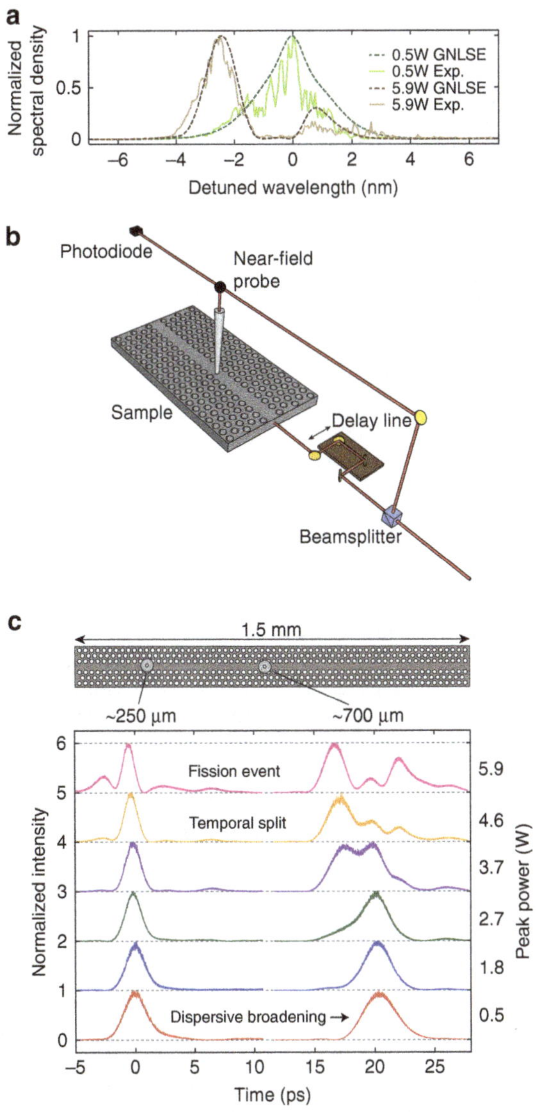

Figure 1 | Spectral transmission and time-resolved near-field microscopy of soliton fission. (**a**) Spectral transmission properties of the optical pulse measured at the waveguide output. (**b**) Time-resolved near-field optical microscope (NSOM) apparatus used in the experiment. (**c**) Experimental cross-correlation measurements as a function of power (vertical axis) at two spatial positions along the nanostructured photonic waveguide. It is clear that as the power is increased a break up of the pulse occurs as it propagates.

The free-carrier plasma modifies the nonlinear pulse evolution with both dispersive (FCD, n_{FC}) and absorptive (FCA, σ) contributions leading to non-trivial dynamics unavailable in other optical systems. While in the spectral domain optical pulses undergo a spectral blueshift due to FCD[8–10], in contrast the temporal properties are governed by the dynamic interaction of FCD and dispersion together leading to, for example, nonlinear pulse temporal broadening[11]. These free-carrier effects can also interplay with and modulate the classical soliton evolution. Temporal solitons in free-carrier media have been shown[12–14] including soliton self-frequency blueshift[15] and soliton acceleration[16,17]. While recent numerical simulations suggest that free carriers could cause soliton fission[15], both a theoretical formulation and direct temporal measurements establishing a causal link remain open challenges to the field.

Here we provide both an experimental demonstration and a theoretical explanation of the physics underpinning soliton fission induced by a free-carrier perturbation. Using an interferometric near-field scanning optical microscope (NSOM), we observe both the spatial and temporal pulse evolution *in situ* along a semiconductor waveguide. This direct measurement is essential to unraveling the localized nonlinear dynamics in nanophotonic waveguides as traditional cut back methods used for macroscopic devices are impractical at these length scales. From the theoretical side, we derive an analytic formalism to reveal the physical parameters governing the system. With this new formalism we determine a quantitative threshold required to observe soliton fission induced by FCD and show that our experimental conditions exceed the threshold by an order of magnitude. In our experiment, the fission occurs on a length scale as small as 160 μm due to a slow-light enhancement of the optical field in the photonic crystal waveguide (PhCWG) device. This value represents the shortest fission length we could find reported in the literature. We confirm these results with a numerical model based on the generalized nonlinear Schrödinger equation (GNLSE) incorporating the higher-order effects.

Results

Near-field measurements of nonlinear pulse propagation. The structure under study is a two-dimensional PhCWG made of air-holes etched in a GaInP slab (see Methods, Supplementary Note 1 and Supplementary Table 1 for additional details). These structures are known to enhance the nonlinear optical properties due to slow light in the periodic medium[18]. We note the increased group index $n_g = 15.1$ is achieved using the dispersion-engineered design outlined in ref. 19 in a region away from the band edge so as to avoid scattering losses[20] and minimize TOD[21]. The earliest investigations of nonlinear evolution of optical pulses in PhCWGs examined the pulse spectra after the pulse propagated through the waveguide[22]. Figure 1a shows the measured spectral transmission in our current experiment (solid) at the waveguide output for low and high power levels for the optical pulse of 2.2 ps (T_{FWHM}, full-width at half-maximum of a hyperbolic secant). Note that the oscillations in the measured spectra arise from disorder in the periodic media[20]. The measured waveguide transmission spectrum is shown as Supplementary Fig. 1. The dashed curves are the result of model calculations detailed below. Spectra measured at different power levels are shown in Supplementary Fig. 2 and described in Supplementary Note 2. We observe a clear spectral blueshift at high power due to FCD[7], as well as a less intense satellite peak. Such satellite peaks have in the past been attributed to soliton fission in fibres, though no similar observations in semiconductor waveguides have been reported to date.

To determine the origin of the satellite peak it is highly desirable to investigate the pulse evolution as it occurs.

Traditionally, this is done through cut back of an optical fibre, wherein measurements are taken at multiple spatial points albeit at the cost of device destruction[23]. This method is impractical for nanoscale devices without high risk of damage to the sample. Fabricating devices of different lengths overcomes this limitation, though with the drawback of device-to-device variation. Non-destructive techniques such as NSOM[24] or photomodulation spectroscopy[25] are well suited to evaluate the propagation dynamics of sub-wavelength structures.

Figure 1b illustrates the time-resolved NSOM we used to measure the pulse evolution in the waveguide[21]. With this set-up we are able to measure the temporal dynamics of the propagating pulse inside the waveguide at the position of the near-field probe. In detail, we measure a temporal electric-field cross-correlation between the pulse in the sample and a pulse in the reference branch of the interferometric set-up. Details about the working principle of the NSOM can be found in the Methods. This cross-correlation contains all crucial information about the evolution of the temporal pulse envelope of the electric field. For example, it has been shown that the temporal broadening due to group-velocity dispersion (GVD, β_2)[26] or the reshaping due to higher-order dispersion[21] transfers directly from the temporal pulse envelope to the measured cross-correlation. We utilize this relation between the cross-correlation and temporal pulse envelope to describe the results in this work.

Figure 1c shows a summary of the NSOM measurements of the temporal pulse dynamics as a function of coupled peak power P_o. The horizontal direction indicates two spatial positions that we measured along the device with the near-field probe: 250 μm (left-hand side) to 700 μm (right-hand side). A clear modulation of the pulse dynamics is seen as a function of P_o in the vertical direction. The soliton number $N^2=\frac{L_D}{L_{NL}}$ indicates the relative balance of the characteristic length scales for linear dispersion L_D and the nonlinear Kerr effect L_{NL} and determines the pulse propagation regime. These lengths will be defined as they are used in the text. In the linear regime the soliton number is $N=0.5$ ($P_0=0.5$ W) and temporal broadening due to GVD (β_2) dominates the propagation from 250 to 700 μm (ref. 27). This makes sense given the dispersion length of 410 μm ($L_D=0.322\frac{T^2_{FWHM}}{\beta_2}$) and a sample length of $L=1.5$ mm. The power-dependent behaviour at the two spatial locations indicates noticeably different evolution patterns. At 250 μm, the pulse narrows with increasing power, indicative of higher-order soliton temporal compression[28]. In contrast, at 700 μm distinct solitons have formed and separated in time for the initially injected $N\approx2$ soliton ($P_0=5.9$ W). This temporal separation is the essence of soliton fission.

To understand the physical origin of this separation from an intuitive perspective we first recall that a change in frequency (spectral shift Ω) in a dispersive medium corresponds to a change in group velocity. This ultimately translates into a shift in temporal position according to the moment evolution equation $\frac{dT_c}{dz}=\beta_2\Omega$ (ref. 29). Since all solitons have anomalous GVD ($\beta_2<0$), and here Ω is blueshifted (positive), the result is a temporal advance. This is opposite to the well-known case of solitons in a Raman medium which redshift and therefore slow down[30–32]. In the context of soliton fission, it has been shown that the fissioned constituents have well predicted and very different energies and power levels[33]. As a consequence, the constituent solitons with larger peak power experience a greater self-frequency shift and a larger temporal advance compared with smaller amplitude solitons. Notice in our experiment that the more energetic main soliton is advanced in time due to FCD and dispersion[16,29].

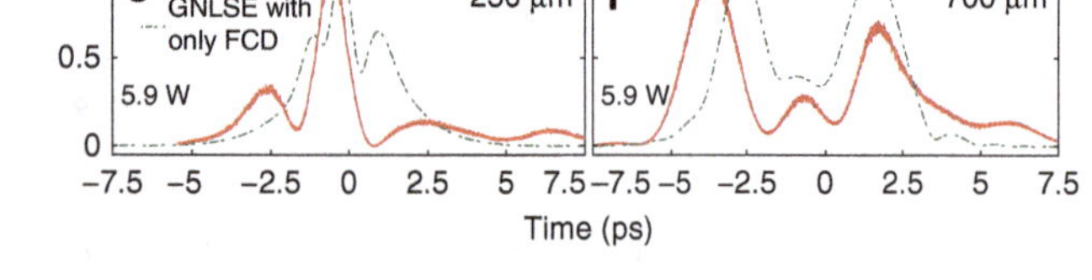

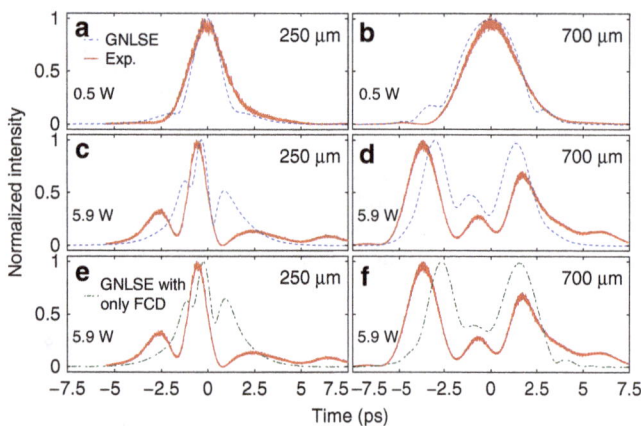

Figure 2 | Comparison of experiment and model of the nonlinear pulse propagation. (a,b) Time-resolved NSOM measurements and GNLSE modelling at a peak power of at a peak power of 0.5 W at a propagation distance of **(a)** 250 μm and **(b)** 700 μm. Temporal broadening of the pulse envelope due to GVD is visible in experiment (red line) and the model (blue line). **(c,d)** Same as above with a peak power of 5.9 W. The multiple peaks characteristic of soliton fission are clearly observable in both theory and experiment. To illustrate that the main features observed in the experiment are related to free-carrier generation, **(e,f)** compare the experimental results with GNLSE modelling results (green line) taking only the soliton terms and FCD/3PA into account, which still results in a good agreement. Note here we show the cross-correlation of the electric field of the temporal pulse envelope for the modelling as well as the experimental results as defined in the Methods.

Confirmation of free-carrier induced fission by modeling. The nonlinear pulse propagation in the GaInP semiconductor waveguide can be described by a GNLSE model (Supplementary Note 3). The nonlinear dynamics here are dominated by the $\chi^{(3)}$ optical Kerr effect (nonlinear parameter γ) with free carriers generated by nonlinear three-photon absorption (3PA, α_3) acting as a perturbation in the wide-gap material ($E_g=1.9$ eV) for our 1,553 nm (~0.8 eV) pulses (ref. 12). Figure 2a–f show detailed GNLSE modelling (dashed blue and green) with the experimental data (solid red) from Fig. 1 superimposed. In particular, we highlight (a),(b) low and (c),(d) high power at the propagation distance of 250 and 700 μm, respectively. The temporal shapes are in good quantitative agreement and excellent qualitative agreement with the experimental data and capture the essential physics of the nonlinear pulse propagation in the nanophotonic waveguide. The good agreement between the experiment and the GNLSE model is even conserved if only the free-carrier effects are included as perturbation to the soliton propagation, as presented in Fig. 2e,f. These results indicate FCD is the dominant perturbation and the cause of the soliton fission. We now perform additional GNLSE modelling to verify this observation and to examine the physical origin of the fission.

Figure 3 summarizes our GNLSE modelling and confirmation that the fission event is triggered by FCD. In particular, we show the modelled pulse temporal $P(t)$ profile along the waveguide. As a baseline, Fig. 3a,b show the GNLSE model in the linear ($P_0=0.5$ W) and nonlinear ($P_0=5.9$ W) regimes, respectively, with identical conditions to Fig. 2. The highest power level results in $L_{NL}=(\gamma P_o)^{-1}=90$ μm. The dashed white lines correspond to the two experimental spatial locations. We observe the pulses already split after ~160 μm. We attribute the short fission length to a slow-light enhancement in the photonic crystal waveguide[18]. We now discern the roles of the different effects by switching them on and off independently in the model.

Figure 3c shows the case where we neglect free carriers by setting the carrier density $N_c=0$ and include only TOD (β_3)

and the soliton terms (Kerr and GVD). The pulse clearly does not undergo fission but rather periodic recurrence as expected from soliton theory in the absence of perturbations[28,31]. This is not surprising due to the small relative magnitude of the TOD effect[3]. A similar result holds for SS. We conclude TOD and SS cannot be the fission mechanisms here. In contrast, Fig. 3d shows that setting $\beta_3 = 0$ (including only FCD, 3PA and the soliton terms) yields a profile in which the main soliton advances in time with the smaller amplitude fissioned pulse trailing behind. The notable qualitative similarity of this result with both the full model (Fig. 3b) and the experimental result (Fig. 2d,f) confirms that FCD is the physical origin of the fission event. We note FCD scales as P_0^3, whereas 3PA scales as P_0^2, thus the reason for the strong FCD effect. Further, FCA is essentially negligible as shown by the ratio of the two effects $\frac{2k_0 n_{FC}}{\sigma} \approx 10$. Now that we have established that FCD is the dominant perturbation in our system, we develop an analytic description to obtain deeper insight.

Derivation of the free-carrier perturbation. It is common to write the GNLSE in a non-dimensional form to analyse the pulse evolution[27]. Here for our case of solitons for a free-carrier perturbation generated by 3PA this is:

$$i\frac{\partial U}{\partial \xi} - \frac{\mathrm{sign}(\beta_2)}{2}\frac{\partial^2 U}{\partial \tau^2} + N^2|U|^2U = N^2\kappa_{FC}^{(3)} U \int_{-\infty}^{\tau} |U|^6 d\tau', \quad (1)$$

where ξ, τ and U are the dimensionless parameters for propagation distance, time and pulse envelope, respectively (Supplementary Note 4). The terms on the left-hand side of the equation are related to soliton propagation. The right-hand side is reserved for perturbations, where the magnitude of the non-dimensional parameter governs the conditions to trigger soliton fission. A higher-order soliton will break apart when the magnitude of these parameters exceeds a minimum threshold. Conversely, when the parameter is below the threshold, the higher-order soliton remains intact and recurrent behaviour is retained. The numerical value of the minimum threshold depends on the specific physical effect causing the fission (that is, TOD, SS and FCD). An important additional property is that the minimum threshold to break a soliton decreases with increasing soliton number N, a topic we will treat in further detail below. Importantly, we have introduced the new term $\kappa_{FC}^{(3)}$ to elucidate the role of the FCD perturbation:

$$\kappa_{FC}^{(3)} = \frac{L_{NL}}{L_{FCD}^{(3)}} = \left(\frac{1}{\gamma P_o}\right) k_0 |n_{FC}| N_c^o$$
$$= \left(\frac{1}{\gamma P_o}\right) k_o |n_{FC}| \rho_{FC}^{(3)} P_o^3 T_o. \quad (2)$$

We have also defined $L_{FCD}^{(3)}$, the FCD length for the free-carrier density generated dynamically from intrapulse 3PA with a peak carrier amplitude $N_c^o = \rho_{FC}^{(3)} P_o^3 T_o$, the free-carrier generation efficiency $\rho_{FC}^{(3)}$ (ref. 11), and $T_o = T_{FWHM}/1.76$ for hyperbolic secant pulses. The physical interpretation of $\kappa_{FC}^{(3)}$ is the relative nonlinear phase shift due to the Kerr effect compared with FCD per unit length.

In terms of characteristic physical scaling, we see that $\kappa_{FC}^{(3)} \propto P_o^2 T_o$, with the material contributing via constants. The power dependence comes from the nonlinear 3PA carrier generation, whereas the T_o term arises due to the fact that free carriers accumulate over the pulse duration, as represented by the integral in Equation (1). It is worth highlighting that the exact scaling of κ_{FC} depends on the specific nonlinear mechanism generating the free carriers (for example, two-photon absorption, ionized gas tunneling and so on). We describe this point further in the Discussion. Note that this carrier perturbation has a completely different form to perturbations caused by TOD, Raman and SS which scale as $\frac{1}{T_o}$ due to a derivative term $\frac{\partial}{\partial \tau}$ in the GNLSE, indicating these effects scale with the local pulse shape, rather than the non-local free-carrier effects[31]. Figure 4a shows the calculated $\kappa_{FC}^{(3)}$ parameters as a function of coupled peak power for our experimental conditions. We have also included

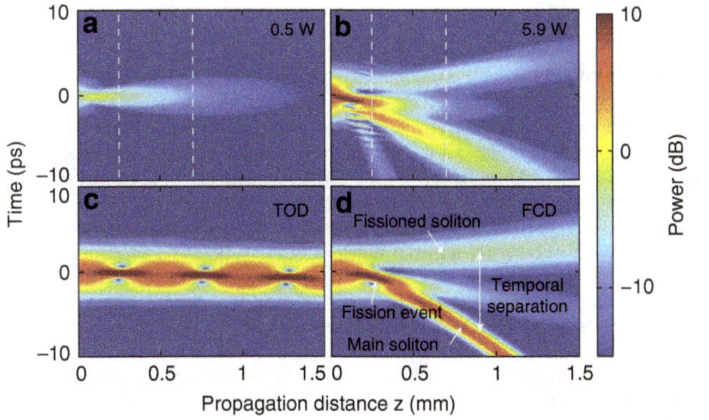

Figure 3 | Time-space propagation maps from a generalised nonlinear Schrödinger equation model. (a–d) A GNLSE model of the pulse dynamics confirms the fission originates from free-carrier dispersion. The dashed lines indicate positions we measured along the waveguide. Note here we show the temporal power $P(t)$ in a dB-scale relative to 1 W, whereas in Figs 1c and 2 we presented the cross-correlation of the electric field $E(t)$, which is the quantity that we measure in the experiment. **a,b** correspond to the experimental conditions with low (**a**) and high (**b**) power, respectively. (**c**) The case modelled with solitons and a TOD perturbation. (**d**) Shows the case modelled with solitons, 3PA and a FCD perturbation.

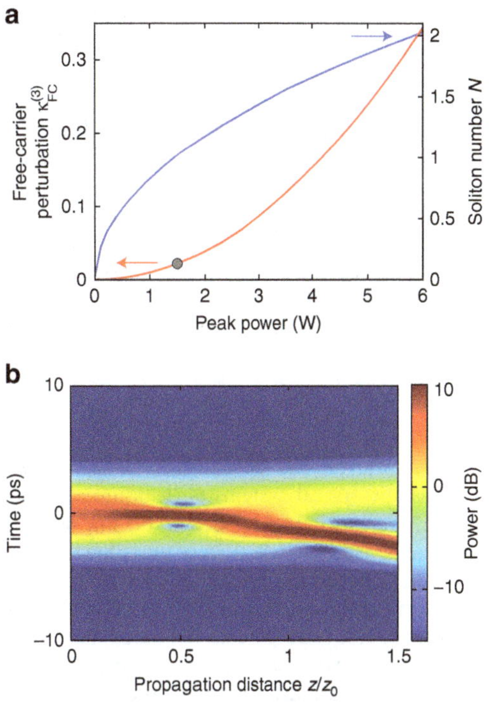

Figure 4 | Analysis of the free-carrier perturbation generated from three-photon absorption. (**a**) Plot of the $\kappa_{FC}^{(3)}$ perturbation and the soliton number N versus power indicating the different scalings for each ($\kappa^{(3)} \propto P_0^2$ and $N \propto \sqrt{P_0}$). (**b**) GNLSE simulation showing the case with the minimum free-carrier dispersion perturbation $\kappa_{FC}^{(3)}$=0.029 required for fission of a $N = 2$ soliton. Note here we show the temporal power $P(t)$ in a dB-scale relative to 1 W.

the scaling of the soliton number to show the comparative evolution of these two parameters ($N \propto \sqrt{P_0}$ and $\kappa_{FC}^{(3)} \propto P_0^2$).

Analytic estimate of FCD-induced fission threshold. We now predict the minimum threshold of $\kappa_{FC}^{(3)}$ required to observe a fission event using this formalism and the characteristic scales for soliton period (z_o) and time duration (T_o) from the literature. First, we define the criteria to call an event a soliton fission. This is non-trivial as the fission is an adiabatic process characterized as a continuous spectral and temporal walk-off of the two constituent solitons[31,33]. We consequently define a clean fission to be a separation of T_0 between the two pulses. Since even for the weakest FCD perturbation this separation can occur at very long distances, we further imposed the condition that the fission must occur within one soliton period z_0 (ref. 3). Under these constraints for a soliton of order $N=2$ we derived an analytic threshold of $\kappa_{FC,min}^{(3)}$(analytic)$= 0.039$ employing a moments method formalism and the equations in the text[29]. We show the full derivation in Supplementary Note 5. The maximum experimental value of $\kappa_{FC,exp}^{(3)}= 0.35$ is approximately an order of magnitude larger than this minimum threshold and clearly of significant strength in the experiments. Note that one can choose arbitrary lengths, temporal separation and soliton number in these equations for desired experimental conditions (see equation 18 in Supplementary Note 5). For higher-order solitons one must consult the analytic relations for the constituent soliton powers following ref. 34 and substitute the appropriate values into the derived equations.

To confirm our analytic theory, we performed GNLSE simulations and varied the strength of $\kappa_{FC}^{(3)}$. We did this by numerically reducing n_{FC} so as not to modify the relationship between the soliton number and $L_{FCD}^{(3)}$. Figure 4b shows the simulation with $\kappa_{FC,min}^{(3)}$(GNLSE)$= 0.029$, which is the minimum strength to meet our criteria. This is on the same order as the analytic theory with the difference attributed to momentum conservation from soliton recoil[4]. This is expected since our analytic formalism treats the constituents independently and neglects soliton interactions. Similar to the known behaviour for other perturbations, we found larger soliton numbers require smaller perturbations to break up the higher-order soliton[3]. This observation is supported by our analytic theory which shows $\kappa_{FC,min}^{(3)}$ scales as $\frac{1}{N^2}$ (see equation 21 in Supplementary Note 5). We compare the FCD perturbation strength derived here with known perturbation mechanisms such as TOD and SS in Supplementary Note 6.

Discussion

From a fundamental physics perspective, these results apply to the general class of optical systems with nonlinear photocarrier (photoelectron) generation. Knowledge of the specific carrier generation mechanism is critical as the physical parameters governing the κ_{FC} perturbation scale differently in the tunneling and multiphoton ionization regimes[14–16,35]. For example, in the case of semiconductor waveguides, a related derivation of the plasma length $L_{FCD}^{(2)}$ was shown for silicon (a TPA-limited material at our wavelength) though soliton perturbation was not addressed in that case[11]. We derive $\kappa_{FC}^{(2)}$ for TPA in Supplementary Note 7 and show that it scales linearly with power. Supplementary Figure 3 compares the power evolution of the 3PA and TPA cases. In the case of ionized gases, an equivalent plasma length for static ionized gases was provided in ref. 36 and we expect a κ parameter could be defined for dynamic nonlinear ionization based on the formalism in ref. 15 for the carrier generation term N_c.

An important application of soliton fission is the generation of ultrabroad coherent light known as supercontinuum (SC)[31,37–39]. The demonstration of SC generation in photonic crystal fibres in 2000 (ref. 40) led to rapid adoption of SC sources in many fields including breakthrough experiments in metrology[41], optical coherence tomography[42] and optical frequency combs[43]. The utility of supercontinuum generation in fibre waveguides has led to significant interest in developing broadband light sources in integrated platforms[44–50]. Examining a recent investigation on supercontinuum in silicon, we computed a value of $\kappa_{FC,exp}^{(2)}$ more than two orders of magnitude larger than our predicted threshold $\kappa_{FC}^{(2)}$(analytic), indicating that the FCD perturbation is required to explain their results[48] (Supplementary Note 7). We expect these observations will facilitate improved SC sources in integrated photonic chips envisioned for future on-chip optical communications systems[51] and lab-on-a-chip spectroscopic tools[52].

In summary, we demonstrated that free-carrier dispersion can induce soliton fission with both theoretical and experimental approaches. Our near-field microscopy measurements enabled the direct observation of the temporal and spatial evolution in the nanoscale waveguide, thereby providing a key new measurement technique for characterizing nonlinear pulses in sub-wavelength structures. We derived an analytic formulation and characteristic parameter $\kappa_{FC}^{(3)}$ for the FCD perturbation and showed that our experimental values were an order of magnitude larger than the minimum required threshold. We supported these results with a GNLSE model confirming both the experiments and theory. These observations elucidate the fundamental physical scaling and dynamics of soliton fission in free-carrier media and could find applications in improved supercontinuum sources in integrated photonic chips and gas-filled microstructured fibres.

Methods

Sample description and material parameters. A scanning electron micrograph and detailed fabrication parameters of the $L = 1.5$ mm air-suspended GaInP waveguide can be found in Supplementary Note 1 and Supplementary Table 1. Our sample includes integrated mode-adapters which reduce the total insertion losses in the linear regime to ~ 17 dB (including propagation loss) and suppress Fabry–Perot oscillations at the end facets. The output coupling from the chip to the 0.4 numerical aperture (NA) lensed fibre (OzOptics) is -2.5 dB in agreement with our earlier work[17]. Due to the need to approach the NSOM tip near to the sample input, coupling is achieved with a 0.4 NA aspheric lens (Newport) with an estimated coupling efficiency of -7 dB which we suspect is due a mode-field size mismatch between the beam waist and the lens. We report the measurements of the sample properties, material parameters and the GNLSE model in Supplementary Notes 1 and 3.

Experimental set-up. For the nonlinear measurements, we employed a mode-locked fibre laser (PriTel) delivering hyperbolic-secant pulses at 1,553 nm with a temporal duration $T_{FWHM} = 2.2$ ps as measured by autocorrelation. The repetition rate is 20 MHz and the laser light is coupled to the waveguide with electric-field polarized in-plane with the slab (TE). The pulses are slightly chirped as confirmed by autocorrelation measurements of the pulse input. For the nonlinear pulse transmission measurements, we used an optical spectrum analyser to measure the pulse spectrum as a function of input power. Two such traces are shown in Fig. 1a with additional traces in Supplementary Fig. 2.

To measure the temporal dynamics of the pulse propagating inside the waveguide we employ a homebuilt time-resolved NSOM[53]. In the set-up the entire microscope, including the sample, is included in one branch of a Mach–Zehnder interferometer. The near-field probe is brought in close proximity (circa 20 nm) of the waveguide where it collects the evanescent tail of the guided mode. As a result, a minute fraction of the guided light is transformed into far-field radiation by the near-field probe and is interferometrically mixed with light from a reference branch. The interference is detected on a photodiode with a heterodyne detection scheme. By scanning an optical delay line and using a pulsed laser source we measure a temporal cross-correlation of the electric field of the pulses propagating in the reference branch of the interferometer and in the waveguide. The measured temporal cross-correlation is described by the following equation:

$$C(z,\tau) \propto \int E_s(z,t-\tau)E_r(t)\mathrm{d}t \quad (3)$$

where $C(z,\tau)$ is the cross-correlation function, z the spatial location of the near-field probe, τ the delay time and $E_s(z,t-\tau)$ and $E_r(t)$ the electric field of the pulse propagating in the sample and the reference branch of the set-up, respectively. Correspondingly, the following equation holds in the frequency domain:

$$C(z,\omega) \propto E_s(z,\omega) \cdot E_r^*(\omega), \quad (4)$$

where $C(z, \omega)$, $E_s(z, \omega)$ and $E^*_r(\omega)$ are the frequency spectra of the temporal cross-correlation function and the electric field of the pulse in the sample and the reference branch, respectively.

It has been shown that various changes of the temporal pulse envelope transfer to the cross-correlation function. For example, the time of flight can be directly extracted from the observed time delay in the experiment[54]. Furthermore, symmetric temporal broadening due to GVD[26], as well as asymmetrical TOD[21], exhibit similar features in the cross-correlation function as in the temporal pulse envelope. However, the cross-correlation function will only directly represent the temporal envelope of the pulse propagating in the sample if the pulse in the reference branch is extremely short in time, ideally a Dirac delta function, and its spectrum extremely broad and constant. Therefore, we show the temporal cross-correlation function in our manuscript where we discuss the experimental measurements (that is, in Figs 1c and 2).

To observe the nonlinear evolution of the pulse we repeat the cross-correlation measurements at different input powers which are controlled by a set of neutral density filtres. Further, to track the changes of the temporal pulse envelope in space we position the near-field probe at different locations along the waveguide and repeat the cross-correlation measurements. This measurement procedure allows, for example, to gain information of the time of flight of the pulse or the reshaping of the pulse envelope[21]. While there are a number of spatially resolved studies in the linear regime[21,55], there are few investigations of nonlinear dynamics with NSOM[24] or complementary techniques[25] and, to our knowledge, no investigations of soliton dynamics.

References

1. Zakharov, V. & Shabat, A. Exact theory of two-dimensional self-focusing and one-dimensional self-modulation of waves in nonlinear media. *J. Exp. Theor. Phys.* **34,** 62–69 (1972).
2. Golovchenko, E. A., Dianov, E. M., Prokhorov, A. M. & Serkin, V. N. Decay of optical solitons. *J. Exp. Theor. Phys.* **42,** 87–91 (1985).
3. Wai, P. K. A., Menyuk, C. R., Lee, Y. C. & Chen, H. H. Nonlinear pulse propagation in the neighborhood of the zero-dispersion wavelength of monomode optical fibers. *Opt. Lett.* **11,** 464–466 (1986).
4. Tai, K., Bekki, N. & Hasegawa, A. Fission of optical solitons induced by stimulated Raman effect. *Opt. Lett.* **13,** 392–394 (1988).
5. Beaud, P., Hodel, W., Zysset, B. & Weber, H. P. Ultrashort pulse propagation, pulse breakup, and fundamental soliton formation in a single-mode optical fiber. *IEEE J. Quant. Electron.* **23,** 1938–1946 (1987).
6. Grudinin, A. B. *et al.* Decay of femtosecond pulses in single-mode optical fibers. *J. Exp. Theor. Phys.* **46,** 221–225 (1987).
7. Lin, Q., Painter, O. J. & Agrawal, G. P. Nonlinear optical phenomena in silicon waveguides: modeling and applications. *Opt. Express* **15,** 16604–16644 (2007).
8. Fedotov, A. B., Serebryannikov, E. E. & Zheltikov, A. M. Ionization-induced blueshift of high-peak-power guided-wave ultrashort laser pulses in hollow-core photonic-crystal fibers. *Phys. Rev. A* **76,** 053811 (2007).
9. Wood, W. M., Siders, C. W. & Downer, M. C. Measurement of femtosecond ionization dynamics of atmospheric density gases by spectral blueshifting. *Phys. Rev. Lett.* **67,** 3523–3526 (1991).
10. Rieger, G. W., Virk, K. S. & Young, J. F. Nonlinear propagation of ultrafast 1.5 µm pulses in high-index-contrast silicon-on-insulator waveguides. *Appl. Phys. Lett.* **84,** 900–902 (2004).
11. Blanco-Redondo, A. *et al.* Controlling free-carrier temporal effects in silicon by dispersion engineering. *Optica* **1,** 299–306 (2014).
12. Colman, P. *et al.* Temporal solitons and pulse compression in photonic crystal waveguides. *Nat. Photon.* **4,** 862–868 (2010).
13. Ding, W. *et al.* Time and frequency domain measurements of solitons in subwavelength silicon waveguides using a cross-correlation technique. *Opt. Express* **18,** 26625–26630 (2010).
14. Travers, J. C., Chang, W., Nold, J., Joly, N. Y. & St J Russell, P. Ultrafast nonlinear optics in gas-filled hollow-core photonic crystal fibers [invited]. *J. Opt. Soc. Am. B* **28,** A11–A26 (2011).
15. Saleh, M. F. *et al.* Theory of photoionization-induced blueshift of ultrashort solitons in gas-filled hollow-core photonic crystal fibers. *Phys. Rev. Lett.* **107,** 203902 (2011).
16. Husko, C. A. *et al.* Soliton dynamics in the multiphoton plasma regime. *Sci. Rep.* **3,** 1100 (2013).
17. Blanco-Redondo, A. *et al.* Observation of soliton compression in silicon photonic crystals. *Nat. Commun.* **5,** 3160 (2014).
18. Bhat, N. A. R. & Sipe, J. E. Optical pulse propagation in nonlinear photonic crystals. *Phys. Rev. E* **64,** 056604 (2001).
19. Colman, P., Combrié, S., Lehoucq, G. & De Rossi, A. Control of dispersion in photonic crystal waveguides using group symmetry theory. *Opt. Express* **20,** 13108–13114 (2012).
20. Hughes, S., Ramunno, L., Young, J. F. & Sipe, J. E. Extrinsic optical scattering loss in photonic crystal waveguides: role of fabrication disorder and photon group velocity. *Phys. Rev. Lett.* **94,** 033903 (2005).
21. Engelen, R. J. P. *et al.* The effect of higher-order dispersion on slow light propagation in photonic crystal waveguides. *Opt. Express* **14,** 1658–1672 (2006).
22. Monat, C. *et al.* Slow light enhancement of nonlinear effects in silicon engineered photonic crystal waveguides. *Opt. Express* **17,** 2944–2953 (2009).
23. Dudley, J. M., Barry, L. P., Bollond, P. G., Harvey, J. D. & Leonhardt, R. Characterizing pulse propagation in optical fibers around 1550 nm using frequency-resolved optical gating. *Opt. Fiber Technol.* **4,** 237–265 (1998).
24. Wulf, M., Beggs, D. M., Rotenberg, N. & Kuipers, L. Unravelling nonlinear spectral evolution using nanoscale photonic near-field point-to-point measurements. *Nano. Lett.* **13,** 5858–5865 (2013).
25. Bruck, R. *et al.* Device-level characterization of the flow of light in integrated photonic circuits using ultrafast photomodulation spectroscopy. *Nat. Photon.* **9,** 54–60 (2015).
26. Gersen, H., Korterik, J. P., van Hulst, N. F. & Kuipers, L. Tracking ultrashort pulses through dispersive media: Experiment and theory. *Phys. Rev. E* **68,** 026604 (2003).
27. Agrawal, G. P. *Nonlinear Fiber Optics* 5th edn (Academic Press, 2013).
28. Mollenauer, L. F., Stolen, R. H. & Gordon, J. P. Experimental observation of picosecond pulse narrowing and solitons in optical fibers. *Phys. Rev. Lett.* **45,** 1095–1098 (1980).
29. Lefrancois, S., Husko, C., Blanco-Redondo, A. & Eggleton, B. J. Nonlinear silicon photonics analyzed with the moment method. *J. Opt. Soc. Am. B* **32,** 218–226 (2015).
30. Gordon, J. P. Theory of the soliton self-frequency shift. *Opt. Lett.* **11,** 662–664 (1986).
31. Dudley, J. M., Genty, G. & Coen, S. Supercontinuum generation in photonic crystal fiber. *Rev. Mod. Phys.* **78,** 1135–1184 (2006).
32. Serkin, V. N. 'colored' envelope solitons in fiber-optic waveguides. *Sov. Tech. Phys. Lett.* **13,** 320–321 (1987).
33. Kodama, Y. & Hasegawa, A. Nonlinear pulse propagation in a monomode dielectric guide. *IEEE J. Quant. Electron.* **23,** 510–524 (1987).
34. Satsuma, J. & Yajima, N. B. Initial Value Problems of One-Dimensional Self-Modulation of Nonlinear Waves in Dispersive Media. *Progress Theor. Phys. Suppl.* **55,** 284–306 (1974).
35. Keldysh, L. V. Ionization in the field of a strong electromagnetic wave. *Sov. Phys. J. Exp. Theor. Phys.* **20,** 1307–1314 (1965).
36. Wagner, N. L. *et al.* Self-compression of ultrashort pulses through ionization-induced spatiotemporal reshaping. *Phys. Rev. Lett.* **93,** 1–4 (2004).
37. Herrmann, J. *et al.* Experimental evidence for supercontinuum generation by fission of higher-order solitons in photonic fibers. *Phys. Rev. Lett.* **88,** 173901 (2002).
38. Alfano, R. R. (ed.) *The Supercontinuum Laser Source* 3rd edn (Springer, 2016).
39. Dudley, J. M. & Taylor, J. R. *Supercontinuum Generation in Optical Fibers* (Cambridge Univ. Press, 2010).
40. Ranka, J. K., Windeler, R. S. & Stentz, A. J. Visible continuum generation in air-silica microstructure optical fibers with anomalous dispersion at 800 nm. *Opt. Lett.* **25,** 25–27 (2000).
41. Udem, T. H., Reichert, J., Holzwarth, R. & Hänsch, T. W. Accurate measurement of large optical frequency differences with a mode-locked laser. *Opt. Lett.* **24,** 881–883 (1999).
42. Povazay, B. *et al.* Submicrometer axial resolution optical coherence tomography. *Opt. Lett.* **27,** 1800–1802 (2002).
43. Cundiff, S. T. & Ye, J. Colloquium: Femtosecond optical frequency combs. *Rev. Mod. Phys.* **75,** 325 (2003).
44. Hsieh, I.-W. *et al.* Supercontinuum generation in silicon photonic wires. *Opt. Express* **15,** 15242–15249 (2007).
45. Yeom, D.-I. *et al.* Low-threshold supercontinuum generation in highly nonlinear chalcogenide nanowires. *Opt. Lett.* **33,** 660–662 (2008).
46. Duchesne, D. *et al.* Supercontinuum generation in a high index doped silica glass spiral waveguide. *Opt. Express* **18,** 923–930 (2010).
47. Lau, R. K. W. *et al.* Octave-spanning mid-infrared supercontinuum generation in silicon nanowaveguides. *Opt. Lett.* **39,** 4518–4521 (2014).
48. Leo, F. *et al.* Dispersive wave emission and supercontinuum generation in a silicon wire waveguide pumped around the 1550 nm telecommunication wavelength. *Opt. Lett.* **39,** 3623–3626 (2014).
49. Singh, N. *et al.* Midinfrared supercontinuum generation from 2 to 6 μm in a silicon nanowire. *Optica* **2,** 797–802 (2015).
50. Yin, L., Lin, Q. & Agrawal, G. P. Soliton fission and supercontinuum generation in silicon waveguides. *Opt. Lett.* **32,** 391–393 (2007).
51. Nakasyotani, T., Toda, H., Kuri, T. & Kitayama, K.-I. Wavelength-division-multiplexed millimeter-waveband radio-on-fiber system using a supercontinuum light source. *J. Lightwave Technol.* **24,** 404–410 (2006).
52. Lindfors, K., Kalkbrenner, T., Stoller, P. & Sandoghdar, V. Detection and spectroscopy of gold nanoparticles using supercontinuum white light confocal microscopy. *Phys. Rev. Lett.* **93,** 037401 (2004).

53. Rotenberg, N. & Kuipers, L. Mapping nanoscale light fields. *Nat. Photon.* **8,** 919–926 (2014).
54. Balistreri, M., Gersen, H., Korterik, J., Kuipers, L. & Van Hulst, N. Tracking femtosecond laser pulses in space and time. *Science* **294,** 1080–1082 (2001).
55. Gersen, H. *et al.* Real-space observation of ultraslow light in photonic crystal waveguides. *Phys. Rev. Lett.* **94,** 073903 (2005).

Acknowledgements

This research was supported by the Australian Research Council (ARC) Center of Excellence CUDOS (CE110001018), ARC Laureate Fellowship (FL120100029), ARC Discovery Early Career Researcher Award (DECRA DE120102069), the Netherlands Foundation for Fundamental Research on Matter (FOM) and the Netherlands Organization for Scientific Research (NWO). L.K. acknowledges funding from ERC Advanced Investigator Grant (no. 240438-CONSTANS). A.D.R, S.C., and G.L. acknowledge financial support from the ERC-Pharos programme lead by A. P. Mosk. C.H. graciously thanks AMOLF for hosting him to conduct the experiments with M.W. and L.K.

Author contributions

L.K., B.J.E., M.W. and C.H. planned the NSOM experiment. M.W. and C.H. performed the experiments. C.H. conceived the idea for free-carrier induced fission. S.L. and C.H. derived the analytic theory. C.H. performed the modelling with support from A.D.R. G.L., S.C. and A.D.R. designed and fabricated the sample. L.K. and B.J.E. supervised the project. C.H. and M.W. wrote the paper. All authors discussed the results and commented on the manuscript.

Additional information

Competing financial interests: The authors declare no competing financial interests.

26

4Pi-RESOLFT nanoscopy

Ulrike Böhm[1], Stefan W. Hell[1] & Roman Schmidt[1]

By enlarging the aperture along the optic axis, the coherent utilization of opposing objective lenses (4Pi arrangement) has the potential to offer the sharpest and most light-efficient point-spread-functions in three-dimensional (3D) far-field fluorescence nanoscopy. However, to obtain unambiguous images, the signal has to be discriminated against contributions from lobes above and below the focal plane, which has tentatively limited 4Pi arrangements to imaging samples with controllable optical conditions. Here we apply the 4Pi scheme to RESOLFT nanoscopy using two-photon absorption for the on-switching of fluorescent proteins. We show that in this combination, the lobes are so low that low-light level, 3D nanoscale imaging of living cells becomes possible. Our method thus offers robust access to densely packed, axially extended cellular regions that have been notoriously difficult to super-resolve. Our approach also entails a fluorescence read-out scheme that translates molecular sensitivity to local off-switching rates into improved signal-to-noise ratio and resolution.

[1] Department of NanoBiophotonics, Max Planck Institute for Biophysical Chemistry, Am Fassberg 11, Göttingen 37077, Germany. Correspondence and requests for materials should be addressed to S.W.H. (email: stefan.hell@mpibpc.mpg.de) or to R.S. (email: roman.schmidt@mpibpc.mpg.de).

The three to seven fold improved axial resolution provided by 4Pi microscopy[1–3] in the 1990s marked a first step in the quest for radically improving the resolution in far-field fluorescence microscopy. Yet the resolution provided by 4Pi microscopy remained diffraction-limited, because by jointly using two opposing lenses for focusing the excitation and/or the fluorescence light, this method just optimized the focusing conditions for feature separation. Modern far-field fluorescence nanoscopy[4], or superresolution microscopy, such as the methods called stimulated emission depletion (STED), reversible fluorescent saturable optical transition (RESOLFT) and later also photoactivated localization microscopy (PALM)/stochastical optical reconstruction microscopy (STORM) fundamentally departed from such early superresolution concepts by discerning features through a molecular state transition. The use of a state transition for feature separation, typically a transition between a fluorescent (ON) and a non-fluorescent (OFF) state, opened the road to lens-based fluorescence microscopy with resolution that is conceptually not limited by diffraction.

Yet diffraction plays a role in these 'diffraction-unlimited' techniques because the resolution of these 'nanoscopy' methods still benefits strongly from focusing the light as sharply as possible. While in STED and RESOLFT, it is the focusing of the illumination light in sample space that matters, in PALM/STORM it is the focusing of the emitted light at the detector. Therefore, the optimization of focusing remains very timely. 4Pi arrangements can also facilitate the doubling of the detected fluorescence without compromising the resolution in the focal plane (x,y), and offer significantly sharper axial (z) intensity gradients than single lenses for both the illumination and the detected light. Consequently, the combination of 4Pi with STED, RESOLFT and PALM/STORM approaches currently offers the most powerful optical setting for three-dimensional (3D) fluorescence nanoscopy[5–7].

Yet 4Pi-type super-resolution arrangements are scarcely reported for STED and PALM and entirely unexplored for RESOLFT, a STED-derivative that typically uses reversibly switchable fluorescent proteins (RSFP) for providing the mandatory ON and OFF states. RSFP-based RESOLFT is particularly attractive because it operates with low light levels, making it gentle to living cells[8].

The difficulties of realizing a 4Pi setup are commonly attributed to the counter-alignment of the two high numerical aperture (NA) lenses. In practice, however, the alignment can be controlled and stabilized over many hours. Instead, a far more general problem that is inherent to all fluorescent imaging modalities comes to the fore. The fluorescence signal (that emanates from each sub-diffraction pixel volume under investigation) needs to be discriminated against 'background' signal from outside of this volume. This 'background' largely stems from optical aberrations that preclude precise spatial control of the illumination or fluorescence beam positions and, in case of STED, RESOLFT or PALM/STORM, from imperfections of the ON/OFF-state transfer (switching) process. Discrimination against this 'background' signal is most challenging along the optical (z) axis, especially when the probed volume is of sub-diffraction dimensions. Lack of sufficient discrimination along the z-axis (optical sectioning) manifests itself as artifacts in the image, particularly as 'ghost features' above and below the real features.

When describing the imaging process in the spatial frequency domain, the appearance of axial lobes corresponds to local depressions in the amplitude of the optical transfer function, that is, the modulation transfer function (MTF) of the microscope. Structural information of the sample can only be retrieved in those spatial frequency bands where the MTF is strong enough to convey a signal that sufficiently exceeds the local noise level. In a 4Pi microscope, MTF depressions are typically restricted to sharp local minima at the so-called critical frequencies[9]. As their amplitude strongly depends on the aperture angle α of the objective lenses used[9], combinations of 4Pi with diffraction-unlimited super-resolution/nanoscopy methods such as isoSTED[5,10,11] and iPALM[6,12], have unfortunately been limited to imaging fixed samples that are more easily accessible with high angle lenses ($\alpha \geq 74°$, as for oil immersion lenses with NA $\geq$ 1.46). Furthermore, the imaged objects were rather thin and labelled very sparsely, as both properties alleviate the requirements on optical sectioning, that is, on suppressing ('background') signal from above and below the focal plane. Fortunately, in a coordinate-targeted nanoscopy method such as RESOLFT, the signal received from the targeted nanosized pixel volume scales with the average number of molecules located within this volume, allowing for tailoring of the pixel volume[11], and hence the resolution and the signal, to the actual imaging conditions to render the 'background' (mathematically) treatable.

Here we report the realization of 4Pi-RESOLFT nanoscopy, that is, of a conceptually diffraction-unlimited resolving method which, by virtue of 4Pi microscopy, provides spatially uniform 3D resolution for imaging (living) cells at the nanometre scale, offers strong optical sectioning due to multiple background suppressing mechanisms and operates at low light levels in 3D.

Results

On-switching order and optical sectioning. The effective point-spread-function (PSF) $h_{ef}(\mathbf{r})$ of a coordinate-targeted super-resolution imaging modality ultimately quantifies the 3D-coordinate range where the fluorescent molecules are allowed to yield measurable signal. If the fluorophores from a certain range are imaged onto a (confocal point) detector, $h_{ef}(\mathbf{r})$ is given by the normalized distribution $S^{ON}(\mathbf{r})$ showing where a molecule is allowed to be in the ON-state at the time of read-out, multiplied by a normalized function $h^{det}(\mathbf{r})$ that describes the detection probability:

$$h_{ef}(\mathbf{r}) = S^{ON}(\mathbf{r}) \cdot h^{det}(\mathbf{r}) \qquad (1)$$

$S_{ON}(\mathbf{r})$ is proportional to a product of normalized terms h^{on} and $\tilde{h}^{off}$ that describe the generation of the ON-state by the use of on- and off-switching processes, respectively. h^{on} describes the spatial probability to assume the ON-state in the absence of off-switching light. It can typically be written as a product of terms h_i^{on} that each express the relative probability for absorption of a single photon that drives a transition to a (virtual) state, and therefore scales with the intensity of the light patterns used (for example, $h_{exc}^{on} \sim I_{exc}^{on}$ for single-photon excitation with intensity I_{exc}^{on}; in case of two-photon excitation: $h_{2phexc}^{on} = h_{exc}^{on} \cdot h_{exc}^{on}$). $\tilde{h}^{off}$ describes the effect of the off-switching light on a potential ON-state distribution; $\tilde{h}^{off} = 1$ where molecules are always allowed to assume the ON-state, and $\tilde{h}^{off} = 0$ where they are forced to stay in an OFF-state. Due to the forced assumption of an OFF-state by 'saturating' off-switching, $\tilde{h}^{off}$ is usually much sharper than the off-switching light intensity patterns.

We formally define h^{det} as the first on-switching term $h_1^{on} \equiv h^{det}$ (because of its similar effect on h_{ef}), drop any h_i^{on} that does not significantly sharpen h_{ef} (for example, widefield-detection or sample-wide switching) and obtain:

$$h_{ef}(\mathbf{r}) = \prod_{i=1}^{O^{on}} h_i^{on}(\mathbf{r}) \cdot \tilde{h}^{off}(\mathbf{r}) \qquad (2)$$

Here the number of on-factors O^{on} denotes the on-(switching) order of h_{ef}, for example, $O^{on} = 2$ (excitation by single-photon absorption and confocal detection) for a typical confocal (STED) microscope. Optical sectioning can generally be improved by engineering h^{off} such that molecules in out-of-focus areas are switched off more effectively (Supplementary Fig. 1), or by requiring the absorption of multiple photons for the occupation of the ON-state, which increases O^{on}. The latter can be realized directly through standard two-photon absorption[13–17], or by requiring the sequential occupation of multiple real states to reach the ON-state[18]. Such sequential state occupation is easily realized using the switching steps offered by RSFP (that are central to the RSFP-based RESOLFT concept[4,19]).

The 4Pi-RESOLFT modality. In this study, we devised a coordinate-targeted 4Pi-RESOLFT modality that utilizes negative-switching RSFP (that is, those that are switched off at a wavelength that is also used for generating the fluorescent signal, Fig. 1a) and that resorts to all the processes mentioned above for strong optical sectioning. Concretely, we opted for the RSFP Dronpa-M159T[20–22], which stands out by relatively fast switching kinetics with comparatively low background. At each scanning position, the local RSFP molecules were cycled through their ON- and OFF-states by consecutive light pulses that defined our RESOLFT imaging sequence (Fig. 1a). In the initial step ('activation' pulse), we applied a µs-long train of 170-fs pulses at 90 MHz/780 nm in a focal pattern h_{ac} to (partially) transfer ('activate') local RSFP to their meta-stable 'active'-state by two-photon absorption, as described by the activation distribution $S^B(\mathbf{r})$. Subsequently, we applied a µs–ms-long 'deactivation' pulse of continuous-wave (CW) irradiation at 488 or 491 nm, focused to form a hollow deactivation pattern (for example, a 'z-donut' h_{zd}, Fig. 1a, Supplementary Methods). This drove active RSFP outside the targeted pixel volume (for example, above and below the focal plane) back to their inactive state, which effectively denied them a further excitation to the fluorescent ON-state and thus improved the spatial ON/OFF-contrast during the following 'read-out' pulse. We finally probed the remaining active RSFP by a second µs–ms-long CW pulse at 491 nm with a focal pattern h_{ro}, which transferred them to their fluorescent (ON) state, and detected the fluorescence through a confocal pinhole.

Our scheme thus entails a number of advantages for live-cell 3D-imaging. First, RSFP are inherently live-cell compatible protein markers, and selection of sufficiently bright and stable RSFP is readily available[8,20,23]. Second, optical sectioning benefits from the additional switching step (activation) involved in the RSFP switching cycle with respect to modalities that do not make use of a meta-stable state. This additional switching becomes especially powerful if it is mediated by two-photon absorption in a 4Pi configuration, as O^{on} rises to 4 and the activation and read-out patterns (h_{ac}, h_{ro}) can be setup to a limited zone of overlap in the focal region (Fig. 1a). While overlapping several pattern h_i^{on} also forms the basis of 4Pi microscopy of type C using two-photon excitation[24], here we do not require coherent double-lens (4Pi) detection of the emitted fluorescence, and therefore do not need broad-band intra-cavity dispersion compensation. The scheme presented here thus acts to the same effect with much less technical complexity. Finally, activation by two-photon absorption entails much less photo damage than two-photon excitation, as it takes place at a time during the switching cycle when virtually no markers can assume their excited fluorescent state.

Under ideal conditions, the effective PSF h_{ef} of such a system is virtually free of axial lobes (Fig. 1a) even without a deactivation pulse. In practice, incomplete deactivation and optical aberrations

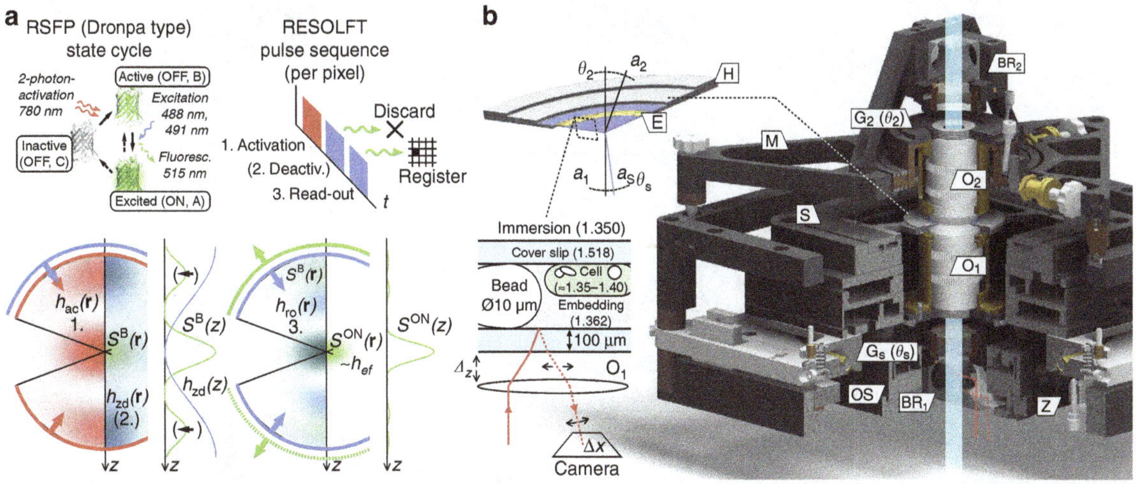

Figure 1 | 4Pi-RESOLFT principle and sample optics. **(a)** Coherent double-lens illumination cycles RSFP markers between dark (OFF) and bright (ON) states to generate spatial ON/OFF-contrast. For each pixel, an activation light pulse (focal pattern h_{ac}) induces two-photon activation of RSFP (state transition C->B) in a pattern $S^B(\mathbf{r})$ with axial side-maxima (lobes) that are optionally suppressed by a subsequently applied deactivation pulse (h_{zd}, B<->A->C). Fluorescence generated by the ON-state A is detected during read-out (B<->A->C) by a pattern h_{ro}. Its mutual overlap with $S^B(\mathbf{r})$ is constrained to the focal centre, resulting in an effective PSF $h_{ef} \sim S^{ON}$ that exhibits ≈100 nm axial FWHM and exceptionally low side-maxima. Profiles show on-axis values. **(b)** The upright 4Pi unit of the microscope. Cells are mounted on a ring-shaped sample holder (H), between two cover glasses fixed at 10 µm distance by spacer beads and epoxy resin (E). The set of refractive indices (in brackets) of the immersion and embedding medium, cover slip thickness and correction collar settings of the objective lenses (O_1, O_2) diminishes aberrations from the sample. The sample stage (S) is mounted on a vertically movable (Z) goniometer (G_S), accepts the sample holder (H) and provides five degrees-of-freedom for coarse *xyz*-positioning and *z*-scanning of the sample, as well as tip-/tilt-alignment (θ_S) of the cover slip normal (a_S) to the optic axis of O_1 (a_1). O_1 itself is mounted on a *xyz*-piezo stage (OS) that provides online fine control over the displacement of both foci. A triangular mount (M) allows for tip/tilt-(θ_2) and coarse *xyz*-alignment of O_2 (axis a_2) with respect to O_1, and can conveniently be detached to change the sample. Two polarizing beam splitter/quarter-wave retarder pairs ($BR_{1,2}$) clean up and tune the polarization of the incident beam pairs to opposing circular states. One beam splitter furthermore serves as a port for an alignment laser beam that provides optical feedback for online-stabilization of the axial sample position (Δz); the beam traverses the respective objective lens off-axis (solid red path), gets reflected at the embedding medium interface and is imaged onto a camera (dotted red path).

may give rise to lobe amplitudes that are still relevant. To counteract these effects, we applied dedicated lobe deactivation by h_{zd} and developed low-aberration[25], live-cell 4Pi optics (Fig. 1b, Methods). These measures enabled volume scans of over 5-μm-thick mammalian cells without noticeable bleaching at an axial (z) resolution in the 100 nm range and axial lobes of only ~15% of the main peak of the z-response, that is, measured on laterally (xy) integrated data (Fig. 2). The base acquisition time of 7–21 s μm^{-3} (depending on the brightness of the labelled structure) was short enough to capture the subtly moving cytoskeleton of a living cell (Fig. 2b,c, total acquisition times incl. drift correction overhead b: 115 min per 703 μm^3, c: 160 min per 400 μm^3). For highly mobile organelles, such as mitochondria (Fig. 2a), fixation of the sample by paraformaldehyde incubation (Supplementary Methods) offered a means to prevent motion blur. While this treatment irreversibly arrests the cell, its potential to introduce structural artifacts is very low with respect to staining/embedding protocols that involve membrane permeabilization.

To resolve smaller features, we implemented a second switching pattern for deactivation of RSFP around the focal centre: A hollow '3D-donut' h_{3d} (Fig. 3a), created by a single focused 4Pi beam pair[26] (Methods), allowed us to tune h_{ef} to a near-isotropic resolution below the diffraction limit (Fig. 3a). Calculations using a vectorial diffraction theory[27] predicted on-axis MTF values of over 40% of the MTF maximum within the MTF bandwidth up to a resolution of 30 nm. This feature keeps the signal well over the noise level in most applications and exemplifies the improvement brought about by higher order on-switching in comparison to modalities of second order such as those reported in isoSTED microscopy[5] (Fig. 3b). Furthermore, the confinement of the fluorescent on-state, that is, of h_{ef}, to sub-diffraction 3D volumes means that fewer fluorophores are interrogated at any point in time. This reduction in number of interrogated molecules (that are inherently co-localized) greatly facilitates the quantitative assessment of the properties of the fluorescent labels as they vary in the sample. We found that in time-resolved recordings, the on/off separation contrast decayed over time, hinting to the contribution of multiple deactivation rates. Thus, we introduced a 'rate-gated' RESOLFT detection scheme that improved both the signal-to-noise ratio in the image and the resolution by discriminating individual signal components (Fig. 3c,d, Methods).

Following this approach, we recorded images of Lifeact-Dronpa-M159T-expressing cells and adjusted rate-gating and the RESOLFT pulse sequence for target resolutions of 30-50 nm; the parameters were established by a PSF simulation using measured rate kinetics. Optical xz-sections taken perpendicular to the run of solitary actin fibre bundles confirmed the effectiveness of rate-gating (Fig. 3c,d) and the overall shape of the effective PSF (Fig. 3e). Illumination with the z-donut-shaped (h_{zd}) focus for 1 ms at an average light power of 1.8 μW (488 nm, CW) was sufficient to virtually eliminate lobe background from the image (Fig. 3e, $+h_{zd}$), while the low gradients around the central zero of h_{zd} with respect to h_{3d} facilitated the mutual alignment of these patterns (Supplementary Fig. 1). Turning to the finer structured actin network inside the cell body, we measured apparent feature sizes well below 40 nm (Fig. 4a–c). At a relaxed target resolution of 50 nm and an acquisition time of 3.3 min μm^{-2}, we observed the time evolution of the actin scaffold at a vertical contact region of two neighbouring cells (Fig. 4e–g).

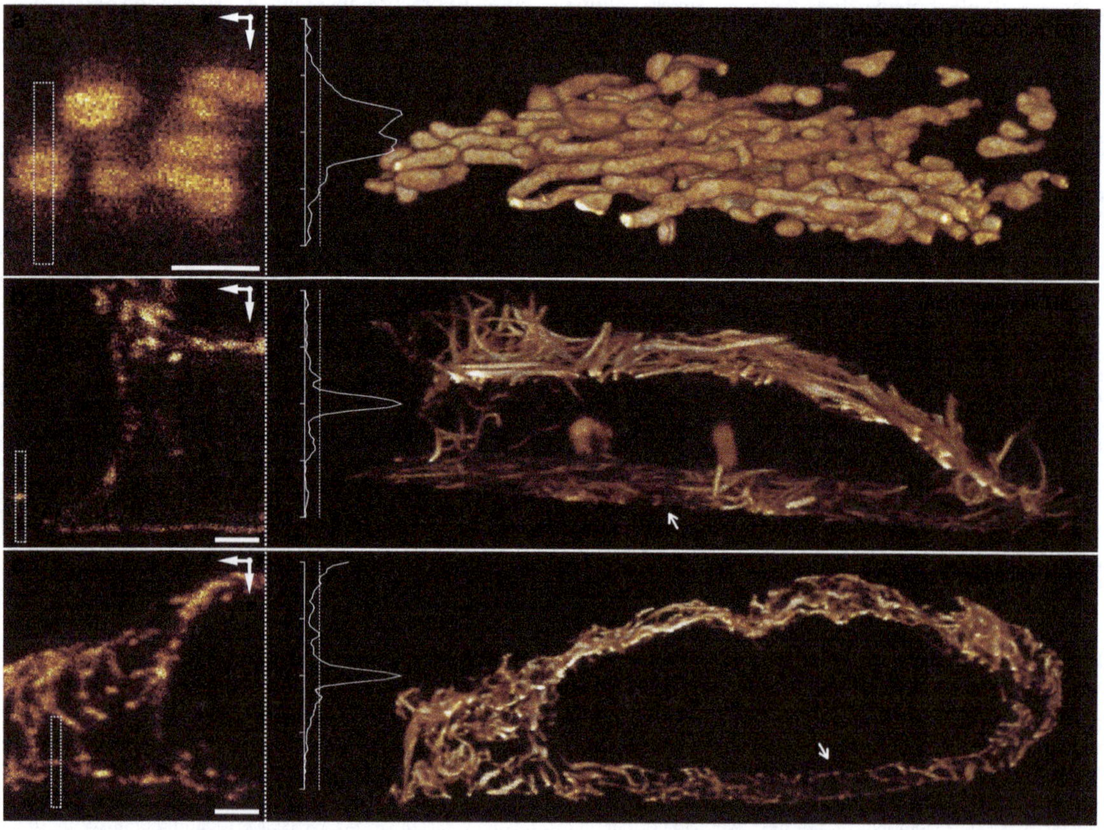

Figure 2 | 4Pi-RESOLFT imaging exhibits only minor axial lobes. 4Pi-RESOLFT raw data (left) and volume renderings (right) of Dronpa-M159T targeted to (**a**) the lumen of mitochondria, (**b**) actin microfilaments and (**c**) intermediate filaments of the cytoskeleton. The sample in **a** was subject to PFA fixation to freeze the motion of mitochondria; the filament networks in **b,c** were recorded from living cells, and exhibit regions of reduced density adjacent to the cover slip (arrows). Estimates of the z-response (insets), measured as box profiles over extended structures, exhibit only minor axial lobes in the 15 % range. Fast-to-slow order of scan axes, xzy. Pulse parameters, E_{ac}, E_{zd}, E_{ro} = 1.6 mW · 50 μs, 18 μW · 50 μs, 3.1 μW · 50 μs. Scale bars, 1 μm.

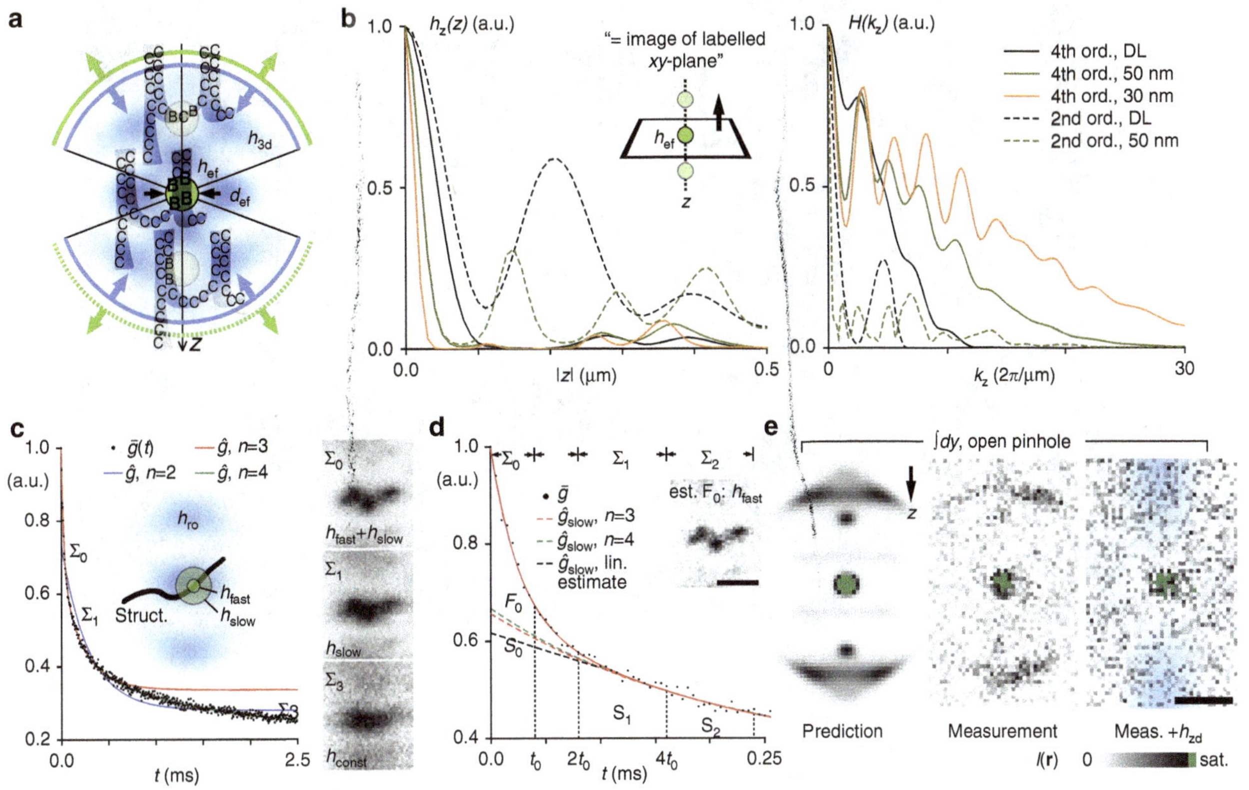

Figure 3 | 4Pi-RESOLFT image formation with <100 nm isotropic resolution. (**a**) A hollow switching pattern h_{3d} confines the central effective PSF to a spot with diameter d_{ef} by switching activated markers (B) back to their inactive state (C). Side-lobes due to inefficient switching at low off-centre h_{3d} amplitudes rise in relative strength as d_{ef} is reduced. μ, labelled structure. (**b**) Simulated z-response $h_z(z)$ (laterally integrated h_{ef}) and axial MTF profile $H(k_z)$ of the 4Pi-RESOLFT microscope (fourth on-order switching, solid lines) at different target resolutions d_{ef}. DL, diffraction limit. Graphs for an isoSTED microscope under similar conditions are included for reference (second on-order, dotted lines). (**c**) Normalized time-resolved, mean fluorescence signal $\bar{g}(t)$ collected from an xz-section through an actin fibre bundle (struct.) in a cell expressing Lifeact-Dronpa-M159T. Target resolution 50 nm, read-out pattern h_{ro} with a total power of $P_{ro} = 3.1\,\mu W$ incident on the sample. An n-component multi-exponential fit to the data corresponds to n apparent switching speeds $\hat{\lambda}_i$. A minimum of $n = 3$ is required to adequately represent the data from the beginning of the read-out pulse $t = 0$ up to 0.5 ms, $\hat{\lambda}_i = 40.5, 4.3, 0.0\,ms^{-1}$ (for up to 2.5 ms: $n = 4$, $\hat{\lambda}_i = 41, 5.9, 0.85, 0.0\,ms^{-1}$). Images $\Sigma_{0,1,3}$ integrated over time regimes that are dominated by fast (h_{fast}), slow (h_{slow}) and about constant PSF components (h_{const}) exhibit a declining resolution. (**d**) Rate-gated 4Pi-RESOLFT. Extrapolation of the initial contribution of h_{slow} ($= S_0$), based on integrated images Σ_1 ($\approx S_1$) and Σ_2 ($\approx S_2$), $t_0 = 40\,\mu s$, provides an estimate of the partial image generated by h_{fast} ($F_0 \approx \Sigma_0 - S_0$, inset), improving resolution and image fidelity over Σ_0. Details are provided in Methods. (**e**) Rate-gated xz-sections through actin fibres, recorded with open pinhole to boost out-of-focus signal. The measured (y-integrated) side-lobe structure closely resembles the numerical prediction and can be further suppressed (right) by an additional z-donut h_{zd} (overlay, $E_{zd} = 1.8\,\mu W \cdot 1.0\,ms$). Simulation parameters, numerical aperture 1.20, refractive index 1.362, pinhole diameter 0.5 airy units (**e**: open pinhole). Pulse parameters, E_{ac}, E_{3d}, $E_{ro} = 1.6\,mW \cdot 0.2\,ms$, $1.3\,\mu W \cdot 1.6\,ms$, $3.1\,\mu W \cdot 2.5\,ms$ (**e**: 0.5 ms). Scale bars, 250 nm.

Discussion

Using the current RSFP Dronpa-M159T, rate-gating allowed us to obtain images based on-switching speeds (switch-off half-time $T^{1/2} = 10$–$17\,\mu s$ at $11.5\,kW\,cm^{-2}$ illumination intensity, Fig. 3c, Supplementary Fig. 2, Supplementary Table 1) that were over an order of magnitude faster than the previously reported corresponding values for rapid switching RSFP ($T^{1/2}$, rsEGFP2: 250 μs (ref. 23), Dronpa-M159T: 230 μs (ref. 28)). Still the recording speed of our 4Pi-RESOLFT nanoscopy scheme can be made substantially faster by parallelization using a multi-spot scanning arrangement. In this case, the recording time of a certain sample area or volume would be cut down by the number of individual recording channels, that is, by the degree of parallelization.

In this study, we opted for cellular structures that are more demanding for 3D-superresolution imaging due to their high spatial density and wide axial extension. Under conditions exacerbated by the optical inhomogeneity of living cells, the signal from a (sub-diffractive) ensemble is easily buried in background (lobe) fluorescence beyond recovery. Nevertheless, owing to the consistently robust MTF of our 4Pi-RESOLFT scheme (Fig. 3b), we obtained raw (Fig. 2, insets) and rate-gated image data (Fig. 4e–g) that were conclusive without the mathematical post-processing (that is, deconvolution) dedicated to lobe-removal that is usually applied in 4Pi-based methods. The actin network seen in the exemplary time-lapse recording (Fig. 4e–g) appeared particularly crowded and extended over 8 μm along the optic axis, which forced the light to pass through several micrometres of cellular material from all angles. Still, the rearrangement of the entwined actin fibres could be traced in great detail, which was possible because the obtained images were practically devoid of axial lobes.

Notably, our scheme of reducing the global refractive index (RI) variance (Fig. 1b) turned out to sufficiently mitigate sample-induced aberrations without adding the complexity associated with adaptive optical elements. The most prominent residual aberration effect was a position-dependent 4Pi phase offset that stemmed from the uneven thickness of the cell layer; it has been accounted for during our recordings by the automated correction mechanism that also counteracted thermal drift (Supplementary Methods).

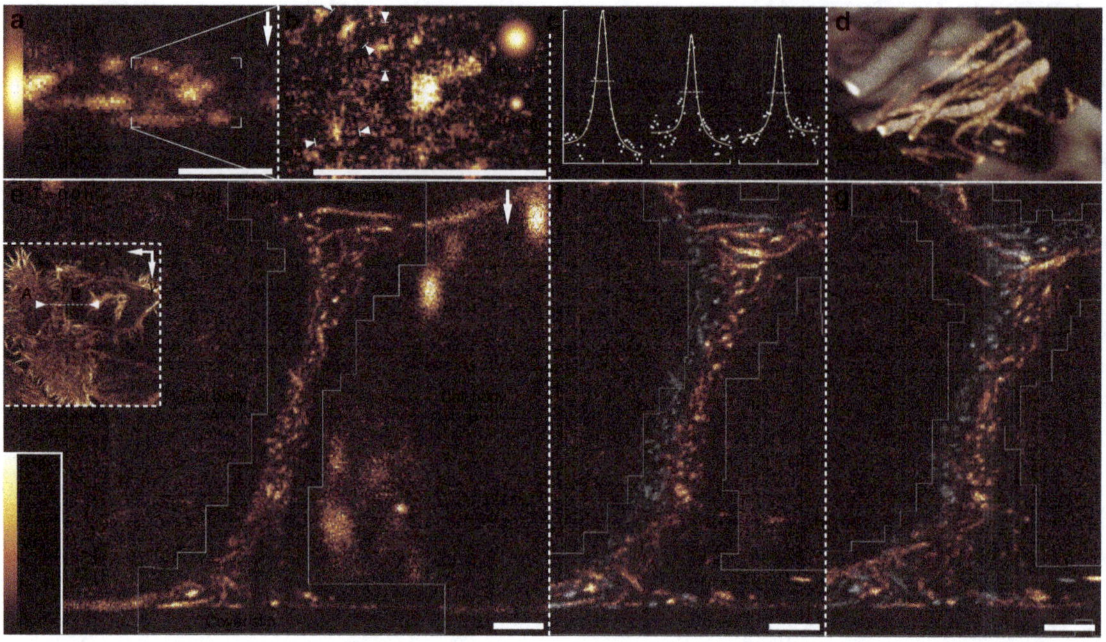

Figure 4 | 3D nanoscopy with strong optical sectioning. *xz*-sections of live HeLa cells expressing Lifeact-Dronpa-M159T. (**a**) Overview (optical *xz*-section) of actin fibre bundles at an axial base resolution in the 100 nm range. Left inset, confocal reference. (**b**) Addition of a 3D deactivation donut ($+h_{3d}$, $E_{3d} = 2.6\,\mu W \cdot 3.2\,ms$) to the RESOLFT pulse sequence reveals Dronpa patterns with apparent feature sizes well below 40 nm (inset, Gaussian reference spheres); (**c**) Lorentzian fits, plus a linear local background, to box profiles p1-3 over marked features in **b** along different directions. Numbers indicate full widths at half maximum (FWHM) over background. (**d**) Rendering of the volume surrounding **a**. (**e-g**) Time (T) evolution of an 8-µm-thick, densely labelled, vertical contact region between two adjacent cells (*xz*-section as marked in the *xy*-overview). Grayscale overlays of the preceding time step (**f**,**g**) aid in the tracking of individual features. A narrowed region of interest was generated online from initial overview scans ($-h_{3d}$) at each time frame and imaged at 50 nm target resolution ($+h_{3d}$, grey outline, E_{3d}, $E_{zd} = 1.3\,\mu W \cdot 1.6\,ms$, $1.8\,\mu W \cdot 0.5\,ms$). Despite the challenging imaging conditions, stacked actin structures are unambiguously resolved across the full axial extent of the cell layer. *xz*-panels depict rate-gated 4Pi-RESOLFT raw data, solely subjected to noise reduction. Fluorescence intensities $I(r)$. Common pulse parameters (**b**,**e-g**), E_{ac}, $E_{ro} = 1.6\,mW \cdot 0.2\,ms$, $3.1\,\mu W \cdot 0.5\,ms$. Scale bars, 1 µm.

In conclusion, by realizing 4Pi-RESOLFT nanoscopy based on RSFPs, we have demonstrated exceptional optical sectioning in coordinate-targeted far-field fluorescence nanoscopy, which greatly facilitates nanometre scale 3D fluorescence imaging in living cells. Many accepted constraints to the sample can be lifted, which opens up an imaging regime that has so far been systematically avoided.

Methods

4Pi sample optics for live-cell imaging. In a 4Pi arrangement, the RI (n) difference between the material forming a living mammalian cell ($n \approx 1.35$–1.40) and the surrounding culture medium (typically $n \approx 1.33$) is a major source of optical aberrations. Aberrations generally reduce the attainable signal-to-noise ratio (*S/N*) by blurring the light intensity distributions in the focal region and by raising the intensity of the central minimum of the off-switching light patterns (Figs 1a and 3a). We therefore devised a mounting procedure that minimizes optical aberrations by raising the RI of a standard cell culture medium to $n = 1.362$ (Supplementary Fig. 3, Supplementary Methods) and by designing the optical setup accordingly (Fig. 1b, Supplementary Figs 4 and 5): The 4Pi foci are jointly created by two 1.20 NA water immersion objectives that are outfitted with individual tip/tilt-correction to prevent aberrations that arise from lens-coverslip misalignment[29]. The refractive indices of the embedding ($n = 1.362$) and immersion media ($n = 1.350$), the correction collar settings and the cover slip thickness were chosen to minimize spherical aberrations[25] over at least 10 µm of sample depth. The changes in the optical path lengths of the two 4Pi-interferometer arms due to z-scanning of the sample were compensated[30] by synchronous position adjustment of the main beam splitter cube (Supplementary Fig. 4).

A single-focus 3D light pattern for deactivation. To resolve features smaller than 100 nm, we added a RSFP deactivation beam to the microscope. It was imprinted with a circular phase ramp that was subsequently imaged into the back pupil planes of both objective lenses. In contrast to the configuration of a single-lens RESOLFT setup, the direction of rotation of the phase ramp was oriented in countersense with respect to the circular beam polarization at the back pupil planes, which produced a 4Pi off-switching pattern h_{3d} that completely surrounded a central zero[26] (Fig. 3a). Applying h_{3d} during the deactivation phase of the RESOLFT switching cycle squeezed the central full width at half maximum (FWHM) d_{ef} of h_{ef} and allowed us to tune h_{ef} to a resolution below the diffraction limit.

RESOLFT imaging with rate-gated detection. In the present case of RESOLFT imaging of negative-switching RSFP, deactivation of fluorophores during read-out gives rise to a time-dependent signal and hence a time-dependent effective PSF $h_{ef}(\mathbf{r},t)$. The time-resolved image $g(\mathbf{r},t)$ obtained by imaging a structure $s(\mathbf{r})$ is thus given by

$$g(\mathbf{r},t) = s(\mathbf{r}) \otimes h_{ef}(\mathbf{r},t). \tag{3}$$

We simplistically assume a deactivation rate $\lambda(\mathbf{r})$ that only depends on the read-out intensity $I_{ro}(\mathbf{r})$ and therefore obtain

$$h_{ef}(\mathbf{r},t) = h_{ef}(\mathbf{r}) \cdot e^{-\lambda(\mathbf{r})\cdot t} = h_{ef}(\mathbf{r}) \cdot e^{-\lambda \cdot I_{ro}(\mathbf{r})\cdot t} \tag{4}$$

with t denoting time relative to the start of the read-out pulse. The deactivation pattern $h_{3d}(\mathbf{r})$ typically confines the effective volume from which fluorescence is collected to a region of FWHM d_{ef} around the primary zero of the deactivation pattern. This region is much narrower than the FWHM d_{ro} of the diffraction pattern used for read-out:

$$d_{ef} \ll d_{ro} \tag{5}$$

The read-out intensity can then be considered constant, and $h_{ef}(\mathbf{r},t)$ follows a mono-exponential decay at a rate that only depends on the peak read-out intensity $I^0_{ro} := I_{ro}(0)$. Thus, $g(\mathbf{r},t)$ becomes separable and transforms into:

$$g(\mathbf{r},t) = g(\mathbf{r}) \cdot g(t) := (s(\mathbf{r}) \otimes h_{ef}(\mathbf{r})) \cdot e^{-\lambda \cdot I^0_{ro} \cdot t} \tag{6}$$

Data from test structures, however, exhibit a distinct multi-exponential behaviour that requires additional components for a proper fit (Fig. 3c):

$$\widehat{g}(\mathbf{r},t) = \sum_{i=0}^{n-1} \widehat{c}_i(\mathbf{r}) \cdot e^{-\widehat{\lambda}_i(\mathbf{r})\cdot t} \tag{7}$$

where n components with ordered switching rates $\widehat{\lambda}_i$, $\widehat{\lambda}_i > \widehat{\lambda}_{i+1}$ and coefficients $\widehat{c}_i$ are fitted to the data ('^' marks fit results). While we attribute the fastest rate $\widehat{\lambda}_0$ to signal from unimpaired RSFP at the focal centre, the presence of additional rates suggests the co-existence of RSFP species that exhibit significantly slower switching kinetics. A slowed-down switching observed after fixation supports this notion.

Unintended processes during image recording also potentially contribute to the observed signal behaviour, for example, the re-activation of RSFP by the read-out light which generates a constant background.

Without loss of generality, we assume a position independent mixture of n species with discrete switching rates λ_j. Since the deactivation pattern h_{3d} shrinks the effective PSF by a λ-dependent factor, $h_{ef}(\mathbf{r},t)$ has to be generalized to a superposition of n individual h^j_{ef} that each correspond to a λ_j. Furthermore, we assume our experimental parameters are chosen such that the in-focus part h_{fast} of h^0_{ef} obeys the analogue to equation (5), and pool the remaining contributions in h_{slow}:

$$h_{ef}(\mathbf{r},t) = h_{fast}(\mathbf{r},t) + h_{slow}(\mathbf{r},t) := h_{fast}(\mathbf{r}) \cdot e^{-\lambda_0 \cdot I^0_{ro} \cdot t} + \sum_{j=1}^{n-1} h^j_{slow}(\mathbf{r}) \cdot e^{-\lambda_j \cdot I_{ro}(\mathbf{r}) \cdot t} \tag{8}$$

Consequently, the apparent resolution of the acquired image is less than the potential resolution provided by h_{fast} and declines over time, as faster components vanish first (Fig. 3c).

To access the full image information that is mediated by h_{fast}, we implemented an unmixing scheme that isolates the fastest switching signal component (rate λ_0), and that we hence termed 'rate-gating': according to equation (3), the image generated by a PSF equation (8) takes on the form

$$g(\mathbf{r},t) = g_{fast}(\mathbf{r},t) + g_{slow}(\mathbf{r},t) := g_0(\mathbf{r}) \cdot e^{-\lambda_0 \cdot I^0_{ro} \cdot t} + \int_0^{\lambda_1} c(\mathbf{r},\Lambda) \cdot e^{-\Lambda t} \mathrm{d}\Lambda \tag{9}$$

whereby g_{slow} is represented by a continuum of exponentials with switching speeds $\Lambda \in [0,\lambda_1]$ and coefficients $c(\mathbf{r},\Lambda)$ as the result of $s \otimes h_{slow}$.

Hence, a fit $\widehat{g}(\mathbf{r},t)$ to an imaged structure according to equation (7), in principle, provides a position invariant estimate for $\lambda_0 = \widehat{\lambda}_0 / I^0_{ro}$ by $\widehat{g}_{fast}$, on top of a local approximation of $g_{slow}(\mathbf{r},t)$. In practice however, an insufficient photon count often prohibits local fitting of equation (9). We therefore implemented a robust approximation scheme for $g_{fast}(\mathbf{r})$ that only relies on parameters that can be extracted from a fit $\widehat{g}(t)$ to the global (that is, from a region much larger than the corresponding diffraction limit) spatial average $\overline{g}(t)$ of the measured data:

First, we estimated the time t_0 at which the integrated signal exhibits the maximum S/N with respect to g_{fast}:

$$t_0 := \operatorname{argmax}\left(\frac{\int_0^T \overline{g}_{fast}(t)\mathrm{d}t}{\sqrt{\int_0^T \overline{g}_{fast}(t) + \overline{g}_{slow}(t)\mathrm{d}t}} \right) \tag{10}$$

Locally calculated values for t_0 would slightly differ, but as this only affects the statistical error of the result, equation (10) is usually sufficiently precise. Second, we determined a cut-off time t_1 such that

$$g_{fast}(t) \ll g_{slow}(t) \quad (t > t_1) \tag{11}$$

which is usually the case for $t_1 = 2t_0$. By further choosing t_2 and t_3 such that $t_i > t_{i+1} (i = 0..3)$, we partition the measured signal into time bins $\sum_{0,1,2}$ (Fig. 3d),

$$\sum\nolimits_{0,1,2}(\mathbf{r}) := \int_{0,t_1,t_2}^{t_0,t_2,t_3} g(\mathbf{r},t)\mathrm{d}t \approx F(\mathbf{r}) + S_0(\mathbf{r}), S_1(\mathbf{r}), S_2(\mathbf{r}) \tag{12}$$

with $F(\mathbf{r})$ and $S(\mathbf{r})$ denoting the time integrals over the fast and slow components:

$$F(\mathbf{r}) := \int_0^{t_0} g_{fast}(\mathbf{r},t)\mathrm{d}t, \quad S_{0,1,2} := \int_{0,t_1,t_2}^{t_0,t_2,t_3} g_{slow}(\mathbf{r},t)\mathrm{d}t \tag{13}$$

Finally, we estimated $F(\mathbf{r})$, and thereby $g_0(\mathbf{r})$, by linear extrapolation in either zeroth or first order:

$$F(\mathbf{r})^{0th} := \sum\nolimits_0(\mathbf{r}) - u \cdot \sum\nolimits_1(\mathbf{r}) \tag{14}$$

$$F(\mathbf{r})^{1st} := F(\mathbf{r})^{0th} - v \cdot \left(\sum\nolimits_1(\mathbf{r}) - \sum\nolimits_2(\mathbf{r}) \right) \tag{15}$$

with $u,v = u,v(t_{0..3})$ denoting geometrical factors that account for the particular choice of the $t_{0..3}$. Narrowing the integration intervals defined by $t_{0..3}$ and moving them closer to $t = 0$, just as the inclusion of the first extrapolation order, reduces the systematic error, but also raises the statistic error due to a reduced photon count. To mitigate this effect, and owing to equation (11), we substituted $\sum_{1,2}$ with their respective resolution-neutral local averages, for example, by applying a Gaussian filter with a FWHM sufficiently far below the FWHM of $h^{1,2}_{slow}$.

References

1. Hell, S. W. Double-confocal scanning microscope. European Patent 0491289 (1992).
2. Hell, S. W. & Stelzer, E. H. K. Properties of a 4pi confocal fluorescence microscope. *J. Opt. Soc. Am. A Opt. Image Sci. Vis.* **9,** 2159–2166 (1992).
3. Gustafsson, M. G. L., Agard, D. A. & Sedat, J. W. Sevenfold improvement of axial resolution in 3D widefield microscopy using two objective lenses. *Proc. Soc. Photo-Opt. Instrum. Eng.* **2412,** 147–156 (1995).
4. Eggeling, C., Willig, K. I., Sahl, S. J. & Hell, S. W. Lens-based fluorescence nanoscopy. *Q. Rev. Biophys.* **48,** 178–243 (2015).
5. Schmidt, R. *et al.* Spherical nanosized focal spot unravels the interior of cells. *Nat. Methods* **5,** 539–544 (2008).
6. Shtengel, G. *et al.* Interferometric fluorescent super-resolution microscopy resolves 3D cellular ultrastructure. *Proc. Natl Acad. Sci. USA* **106,** 3125–3130 (2009).
7. Hell, S. W., Schmidt, R. & Egner, A. Diffraction-unlimited three-dimensional optical nanoscopy with opposing lenses. *Nat. Photon.* **3,** 381–387 (2009).
8. Grotjohann, T. *et al.* Diffraction-unlimited all-optical imaging and writing with a photochromic GFP. *Nature* **478,** 204–208 (2011).
9. Lang, M. C., Engelhardt, J. & Hell, S. W. 4Pi microscopy with linear fluorescence excitation. *Opt. Lett.* **32,** 259–261 (2007).
10. Schmidt, R. *et al.* Mitochondrial cristae revealed with focused light. *Nano Lett.* **9,** 2508–2510 (2009).
11. Ullal, C. K., Schmidt, R., Hell, S. W. & Egner, A. Block copolymer nanostructures mapped by far-field optics. *Nano Lett.* **9,** 2497–2500 (2009).
12. Kanchanawong, P. *et al.* Nanoscale architecture of integrin-based cell adhesions. *Nature* **468,** 580–584 (2010).
13. Denk, W., Strickler, J. H. & Webb, W. W. 2-photon laser scanning fluorescence microscopy. *Science* **248,** 73–76 (1990).
14. Schneider, M., Barozzi, S., Testa, I., Faretta, M. & Diaspro, A. Two-photon activation and excitation properties of PA-GFP in the 720-920-nm region. *Biophys. J.* **89,** 1346–1352 (2005).
15. Glaschick, S. *et al.* Axial resolution enhancement by 4Pi confocal fluorescence microscopy with two-photon excitation. *J. Biol. Phys.* **33,** 433–443 (2007).
16. Moneron, G. & Hell, S. W. Two-photon excitation STED microscopy. *Opt. Express* **17,** 14567–14573 (2009).
17. York, A. G., Ghitani, A., Vaziri, A., Davidson, M. W. & Shroff, H. Confined activation and subdiffractive localization enables whole-cell PALM with genetically expressed probes. *Nat. Methods* **8,** 327–333 (2011).
18. Hell, S. W. Improvement of lateral resolution in far-field light microscopy using two-photon excitation with offset beams. *Opt. Commun.* **106,** 19–24 (1994).
19. Hell, S. W. Toward fluorescence nanoscopy. *Nat. Biotechnol.* **21,** 1347–1355 (2003).
20. Stiel, A. C. *et al.* 1.8 angstrom bright-state structure of the reversibly switchable fluorescent protein dronpa guides the generation of fast switching variants. *Biochem. J.* **402,** 35–42 (2007).
21. Willig, K. I., Stiel, A. C., Brakemann, T., Jakobs, S. & Hell, S. W. Dual-label STED nanoscopy of living cells using photochromism. *Nano Lett.* **11,** 3970–3973 (2011).
22. Testa, I. *et al.* Nanoscopy of living brain slices with low light levels. *Neuron* **75,** 992–1000 (2012).
23. Grotjohann, T. *et al.* rsEGFP2 enables fast RESOLFT nanoscopy of living cells. *eLife* **1,** e00248 00241-00214 (2012).
24. Gugel, H. *et al.* Cooperative 4pi excitation and detection yields sevenfold sharper optical sections in live-cell microscopy. *Biophys. J.* **87,** 4146–4152 (2004).
25. Wan, D. S., Rajadhyaksha, M. & Webb, R. H. Analysis of spherical aberration of a water immersion objective: application to specimens with refractive indices 1.33-1.40. *J. Microsc.* **197,** 274–284 (2000).
26. Schmidt, R. *3D Fluorescence Microscopy with Isotropic Resolution on the Nanoscale* (PhD thesis, Univ. Heidelberg, 2008).
27. Richards, B. & Wolf, E. Electromagnetic diffraction in optical systems. II. structure of the image field in an aplanatic system. *Proc. R. Soc. Lond. A* **253,** 358–379 (1959).
28. Testa, I., D'Este, E., Urban, N. T., Balzarotti, F. & Hell, S. W. Dual channel RESOLFT nanoscopy by using fluorescent state kinetics. *Nano Lett.* **15,** 103–106 (2015).
29. Arimoto, R. & Murray, J. M. A common aberration with water-immersion objective lenses. *J. Microsc.* **216,** 49–51 (2004).
30. Hell, S. W., Schrader, M. & VanderVoort, H. T. M. Far-field fluorescence microscopy with three-dimensional resolution in the 100-nm range. *J. Microsc.* **187,** 1–7 (1997).

Acknowledgements

We thank C. Gregor for providing the plasmid Lifeact-Dronpa-M159T, T. Gilat and E. Rothermel for cloning support and assistance with cell culture preparation, B. Thiel for Imspector software support and A. Pucher-Diehl for precision mechanic support for custom microscope parts. S.W.H. and R.S. acknowledge a grant by the Deutsche Forschungsgemeinschaft within SFB 775.

Author contributions

U.B. prepared the samples and carried out the measurements. U.B. and R.S. designed and built the RESOLFT microscope and evaluated the data. R.S. designed and implemented the 4Pi unit and the data acquisition and rate-gating algorithms. The project was defined and supervised by R.S. and S.W.H. All authors discussed the project as it evolved and edited the final version of the paper. The paper was written by R.S. and S.W.H. with contributions by U.B.

Additional information

Competing financial interests: The authors declare no competing financial interests.

27

Manipulation of charge transfer and transport in plasmonic-ferroelectric hybrids for photoelectrochemical applications

Zhijie Wang[1,2,*], Dawei Cao[1,*], Liaoyong Wen[1], Rui Xu[1], Manuel Obergfell[3,4], Yan Mi[1], Zhibing Zhan[1], Nasori Nasori[1], Jure Demsar[1,4] & Yong Lei[1]

Utilizing plasmonic nanostructures for efficient and flexible conversion of solar energy into electricity or fuel presents a new paradigm in photovoltaics and photoelectrochemistry research. In a conventional photoelectrochemical cell, consisting of a plasmonic structure in contact with a semiconductor, the type of photoelectrochemical reaction is determined by the band bending at the semiconductor/electrolyte interface. The nature of the reaction is thus hard to tune. Here instead of using a semiconductor, we employed a ferroelectric material, $Pb(Zr,Ti)O_3$ (PZT). By depositing gold nanoparticle arrays and PZT films on ITO substrates, and studying the photocurrent as well as the femtosecond transient absorbance in different configurations, we demonstrate an effective charge transfer between the nanoparticle array and PZT. Most importantly, we show that the photocurrent can be tuned by nearly an order of magnitude when changing the ferroelectric polarization in PZT, demonstrating a versatile and tunable system for energy harvesting.

[1] Institut für Physik & IMN MacroNano (ZIK), Technische Universität Ilmenau, 98693 Ilmenau, Germany. [2] Key Laboratory of Semiconductor Materials Science, Institute of Semiconductors, CAS, 100083 Beijing, China. [3] Physics Department, University of Konstanz, 78457 Konstanz, Germany. [4] Institute of Physics, Johannes Gutenberg-University Mainz, 55128 Mainz, Germany. * These authors contributed equally to this work. Correspondence and requests for materials should be addressed to Y.L. (email: yong.lei@tu-ilmenau.de).

For photoelectrochemical (PEC) systems based on plasmonics, in addition to the scattering effect in metallic nanostructures[1,2], three factors play decisive roles: the Schottky junction at the interface between the metallic nanoparticle and the semiconductor, enabling the capture of hot electrons generated in photon-stimulated nanometals to semicondcutors[3–5]; the interface between the semiconductor and electrolyte, governing the transfer of the hot carriers from the semiconductor to the electrolyte; and the transport of hot carriers between the two interfaces. Considering that the properties of the Schottky or Ohmic junction are fixed for a given combination of a metal and a semiconductor, the other two factors are crucial for adjusting the PEC performance. Particularly, the semiconductor/electrolyte interface is important, since the band bending is either upward (from semiconductor to electrolyte) for an easy hole transfer or downward to facilitate electron transfer to the electrolyte[6,7]. As a conventional PEC semiconductor, TiO_2 has been widely used in water splitting for collecting hot electrons from plasmonic nanostructures[2,8,9]. However, Pt nanoparticles or other catalysts have to be adopted to adjust the upward band bending at the TiO_2/electrolyte interface that inhibits the transfer of electrons in the conduction band of TiO_2 to the electrolyte[10,11]. Though different approaches have been followed to tune the plasmonic properties of metallic nanostructures to enhance the PEC performance[12–14], insightful mechanism and technique proposed for tailoring both the band bending at the semiconductor/electrolyte interface and the transport of hot carriers in the PEC film, preferably at the same time, have so far been lacking.

Here we present an approach where a conventional semiconductor has been replaced by ferroelectric $Pb(Zr,Ti)O_3$ (PZT), which possesses a large, stable and manipulable remnant polarization[15–18]. The associated depolarization electric field (E_{DP}), extending over the entire thin film volume, enables tuning the band bending at the ferroelectric/electrolyte interface by poling pretreatments and thus adding extra functionality for scavenging and conducting the excited charges. We report on manipulation of the charge transfer and transport in nano-Au/PZT hybrids by placing a nano-Au array in different positions within ITO/PZT and by poling the PZT films with different potentials. Among the PEC electrodes (as grown), the structure of ITO/nano-Au/PZT provides the best performance among the three structures. On the other hand, using the ITO/PZT/nano-Au/PZT electrodes, we demonstrate tuning of the short-circuit photocurrent by nearly an order of magnitude, when the pre-poling bias is switched from $+10$ to -10 V. The transport studies are accompanied by femtosecond transient absorbance study to track the dynamics of the hot-charge transfer from Au nanoparticles to PZT. Such simultaneous manipulation of the charge transfer and interface-related PEC phenomena within a given nano-metal/PEC film/electrolyte system presents a route to optimally and flexibly manipulate the photoexcited charges for PEC energy conversion (for example, solar water splitting).

Results

Characteristics of the nano-Au array. Figure 1 summarizes the fabrication processes used herein for preparing different nano-Au/PZT hybrids as PEC photoelectrodes. Well-ordered nano-Au array was prepared utilizing a well-established ultra-thin alumina mask (UTAM) technique[19–21]. A representative top-view scanning electron microscope (SEM) image is shown in Fig. 2a, where the spacing between the square Au dots is $\sim$130 nm and the dot dimension is gauged at $\sim 270 \times 270\,nm^2$ (Fig. 2a, inset). In combination with atomic force microscopic analysis illustrated in Supplementary Fig. 1, the thickness of the Au dots is determined to be $\sim$60 nm. These parameters govern the characteristic absorbance spectrum of the nano-Au array, as shown in Fig. 2b. The spectrum exhibits a main absorbance peak at $\sim$800 nm, which can be attributed to the localized surface plasmon resonance (LSPR) of the square-like Au dots with the lateral dimension of 270 nm (ref. 22). Indeed, the simulated spectra obtained by the finite difference time domain (FDTD) method, qualitatively agree well with experimental data (Supplementary Fig. 2a). Moreover, we simulated the spatial distribution of the electric field intensity around the square dot illuminated by light at 800 nm (Fig. 2c,d). The hot spots are located at the edge of the square dot and may amplify the probability of the hot-charge transfer. The advantages of such periodic nano-Au pattern are elaborated in the Supplementary Note 1 (Supplementary Fig. 3).

Performance of plasmonic-ferroelectric hybrids. Rather than growing PZT films epitaxially in high vacuum conditions, we adopted a cost-effective spin-coating technology for obtaining

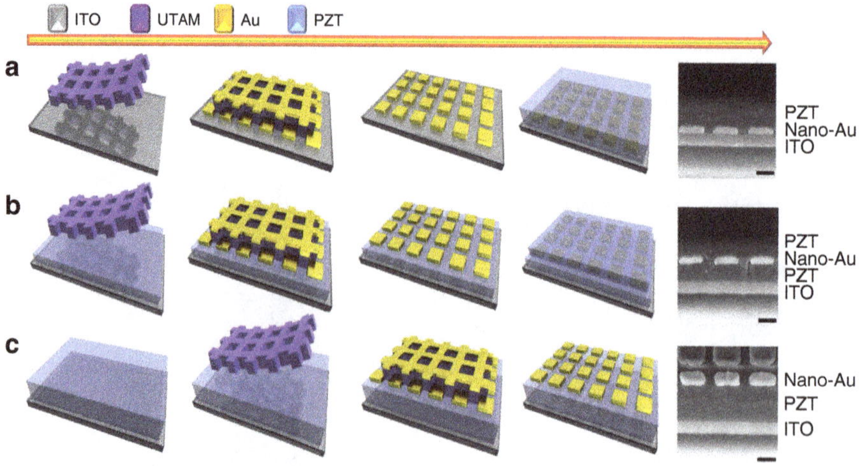

Figure 1 | Schematic of fabrication processes for preparing the plasmonic-ferroelectric hybrids. (**a**) The procedure for fabricating the nano-Au array at the interface of ITO/PZT (structure: ITO/nano-Au/PZT). (**b**) The procedure for embedding the nano-Au array in the PZT films (structure: ITO/PZT/nano-Au/PZT), poling treatment is performed on this structure for investigating the impact of the orientation of ferroelectric polarization on the PEC performance. (**c**) The procedure for making nano-Au array on top of ITO/PZT (structure: ITO/PZT/nano-Au). Cross-sectional image of scanning electron microscope for the corresponding structure is given on the right hand. All scale bars, 200 nm.

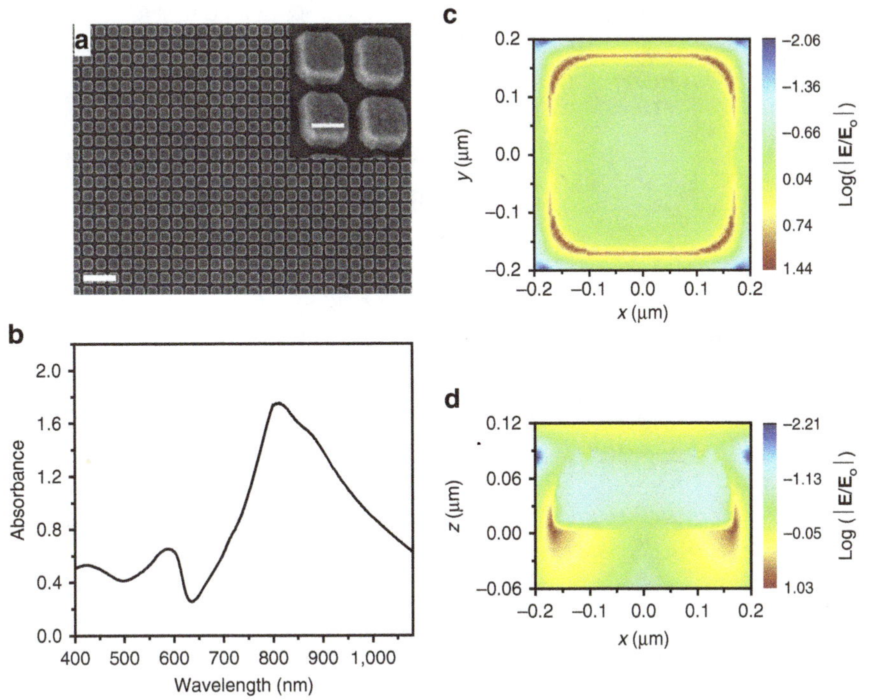

Figure 2 | Morphology and optical absorbance of nano-Au array. (**a**) Top-view SEM images of nano-Au array fabricated on ITO glass using a UTAM template; scale bar, 1 μm. The inset zooms in one part of the SEM; scale bar, 200 nm. (**b**) Ultraviolet-visible absorbance spectrum (absorbance = − $\log_{10}$ (transmission)) of ITO/nano-Au array. (**c**,**d**) FDTD-simulated spatial distribution of the electric field intensity around the square dot ((**c**) planar FDTD; (**d**) cross-sectional FDTD) illuminated by 800 nm.

high quality PZT films[18,23]. The X-ray diffraction patterns of the samples shown in Supplementary Fig. 4a illustrate a pure PZT phase (perovskite structure). As shown in Supplementary Fig. 4b, a typical polarization–voltage hysteresis loop yields a coercive field of ~170 kV cm^{-1}. Thus, an applied potential of 10 V is sufficient to switch the ferroelectric domains in the films of 300 nm thickness. Considering the significant role of the ITO/PZT contact on the charge transfer process, the ITO/PZT junction was elaborately investigated and the Schottky barrier was determined to be ~1.03 eV (Supplementary Fig. 5; Supplementary Note 2).

Figure 3a presents the representative steady-state external quantum efficiency (EQE) spectra of PEC electrodes of ITO/PZT and ITO/nano-Au/PZT (note that the corresponding internal quantum efficiency would be at least an order of magnitude higher). Compared with the bare (intrinsic) PZT photoelectrode on ITO substrate, the nano-Au/PZT photoelectrode exhibits a distinctive EQE for photon energies below the absorption threshold of PZT ($E_g = 3.6$ eV, Supplementary Fig. 4c). The spectrum of the EQE qualitatively matches the absorbance spectrum of nano-Au/PZT (Supplementary Fig. 6a; Supplementary Note 3), demonstrating the occurrence of hot-electron injection from the excited nano-Au to PZT. Photocurrent–potential profiles were measured by soaking the photoelectrodes into 0.1 M Na_2SO_4 aqueous solutions. Each plot represents typical photoresponse obtained by illumination with a standard 300 W Xe lamp (Newport). To get the photocurrent signal from nano-Au array solely, a 455-nm-low pass optical filter was used to avoid the excitation of PZT. The light intensity was characterized as 100 mW cm^{-2}. As illustrated in Fig. 3b, the ITO/nano-Au/PZT electrode possesses a distinct PEC performance with a short-circuit current ~10 μA cm^{-2} and an open circuit potential close to 0.6 V versus Ag/AgCl. The photocurrent direction is cathodic, demonstrating that it is the hot electrons that have been transferred from the nano-Au to the PZT/electrolyte and hence initiate the PEC reactions. To attribute the hot-electron collection to the presence of PZT, we also made the electrode based on ITO/nano-Au and cannot observe any EQE signal at all, as shown in Supplementary Fig. 7.

Two strategies were adopted to manipulate the hot-electron injection efficiency and to optimize the PEC performance: adjusting the positions of nano-Au within the ITO/PZT and tuning the ferroelectric polarization in PZT films with external potential. First, the nano-Au array was placed in varied positions: at the interface of ITO/PZT, in the middle of PZT films and on the top of PZT films, respectively (as shown in Fig. 1). PEC results in Fig. 3c,d show that the electrodes (as grown) with nano-Au array at the ITO/PZT interface have the best performance among these three structures. The Schottky barrier of ITO/PZT is 1.03 eV (Supplementary Fig. 5). When the nano-Au array is placed in the depletion region of such Schottky contact, with continuously varying band bending, the collection and conduction of hot electrons injected into PZT should be more efficient. Importantly, the work function of Au is larger than that of ITO[24], which supports the transfer of photogenerated holes in nano-Au into ITO. Considering the fact that the valence band position of PZT is almost 1.5 eV below the work function of Au[18], it is on the other hand hard for the remaining holes in the Au to overcome the barrier at the Au/PZT interface and be collected by the external circuit when the nano-Au array is sandwiched within the PZT films. This can be relaxed if the ferroelectric domain structure is optimized by poling treatment[25]. As to the electrodes with the nano-Au array located on the top of the PZT, even though hot electrons can be injected into the PZT, the 1.03 eV Schottky barrier at the ITO/PZT interface prevents the electrons from being transferred to the external circuit.

The tunability of the E_{DP} in PZT films offers another opportunity to manipulate hot-electron injection and transfer. Experiments were conducted by poling the electrodes with different potentials in a propylene carbonate solution. For

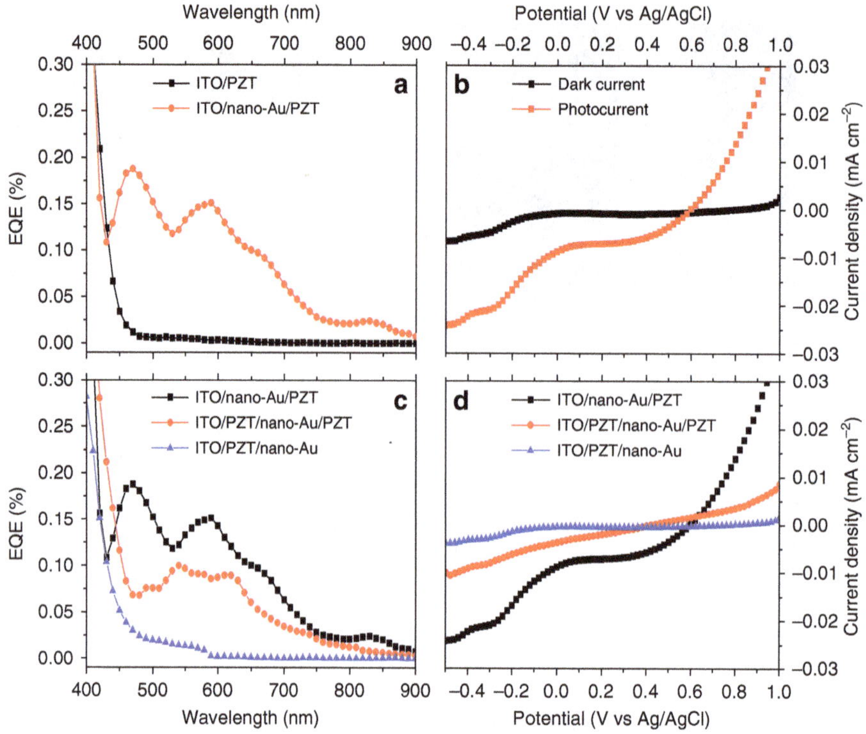

Figure 3 | Comparison of the PEC responses from different photoelectrodes. (**a**) EQE spectra of the PEC electrodes of ITO/PZT and ITO/nano-Au/PZT, respectively. (**b**) Current density-potential curves of the ITO/nano-Au/PZT photoelectrodes under white-light excitation (filtered, $\lambda > 450$ nm), compared with dark measurement. (**c**) EQE spectra of the electrodes of ITO/nano-Au/PZT(black), ITO/PZT/nano-Au/PZT (red) and ITO/PZT/nano-Au (blue). (**d**) Photocurrent-potential measurements of the samples in **c** under the white-light excitation (filtered, $\lambda > 450$ nm).

ITO/nano-Au/PZT, degradation of the performance was observed after poling treatments (Supplementary Fig. 8); the relevant discussion is given in Supplementary Note 4. In ITO/PZT/nano-Au/PZT electrodes, with a lower as-grown PEC performance than the ITO/nano-Au/PZT electrodes, the process was reversible and no deterioration was observed with cycling. Following poling, steady-state PEC and transient absorbance measurements were performed. Noteworthy, we demonstrate that these electrodes exhibit a high tuning capability in terms of the PEC performance. As shown in Fig. 4a, +10 V poling pretreatment results in the highest EQE compared with the same electrodes undergone -10 V poling and no poling treatments, respectively. This EQE value is even higher than that of the ITO/nano-Au/PZT electrodes, indicating that the poling condition in PZT is crucial for optimizing the PEC performance. The -10 V poling treatment strongly suppresses the EQE, while the as-grown sample shows an intermediate EQE, suggesting that the ferroelectric domains in the as-grown polycrystalline PZT films are randomly distributed. Correspondingly, the photocurrent-potential plots, displayed in Fig. 4b, demonstrate the same tendency. The short-circuit current can thus be tuned from 2.4 to 16.7 μA cm^{-2} (for white-light excitation density of 100 W cm^{-2}) just by switching the poling conditions from -10 to $+10$ V. As shown in Supplementary Fig. 9, the poling does not affect the absorbance of this structure.

The randomly oriented ferroelectric domains in the as-grown PZT films can be poled using electric fields larger than the coercive field[26]. In this way, the direction of E_{DP} can be correspondingly switched[27,28]. The +10 V poling potential induces an E_{DP} with the direction pointing towards the ITO substrate and a downward band bending at the PZT/electrolyte interface (Fig. 4c). This configuration is favourable for the injected hot electrons being transferred to the interface and driving the PEC reactions. The optimized E_{DP} across the entire PZT films could also be helpful for transferring the excited holes to ITO electrode[25]. The -10 V poling potential, however, switches the direction of the E_{DP}, which points towards the PZT/electrolyte interface and renders an upward band bending at the PZT/electrolyte interface (Fig. 4d). In this case, the hot electrons injected into the PZT cannot be transferred to the PZT/electrolyte interface and get trapped in the bulk of the PZT film.

We have characterized the stability of the PEC performance of our hybrid structures. As shown in Supplementary Fig. 10 and elaborated in Supplementary Note 4, the ITO/PZT/nano-Au/PZT structure shows high stability and reproducibility, indicating that the depolarization electric field in the PZT film is stable and can provide a sustainable driving force to conduct the charge carriers toward a certain direction, consistent with the previous reports[18,29].

Hot-charge transfer dynamics. To shed light on the hot-electron transfer, we performed broadband transient absorbance measurements. For photoexcitation, we used 70 fm optical pulses at 400 nm central wavelength, with the excitation density of ~1 mJ cm^{-2}. The photoexcitation photon energy (3 eV) is lower than the band gap of the PZT, yet high enough to induce the hot-electron transfer from the stimulated nano-Au to PZT. Photoinduced changes in transmission for wavelengths between 950 and 440 nm (1.3–2.8 eV) were recorded using (~100 fs) white-light continuum pulses.

Figure 5a presents the time evolution of the relative transmission changes ($\Delta T/T$) recorded on the nano-Au array on ITO/glass. Here two distinct, spectrally well-separated components can be identified (Fig. 5d), the enhanced transmission peaked at ~1.65 eV (~750 nm) and the reduced transmission peaked at ~2.5 eV. The former can be linked to the

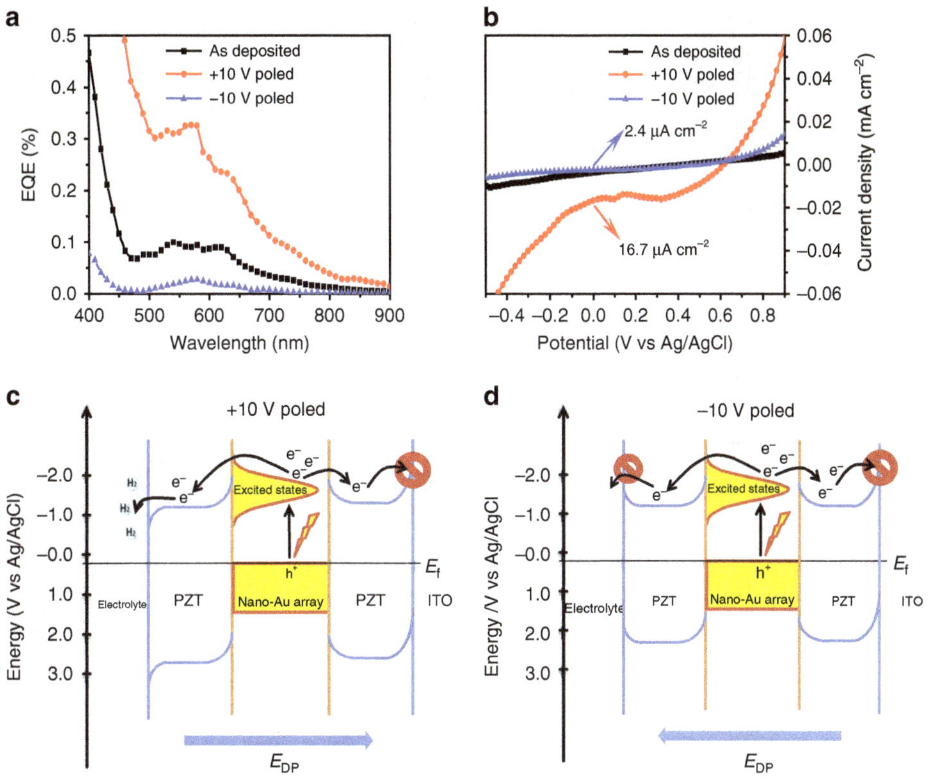

Figure 4 | Polarization switching behavior of the ITO/PZT/nano-Au/PZT photoelectrode. (**a**,**b**) EQE spectra and photocurrent-potential measurements (under the filtered white-light excitation) of the as-grown (black), +10 V (red) and −10 V (blue) poled samples. (**c**,**d**) Schematic electronic band structure and the mechanisms for the injected hot-electron transfer from PZT films to the electrolyte for the two poling configurations.

photoinduced changes in absorbance due to the photoinduced changes in the LSPR centred at ~1.5 eV (Fig. 2b). Photo-excitation, the resulting electron–electron and electron–phonon thermalization result in broadening of the LSPR due to the enhanced scattering[30,31]. The strongest photoinduced changes in transmission, caused by the broadening of the LSPR, may be expected near the LSPR for photon energies where the linear transmission strongly varies with the photon energy. Apart from the broadening of the LSPR, the photoinduced shift of the central frequency may be expected due to the photoinduced expansion, particularly for longer time delays[30].

Even more pronounced is the photoinduced decrease of transmission, peaked at about 2.5 eV. We attribute this peak to a bulk-like response of Au, governed by the photoinduced changes in the joint density of states for the optical transition between the d-band and Fermi level (E_f). Indeed, the spectral shape of the induced change in transmission (Fig. 5d, inset) matches well with the results obtained on extended thin films[32]. Unlike in gold nanoparticles with lateral dimensions on the 10-nm scale, where the LSPR spectrally overlaps with the d-band to E_f transition[30], the two spectral features are well separated in our case.

The photoinduced transmission spectra recorded on ITO/nano-Au/PZT (Fig. 5b) and ITO/PZT/nano-Au/PZT (Fig. 5c) are much more complicated. This can be linked to the complicated linear transmission spectra (Supplementary Fig. 6) caused by Fabry-Perot interference due to the additional PZT layer(s). Nevertheless, the reduced transmission in the high-frequency range and bleaching absorbance in the low-frequency part of the spectra are still recognizable. To rule out the influence of PZT on the transient dynamics in the visible range, we performed transient absorbance measurement on ITO/PZT. As shown in Supplementary Fig. 11, the contribution of PZT to the transient changes in transmission dynamics can be ignored.

Let us now address the dynamical aspect of the data. The rise-time (20 − 80%) is resolution limited (~120 fs) for all samples. Plotting $\Delta T/T(t)$ in ITO/nano-Au for photon energies at the two spectral peaks we find, however, that their respective $\Delta T/T$ decays with considerably different time constants. To avoid artefacts that can arise due to the time-dependent spectral shifts (particularly critical for data with narrow spectral features as in Fig. 5b,c), we analysed the time evolution of the low-frequency ($1.3\,\mathrm{eV} < h\nu < 2\,\mathrm{eV}$) and high-frequency ($2\,\mathrm{eV} < h\nu < 2.8\,\mathrm{eV}$) responses separately. Applying singular value decomposition on both spectral ranges (Supplementary Fig. 12; Supplementary Note 5), we demonstrate that the photoinduced transient spectra in each of the spectral ranges can be well reproduced by single components (their spectral weights are ~90% of the entire signal). In other words, the time evolution of the induced change in transmission, $\Delta T/T(h\nu,t)$, in both spectral ranges can be well reproduced by $\Delta T/T(h\nu,t) = \Delta T/T(h\nu) \times S(t)$, where $\Delta T/T(h\nu)$ is the spectrum and $S(t)$ is respective temporal evolution. Figure 5d presents the decomposition of $\Delta T/T(h\nu)$ on ITO/nano-Au. The main panel presents the temporal evolutions for the two spectral ranges, while the inset shows the corresponding $\Delta T/T(h\nu)$. It should be noted that, while the two spectral ranges are Kramers–Kronig connected, the fact that the underlying excitations (LSPR and the interband transition) are well spectrally separated justifies this approach.

There are two noteworthy observations as far as the dynamics is concerned. First, the relaxation of the high-frequency part is substantially slower. Second, while the dynamics of the low-frequency part is well described by an exponential decay, the high-frequency part clearly displays a non-exponential relaxation, which can be well seen on the semi-log plot. Considering the different natures of the two processes, the observation may not be too surprising. The relaxation of the

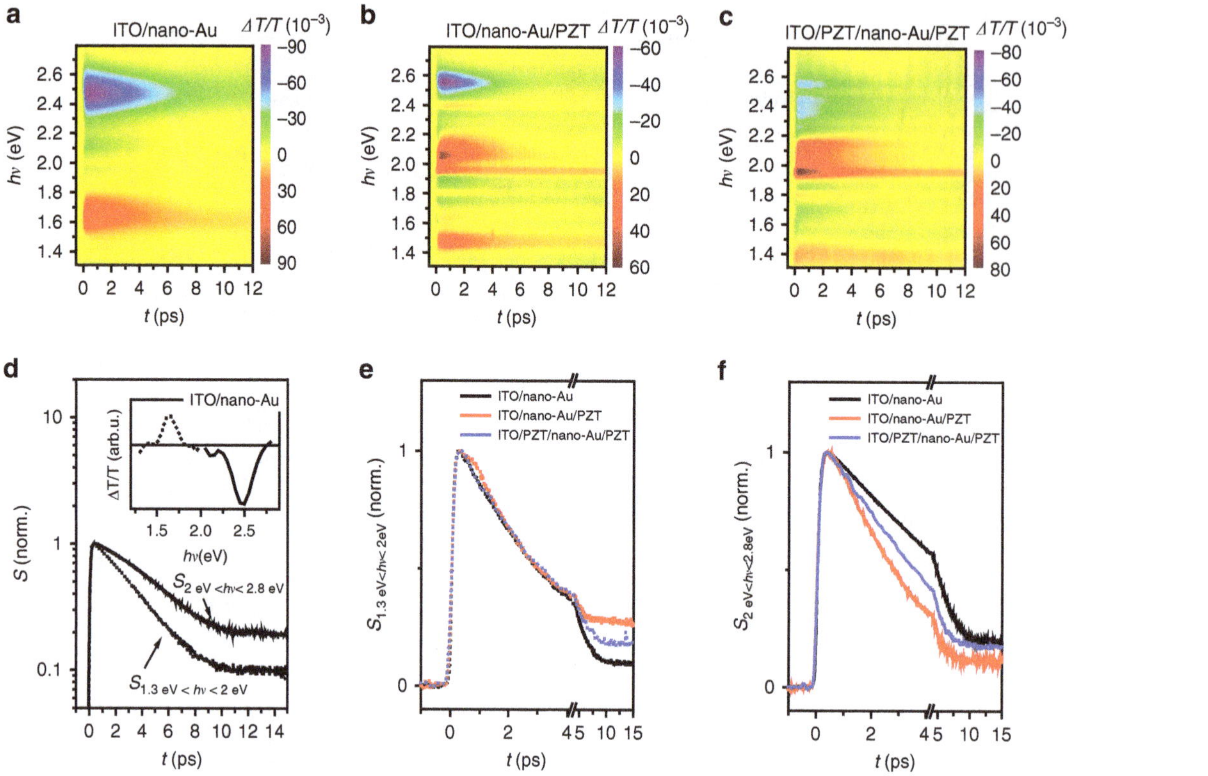

Figure 5 | Transient absorbance spectroscopy of the plasmonic-ferroelectric hybrids. Photoinduced transmission, $\Delta T/T(h\nu,t)$ spectra recorded on (**a**) ITO/nano-Au, (**b**) ITO/nano-Au/PZT and (**c**) ITO/PZT/nano-Au/PZT, following photoexcitation with 70 fs ultraviolet pulses at 400 nm. (**d**) The time evolution $\Delta T/T(h\nu,t)$ in ITO/nano-Au that can be reproduced by $\Delta T/T(h\nu,t) = \Delta T/T(1.3\,\text{eV} < h\nu < 2\,\text{eV}) \times S_{1.3\,\text{eV}<h\nu<2\text{eV}}(t) + \Delta T/T(2\,\text{eV} < h\nu < 2.8\,\text{eV}) \times S_{2\,\text{eV}<h\nu<2.8\,\text{eV}}(t)$. The time evolutions, $S(t)$, are shown in the main panel, while their respective spectra are shown in inset. The dotted line presents the time evolution of the low-frequency range ($1.3\,\text{eV} < h\nu < 2\,\text{eV}$), while the solid line corresponds to the high frequency ($2\,\text{eV} < h\nu < 2.8\,\text{eV}$). Comparison of the low-frequency (**e**) and high-frequency (**f**) dynamics between the three nano-Au hybrids. The high-frequency response demonstrates the speeding up of relaxation in nano-Au/PZT hybrids, consistent with the existence of an additional relaxation channel (Au-PZT charge transfer).

LSPR is governed by the time evolution of the (collective) plasma scattering rate, while the interband transition is governed by the photoexcited quasiparticle density and their distribution. Indeed, for the high-frequency part, the slope of $S(t)$ changes with time, suggesting the presence of multiple decay channels[33]. Since this component corresponds to a bulk-like interband transition, the carrier diffusion (ballistic transport) into the 60-nm-thick Au nanoparticles, competing with electron–phonon thermalization, could be the origin of the observed functional form of $S(t)$[33].

Further evidence for the above suggestion comes from the comparison of relaxation dynamics between different samples. While no measurable changes are observed for the low-frequency part (Fig. 5e), the high-frequency part (Fig. 5f) shows a pronounced variation of relaxation rates, recovering substantially faster in nano-Au/PZT hybrids. The relaxation rates show a clear trend: $\tau^{-1}_{\text{ITO/nano-Au/PZT}} > \tau^{-1}_{\text{ITO/PZT/nano-Au/PZT}} > \tau^{-1}_{\text{ITO/nano-Au}}$. Since photoconductivity data demonstrate a substantial charge transfer from nano-Au to PZT, with the highest efficiency in ITO/nano-Au/PZT (Fig. 3c), we suggest that the nano-Au – PZT charge transfer may be responsible for speeding up the relaxation of the high-frequency spectral component in nano-Au/PZT hybrids.

We have further performed studies of photoinduced transmission on the poled ITO/PZT/nano-Au/PZT samples and some of the data (Supplementary Fig. 13) do show slight changes in the relaxation dynamics between samples poled with + 10 and –10 V. While this is consist with the PEC analysis, the changes in relaxation timescales were within the scatter between timescales recorded on different (as grown) samples from the same batch.

Discussion

Here we have studied and manipulated hot-electron transfer in nano/PZT hybrids, both from a solid-state PEC view and from the ultrafast dynamic point of view, where direct evidence of picosecond charge transfer from Au nanoparticles into PZT was obtained. To make a solid comparison of the nano-Au/PZT system with systems based on conventional semiconductors, we fabricated ITO/nano-Au/TiO_2 photoelectrodes and tested their PEC properties. As elaborated in Supplementary Note 4, due to the n-type nature of TiO_2 the hydrogen evolution catalyst (Pt nanoparticles) had to be deposited on TiO_2 to adjust the band bending. The EQE of this device, which is consistent with the values summarized by Pu *et al.*[34], is lower than in the nano-Au/PZT hybrids (Supplementary Fig. 14). After optimizing the redox couples in the electrolyte and the structural design of the plasmonic structure, the EQE of the metallic nanoparticles/semiconductor system could be enhanced to over 1% (refs 35,36). We believe the similar could be achieved by optimizing the two factors in metallic nanoparticles/ferroelectric system.

These cumulative evidences point out that the hot-electron injection from excited nano-Au to the ferroelectric material mirrors the hot-electron transfer in nano-Au/semiconductor structures in terms of hot-electron collection. The employment of ferroelectric material in plasmonic hybrids, however, introduces another dimension to effectively manipulate the PEC properties. The tunable electric polarization offers a flexible platform to freely utilize the optical energy collected by the plasmonic nanostructure. In particular, the hot electrons could be either conducted to the ferroelectric/electrolyte interface to drive PEC reduction reactions or be transferred to the bulk of the ferroelectric material leaving the holes to initiate PEC oxidation reactions, just by switching the direction of the depolarization field in the ferroelectric films. Moreover, adding an additional finger-type electrode on top of the device would enable *in situ* control of the device performance. This concept could have a great impact on the field of solar fuel generation by water splitting or carbon dioxide reduction.

Methods

Fabrication of ultra-thin alumina masks and Au nanoparticle arrays. The UTAMs were prepared in a standard prepatterned anodization process. High-purity (99.99 + %) aluminum foil with ~0.22-mm thickness was used as the starting material. The aluminum foil was first degreased with acetone then annealed at 400 °C for 4 h under vacuum conditions to remove mechanical stresses. After electro-polishing in a 1:7 solution of perchloric acid and ethanol, a Ni stamp with 400 nm nanopillar array was placed on the electropolished Al foil. The pressing was carried out using an oil press system under a pressure of ~10.0 MPa for 3 min. After that, the Ni imprinted stamp was carefully detached from the patterned Al foil and reused. The imprinting process generated an array of highly ordered concaves on the surface of the Al foil. The pre-patterned Al was anodized for 8 min under a constant voltage of 160 V in a 0.3 M phosphoric acid at 15 °C. A PMMA (poly(methylmethacrylate)) layer was subsequently deposited on the top of the alumina layer from a 6% PMMA/chlorobenzene solution to support the UTAM and baked at 100 °C for ~25 min. The Al layer on the backside of the UTAM/PMMA was removed in a mixture solution of $CuCl_2$ and HCl. The PMMA layer on the top of UTAM was removed by acetone. With the aid of a plastic strainer, the UTAM was transferred to a H_3PO_4 solution (5 wt %) at 30 °C for 2 h to remove the barrier layer and widen the size of the pores. Then the UTAM with uniform opened pores was transferred into deionized water from the H_3PO_4 solution, and the clean UTAM was carefully mounted on the substrate (ITO/glass or PZT films coated ITO/glass) in deionized water. Subsequently, the substrates with UTAMs were taken out and dried, and then Au nanoparticles were deposited into highly ordered nanopores of the UTAM by the electron beam evaporation method (Kurt J. Lesker). During the deposition process, substrates were kept in rotation at 20 rounds per minute. Finally, the UTAM was peeled off by Scotch tape, leaving a perfectly ordered nanoparticle array on the surface of the substrate.

Preparation of polycrystalline PZT films. The PZT films with a stoichiometry of $Pb(Zr_{0.20}Ti_{0.80})O_3$ were deposited on ITO/glass (or ITO/glass with ordered nano-Au array) by a sol-gel method. The precursor solution for the coating was prepared by dissolving an appropriate amount of lead acetate ($Pb(CH_3COO)_2 \cdot 5H_2O$) in acetic acid at room temperature in air. A stoichiometric amount of titanium isopropoxide ($Ti((CH_3)_2CHO)_4$) and zirconium isopropoxide ($Zr((CH_3)_2CHO)_4$) was slowly added to the precursor solution. Subsequently, 2-methoxyethanol was added to adjust the concentration until a clear yellow sol with a molar concentration of 0.2 mol l^{-1} was obtained. A 10 mol% excess amount of lead acetate was used to compensate the Pb evaporation during annealing. The wet films were dried at 150 °C for 5 min in air and annealed at 400 °C for 10 min. Finally, the films were crystallized in air atmosphere under 550 °C for 2 h.

Fabrication of PZT photoelectrodes. A strip of conductive copper tape was adhered on the exposed ITO part of the ITO/PZT (or ITO/nano-Au/PZT) and threaded through a glass tube, and then sealed with an insulating epoxy. Electrode areas were optically measured as 0.5 cm^2. For measuring the hysteresis loop of the PZT films and evaluation of the Schottky barrier at the ITO/PZT, ~40-nm-thick Pt dots with diameters of 0.28 mm were deposited onto the PZT films using physical vapour deposition.

PEC measurements. Poling pretreatment was conducted in a quartz electrochemical cell with PZT or nano-Au/PZT photoelectrode as the working electrode and Pt plate as the counter electrode, respectively. Due to the large electrochemical windows, propylene carbonate solution containing 0.1 M $LiClO_4$ was chosen as the electrolyte for poling pretreatments. The poling bias was controlled in the range between + 10 and − 10 V; the poling time was 10 s. Current–potential curves were measured using the digital BioLogic potentiostat (SP-200) and 0.1 M Na_2SO_4 aqueous solution served as the electrolyte. A Pt counter electrode and Ag/AgCl reference electrode were used during the measurements and standard 300 W Xe lamp (Newport) served as the light source; the illumination was filtered by a 455-nm filter (Newport: 20CGA–455). The intensity was 100 mW cm^{-2} determined by a Si photodiode (Newport). EQE was measured with an Oriel 150 W Xe arc lamp (Newport) and a quarter-turn single-grating monochromator (Newport). Measurements were recorded with chopped illumination (20 Hz) and no external bias was applied during the measurements to get a pure photocurrent signal. The output current signal was connected to a Merlin digital lock-in radiometry system and the output signal from the lock-in amplifier was fed into a computer controlled by TRACQ BASIC software.

Characterizations. X-ray diffraction measurement was recorded on Bruker D8 Advance equipped with graphite monochromatized high-intensity Cu Kα radiation ($\lambda = 1.54178$ Å). The SEM images were obtained by Auriga Zeiss focused ion beam SEM. Transient absorbance spectroscopy was carried out by 70 fs optical pulses at 400 nm central wavelength, with the excitation density of 1 mJ cm^{-2}. The wavelength of the pump pulse was selected as 400 nm to avoid exciting PZT. Room-temperature ultraviolet–visible absorbance spectroscopy was carried out on Varian Cary 5,000 UV–vis-NIR spectrophotometer. Polarization–voltage hysteresis loops were examined using a precision ferroelectric analyzer from Radiant Technology. The dark leakage current–voltage (*J–V*) curves of ITO/PZT/Pt were recorded by Keithley 4,200. FDTD simulations were performed using FDTD Solutions by Lumerical Computational solutions, Inc. Atomic force microscopic measurements were carried out using NTEGRA Probe NanoLaboratory (NT-MDT).

References

1. Burda, C., Chen, X., Narayanan, R. & El-Sayed, M. A. Chemistry and properties of nanocrystals of different shapes. *Chem. Rev.* **105,** 1025–1102 (2005).
2. Warren, S. C. & Thimsen, E. Plasmonic solar water splitting. *Energy Environ. Sci.* **5,** 5133–5146 (2012).
3. Knight, M. W., Sobhani, H., Nordlander, P. & Halas, N. J. Photodetection with active optical antennas. *Science* **332,** 702–704 (2011).
4. Sobhani, A. *et al.* Narrowband photodetection in the near-infrared with a plasmon-induced hot electron device. *Nat. Commun.* **4,** 1643 (2013).
5. Wang, F. & Melosh, N. A. Plasmonic energy collection through hot carrier extraction. *Nano Lett.* **11,** 5426–5430 (2011).
6. Chitambar, M., Wang, Z., Liu, Y., Rockett, A. & Maldonado, S. Dye-sensitized photocathodes: Efficient light-stimulated hole injection into p-GaP under depletion conditions. *J. Am. Chem. Soc.* **134,** 10670–10681 (2012).
7. Walter, M. G. *et al.* Solar water splitting cells. *Chem. Rev.* **110,** 6446–6473 (2010).
8. Silva, C. G., Juárez, R., Marino, T., Molinari, R. & García, H. Influence of excitation wavelength (UV or visible light) on the photocatalytic activity of titania containing gold nanoparticles for the generation of hydrogen or oxygen from water. *J. Am. Chem. Soc.* **133,** 595–602 (2011).
9. Liu, Z., Hou, W., Pavaskar, P., Aykol, M. & Cronin, S. B. Plasmon resonant enhancement of photocatalytic water splitting under visible illumination. *Nano Lett.* **11,** 1111–1116 (2011).
10. Mubeen, S. *et al.* An autonomous photosynthetic device in which all charge carriers derive from surface plasmons. *Nat. Nanotechnol.* **8,** 247–251 (2013).
11. Lee, J., Mubeen, S., Ji, X., Stucky, G. D. & Moskovits, M. Plasmonic photoanodes for solar water splitting with visible light. *Nano Lett.* **12,** 5014–5019 (2012).
12. Linic, S., Christopher, P. & Ingram, D. B. Plasmonic-metal nanostructures for efficient conversion of solar to chemical energy. *Nat. Mater.* **10,** 911–921 (2011).
13. Thomann, I. *et al.* Plasmon enhanced solar-to-fuel energy conversion. *Nano Lett.* **11,** 3440–3446 (2011).
14. Hou, W. & Cronin, S. B. A review of surface plasmon resonance-enhanced photocatalysis. *Adv. Funct. Mater.* **23,** 1612–1619 (2013).
15. Zheng, F., Xu, J., Fang, L., Shen, M. & Wu, X. Separation of the Schottky barrier and polarization effects on the photocurrent of Pt sandwiched $Pb(Zr_{0.20}Ti_{0.80})O_3$ films. *Appl. Phys. Lett.* **93,** 172101 (2008).
16. Qin, M., Yao, K. & Liang, Y. C. High efficient photovoltaics in nanoscaled ferroelectric thin films. *Appl. Phys. Lett.* **93,** 122904 (2008).
17. Qin, M., Yao, K., Liang, Y. C. & Gan, B. K. Stability of photovoltage and trap of light-induced charges in ferroelectric WO_3-doped $(Pb_{0.97}La_{0.03})(Zr_{0.52}Ti_{0.48})O_3$ thin films. *Appl. Phys. Lett.* **91,** 092904 (2007).
18. Cao, D. *et al.* High-efficiency ferroelectric-film solar cells with an n-type Cu_2O cathode buffer layer. *Nano Lett.* **12,** 2803–2809 (2012).
19. Lei, Y., Cai, W. & Wilde, G. Highly ordered nanostructures with tunable size, shape and properties: A new way to surface nano-patterning using ultra-thin alumina masks. *Prog. Mater. Sci.* **52,** 465–539 (2007).
20. Lei, Y., Yang, S., Wu, M. & Wilde, G. Surface patterning using templates: concept, properties and device applications. *Chem. Soc. Rev.* **40,** 1247–1258 (2011).
21. Wu, M. *et al.* Ultrathin alumina membranes for surface nanopatterning in fabricating quantum-sized nanodots. *Small* **6,** 695–699 (2010).
22. Ringe, E. *et al.* Plasmon length: A universal parameter to describe size effects in gold nanoparticles. *J. Phys. Chem. Lett.* **3,** 1479–1483 (2012).
23. Cao, D. *et al.* Interface effect on the photocurrent: A comparative study on Pt sandwiched $(Bi_{3.7}Nd_{0.3})Ti_3O_{12}$ and $Pb(Zr_{0.2}Ti_{0.8})O_3$ films. *Appl. Phys. Lett.* **96,** 192101 (2010).
24. Park, Y., Choong, V., Gao, Y., Hsieh, B. R. & Tang, C. W. Work function of indium tin oxide transparent conductor measured by photoelectron spectroscopy. *Appl. Phys. Lett.* **68,** 2699–2701 (1996).
25. Yang, X. *et al.* Enhancement of photocurrent in ferroelectric films via the incorporation of narrow bandgap nanoparticles. *Adv. Mater.* **24,** 1202–1208 (2012).

26. Cao, D. *et al.* Understanding the nature of remnant polarization enhancement, coercive voltage offset and time-dependent photocurrent in ferroelectric films irradiated by ultraviolet light. *J. Mater. Chem.* **22,** 12592–12598 (2012).
27. Cao, D. *et al.* Switchable charge-transfer in the photoelectrochemical energy-conversion process of ferroelectric $BiFeO_3$ photoelectrodes. *Angew. Chemie Int. Ed. Engl.* **53,** 11027–11031 (2014).
28. Choi, T., Lee, S., Choi, Y. J., Kiryukhin, V. & Cheong, S.-W. Switchable ferroelectric diode and photovoltaic effect in $BiFeO_3$. *Science* **324,** 63–66 (2009).
29. Wang, C. *et al.* Photocathodic behavior of ferroelectric $Pb(Zr,Ti)O_3$ films decorated with silver nanoparticles. *Chem. Commun.* **49,** 3769–3771 (2013).
30. Voisin, C. *et al.* Ultrafast electron dynamics and optical nonlinearities in metal nanoparticles. *J. Phys. Chem. B* **105,** 2264–2280 (2001).
31. Link, S. & El-Sayed, M. A. Spectral properties and relaxation dynamics of Surface plasmon electronic oscillations in gold and silver nanodots and nanorods. *J. Phys. Chem. B* **103,** 8410–8426 (1999).
32. Sun, C. K., Vallée, F., Acioli, L. H., Ippen, E. P. & Fujimoto, J. G. Femtosecond-tunable measurement of electron thermalization in gold. *Phys. Rev. B* **50,** 15337–15348 (1994).
33. Demsar, J. *et al.* Hot electron relaxation in the heavy-fermion $Yb_{1-x}Lu_xAl_3$ compound using femtosecond optical pump-probe spectroscopy. *Phys. Rev. B* **80,** 085121 (2009).
34. Pu, Y. & Zhang, J. Z. Mechanisms behind plasmonic enhancement of photocurrent in metal oxides. *Austin J. Nanomed. Nanotechnol.* **2,** 1030 (2014).
35. Chen, Y.-S. & Kamat, P. V. Glutathione-capped gold nanoclusters as photosensitizers. Visible light-induced hydrogen generation in neutral water. *J. Am. Chem. Soc.* **136,** 6075–6082 (2014).
36. Nishijima, Y., Ueno, K., Yokota, Y., Murakoshi, K. & Misawa, H. Plasmon-assisted photocurrent generation from visible to near-infrared wavelength using a Au-nanorods/TiO_2 electrode. *J. Phys. Chem. Lett.* **1,** 2031–2036 (2010).

Acknowledgements

This work was supported by European Research Council Grant (Three-D Surface: 240144), Federal Ministry of Education and Research in Germany (BMBF: ZIK-3DNano-Device: 03Z1MN11), Alexander von Humboldt Foundation, Volkswagen-Stiftung (Herstellung funktionaler Oberflächen: I/83 984), Carl Zeiss Stiftung, National Natural Science Foundation of China (21503209) and the Hundred-Talent Program (Chinese Academy of Sciences). D.C. thanks Y. Yang for her support and guidance of FDTD Solutions. We acknowledge valuable discussions with C. Lienau, E. Runge and K.F. Domke.

Author contributions

Z.W. and D.C. contributed equally to this work. Z.W., D.C. and Y.L. conceived the research plan. Z.W. and D.C. designed the experiments. Z.W., D.C., L.W., R.X., N.N., Z.Z. and Y.M. performed the experiments and analysed the data. D.C. carried out the FDTD simulation. M.O. and Z.W. performed the transient absorbance measurements, M.O., Z.W. and J.D. analysed the data. Z.W., D.C., J.D. and Y.L. co-wrote the paper. Y.L. supervised the project.

Additional information

Competing financial interests: The authors declare no competing financial interests.

Exciton-exciton annihilation and biexciton stimulated emission in graphene nanoribbons

Giancarlo Soavi[1,†], Stefano Dal Conte[2], Cristian Manzoni[2], Daniele Viola[1], Akimitsu Narita[3], Yunbin Hu[3], Xinliang Feng[3], Ulrich Hohenester[4], Elisa Molinari[5,6], Deborah Prezzi[6], Klaus Müllen[3] & Giulio Cerullo[1,2]

Graphene nanoribbons display extraordinary optical properties due to one-dimensional quantum-confinement, such as width-dependent bandgap and strong electron-hole interactions, responsible for the formation of excitons with extremely high binding energies. Here we use femtosecond transient absorption spectroscopy to explore the ultrafast optical properties of ultranarrow, structurally well-defined graphene nanoribbons as a function of the excitation fluence, and the impact of enhanced Coulomb interaction on their excited states dynamics. We show that in the high-excitation regime biexcitons are formed by nonlinear exciton-exciton annihilation, and that they radiatively recombine via stimulated emission. We obtain a biexciton binding energy of $\approx$250 meV, in very good agreement with theoretical results from quantum Monte Carlo simulations. These observations pave the way for the application of graphene nanoribbons in photonics and optoelectronics.

[1] Dipartimento di Fisica, Politecnico di Milano, Piazza Leonardo Da Vinci 32, Milano 20133, Italy. [2] Istituto di Fotonica e Nanotecnologie, CNR, Piazza Leonardo Da Vinci 32, Milano 20133, Italy. [3] Max Planck Institute for Polymer Research, Ackermannweg 10, Mainz 55128, Germany. [4] Institute of Physics, University of Graz, Universitätsplatz 5, Graz 8010, Austria. [5] Dipartimento di Scienze Fisiche, Informatiche e Matematiche, Università di Modena e Reggio Emilia, Modena 41125, Italy. [6] Istituto Nanoscienze, CNR, via G. Campi 213/a, Modena 41125, Italy. † Present address: Cambridge Graphene Centre, University of Cambridge, Cambridge CB3 0FA, UK. Correspondence and requests for materials should be addressed to G.S. (email: gs544@cam.ac.uk) or to D.P. (email: deborah.prezzi@nano.cnr.it) or to G.C. (email: giulio.cerullo@polimi.it).

Slicing graphene into nanoribbons (GNRs) allows to open a bandgap in the graphene electronic structure, owing to the quasi-one-dimensional confinement[1]. This has important implications for electronic devices, such as GNR transistors[2], and forms the basis of the emerging field of graphene nanoplasmonics[3,4]. Especially appealing is also the possibility to additionally tailor specific properties through edge-structure engineering, such as magnetic ordering in zigzag terminated GNRs[5,6]. Further improvements along these lines are envisaged on the basis of recent advances in fabrication[7–10] and processing routes[11,12]. In particular, the bottom-up synthesis[7,8] of GNRs based on molecular precursors designed on purpose has proven capable of reaching nanometric widths with atomically precise edges, a regime where the GNR properties are widely tunable[1]. While single-walled carbon nanotubes (SWNTs) cannot be prepared with single chirality and require further processing with surfactants for the sorting[13,14], the bottom-up synthesis directly affords GNRs with a uniform chemical structure with 100% selectivity. Thus, prepared GNRs show well-defined electronic and optical properties, which are fully determined by their specific structure and can be further tuned by modulation of their width and edge configuration[10,15,16] as well as by atomically controlled doping[12,17]. All of this holds promise for application in next-generation optoelectronic and photonic devices, as recently suggested by the realization of all-GNR-based heterojunctions[12,18] and by other proposals for photovoltaic applications[16,19,20].

In spite of this interest, the field of GNRs is still in its infancy and little is known about their photophysical properties, especially in the non-equilibrium regime. Extraordinary optical properties were predicted[21–23], such as width-dependent bandgap and the formation of excitons with extremely high binding energies, which have been only recently demonstrated in bottom-up GNRs[15,24–26]. In particular, the pronounced excitonic effects[26] are accompanied by a significant increase of the optical absorbance, as compared with graphene, for light that is linearly polarized along the ribbon axis. At this stage, the understanding of the excited-state relaxation dynamics of GNRs would offer not only a deeper insight into the fundamental physics of these ideal one-dimensional systems but also a benchmark for their integration in advanced optoelectronic devices[27].

In the following, we apply resonant ultrafast pump–probe spectroscopy to nanometre-wide atomically precise GNRs obtained by a bottom-up solution synthesis[8] to study the kinetics of excitons and their interactions in the saturation (that is, nonlinear) excitation regime. For high-excitation fluences, we observe bimolecular exciton annihilation and the concomitant ultrafast (≈ 1 ps) buildup of a stimulated emission (SE) signal from an excited biexciton state that is populated via a nonlinear process, a result that is extremely promising in view of applications of GNRs as tuneable active materials in lasers and light-emitting diodes.

Results

Linear absorption of GNRs. The GNRs studied here, characterized by cove-shaped edge morphology and hereafter labelled 4CNR (following the notation in ref. 16), were chemically synthesized as described in ref. 8. Their aromatic core structure, displayed in Fig. 1a (inset), features a modulated width of 0.7–1.1 nm, and is functionalized with long and branched alkyl chains (2-decyltetradecyl) at the outer benzene rings to guarantee dispersibility in organic solvents. The linear absorption spectrum of the GNR sample in tetrahydrofuran (THF) dispersion is shown in Fig. 1a (black curve), and compared with the simulated gas-phase spectrum obtained from *ab initio* *GW* plus Bethe–Salpeter (*GW*-BS) calculations (blue arrows), performed on H-passivated 4CNR (details concerning the calculations are reported in the Methods section). THF was chosen as a solvent to minimize aggregation of GNRs, which would significantly alter their optical properties. The effect of the solvent on the spectrum is instead expected to be minor (see, for example, ref. 28). The experimental spectrum is dominated by an optical transition of excitonic origin located at ≈ 570 nm, in good agreement with simulations. According to our *GW*-BS results, the first two excitons both arise from the linear combination of transitions among the two highest valence and two lowest conduction bands around the Γ point (mainly E_{12} and E_{21} transitions for the first and second exciton, respectively, as indicated in the band structure of Fig. 1b). In vacuum, excitons are tightly bound, with a giant binding energy of ~ 1.5 eV (defined as the difference between the quasi-particle gap and the energy of the excitonic states), which is expected to diminish in presence of a dielectric environment.

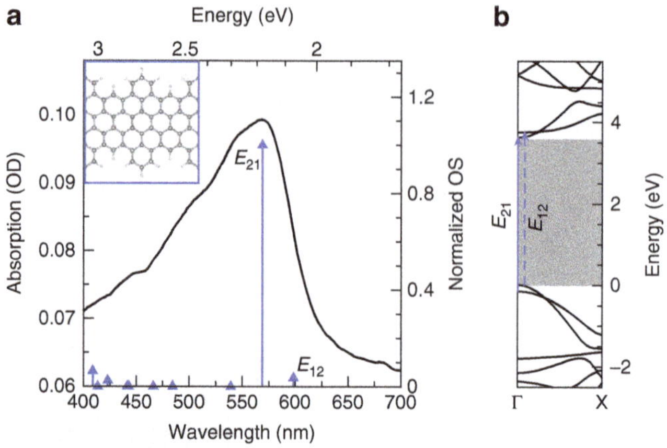

Figure 1 | Linear absorption and excitons in GNRs. (**a**) Linear absorption spectrum of the 4CNR sample in THF solution (black curve). A ball-and-stick model of the GNR without alkyl chains at the edges is shown in the inset. The experimental spectrum is compared with the result of *GW*-BS calculations, with excitonic transitions indicated by blue arrows. (**b**) *GW* quasi-particle band structure. The lines indicate the transitions that are mainly contributing to the first and second exciton. The 1.5 eV difference between the *GW* gap (**b**; grey area) and the excitonic transition reported in **a** defines the exciton binding energy in vacuum.

Ultrafast pump-probe spectroscopy of GNRs. We performed broadband pump–probe spectroscopy of the 4CNR samples using resonant excitation at 570 nm and white-light probing covering the 500–700 nm range, with an overall temporal resolution of ≈ 100 fs (see Methods for details of the experimental setup). Figure 2 shows the differential transmission ($\Delta T/T$) spectra for different pump–probe delays (Fig. 2a), and the $\Delta T/T$ dynamics at 600 nm probe wavelength (Fig. 2b), when the sample is excited with a low fluence of $\approx 100\ \mu\mathrm{J\,cm}^{-2}$. In the $\Delta T/T$ spectra we clearly distinguish two bands: (i) an intense photo-bleaching (PB) of the excitonic transition peaked at ≈ 570 nm, which we assign to ground-state depletion and/or phase space filling of the excited state; (ii) a relatively weak red-shifted photo-induced absorption (PA) band starting from ≈ 620 nm. Since these two bands have the same decay kinetics (Fig. 2b), we attribute the PA signal to excited-state absorption from the exciton to higher energy states or to the e–h continuum. The comparison of the $\Delta T/T$ spectra (Fig. 2a) for different pump–probe delays (from 500 fs to 50 ps) also highlights that the signal decays without any significant

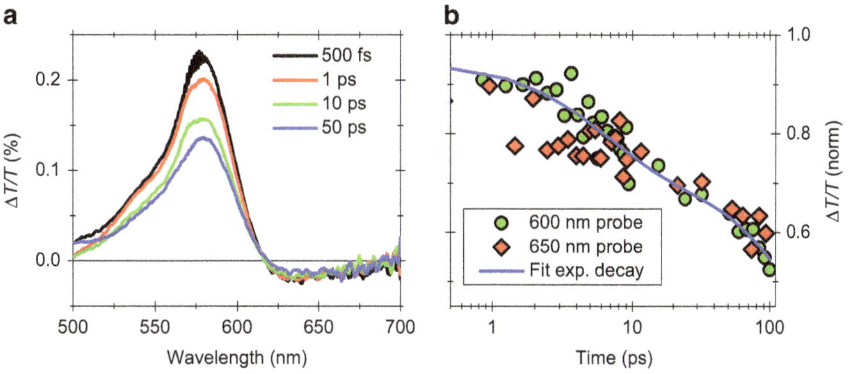

Figure 2 | Transient absorption spectra and dynamics at low fluence. (**a**) $\Delta T/T$ spectra of 4CNRs at different pump-probe delays and (**b**) decay dynamics at 600- (green circles) and 650- (red diamonds) nm probe wavelengths for an excitation fluence of $\approx 100\,\mu J\,cm^{-2}$. The fit (blue line) in **b** correspond to a bi-exponential function with time constants $\tau_1 \approx 6$ ps and $\tau_2 \approx 330$ ps.

spectral evolution at this excitation fluence, thus indicating a simple relaxation mechanism of the excited state (either radiative via emission of photons and/or non-radiative via interaction with phonons).

To understand the carrier relaxation process, we can take as a reference the large amount of experimental results on SWNTs, since we expect them to display similar recombination dynamics. SWNTs also show complex, multi-component decay kinetics. In the low-excitation fluence regime, the long-lived (>5 ps) decay components in SWNTs have been reproduced with a bi-exponential model, which includes both the radiative and non-radiative lifetime[29–32]. Other studies describe these decay dynamics by a model for a diffusion-limited regime[33,34] or geminate e–h recombination in one dimension[35], thus leading to a power law ($\approx t^{-0.5}$) kinetics. Since a detailed study of the dynamics of these long-lived photoexcitations for GNRs is beyond the scope of this paper, we fit the kinetics of the PB signal at 600 nm probe wavelength with a simple bi-exponential decay model (Fig. 2b), which gives us the timescale of the relaxation processes. From the fit we obtain $\tau_1 \approx 6$ ps and $\tau_2 \approx 330$ ps, in good agreement with the results obtained for SWNTs[30].

Instead, we concentrate on the ultrafast (<5 ps) decay dynamics, and in particular on their dependence on the excitation fluence, as we show in Fig. 3. The normalized $\Delta T/T$ spectra at different excitation fluences for a 1 ps pump-probe delay (Fig. 3a) display very similar, fluence-independent PB and PA spectral features, while for a 5 ps pump-probe delay (Fig. 3b) we unambiguously observe the buildup, with increasing fluence, of a positive and red-shifted $\Delta T/T$ peak, at ≈ 650 nm. From a general point of view, a positive $\Delta T/T$ signal in pump-probe experiments describes either a PB or a SE process. We here assign the peak at ≈ 650 nm to a SE process, since (i) it does not correspond to any resonant feature in the linear absorption spectrum (Fig. 1a), as also confirmed by simulations, being instead red-shifted with respect to the main excitonic transition (at ≈ 570 nm); and (ii) it appears with a $\approx$ps delay with respect to the pump pulse and only at high-excitation fluences. We note that photons produced by SE are identical (phase, energy and momentum) to the probe photons and thus can be detected in pump-probe experiments, at variance with those produced by spontaneous emission.

Exciton–exciton annihilation and biexciton formation in GNRs. To clarify the origin of this SE signal, we concentrate on the fluence-dependent dynamics at the probe wavelengths of 600 nm, that is, the PB signal, and 650 nm, that is, the SE signal (Fig. 3c and d, respectively). We immediately observe that, for increasing fluence, the PB signal (Fig. 3c) displays a faster decay, while, correspondingly, the signal at 650 nm (Fig. 3d) undergoes a clear change in sign (from negative to positive) that corresponds to the delayed formation of the SE signal. First, let us analyse the fast PB decay at 600-nm probe wavelength (Fig. 3c). In semiconducting SWNTs the appearance of an ultrafast fluence-dependent decay component is explained by exciton–exciton annihilation, a two-exciton interaction process, in which one exciton recombines to the ground state and the other either dissociates into a free e–h pair or is promoted into a higher energy level[36–38]. Such process is also commonly observed in other one-dimensional semiconductors, such as conjugated polymers[39,40], and it has been recently observed also in monolayer MoS_2 (ref. 41). Theoretical calculations[42] also predict that an Auger-like mechanism occurs in semiconducting armchair GNRs because of effectively enhanced Coulomb interaction. In their work, Konabe *et al.*[42] find an exciton–exciton annihilation time in the order of few ps for 1.2–2.5 nm wide GNRs. After the initial ultrafast nonlinear decay process, the dynamics at all pump fluences are instead the same. This can be noticed by comparing the kinetics of the PB signal at high and low fluences over the full temporal range (inset in Fig. 3c). Being a two-body interaction process, exciton–exciton annihilation is expected to display a nonlinear dependence on the exciton density (see Methods for details) and/or the excitation fluence[37], in agreement with our experimental results (inset in Fig. 3a).

Second, we need to understand the origin of the delayed formation of the SE signal at 650 nm (Fig. 3d). As we have already discussed, following exciton–exciton annihilation both free e–h pairs and/or higher energy-excited states can be formed. The first scenario has been observed in SWNTs, where the creation of charges in the high-excitation regime leads to the formation of trions[43,44], which are detected as a negative (PA) and red-shifted $\Delta T/T$ signal. Clearly, this scenario is in contrast with our experimental results, which present a positive (SE) signal. Instead, the formation of a delayed and red-shifted SE signal in the high-excitation regime was observed in semiconducting quantum-dots (QDs)[45–48], another prototype of quantum-confined systems. In the case of QDs, the SE signal was explained in terms of emission from biexcitons and the energy distance between the main exciton PB signal and the biexciton SE signal gives the biexciton binding energy, which is typically of the order of few tens of meV. For our GNRs, the SE peak corresponds to a much larger value for the biexciton binding energy, that is, $E_b \approx 250$ meV. This value is in accordance with the comparably larger exciton binding energies in GNRs, that are at least one order of magnitude larger than in QDs due to the reduced screening as well as the extreme two-dimensional and transversal confinements. The following

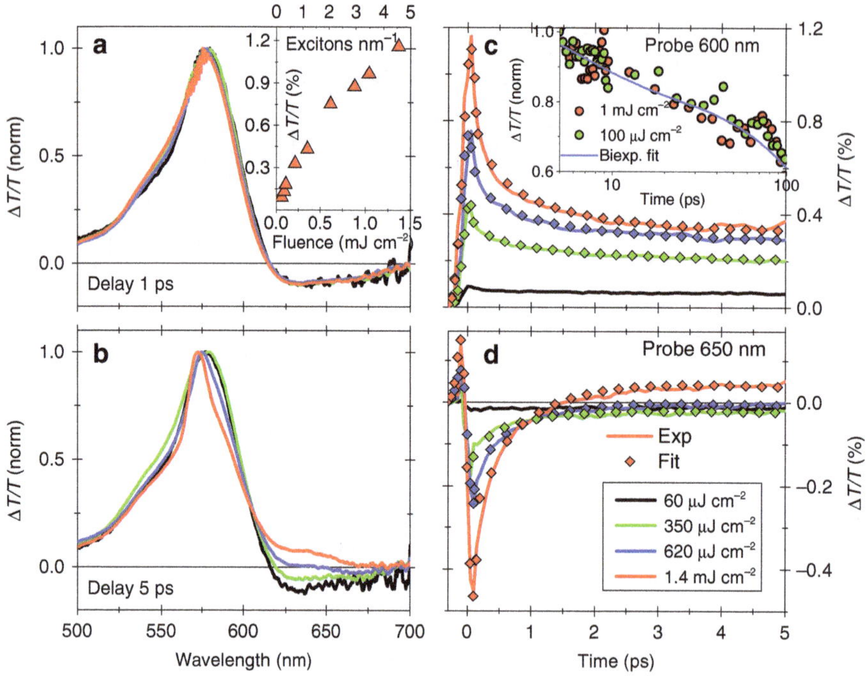

Figure 3 | Exciton-exciton annihilation and biexciton formation. Normalized $\Delta T/T$ spectra of 4CNRs for different excitation fluences at a fixed pump-probe delay of (**a**) 1 ps and (**b**) 5 ps. The inset in **a** reports the peak amplitude of the signal at 600 nm probe wavelength as a function of the excitation fluence (bottom x axis) and the exciton linear density (top x axis). Excitation-fluence-dependent dynamics at (**c**) 600 nm probe wavelength and (**d**) 650 nm probe wavelength. The fit (diamonds) is obtained from the coupled rate-equations (described in the text) based on exciton-exciton annihilation in one dimension. Inset in **c** represents the dynamics on a 100 ps timescale for 100 μJ cm^{-2} (low) and 1 mJ cm^{-2} (high) fluences, together with the bi-exponential fit used in Fig. 2b.

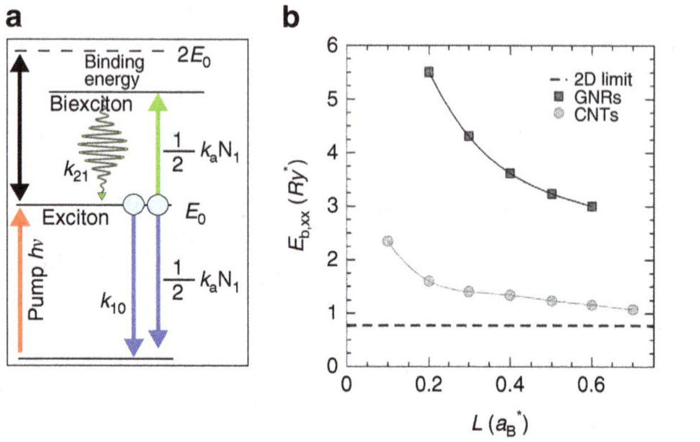

Figure 4 | Photoexcitation scenario in GNRs and biexciton binding energy. (**a**) Sketch of the energetic levels and the kinetic model described in the text. (**b**) Biexciton (XX) binding energy ($E_{b,XX}$) as a function of the lateral dimension (L), that is, width for GNRs (black squares) and diameter for SWNTs (grey circles). The data, obtained by guide-function QMC simulations, are shown in dimensionless exciton units, as detailed in the text. In these units, the binding energy of the 4CNR is $\sim 3.4\,Ry^*$. The dashed line indicates the two-dimensional value of $E_{b,XX} = 0.77\,Ry^*$.

photoexcitation scenario in GNRs thus emerges: at high-excitation fluences, exciton–exciton annihilation leads to the population of a radiative biexciton state, which then undergoes SE to the one-exciton level upon interaction with the probe pulse (Fig. 4a).

Discussion

To support our assignment, we first compute the biexciton binding energy in GNRs by means of guide-function quantum Monte Carlo (QMC) simulations, using an envelope function approach with effective masses, as previously reported in ref. 49 for SWNTs. Full description of the method is reported in the Methods section. In Fig. 4b, we show the biexciton binding energy dependence on the lateral confinement in dimensionless exciton units, that is, distances are measured in units of the effective Bohr radius $a_B^* = a_B\varepsilon/\mu$ and energies in units of the effective Rydberg $Ry^* = \frac{e^2}{2\varepsilon a_B^*}$, where μ is the reduced e–h mass, ε is the dielectric constant, a_B is the Bohr radius and e is the electron charge. By considering an average reduced mass as obtained from *ab initio* calculations (computed as the weighted average relative to the E_{21} and E_{12} transitions, that is, $\mu = 0.22$), the average width of the 4CNR ($w = 0.84$ nm), and the dielectric constant of the solvent ($\varepsilon = 7.5$), we obtain a biexciton binding energy of 180 meV, in good agreement with the experimental result of 250 meV. The discrepancy with respect to experiments is reasonable in view of the simplified description scheme used here, which is not expected to capture all microscopic details.

In Fig. 4b, we also report the biexciton binding energy for SWNTs, calculated by the same approach[49]: for a SWNT of similar lateral dimension, we find a binding energy that is less than one-half of that of the 4CNR in dimensionless units. This can be understood by considering the different biexciton confinement, since in the case of SWNTs the biexciton wavefunction is delocalized over the whole circumference, whereas in GNRs it becomes strongly confined in the transversal direction. The value obtained for GNRs is quite large also in comparison with the biexciton binding energy $E_b \approx 50$ meV of different monolayer transition metal dichalcogenides obtained by recent transient absorption[50] and photoluminescence experiments[51].

To further confirm our interpretation, we finally reproduce the temporal evolution of the exciton (that is, decay of the PB signal) and the biexciton (that is, formation of the SE signal) populations with the following coupled rate-equations[52–54]:

$$\frac{dn_E}{dt} = -k_{10}n_E - k_{e-e}\frac{n_E^2}{\sqrt{t}} + k_{21}n_B \quad (1)$$

$$\frac{dn_B}{dt} = +\frac{1}{2}k_{e-e}\frac{n_E^2}{\sqrt{t}} - k_{21}n_B \quad (2)$$

where n_E and n_B are the exciton and biexciton population, respectively, k_{10} is the decay rate from the exciton to the ground state, k_{21} is the radiative decay rate from the biexciton to the exciton producing the SE signal and k_{e-e} is the exciton annihilation rate constant (see Fig. 4a for the adopted model). The $t^{-0.5}$ dependence of the exciton annihilation rate arises from the one-dimensional diffusion mechanism of excitons, and its divergence for $t \to 0$ is cured by truncating it for times shorter than the width of the instrumental response function (IRF) ($\approx$100 fs in our case). The quality of the fit (Fig. 3c,d) further confirms the validity of our model. From the fit we find that $k_{10} \approx 0$ ps and $k_{21} \approx 0.15\ \text{ps}^{-1}$, meaning that both mechanisms occur on a timescale that is longer with respect to the temporal window (5 ps) that we use for this analysis.

It is worth noting that we can fit our experimental data considering that all the excitons that undergo exciton–exciton annihilation form biexcitons. In particular, we can exclude that the observed SE signal is because of the dissociation of excitons into free e–h carriers, since SE from electrons in the continuum is not expected. Another possibility is that free e–h pairs get trapped into low-energy states; for example, due to defects. Nevertheless, since the selection rules for emission and absorption of photons are the same, the presence of such bright low-energy states should be detectable also in the linear absorption spectrum, and thus this scenario can be also excluded by looking at the absorption spectrum and theoretical simulations in Fig. 1. Thus, although we cannot exclude that free e–h carriers are formed, we can strongly assert that the observed SE arises from biexcitons. Since our model is able to correctly reproduce also the amplitude of the pump–probe signal without additional loss channels, we conclude that biexciton formation upon exciton–exciton annihilation is extremely efficient. Finally, for a fixed pump–probe delay we can also evaluate the annihilation rate $k_a = \frac{k_{e-e}}{\sqrt{t}}$, which corresponds to an initial $k_a \approx 2\ \text{ps}^{-1}$ for a pump–probe delay of 180 fs, close to our temporal resolution, and thus to an annihilation lifetime of $\approx$0.5 ps, in excellent agreement with experimental results on SWNTs[54] and theoretical calculations on GNRs[42].

In conclusion, we studied the transient photophysical properties of ultranarrow structurally well-defined GNRs by means of ultrafast pump-probe spectroscopy. We show that a nonlinear decay channel for the main excitonic transition sets in at high excitation densities, and, correspondingly, we unambiguously observe a red-shifted SE signal. Our experiments demonstrate that exciton–exciton annihilation populates a radiative biexciton state, with an extremely high binding energy $\approx$250 meV, in agreement with estimates from QMC simulations. The high efficiency we find for both exciton–exciton annihilation and biexciton formation is of great importance, not only for gaining fundamental understanding on strongly enhanced quantum effects in low-dimensional materials but also for its implications in GNR-based optoelectronic devices. Indeed, the clear observation of a strong SE is extremely promising in view of using GNRs as active light-amplifying materials in tuneable lasers and light-emitting diodes. Moreover, our results suggest that also multiple-exciton generation[55–57] can be extremely efficient in GNRs since it is governed by the same exciton–exciton annihilation rate. A detailed understanding of the photophysics of biexcitons and the mechanism of multiple-exciton generation will help to improve the efficiency of photovoltaic devices, with GNRs acting as light absorbers.

Methods

Pump-probe experimental setup. The experimental setup used for pump–probe measurements has been described in detail elsewhere[58]. In brief, the setup is based on a regeneratively amplified Ti:sapphire laser (Coherent, Libra) producing 100 fs, 4 mJ pulses at 800 nm wavelength and 1 kHz repetition rate. The probe pulse is obtained by focusing a fraction of the laser beam in a 2-mm-thick sapphire plate to generate a broadband single-filament white-light continuum. The pump pulse, generated by an optical parametric amplifier, is centred at 570 nm (at resonance with the main excitonic transition) with a bandwidth of $\approx$10 nm, corresponding to $\approx$70 fs duration. The probe light transmitted by the sample is dispersed on an optical multichannel analyser equipped with fast electronics, allowing single-shot recording of the probe spectrum at the full 1 kHz repetition rate. By changing the pump–probe delay, we record two-dimensional maps of the differential transmission ($\Delta T/T$) signal as a function of probe wavelength and delay. The temporal resolution (taken as full-width at half-maximum of pump–probe cross-correlation) is $\approx$100 fs over the entire probe spectrum.

Exciton linear density estimation. To estimate the exciton linear density reported in the inset of Fig. 3a, we proceed as follows: (i) we calculate the number of absorbed photons cm^{-2} from the measured pump fluence, the measured absorbance, and the pump photon energy (2.18 eV); (ii) we calculate the number of excitons cm^{-2} by considering an exciton photogeneration efficiency (Quantum Yield) of 96% based on optical pump—THz probe experiments on similar GNRs[59]; and (iii) we multiply this value by the concentration ($0.0021\ \text{g l}^{-1}$) of the dispersion and divide by the mass of a unit length of the GNRs ($0.83 \times 10^{-21}\ \text{g nm}^{-1}$). This estimate gives a density of $\sim$0.2–0.3 excitons nm^{-1} for pump fluences below $100\ \mu\text{J cm}^{-2}$, corresponding to an average distance between excitons of $\sim$4–5 nm. Such a value appears to be sufficient to prevent exciton–exciton interactions.

Coherent artefacts in pump-probe dynamics. The coupled-rate equations used to model the evolution of the exciton and biexciton populations are solved by taking into account both the IRF and possible coherent artefacts present in pump–probe measurements. For our experimental setup the IRF is the cross-correlation of the pump and probe pulses, and can be quite accurately modelled by a Gaussian function with $\approx$100 fs full-width at half-maximum. Possible coherent artefacts in pump–probe experiments are stimulated Raman amplification and cross-phase modulation (XPM)[60]. We fitted the initial 100 fs of the PB signal at 600 nm (Fig. 3c) including the IRF to reproduce the initial buildup and a Gaussian function (the same used for the IRF) to reproduce the initial ultrafast stimulated Raman amplification coherent artefact. For the SE dynamics at 650 nm (Fig. 3d), instead, we include also XPM, which we fit as the first derivative of a Gaussian function (the same used for the cross-correlation). Although extremely simple, this model correctly reproduces not only the evolution but also the initial steps of the pump–probe signal.

Thermal effects and sample heating. Time-resolved experiments were carried out at room temperature, assuming negligible temperature effects on the spectra on the basis of previous results on SWNTs (see, for example, ref. 61). Regarding the sample heating during experiments, we also expect negligible temperature changes upon photoexcitation. In fact, we can estimate a maximum increase in temperature of $\sim$0.1 K, based on a comparison with the work of Abdelsayed *et al.*[62], and by considering the following parameters: volume of the sample (0.3 ml), THF heat capacity ($123\ \text{J mol}^{-1}\ \text{K}^{-1}$), density ($889\ \text{kg m}^{-3}$), molar mass ($72\ \text{g mol}^{-1}$) and laser total energy ($\sim$100 nJ per pulse at 500 Hz for 10 min of irradiation, corresponding to 30 mJ). This indicates that we are working in a perturbative regime for what concerns thermal effects.

***GW*-BS calculations for the static absorption.** The ground-state atomic structure of the 4CNR was optimized by using the PWscf code of the Quantum ESPRESSO package (ref. 63), which is based on a plane-wave pseudopotential implementation of density functional theory. Calculations were performed by employing local density approximation exchange correlation (xc) functional and norm-conserving pseudopotentials, with a cutoff energy for the wavefunctions of 60 *Ry*. The atomic positions within the cell were fully relaxed until forces were $<5 \times 10^{-4}\ Ry\ \text{bohr}^{-1}$, while the lattice constant along the periodic direction was optimized separately. The Brillouin zone was sampled with 16 **k**-points along the periodic direction. The Kohn-Sham band structure obtained for the optimized geometry was improved by introducing many-body corrections within the G_0W_0 approximation for the self-energy operator. Here, the dynamic dielectric function was obtained within the plasmon-pole approximation, by employing a box-shaped truncation of the Coulomb potential[64] to remove the long-range interaction between periodic images. The optical absorption spectrum was then computed as the imaginary part of the macroscopic dielectric function starting from the solution of the BS equation, which allows for the inclusion of e–h interaction effects.

The static screening in the direct term was calculated within the random-phase approximation with inclusion of local field effects; the Tamm–Dancoff approximation for the BS Hamiltonian was employed. The aforementioned *GW*-BS calculations, performed by using the YAMBO code (ref. 65), were carried out for the fully H-passivated 4CNR, that is, by removing the alkyl side chains, in order to make them computationally affordable. We have checked that the different passivation does not affect the band structure properties at the density functional theory and local density approximation level in the energy window of interest for the determination of the optical absorption. Similar results were reported in ref. 66.

The guide function quantum Monte Carlo approach. Atomistic simulations as the ones described above are presently unfeasible for the investigation of biexcitons in realistic systems. Here we resort to an effective model based on guide function QMC simulations, as previously reported in ref. 49 for SWNTs. For the (unnormalized) guide function we use

$$\Psi_T(r_1, r_2, r_a, r_b) = \exp[-(r_{1a} + r_{1b} + r_{2a} + r_{2b})] \tag{3}$$

where $r_{1,2}$ ($r_{a,b}$) are the dimensionless positions of the two electrons (holes), and $r_{i,j} = \left[(x_i - x_j)^2 + (y_i - y_j)^2\right]^{1/2}$ is the distance between particles confined to the two-dimensional nanoribbon. The Monte Carlo simulation approach is the same as in ref. 49 with the only exception that we add a box-like confinement potential

$$V(y) = V_0 / \left[e^{\beta\left(0.5 - \frac{|y|}{w}\right)} + 1\right] \tag{4}$$

along the transversal direction, with $V_0 = 1{,}000$, $\beta = 20$ and w the GNR width. In the simulations we use 20,000 walkers[67], a time step of $\Delta t = 0.25 \times 10^{-4}$, an equilibration interval of 20,000 and a measurement interval of 30,000 time steps. The smoothed confinement potential and the time step were chosen such that for 'typical' paths the inequality $|V(y_t) - V(y_{t+1})| \ll |(y_t - y_{t+1})/(2\Delta t^2)|$ holds[68]. We checked that the biexciton binding energy did not change substantially upon modifying Δt or other simulation parameters. The QMC approach has been able to predict quite accurately the biexciton binding energy of SWNTs. In fact, Colombier *et al.*[69] detected the presence of biexcitons in SWNTs embedded in a gelatine matrix by means of nonlinear optical spectroscopy, reporting a binding of 106 meV energy for the (9, 7) tube. The QMC model predicts a value of the binding energy of ≈ 100 meV assuming a dielectric constant of $\varepsilon = 3$ (instead of 2.3 as appropriate for a gelatin matrix). Such an overestimation in the case of small values of ε is known for phenomenological models, whereas for larger values of ε (like the one considered in this work, i.e., $\varepsilon = 7.5$) the QMC model is expected to give results similar to more refined (though not yet atomistic) approaches, as discussed in ref. 70.

References

1. Castro Neto, A. H., Guinea, F., Peres, N. M. R., Novoselov, K. S. & Geim, A. K. The electronic properties of graphene. *Rev. Mod. Phys.* **81,** 109–162 (2009).
2. Schwierz, F. Graphene transistors. *Nat. Nanotechnol.* **5,** 487–496 (2010).
3. Chen, J. *et al.* Optical nano-imaging of gate-tunable graphene plasmons. *Nature* **487,** 77–81 (2012).
4. Rodrigo, D. *et al.* Mid-infrared plasmonic biosensing with graphene. *Science* **349,** 165–168 (2015).
5. Son, Y.-W., Cohen, M. L. & Louie, S. G. Half-metallic graphene nanoribbons. *Nature* **444,** 347–349 (2006).
6. Magda, G. Z. *et al.* Room-temperature magnetic order on zigzag edges of narrow graphene nanoribbons. *Nature* **514,** 608–611 (2014).
7. Cai, J. *et al.* Atomically precise bottom-up fabrication of graphene nanoribbons. *Nature* **466,** 470–473 (2010).
8. Narita, A. *et al.* Synthesis of structurally well-defined and liquid-phase-processable graphene nanoribbons. *Nat. Chem.* **6,** 126–132 (2014).
9. Jacobberger, R. M. *et al.* Direct oriented growth of armchair graphene nanoribbons on germanium. *Nat. Commun.* **6,** 8006 (2015).
10. Narita, A., Wang, X.-Y., Feng, X. & Müllen, K. New advances in nanographene chemistry. *Chem. Soc. Rev.* **44,** 6616–6643 (2015).
11. Bennett, P. B. *et al.* Bottom-up graphene nanoribbon field-effect transistors. *Appl. Phys. Lett.* **103,** 253114 (2013).
12. Cai, J. *et al.* Graphene nanoribbon heterojunctions. *Nat. Nanotechnol.* **9,** 896–900 (2014).
13. Yang, F. *et al.* Chirality-specific growth of single-walled carbon nanotubes on solid alloy catalysts. *Nature* **510,** 522–524 (2014).
14. Omachi, H., Nakayama, T., Takahashi, E., Segawa, Y. & Itami, K. Initiation of carbon nanotube growth by well-defined carbon nanorings. *Nat. Chem.* **5,** 572–576 (2013).
15. Narita, A. *et al.* Bottom-up synthesis of liquid-phase-processable graphene nanoribbons with near-infrared absorption. *ACS Nano* **8,** 11622–11630 (2014).
16. Osella, S. *et al.* Graphene nanoribbons as low band gap donor materials for organic photovoltaics: quantum chemical aided design. *ACS Nano* **6,** 5539–5548 (2012).
17. Kawai, S. *et al.* Atomically controlled substitutional boron-doping of graphene nanoribbons. *Nat. Commun.* **6,** 8098 (2015).
18. Chen, Y.-C. *et al.* Molecular bandgap engineering of bottom-up synthesized graphene nanoribbon heterojunctions. *Nat. Nanotechnol.* **10,** 156–160 (2015).
19. Yan, X., Cui, X., Li, B. & Li, L. Large, solution-processable graphene quantum dots as light absorbers for photovoltaics. *Nano Lett.* **10,** 1869–1873 (2010).
20. Lee, S. L. *et al.* An organic photovoltaic featuring graphene nanoribbons. *Chem. Commun. (Camb.)* **51,** 9185–9188 (2015).
21. Prezzi, D., Varsano, D., Ruini, A., Marini, A. & Molinari, E. Optical properties of graphene nanoribbons: the role of many-body effects. *Phys. Rev. B* **77,** 041404 (2008).
22. Yang, L., Cohen, M. L. & Louie, S. G. Excitonic effects in the optical spectra of graphene nanoribbons. *Nano Lett.* **7,** 3112–3115 (2007).
23. Prezzi, D., Varsano, D., Ruini, A. & Molinari, E. Quantum dot states and optical excitations of edge-modulated graphene nanoribbons. *Phys. Rev. B* **84,** 041401 (2011).
24. Ruffieux, P. *et al.* Electronic structure of atomically precise graphene nanoribbons. *ACS Nano* **6,** 6930–6935 (2012).
25. Linden, S. *et al.* Electronic structure of spatially aligned graphene nanoribbons on Au(788). *Phys. Rev. Lett.* **108,** 216801 (2012).
26. Denk, R. *et al.* Exciton-dominated optical response of ultra-narrow graphene nanoribbons. *Nat. Commun.* **5,** 4253 (2014).
27. Bonaccorso, F., Sun, Z., Hasan, T. & Ferrari, A. C. Graphene photonics and optoelectronics. *Nat. Photon.* **4,** 611–622 (2010).
28. Ohno, Y. *et al.* Excitonic transition energies in single-walled carbon nanotubes: dependence on environmental dielectric constant. *Phys. Status Solidi B Basic Solid State Phys.* **244,** 4002–4005 (2007).
29. Styers-Barnett, D. J., Ellison, S. P., Park, C., Wise, K. E. & Papanikolas, J. M. Ultrafast dynamics of single-walled carbon nanotubes dispersed in polymer films. *J. Phys. Chem. A* **109,** 289–292 (2005).
30. Chou, S. *et al.* Phonon-assisted exciton relaxation dynamics for a (6,5)-enriched DNA-wrapped single-walled carbon nanotube sample. *Phys. Rev. B* **72,** 195415 (2005).
31. Dyatlova, O. A. *et al.* Ultrafast relaxation dynamics via acoustic phonons in carbon nanotubes. *Nano Lett.* **12,** 2249–2253 (2012).
32. Huang, L., Pedrosa, H. N. & Krauss, T. D. Ultrafast ground-state recovery of single-walled carbon nanotubes. *Phys. Rev. Lett.* **93,** 017403 (2004).
33. Russo, R. M. *et al.* One-dimensional diffusion-limited relaxation of photoexcitations in suspensions of single-walled carbon nanotubes. *Phys. Rev. B* **74,** 041405 (2006).
34. Allam, J. *et al.* Measurement of a reaction-diffusion crossover in exciton-exciton recombination inside carbon nanotubes using femtosecond optical absorption. *Phys. Rev. Lett.* **111,** 197401 (2013).
35. Soavi, G. *et al.* High energetic excitons in carbon nanotubes directly probe charge-carriers. *Sci. Rep.* **5,** 9681 (2015).
36. Ma, Y.-Z., Valkunas, L., Dexheimer, S. L., Bachilo, S. M. & Fleming, G. R. Femtosecond spectroscopy of optical excitations in single-walled carbon nanotubes: evidence for exciton-exciton annihilation. *Phys. Rev. Lett.* **94,** 157402 (2005).
37. Valkunas, L., Ma, Y.-Z. & Fleming, G. R. Exciton-exciton annihilation in single-walled carbon nanotubes. *Phys. Rev. B* **73,** 115432 (2006).
38. Huang, L. & Krauss, T. D. Quantized bimolecular auger recombination of excitons in single-walled carbon nanotubes. *Phys. Rev. Lett.* **96,** 057407 (2006).
39. Maniloff, E. S., Klimov, V. I. & McBranch, D. W. Intensity-dependent relaxation dynamics and the nature of the excited-state species in solid-state conducting polymers. *Phys. Rev. B* **56,** 1876 (1997).
40. Stevens, M. A., Silva, C., Russel, D. M. & Friend, R. H. Exciton dissociation mechanisms in the polymeric semiconductors poly(9,9-dioctylfluorene) and poly(9,9-dioctylfluorene-co-benzothiadiazole). *Phys. Rev. B* **63,** 165213 (2001).
41. Sun, D. *et al.* Observation of rapid exciton–exciton annihilation in monolayer molybdenum disulfide. *Nano Lett.* **14,** 5625–5629 (2014).
42. Konabe, S., Onoda, N. & Watanabe, K. Auger ionization in armchair-edge graphene nanoribbons. *Phys. Rev. B* **82,** 073402 (2010).
43. Santos, S. M. *et al.* All-optical trion generation in single-walled carbon nanotubes. *Phys. Rev. Lett.* **107,** 187401 (2011).
44. Matsunaga, R., Matsuda, K. & Kanemitsu, Y. Observation of charged excitons in hole-doped carbon nanotubes using photoluminescence and absorption spectroscopy. *Phys. Rev. Lett.* **106,** 037404 (2011).
45. Levy, R., Hönerlage, B. & Grun, J. B. Time-resolved exciton-biexciton transitions in CuCl. *Phys. Rev. B* **19,** 2326 (1979).
46. Klimov, V. I. *et al.* Optical gain and stimulated emission in nanocrystal quantum dots. *Science* **290,** 314–317 (2000).
47. Kreller, F., Lowisch, M., Puls, J. & Hennenberger, F. Role of biexcitons in the stimulated emission of wide-gap II-VI quantum wells. *Phys. Rev. Lett.* **75,** 2420 (1995).
48. Klimov, I. V. Spectral and dynamical properties of multiexcitons in semiconductor nanocrystals. *Annu. Rev. Phys. Chem.* **58,** 635–673 (2007).
49. Kammerlander, D., Prezzi, D., Goldoni, G., Molinari, E. & Hohenester, U. Biexciton stability in carbon nanotubes. *Phys. Rev. Lett.* **99,** 126806 (2006).

50. Sie, E. J., Lee, Y.-H., Frenzel, A. J., Kong, J. & Gedik, N. Intervalley biexcitons and many-body effects in monolayer MoS_2. *Phys. Rev. B* **92,** 125417 (2015).
51. You, Y. *et al.* Observation of biexcitons in monolayer WSe2. *Nat. Phys.* **11,** 477–481 (2015).
52. Lüer, L. *et al.* Size and mobility of excitons in (6, 5) carbon nanotubes. *Nat. Phys.* **5,** 54–58 (2009).
53. Martini, I. B., Smith, A. D. & Schwartz, B. J. Exciton-exciton annihilation and the production of interchain species in conjugated polymer films: comparing the ultrafast stimulated emission and photoluminescence dynamics of MEH-PPV. *Phys. Rev. B* **69,** 035204 (2004).
54. Wang, F., Duckovic, G., Knoesel, E., Brus, L. E. & Heinz, T. F. Observation of rapid Auger recombination in optically excited semiconducting carbon nanotubes. *Phys. Rev. B* **70,** 241403 (R) (2004).
55. Schaller, R. D. & Klimov, V. I. High efficiency carrier multiplication in PbSe nanocrystals: implications for solar energy conversion. *Phys. Rev. Lett.* **92,** 186601 (2004).
56. Wang, S., Khafizov, M., Tu, X., Zheng, M. & Krauss, T. D. Multiple exciton generation in single-walled carbon nanotubes. *Nano Lett.* **10,** 2381–2386 (2010).
57. Konabe, S. & Okada, S. Multiple exciton generation by a single photon in single-walled carbon nanotubes. *Phys. Rev. Lett.* **108,** 227401 (2012).
58. Polli, D., Lüer, L. & Cerullo, G. High-time-resolution pump-probe system with broadband detection for the study of time-domain vibrational dynamics. *Rev. Sci. Intrum.* **78,** 103108 (2007).
59. Jensen, S. A. *et al.* Ultrafast photoconductivity of graphene nanoribbons and carbon nanotubes. *Nano Lett.* **13,** 5925–5930 (2013).
60. Lorenc, M. *et al.* Artifacts in femtosecond transient absorption spectroscopy. *Appl. Phys. B* **74,** 19–27 (2002).
61. Fantini, C. *et al.* Optical transition energies for carbon nanotubes from resonant raman spectroscopy: environment and temperature effects. *Phys. Rev. Lett.* **93,** 147406 (2004).
62. Abdelsayed, V. *et al.* Photothermal deoxygenation of graphite oxide with laser excitation in solution and graphene-aided increase in water temperature. *J. Phys. Chem. Lett.* **1,** 2804–2009 (2010).
63. Giannozzi, P. *et al.* QUANTUM ESPRESSO: a modular and open-source software project for quantum simulations of materials. *J. Phys. Condens. Matter* **21,** 395502 (2009).
64. Rozzi, C. A., Varsano, D., Marini, A., Gross, E. K. U. & Rubio, A. Exact Coulomb cutoff technique for supercell calculations. *Phys. Rev. B* **73,** 205119 (2006).
65. Marini, A., Hogan, C., Grüning, M. & Varsano, D. Yambo: an *ab initio* tool for excited state calculations. *Comput. Phys. Commun.* **180,** 1392–1403 (2009).
66. Villegas, C. E. P., Mendoca, P. B. & Rocha, A. R. Optical spectrum of bottom-up graphene nanoribbons: towards efficient atom-thick excitonic solar cells. *Sci. Rep* **4,** 6579 (2014).
67. Thijssen, J. M. *Computational Physics* (Cambridge University Press, 2007).
68. Hüser, M. H. & Berne, B. J. Circumventing the pathological behavior of path-integral Monte Carlo for systems with Coulomb potentials. *J. Chem Phys.* **107,** 571–575 (1997).
69. Colombier, L. *et al.* Detection of a biexciton in semiconducting carbon nanotubes using nonlinear optical spectroscopy. *Phys. Rev. Lett.* **109,** 197402 (2012).
70. Watanabe, K. & Asano, K. Biexcitons in semiconducting single-walled carbon nanotubes. *Phys. Rev. B* **83,** 115406 (2011).

Acknowledgements

This work was funded by MIUR PRIN Grant No. 20105ZZTSE, MIUR FIRB Grant No. RBFR12SWOJ, and MAE Grant No. US14GR12. It has been supported in part by the Austrian Science Fund FWF, under the SFB F49 NextLite, by NAWI Graz, the European Research Council grant on NANOGRAPH, DFG Priority Program SPP 1459 and European Union Projects MoQuaS. G.C., A.N., X.F. and K.M. acknowledge support by the EC under Graphene Flagship (contract no. CNECT-ICT-604391). Computing time was provided by the Center for Functional Nanomaterials at Brookhaven National Laboratory, supported by the US Department of Energy, Office of Basic Energy Sciences, under contract number DE-SC0012704.

Authors contributions

A.N., Y. H., X.F. and K.M. prepared the GNR sample. G.S., S.D.C., D.V., C.M. and G.C. performed the pump–probe experiments. G.S. and D.V. performed the data analysis and simulation with the coupled-rate equations model. D.P., U.H. and E.M. performed the simulations on the exciton and biexciton binding energies. G.S., D.P. and G.C. wrote the manuscript. All authors designed the research and discussed the results.

Additional information

Competing financial interests: The authors declare no competing financial interests.

29

Wavefront shaping through emulated curved space in waveguide settings

Chong Sheng[1,*], Rivka Bekenstein[2,*], Hui Liu[1], Shining Zhu[1] & Mordechai Segev[2]

The past decade has witnessed remarkable progress in wavefront shaping, including shaping of beams in free space, of plasmonic wavepackets and of electronic wavefunctions. In all of these, the wavefront shaping was achieved by external means such as masks, gratings and reflection from metasurfaces. Here, we propose wavefront shaping by exploiting general relativity (GR) effects in waveguide settings. We demonstrate beam shaping within dielectric slab samples with predesigned refractive index varying so as to create curved space environment for light. We use this technique to construct very narrow non-diffracting beams and shape-invariant beams accelerating on arbitrary trajectories. Importantly, the beam transformations occur within a mere distance of 40 wavelengths, suggesting that GR can inspire any wavefront shaping in highly tight waveguide settings. In such settings, we demonstrate Einstein's Rings: a phenomenon dating back to 1936.

[1] National Laboratory of Solid State Microstructures & School of Physics, Collaborative Innovation Center of Advanced Microstructures, Nanjing University, Nanjing, Jiangsu 210093, China. [2] Department of Physics and Solid State Institute, Technion, Haifa 32000, Israel. * These authors contributed equally to this work. Correspondence and requests for materials should be addressed to R.B. (email: beken@tx.technion.ac.il) or to H.L. (email: liuhui@nju.edu.cn).

General electromagnetic (EM) beams propagating through linear homogenous media experience diffraction broadening. However, many applications would greatly benefit from having beams that remain very narrow or shape-invariant for large distances. The past two decades have witnessed remarkable progress in wavefront shaping specifically for the purpose of generating non-diffracting beams, such as shape-preserving Bessel beams[1] and accelerating beams in free space[2–5], in plasmonics[6–9] and even in nonlinear materials[10–15]. The concept of shape-invariant wavepackets was extended beyond EM waves, for example to shaping wavefunctions of electrons[16–19] and generating shape-invariant acoustic beams[20,21], and even accelerating surface water gravity waves[22]. All of these shape-invariant wavepackets are not square integrable (they carry infinite power), hence physically they must be truncated, which implies that they stay non-diffracting only for a finite distance[2]. In a similar vein, there are other kind of beams which are a priori designed to stay shape-invariant only for a finite distance, for example, the cosine-Gauss beams[23] and a class of beams that propagate on arbitrary curved trajectories[5,24,25]. Naturally, all of these beams require wavefront shaping: the launch beam must be shaped in a specific structure (amplitude and phase), to stay non-diffracting for the specified distance.

Wavefront shaping for generating non-diffracting optical beams can be achieved by various methods, ranging from annular slits[1], axicon lenses[26], computer generated holograms[24], spatial light modulators[3,28], gratings[7,23,29], metasurfaces[30–32] and diffraction from nanoparticles[4,33]. Importantly, non-diffracting beams can also be generated in inhomogeneous media such as photonic crystal slabs[34–38], photonic crystals[39,40] and photonic lattices[41]. All these too require wavefront shaping, that is typically done externally, outside the medium within which the beam is propagating. However, wavefront shaping can also be done by shaping the EM environment in which the wave is propagating[42,43]. The fact that the propagation of EM waves in static curved space is analogous to that in inhomogeneous media[42–44] is the underlying principle of emulating general relativity (GR) phenomena in transformation optics[42,43,45–52]. In transformation optics, the permittivities and permeabilities are structured to vary according to the curvature of space[53–59], giving rise to unique trajectories[55–57,60] and controlling the diffraction of light[61,62].

Here, we show that using ideas inspired by GR yields efficient beam shaping in waveguide settings. The concept is general, applicable to many cases where wavefront beam shaping in a waveguide platform is required. First, we fabricate the micro-structured optical waveguide with the specific refractive index emulating the curved space environment generated by a massive gravitational object. This dielectric structure yields a very narrow beam that remains non-diffracting for many Rayleigh lengths. Second, with the same experimental system, we demonstrate the Einstein's rings phenomenon, matching Einstein's 80 years old formula. Finally, we present a general formalism to transform Gaussian beams to considerably narrower shape-invariant beams accelerating (bending) along arbitrary trajectories.

Results

Generating non-diffracting beams through gravitational collimation. The first goal is to create a narrow beam that would propagate in a non-diffracting fashion for a considerable distance in a homogeneous medium. We do that by passing a Gaussian beam through a specific refractive index structure, inspired by the

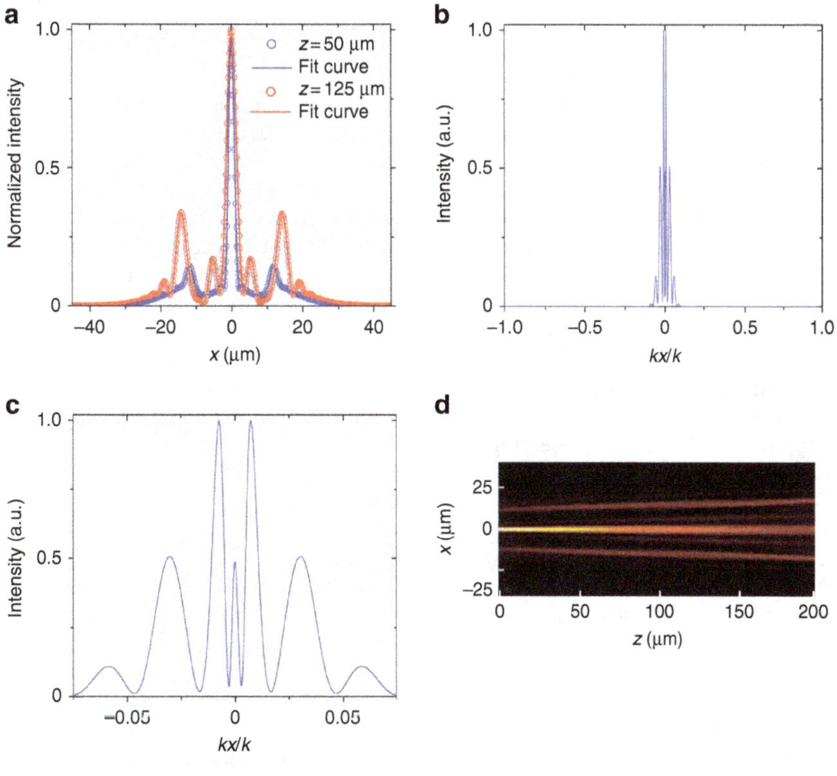

Figure 1 | Calculated propagation of gravitational collimation resulting in a non-diffracting beam. (**a**) The calculated non-diffracting beam fitted to the beam arising from the simulation of the experimental setting. (**b**) Spatial spectrum of the beam displaying two main peaks, as can be seen in **c** showing zoom-in on the central section of the spectrum. The two pronounced peaks correspond to a superposition of non-diffracting cosine and sine distributions, resulting in the narrow non-diffracting beam. (**d**) Simulated propagation of the non-diffracting beam of **a**, for a distance of 200 µm inside a homogenous medium, revealing the non-diffracting property.

gravitational lensing phenomenon occurring around massive stars. We design a specific curvature where the emulated gravitational lensing of the light on the micro-scale can create a very narrow non-diffracting beam. The basic principles of diffraction imply that non-diffracting beams can be constructed when their plane-waves constituents accumulate phase at the same rate. The non-diffracting property of beams depends on the dimensions of the wavepackets, that is, a non-diffracting beam can be a shape-invariant solution to the wave equation in three dimensions (3D) or in two dimensions (2D). In 3D homogeneous media, beams that are structured in both their transverse dimensions exhibit shape-invariant propagation on a straight line in the third dimension include the family of Bessel beams[1]. In 2D, on the other hand, when the beams are structured in a single transverse dimension (for example, when the beam is propagating in a planar waveguide), an ideal non-diffracting beam has a unique shape: two plane waves propagating at opposite symmetric angles with respect to the propagation axis. However, whereas the Bessel beams are localized, that is, they have a main lobe carrying most of the power, the planar case is just an interference grating—which is periodic and cannot be used for applications that require a beam with a single main lobe. Interestingly, providing proper spatial bandwidth to each of the opposite waves in the one-dimensional (1D) case does lead to a localized beam displaying non-diffracting features for some finite distance. More specifically, superimposing two beams whose spectrum in k-space is small compared with the wavenumber, at opposite angles with respect to the propagation axis, gives rise to non-diffracting propagation up to a finite distance, due to the similar rate of phase accumulation of the different modal (plane waves) constituents. Here, we construct such a very narrow non-diffracting beam by drawing on intuition from GR, where it is known that light waves are deflected by the space curvature generated by a massive star[63,64]. We exploit this gravitational lensing effect to construct a field that is a superposition of two beams of a finite spatial bandwidth, propagating at opposite angles with respect to the propagation axis. Such a beam remains non-broadening for a finite distance that can be much larger than the Rayleigh length of its main lobe. An example for such a 1D non-diffracting beam and its spectrum is displayed in Fig. 1a,b, respectively. Figure 1c shows zoom-in on the spectrum, while Fig. 1d presents its simulated propagation—where it is clear that the main lobe remains narrow for a large distance, in spite of the fact that its width is only four wavelength. The two main peaks in the spectrum (Fig. 1c) represent a superposition of cosine/sine distributions, along with a central peak. The width of the spectral peaks is two orders of magnitude smaller than the wavenumber, enabling a non-diffracting property to a finite distance. This structured beam, whose full-width-half-maximum (FWHM) is 2 μm, is approximately shape-preserving for ~200 μm, which corresponds to six Rayleigh lengths (Fig. 1d).

To transform a broad Gaussian beam (FWHM ~30 μm) into this non-diffracting beam in a planar waveguide setting, we fabricate a specific refractive index structure inspired by the concepts of curved space known from GR. Namely, curved space generated by a massive gravitational body leads to gravitational lensing, that can in principle overcome diffraction broadening and cause beam collimation. The planar waveguide has a unique width profile, causing a change in the propagation constant and effectively modifying the refractive index. The structure is shown in Fig. 2a. During the fabrication process, a silver film is deposited on a silica (SiO_2) substrate with a thickness of 80 nm, followed by polymethyl methacrylate (PMMA) microsphere powder scattered on the substrate. The microspheres are distributed on the substrate, with a small density and large separation distance between microspheres. The sample processing includes a stage

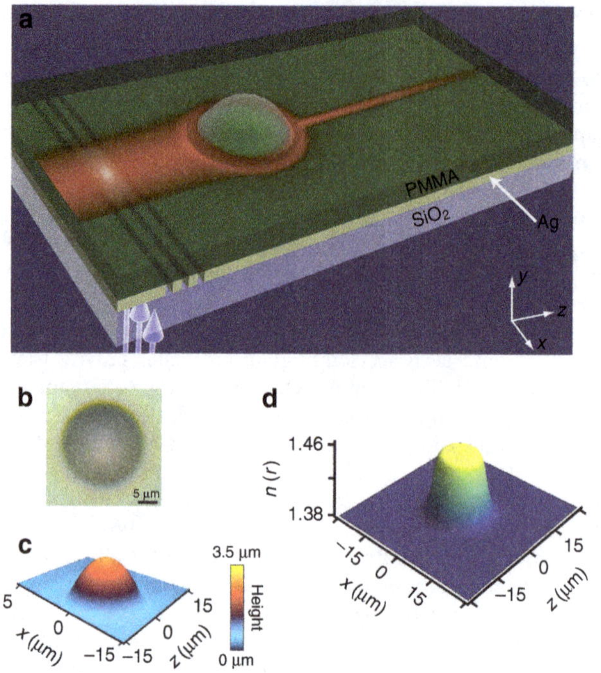

Figure 2 | The sample fabricated for generating a narrow collimated beam. (**a**) Schematic view of the fabricated waveguide: the inhomogeneous planar waveguide with the specifically designed refractive index structure. The structure is fabricated by depositing a thin silver film on a silica (SiO_2) substrate with a thickness of 80 nm, followed by PMMA microsphere powder scattered on the substrate. The blue arrows at the bottom represent the incident 457 nm blue laser light, and the bright spot marks the illumination spot where the light is incident on the grating. (**b**) Top-view optical microscopy image of the microdroplet. (**c**) The surface structure of the microdroplet, as mapped by AFM measurements. (**d**) The effective refractive index structure calculated from **c**, based on waveguide theory.

where the sample is put on the heating table (300 °C) for 30 s. As the melting temperature of PMMA polymer is ~250 °C, the heating process deforms the PMMA microspheres into domes, just as shown in Fig. 2b,c. In this process, the size of resultant PMMA domes is not uniform, and their diameters can vary greatly, from 1 to 100 μm. For the experiment presented here, we work with one of domes that has an appropriate size, as shown in its optical microscope image in Fig. 2b. The structure is shaped as a dome protruding from the plane of the waveguide (Fig. 2a). This is further confirmed by mapping the surface structure with atomic force microcopy (Asylum Research, MFP-3D-SA, USA), as shown in Fig. 2c. Next, a set of gratings with the period 310 nm are drilled on the sliver film around the microdroplet with focused ion beam (FEI Strata FIB 201, 30 keV, 150 pA). These gratings enable to couple the light into the slab waveguide. Next, we spin-coat the sample with a PMMA photoresist mixed with rare earth (Eu^{3+}) to a thickness of ~1 μm, and subsequently dry the sample in the oven at 70 °C for 2 h. The Eu^{3+} rare earth ions are added to the sample to facilitate fluorescence imaging that will reveal the propagation dynamics of the beam. These Eu^{3+} ions absorb the beam propagating in the slab waveguide, whose wavelength (457 nm) is specifically chosen to excite the rare earth ions, that in turn emit fluorescent light at 615 nm wavelength. We note that, although the 1-μm-thick PMMA layer is not single-mode waveguide for the 457 nm beam, the designed grating allows only one mode to be excited inside the waveguide. Here, only the TM3 mode is excited in our experiment (The grating is designed that only one waveguide mode is excited. Hence, plasmonic modes are not excited in the experiment). The resultant 2D structure of the refractive index is displayed in Fig. 2d, together with a 3D illustration of the entire sample (Fig. 2a). Figure 2c shows the width of the PMMA waveguide as mapped by AFM measurements. From this width, we calculate the refractive index structure displayed in Fig. 2d, which is fitted with the function $n(x,z) = n_0 + a/(1+(\sqrt{x^2+z^2}/r_c)^8)$, with $n_0 = 1.37$, $a = 9.22 \times 10^{-2}$, $r_c = 9.69$. Recall that the refractive index of bulk PMMA polymer is 1.49, hence our fabrication process reduces the refractive index according to our design. Specifically, in the region of the dome, the thickness is increased to 3.5 μm, and therefore the effective index of the TM3 waveguide mode is increased from 1.37 to 1.49.

In the experiment, we launch a Gaussian beam of 457 nm wavelength and 11.3 μm FWHM to propagate inside the PMMA layer that acts as a waveguide. The loss in this waveguide is quite small, in spite of the proximity of the thin Ag layer, enabling propagation distances of hundreds of micrometres. The specifically designed refractive index structure focuses the wide beam to a very narrow (2 μm) beam that is subsequently propagating without diffraction for ~200 μm, as expected from the theory. We emphasize that, after passing the 'star', the very narrow beam is propagating in a completely homogeneous medium, hence its non-diffracting property arises solely from the beam structure generated by passing the 'star'. Moreover, whereas most shape-preserving beams are very broad, this beam presents a narrow profile, only 2 μm wide. For comparison, we study the dynamic of a Gaussian beam passing through the same medium numerically and compare it with the experimental results (Fig. 3). We do this by numerically simulating the beam propagation, with the beam propagation method in a medium with the specific refractive index structure conforming to that of the sample used in the experiment (Fig. 2d). In both the experiments and the simulations the transformation of the wide Gaussian beam to a narrow collimated beam is achieve within a very short propagation distance (~20 μm), allowing the use of this scheme in integrated photonics circuits. In Fig. 3, the diameter of the dome is roughly 25 μm. In the experiment, we can fabricate domes with different diameters, always with circular shape. Naturally, domes of different sizes yield collimation for different propagation distances and with different beam widths.

Experiments emulating the Einstein rings phenomenon. Interestingly, we find that besides producing collimated beams, the same planar 'central potential' index structure can also be used to emulate the phenomenon of Einstein's Rings, which is a famous phenomenon predicted by GR and observed in astronomy[65,66]. The Einstein Ring phenomena occurs when light from a point source is deformed by a mass distribution through gravitation lensing that causes the appearance of a ring around the mass distribution. For this case, the beam approaching the 'star' should emulate the radiation originating from a point source, namely, the wave reaching the 'star' should be a spherical wave. To emulate a point source, we fabricate (with focused ion beam) an arc-shaped grating (period of 310 nm) inside the metal film. This is shown in Fig. 4b, where the radius of the arc is 30 μm. When a plane wave is incident (from below) on the arc grating, the grating transforms it into a spherical wave propagating inside the waveguide layer. The region of incidence on the grating acts as a point source, emitting a spherical wave diverging both to the left and to the right of that point (negative and positive z, respectively). In such a setting, the spherical wavefront produced by the arc-grating emulates the wave radiated outwards from a point source located at the centre of grating arc. When this 1D spherical wavefront is passing

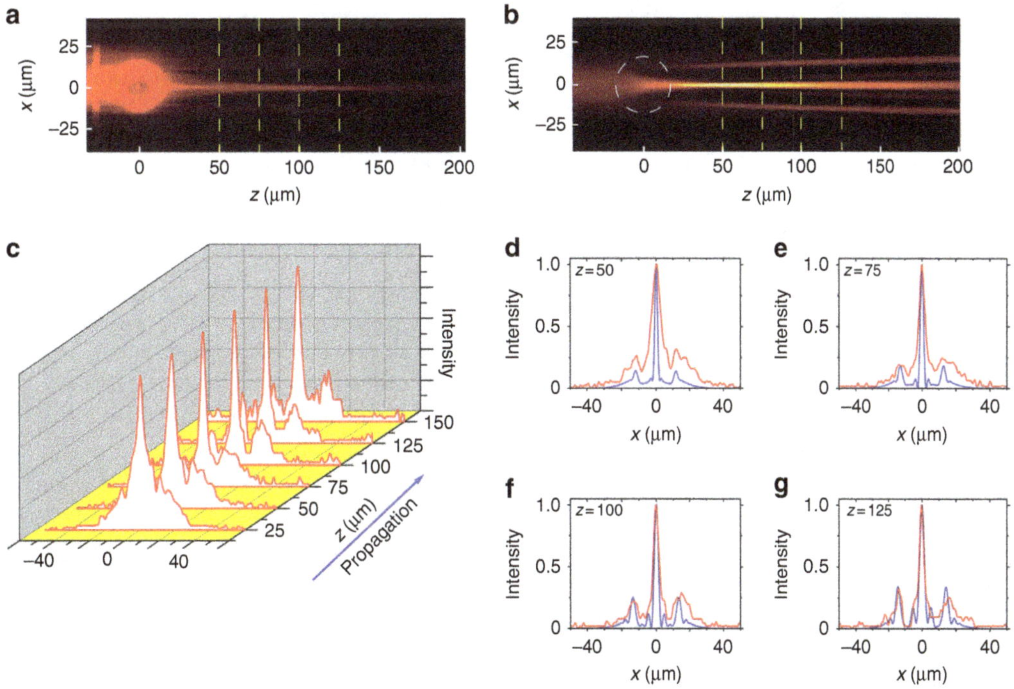

Figure 3 | Experimentally observed propagation dynamics of gravitationally collimated non-diffracting beam. (**a**) Top-view photograph of the experimentally observed results obtained through florescence. A broad Gaussian beam with FWWH 11.3 μm passes through the region of the dome, giving rise to the refractive index profile described in Fig. 2c. The wide Gaussian beam focuses to a narrow collimated beam that is non-diffracting for ~200 μm. The entire beam transformation process occurs within20 μm. (**b**) Simulated results of the same beam showing a similar effect as the experiment. The white dashed circle corresponds to the dome region. (**c**) Normalized intensity profile of the beam for several propagation distances, after passing though the dome region. (**d-g**) Measured (red) and simulated (blue) 1D intensity profiles for $z = 50$ μm, $z = 75$ μm, $z = 100$ μm, $z = 125$ μm, respectively, which correspond to the planes marked by the yellow dashed lines in **a-b**.

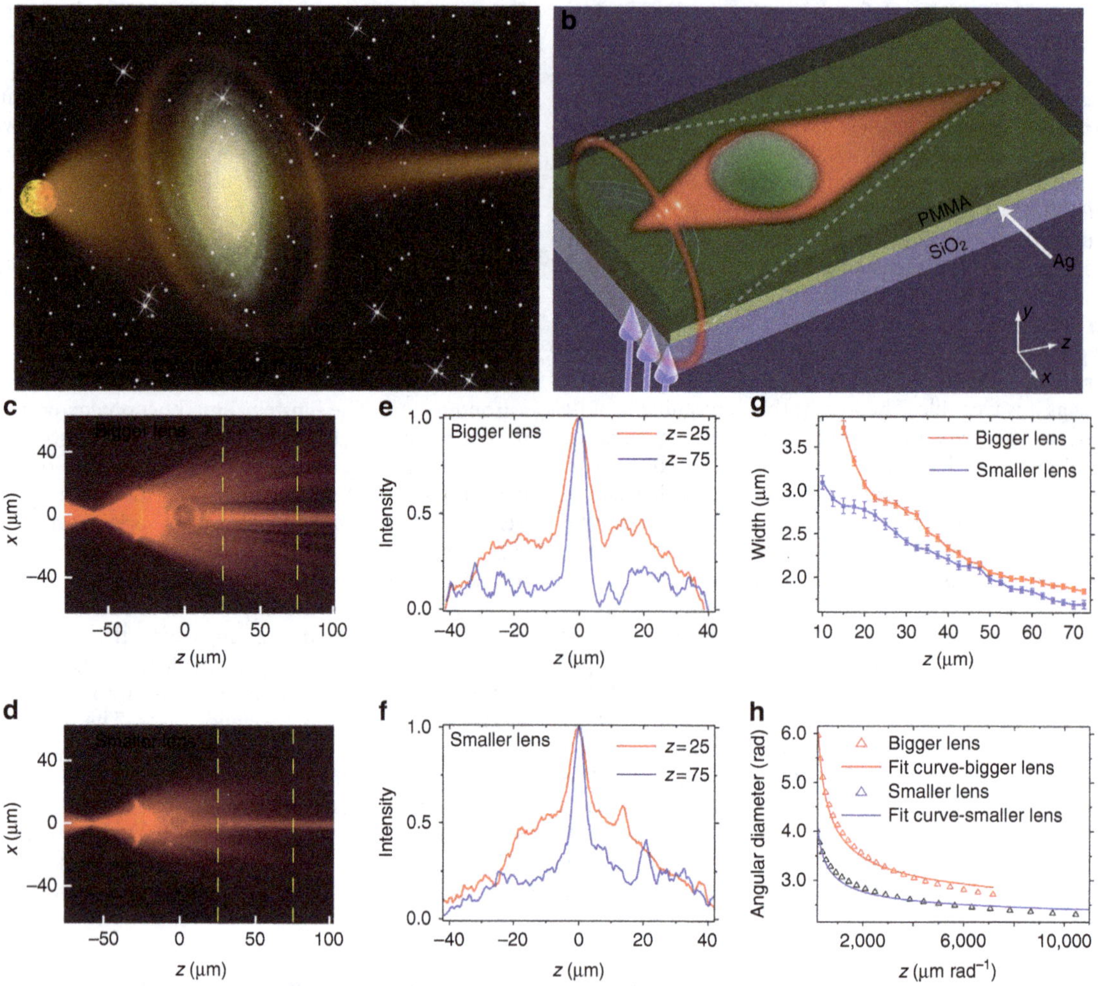

Figure 4 | Experimental emulation of the formation of Einstein's ring. (**a**) Einstein's vision: light from a point source is focused by a gravitational lens, and is subsequently observed as a virtual ring around the mass distribution. (**b**) Schematic view of the fabricated inhomogeneous waveguide. (**c,d**) Experimental results (obtained through florescence) showing a spherical wave passing though the dome region, for two domes of different radii. The inhomogeneous area acts as a gravitational lens on the light. (**e,f**) Measured beam profiles at $z=25$ (red line), $z=75$ (blue line), respectively, which correspond to the locations marked by the yellow dashed lines in **c,d**. (**g**) Measured beam width as a function of the propagation distance in the homogenous medium after the dome region. (**h**) Fit to Einstein's formula for the angular diameter of the Einstein rings. The calculated angular diameter from the experimental measurement is in a very good agreement with the theoretical formula.

by the star—it is focused and the beam width changes as a function of the propagation distance, as extracted from the experimental data. It is important to emphasize that our optical setting represents Einstein's rings formed by a time-harmonic EM waves, hence the entire dynamics is in space (not in time). Typical results for two different 'stars' (microdroplets with two different radii) are displayed in Fig. 4. As the Radius of the 'star' is larger the convergence of the beam is more extreme, but the final beam is wider (Fig. 4). At this point it is intriguing to compare our emulation results with Einstein's prediction. The Einstein Formula for the angular diameter of the virtual ring[64] is given by $\beta = \sqrt{\alpha_0 R_0 / z}$, that depends on the convergence angle α_0, the radius of the mass distribution R_0 and the distance between the centre of the mass distribution to the observation point. We calculate the angular diameter of the Einstein Ring from the measured convergence angle of the beam, for several different observation points (propagation distances). For a given observation point, the focusing angle of the beam after passing the 'star' gives the slope, from which we calculate the angular diameter of the virtual ring that an observer located at this specific distance (from the 'star') will see. To conform with the Einstein formula, we calculate the relative angular radius between the two mass distributions (two samples). Namely, instead of calculating the absolute angular radius as a function of z, we calculate the relative angular radius between the results of each sample. We then fit the curve $\beta\left(\frac{z}{\alpha_0}\right) = \sqrt{c/\left(\frac{z}{\alpha_0}\right)}$ with c as a free parameter and compare the relative constant extracted from the experiment with the constant expected from Einstein's formula. In comparing the ratio and not the absolute number, we avoid the factor 2 between the relativistic Einstein formula and our experiment that represents Newtonian dynamics. As Fig. 4h shows, the experiments agree well with theory, although at large z, the experimental values are slightly lower than the model. This minute discrepancy arises from the difference between the fabricated optical potential (refractive index structure) and the $1/r$ gravitational potential of a point source. Consequently, for large values of z (distances), the focusing angle of the light deviates from Einstein's formula, hence the measured focusing angle is somewhat smaller than the theoretical curve.

Shaping beams accelerating on arbitrary trajectories. Finally, we present a general formalism for transforming broad Gaussian beams to accelerating beams that bend along arbitrary (convex) trajectories in a planar waveguide setting. As above, we do that by passing an incident broad Gaussian beam (11.3 μm FWHM) through a miniature refractive index structure that is designed specifically for this task. Accelerating beams are beams with a well-defined peak intensity that propagates along some non-straight trajectory, depending on the phase of the initial beam[4,5,29]. From the point of view of GR, the peak intensity of the beam does not follow geodesics paths[67], which are the shortest paths that light propagated along (by the Fermat principle). This important property of accelerating beams had been exploited for various applications, such as curved plasma channels[65], manipulating microparticles[68,69] and micromachining[70]. We design accelerating beams by utilizing the formalism suggested in ref. 5, for finding the specific 1D phase $\phi(x)$ required for shaping the wavefront of an accelerating beam that will propagate along a specific trajectory. This ID phase can be achieved by a 2D refractive index structure that the beam passes through, and obeys the relation

$$\phi(x) = k_0 \int_{z_i}^{z_f} n(x,z)dz, \tag{1}$$

under the assumption that the propagation of the beam is in the paraxial regime. Using this method, there is no unique solution for $n(x, z)$. We therefore suggest a simple method that solves equation (1) for one specific refractive index profile to a specified phase, by assuming $n(x, z)$ is constructed from a function that is separable in x, z, namely $n(x, z) = f(x)g(z)$. For simplicity, we take $g(z) = \exp(-z^2/\sigma^2)$, and assume the Gaussian width (in z) is small compared with the propagation distance ($\sigma << z_f - z_i$). This allows setting the boundaries of the integral to infinity which after integrating over z yields:

$$f(x) = \frac{\phi(x)}{k_0\sqrt{\pi\sigma}}. \tag{2}$$

It is important to emphasize that the approximation we used for solving the integral of the phase only, causes additional effects. Due to the 2D refractive index distribution the beam is shifted to some different direction of propagation—$z' = ze^{i\theta}$ while propagating through the inhomogeneous area. Consequently, $n(x, z) = \frac{\phi(x)}{k_0\sqrt{\pi\sigma}}\exp(-z^2/\sigma^2)$. To present an example for this method, we find the refractive index profile required to create the phase for an accelerating beam along the trajectory $f(z) = az'^3$. In this specific case, the propagation of the resulting beam can be solved analytically using the method presented in ref. 5. In more complicated cases, a numerical solution for the ordinary differential equation (ODE) is required. We then use equation (1) to calculate the 2D refractive index structure that will provide the beam with the appropriate phase. By simulating the dynamic of a broad Gaussian beam passing through the designed refractive index structure, we find that the main lobe indeed accelerates along the expected trajectory, for a distance of 20 μm as displayed in Fig. 5. In this regime, it is possible to design a beam that will accelerate beam on an arbitrary trajectory. As any accelerating beam, the structure of such a beam involves a main lobe accompanied by oscillations on one side, and exponential decay on the other side. An example is shown in Fig. 5c, where the beam cross-sections at several propagation distances is displayed. This technique for beam shaping inside a slab waveguide is general, and can be used to shape the wavefront of non-diffracting beams accelerating on any convex trajectory, by designing the refractive index structure using equations (1 and 2), which relates the initial phase front (assumed here to be of a broad Gaussian beam) and the desired phase front $\phi(x)$ to the refractive index structure required for such wavefront shaping.

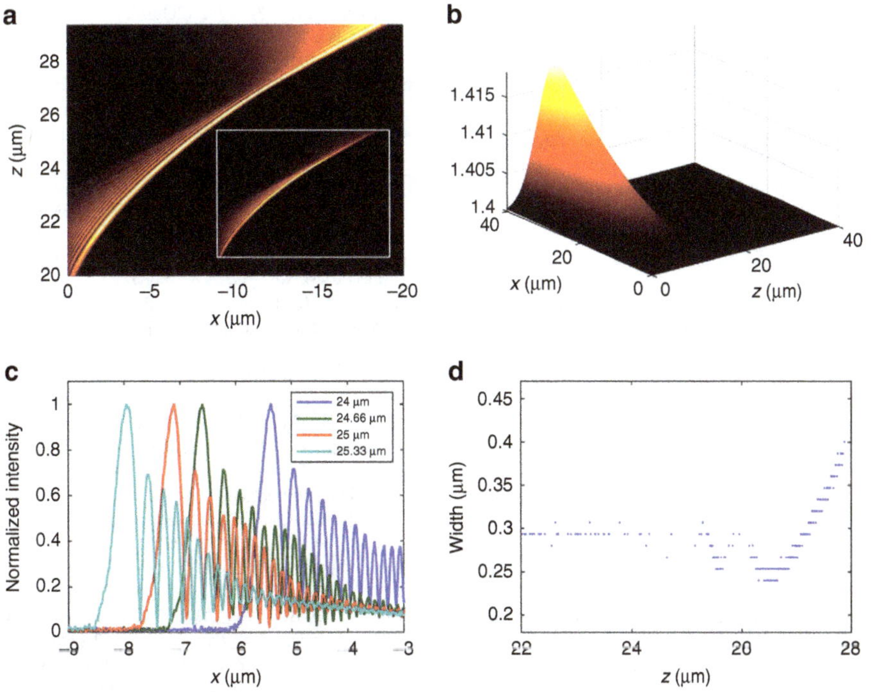

Figure 5 | Accelerating beams propagating along arbitrary trajectories produced by designing the refractive index structure within the initial 10 μm propagation distances in the waveguide layer. (**a**) Simulated evolution of the accelerating beam, where the peak intensity is propagating along the curve $f(z) = az^3$. Inset: the evolution displayed with a non-normalized intensity (**b**) The calculated refractive index structure which transforms a broad Gaussian beam into the narrow non-diffracting accelerating beam of **a**. (**c**) Structure of the accelerating beam for different propagation distance. (**d**) Width of the main lobe as a function of the propagation distance.

Discussion

To conclude, we have presented a method for shaping optical wavefronts in waveguide settings. Our technique is inspired by GR and it provides a platform for emulating the spatial dynamics of EM waves in curved space. This method can be achieved in thin film waveguides and can be implemented in integrated photonics settings. Specifically, we have demonstrated experimentally the construction of a narrow non-diffracting beam, the formation of Einstein's rings, and presented a general method to construct accelerating beans propagating along arbitrary trajectories. This method can be used for shaping any general beam, thereby suggesting a new way of using transformation optics media for beam shaping in waveguide settings with a single dielectric material. In this work, we presented beam shaping in the spatial domain; consequently, our experiments employed only continuous laser beams as our input waves. However, in principle this technique can also be used to shape ultrashort laser pulses with the traditional grating pairs, the lenses and the spatial modulation at the focal plane, all implemented in a slab waveguide geometry with proper design of the planar refractive index structure. This idea will be pursued in our future research.

References

1. Durnin, J., Miceli, J. J. & Eberly, J. H. Diffraction-free beams. *Phys. Rev. Lett.* **58,** 1499–1501 (1987).
2. Siviloglou, G. A. & Christodoulides, D. N. Accelerating finite energy Airy beams. *Opt. Lett.* **32,** 979–981 (2007).
3. Siviloglou, G. A., Broky, J., Dogariu, A. & Christodoulides, D. N. Observation of Accelerating Airy Beams. *Phys. Rev. Lett.* **99,** 213901 (2007).
4. Kaminer, I., Bekenstein, R., Nemirovsky, J. & Segev, M. Nondiffracting accelerating wave packets of Maxwell's equations. *Phys. Rev. Lett.* **108,** 163901 (2012).
5. Greenfield, E., Segev, M., Walasik, W. & Raz, O. Accelerating light beams along arbitrary convex trajectories. *Phys. Rev. Lett.* **106,** 213902 (2011).
6. Salandrino, A. & Christodoulides, D. N. Airy plasmon: a nondiffracting surface wave. *Opt. Lett.* **35,** 2082–2084 (2010).
7. Zhang, P. *et al.* Plasmonic Airy beams with dynamically controlled trajectories. *Opt. Lett.* **36,** 3191–3193 (2011).
8. Minovich, A. *et al.* Generation and near-field imaging of airy surface plasmons. *Phys. Rev. Lett.* **107,** 116802 (2011).
9. Epstein, I. & Arie, A. Arbitrary bending plasmonic light waves. *Phys. Rev. Lett.* **112,** 023903 (2014).
10. Wulle, T. & Herminghaus, S. Nonlinear optics of Bessel beams. *Phys. Rev. Lett.* **70,** 1401–1404 (1993).
11. Kaminer, I., Segev, M. & Christodoulides, D. N. Self-accelerating self-trapped optical beams. *Phys. Rev. Lett.* **106,** 213903 (2011).
12. Lotti, A. *et al.* Stationary nonlinear Airy beams. *Phys. Rev. A* **84,** 021807 (2011).
13. Bekenstein, R. & Segev, M. Self-accelerating optical beams in highly nonlocal nonlinear media. *Opt. Express* **19,** 23706–23715 (2011).
14. Dolev, I., Kaminer, I., Shapira, A., Segev, M. & Arie, A. Experimental observation of self-accelerating beams in quadratic nonlinear media. *Phys. Rev. Lett.* **108,** 113903 (2012).
15. Bekenstein, R., Schley, R., Mutzafi, M., Rotschild, C. & Segev, M. Optical simulations of gravitational effects in the Newton-Schrodinger system. *Nat. Phys.* **11,** 872–878 (2015).
16. Uchida, M. & Tonomura, A. Generation of electron beams carrying orbital angular momentum. *Nature* **464,** 737–739 (2010).
17. Voloch-Bloch, N., Lereah, Y., Lilach, Y., Gover, A. & Arie, A. Generation of electron Airy beams. *Nature* **494,** 331–335 (2013).
18. Grillo, V. *et al.* Generation of nondiffracting electron bessel beams. *Phys. Rev. X* **4,** 011013 (2014).
19. Kaminer, I., Nemirovsky, J., Rechtsman, M., Bekenstein, R. & Segev, M. Self-accelerating Dirac particles and prolonging the lifetime of relativistic fermions. *Nat. Phys.* **11,** 261–267 (2015).
20. Zhang, P. *et al.* Generation of acoustic self-bending and bottle beams by phase engineering. *Nat. Commun.* **5,** 4316 (2014).
21. Bar-Ziv, U., Postan, A. & Segev, M. Observation of shape-preserving accelerating underwater acoustic beams. *Phys. Rev. B* **92,** 100301 (2015).
22. Fu, S., Tsur, Y., Zhou, J., Shemer, L. & Arie, A. Propagation dynamics of airy water-wave pulses. *Phys. Rev. Lett.* **115,** 034501 (2015).
23. Lin, J. *et al.* Cosine-gauss plasmon beam: a localized long-range nondiffracting surface wave. *Phys. Rev. Lett.* **109,** 093904 (2012).
24. Rosen, J. & Yariv, A. Snake beam: a paraxial arbitrary focal line. *Opt. Lett.* **20,** 2042–2044 (1995).
25. Froehly, L. *et al.* Arbitrary accelerating micron-scale caustic beams in two and three dimensions. *Optics Express* **19,** 16455 (2011).
26. Scott, G. & McArdle, N. Efficient generation of nearly diffraction-free beams using an axicon. *Opt. Eng.* **31,** 2640–2643 (1992).
27. Rosen, J. & Yariv, A. Synthesis of an arbitrary axial field profile by computer-generated holograms. *Opt. Lett.* **19,** 843–845 (1994).
28. Zhang, P. *et al.* Nonparaxial mathieu and weber accelerating beams. *Phys. Rev. Lett.* **109,** 193901 (2012).
29. Li, L., Li, T., Wang, S. M. & Zhu, S. N. Collimated plasmon beam: nondiffracting versus linearly focused. *Phys. Rev. Lett.* **110,** 046807 (2013).
30. Bomzon, Z., Kleiner, V. & Hasman, E. Formation of radially and azimuthally polarized light using space-variant subwavelength metal stripe gratings. *Appl. Phys. Lett.* **79,** 1587–1589 (2001).
31. Yu, N. *et al.* Light propagation with phase discontinuities: generalized laws of reflection and refraction. *Science* **334,** 333–337 (2011).
32. Kildishev, A. V., Boltasseva, A. & Shalaev, V. M. Planar photonics with metasurfaces. *Science* **339,** 1232009 (2013).
33. Chen, Z., Taflove, A. & Backman, V. Photonic nanojet enhancement of backscattering of light by nanoparticles: a potential novel visible-light ultramicroscopy technique. *Opt. Express* **12,** 1214–1220 (2004).
34. Yu, X. & Fan, S. Bends and splitters for self-collimated beams in photonic crystals. *Appl. Phys. Lett.* **83,** 3251–3253 (2003).
35. Rakich, P. T. *et al.* Achieving centimetre-scale supercollimation in a large-area two-dimensional photonic crystal. *Nat. Mater.* **5,** 93–96 (2006).
36. Shih, T.-M. *et al.* Supercollimation in photonic crystals composed of silicon rods. *Appl. Phys. Lett.* **93,** 131111 (2008).
37. Hamam, R. E., Ibanescu, M., Johnson, S. G., Joannopoulos, J. D. & Soljacic, M. Broadband super-collimation in a hybrid photonic crystal structure. *Opt. Express* **17,** 8109–8118 (2009).
38. Mocella, V. *et al.* Self-collimation of light over millimeter-scale distance in a quasi-zero-average-index metamaterial. *Phys. Rev. Lett.* **102,** 133902 (2009).
39. Longhi, S. & Janner, D. X-shaped waves in photonic crystals. *Phys. Rev. B* **70,** 235123 (2004).
40. Conti, C. & Trillo, S. Nonspreading wave packets in three dimensions formed by an ultracold bose gas in an optical lattice. *Phys. Rev. Lett.* **92,** 120404 (2004).
41. Manela, O., Segev, M. & Christodoulides, D. N. Nondiffracting beams in periodic media. *Opt. Lett.* **30,** 2611–2613 (2005).
42. Leonhardt, U. Optical conformal mapping. *Science* **312,** 1777–1780 (2006).
43. Pendry, J. B., Schurig, D. & Smith, D. R. Controlling electromagnetic fields. *Science* **312,** 1780–1782 (2006).
44. Laundau, L.D. & Lifshitz, E. M. *The Classical Theory Of Fields* (Butterworth-Heinemann, 1975).
45. Li, J. & Pendry, J. B. Hiding under the carpet: a new strategy for cloaking. *Phys. Rev. Lett.* **101,** 203901 (2008).
46. Alù, A. & Engheta, N. Multifrequency optical invisibility cloak with layered plasmonic shells. *Phys. Rev. Lett.* **100,** 113901 (2008).
47. Smolyaninov, I. I., Smolyaninova, V. N., Kildishev, A. V. & Shalaev, V. M. Anisotropic metamaterials emulated by tapered waveguides: application to optical cloaking. *Phys. Rev. Lett.* **102,** 213901 (2009).
48. Valentine, J., Li, J., Zentgraf, T., Bartal, G. & Zhang, X. An optical cloak made of dielectrics. *Nat. Mater.* **8,** 568–571 (2009).
49. Gabrielli, L. H., Cardenas, J., Poitras, C. B. & Lipson, M. Silicon nanostructure cloak operating at optical frequencies. *Nat. Photon.* **3,** 461–463 (2009).
50. Smolyaninova, V. N., Smolyaninov, I. I., Kildishev, A. V. & Shalaev, V. M. Experimental observation of the trapped rainbow. *Appl. Phys. Lett.* **96,** 211121 (2010).
51. Chen, H., Chan, C. T. & Sheng, P. Transformation optics and metamaterials. *Nat. Mater.* **9,** 387–396 (2010).
52. Zentgraf, T., Liu, Y., Mikkelsen, M. H., Valentine, J. & Zhang, X. Plasmonic luneburg and eaton lenses. *Nat. Nanotechnol.* **6,** 151–155 (2011).
53. Smolyaninov, I. I. Surface plasmon toy model of a rotating black hole. *New J. Phys.* **5,** 147–147 (2003).
54. Leonhardt, U. & Philbin, T. G. General relativity in electrical engineering. *New J. Phys.* **8,** 247 (2006).
55. Genov, D. A., Zhang, S. & Zhang, X. Mimicking celestial mechanics in metamaterials. *Nat. Phys.* **5,** 687–692 (2009).
56. Narimanov, E. E. & Kildishev, A. V. Optical black hole: broadband omnidirectional light absorber. *Appl. Phys. Lett.* **95,** 041106–041106-3 (2009).
57. Cheng, Q., Cui, T. J., Jiang, W. X. & Cai, B. G. An omnidirectional electromagnetic absorber made of metamaterials. *New J. Phys.* **12,** 063006 (2010).
58. Smolyaninov, I. I. & Narimanov, E. E. Metric signature transitions in optical metamaterials. *Phys. Rev. Lett.* **105,** 067402 (2010).
59. Genov, D. A. General relativity: optical black-hole analogues. *Nat. Photon.* **5,** 76–78 (2011).

60. Sheng, C., Liu, H., Wang, Y., Zhu, S. N. & Genov, D. A. Trapping light by mimicking gravitational lensing. *Nat. Photon.* **7,** 902–906 (2013).
61. Batz, S. & Peschel, U. Linear and nonlinear optics in curved space. *Phys. Rev. A* **78,** 043821 (2008).
62. Bekenstein, R., Nemirovsky, J., Kaminer, I. & Segev, M. Shape-preserving accelerating electromagnetic wave packets in curved space. *Phys. Rev. X* **4,** 011038 (2014).
63. Einstein, A. Die Grundlage der allgemeinen relativitätstheorie. *Ann. Phys.* **354,** 769–822 (1916).
64. Einstein, A. Lens-like action of a star by the deviation of light in the gravitational field. *Science* **84,** 506–507 (1936).
65. Hewitt, J. N. *et al.* Unusual radio source MG1131 + 0456: a possible Einstein ring. *Nature* **333,** 537–540 (1988).
66. King, L. J. *et al.* A complete infrared Einstein ring in the gravitational lens system B1938 + 666. *MNRAS* **295,** L41–L44 (1998).
67. Polynkin, P., Kolesik, M., Moloney, J. V., Siviloglou, G. A. & Christodoulides, D. N. Curved plasma channel generation using ultraintense airy beams. *Science* **324,** 229–232 (2009).
68. Baumgartl, J., Mazilu, M. & Dholakia, K. Optically mediated particle clearing using Airy wavepackets. *Nat. Photon.* **2,** 675–678 (2008).
69. Schley, R. *et al.* Loss-proof self-accelerating beams and their use in non-paraxial manipulation of particles' trajectories. *Nat. Commun.* **5,** 5189 (2014).
70. Mathis, A. *et al.* Micromachining along a curve: Femtosecond laser micromachining of curved profiles in diamond and silicon using accelerating beams. *Appl. Phys. Lett.* **101,** 071110–071113 (2012).

Acknowledgements

R.B. gratefully acknowledges the support of the Adams Fellowship Programme of the Israel Academy of Sciences and Humanities. This research was also supported by the ICore Excellence center 'Circle of Light' and a grant from the US Air Force Office for Scientific Research (AFOSR). H.L. gratefully acknowledges the support of the National Natural Science Foundation of China (No's 11321063, 61425018 and 11374151), the National Key Projects for Basic Researches of China (No. 2012CB933501 and 2012CB921500), the Doctoral Programme of Higher Education (20120091140005) and Dengfeng Project B of Nanjing University. C.S. gratefully acknowledge the support of the programme A for Outstanding PhD candidate of Nanjing University.

Author contributions

All authors contributed to all aspects of this work.

Additional information

Competing financial interests: The authors declare no competing financial interests.

Permissions

All chapters in this book were first published in NC, by Nature Publishing Group; hereby published with permission under the Creative Commons Attribution License or equivalent. Every chapter published in this book has been scrutinized by our experts. Their significance has been extensively debated. The topics covered herein carry significant findings which will fuel the growth of the discipline. They may even be implemented as practical applications or may be referred to as a beginning point for another development.

The contributors of this book come from diverse backgrounds, making this book a truly international effort. This book will bring forth new frontiers with its revolutionizing research information and detailed analysis of the nascent developments around the world.

We would like to thank all the contributing authors for lending their expertise to make the book truly unique. They have played a crucial role in the development of this book. Without their invaluable contributions this book wouldn't have been possible. They have made vital efforts to compile up to date information on the varied aspects of this subject to make this book a valuable addition to the collection of many professionals and students.

This book was conceptualized with the vision of imparting up-to-date information and advanced data in this field. To ensure the same, a matchless editorial board was set up. Every individual on the board went through rigorous rounds of assessment to prove their worth. After which they invested a large part of their time researching and compiling the most relevant data for our readers.

The editorial board has been involved in producing this book since its inception. They have spent rigorous hours researching and exploring the diverse topics which have resulted in the successful publishing of this book. They have passed on their knowledge of decades through this book. To expedite this challenging task, the publisher supported the team at every step. A small team of assistant editors was also appointed to further simplify the editing procedure and attain best results for the readers.

Apart from the editorial board, the designing team has also invested a significant amount of their time in understanding the subject and creating the most relevant covers. They scrutinized every image to scout for the most suitable representation of the subject and create an appropriate cover for the book.

The publishing team has been an ardent support to the editorial, designing and production team. Their endless efforts to recruit the best for this project, has resulted in the accomplishment of this book. They are a veteran in the field of academics and their pool of knowledge is as vast as their experience in printing. Their expertise and guidance has proved useful at every step. Their uncompromising quality standards have made this book an exceptional effort. Their encouragement from time to time has been an inspiration for everyone.

The publisher and the editorial board hope that this book will prove to be a valuable piece of knowledge for researchers, students, practitioners and scholars across the globe.